Natural and Social Sciences of Patagonia

Despite being an underpopulated region, Patagonia has attracted the attention of scientists since the very beginning of its settlement. From classical explorers such as Darwin or D'Orbigny, to modern science including nuclear and satellite developments, several disciplines have focused their efforts on unraveling Patagonia's natural and social history. Today, scientific and technological research is shifting from being shaped by northern agendas, towards more locally oriented objectives, such as the management of natural resources, the modernization of energy production and distribution, and the coexistence of rural and cosmopolitan social lifestyles. At the intersection of all these topics, new conflicts concerning the economy, human development, population, and the proper and long-standing planification and management of the landscape and its natural resources have emerged. These conflicts, of course, have also caught the attention of many interdisciplinary research groups.

This series is aimed at describing and discussing various aspects of this complex reality, but also at bridging the gaps between the scientific community and governments, policymakers, and society in general. The respective volumes will analyze and synthesize our knowledge of Patagonian biodiversity at different scales, from alleles, genes and species, to ecosystems and the biosphere, including its multilevel interactions. As humans cannot be viewed as being separate from biodiversity, the series' volumes will also share anthropological, archaeological, sociological and historical views of humanity, and highlight the wide range of benefits that ecosystems provide to humanity including provisioning, regulating and cultural services.

More information about this series at http://www.springer.com/series/16384

Pablo L. Peri
Guillermo Martínez Pastur • Laura Nahuelhual
Editors

Ecosystem Services in Patagonia

A Multi-Criteria Approach for an Integrated Assessment

Springer

Editors
Pablo L. Peri
National Agricultural Technology Institute
Río Gallegos, Argentina

Laura Nahuelhual
Instituto de Economía
Universidad Austral de Chile
Centro de Investigación: Dinámica de
Ecosistemas Marinos de Altas Latitudes
Instituto Milenio en Socio-Ecología Costera
Valdívia, Chile

Guillermo Martínez Pastur
Consejo Nacional de Investigaciones
Científicas y Técnicas
Centro Austral de Investigaciones
Científicas y Técnicas
Ushuaia, Tierra del Fuego, Argentina

ISSN 2662-3463 ISSN 2662-3471 (electronic)
Natural and Social Sciences of Patagonia
ISBN 978-3-030-69165-3 ISBN 978-3-030-69166-0 (eBook)
https://doi.org/10.1007/978-3-030-69166-0

This Springer imprint is published by the registered company Springer Nature Switzerland AG
The registered company address is: Gewerbestrasse 11, 6330 Cham, Switzerland

Contents

Chapter 1
Ecosystem Services as a Tool for Decision-Making in Patagonia

Pablo L. Peri, Laura Nahuelhual, and Guillermo Martínez Pastur

Abstract The Patagonia region that lies within two countries (12% in Chile and 88% in Argentina) has some of the most extensive wilderness areas in the planet (forests, shrublands, grasslands, and wetlands), as well as many low affected coastal-marine ecosystems. However, this "last frontier of capitalism" has started to experience the simultaneous expansion of the green and blue growth sectors, with unknown consequences for biodiversity, ecosystem services (ES) and livelihoods. In this chapter, we (i) present a conceptual framework for the sustainable management of the natural ecosystems in Patagonia that guides the chapters of this book and (ii) highlight the importance of the ES approach for better decision-making. Our conceptual framework starts from a non-dichotomous conception of the relationships between society and ecosystems. This determines a socio-ecosystem composed by a biophysical, economic-productive and socio-political-cultural sub-system. One major challenge in Patagonia is that ES are often unrecognized and undervalued and, therefore, ignored in market transactions, government policies and land and ocean management practices. In this context, this book aims to highlight the multiple ES provided by Patagonian ecosystems, their relation to ecosystems' functioning, how they sustain human well-being and the threats they are subjected to.

P. L. Peri (✉)
National Agricultural Technology Institute, Río Gallegos, Argentina
e-mail: peri.pablo@inta.gob.ar

L. Nahuelhual
Instituto de Economía, Universidad Austral de Chile Centro de Investigación: Dinámica de Ecosistemas Marinos de Altas Latitudes Instituto Milenio en Socio-Ecología Costera, Valdívia, Chile

G. Martínez Pastur
Consejo Nacional de Investigaciones Científicas y Técnicas, Centro Austral de Investigaciones Científicas y Técnicas, Ushuaia, Tierra del Fuego, Argentina

P. L. Peri et al. (eds.), *Ecosystem Services in Patagonia*, Natural and Social Sciences of Patagonia, https://doi.org/10.1007/978-3-030-69166-0_1

Keywords Ecosystem services · Patagonia · Landscape Planning · Sustainable management · Governance

1 The Patagonia Region

Patagonia lies within two countries (12% in Chile and 88% in Argentina) and comprises a diversity of ecosystems and livelihoods (Fig. 1.1). In Argentina, Patagonia includes five provinces (Neuquén, Río Negro, Chubut, Santa Cruz, and Tierra del Fuego) with an area of 197 million ha, which extends from latitudes 37.0° to 52.5° S (Peri et al. 2017).

In Chile, Patagonia does not exist as a political entity, constituting only an imaginary limit. It is most common to differentiate southern Patagonia from the territory located further north of the Corcovado Gulf. Southern Patagonia comprises the province of Palena and the regions of Aysén and Magallanes (excluding the Antarctic Territory). However, from a geological point of view (geological unity of the South Patagonian massif with the North Patagonian massif), the northern limit of Patagonia would be the Huincul fault, thus incorporating the Los Ríos and the Araucanía regions (Schilling et al. 2017). For the purpose of this book, we adopt the later criteria and set the northern limit of the Chilean Patagonia in the Cautín Province, in the Araucanía region.

The Patagonian region as a whole is between the semi-permanent anticyclones of the Pacific and the Atlantic oceans. The strong, constant west winds are dominant across the region. The Andes Mountains play a crucial role in determining climate. The north-south distribution of the mountains imposes an important barrier for humid air masses coming from the Pacific Ocean. Most of the water contained in these maritime air masses falls on the Chilean side, and the air becomes hotter and drier through adiabatic warming as it descends on the Argentinean side of the Andes. From the Andes and eastward, total precipitation decreases exponentially. Most of the central part of Argentinean Patagonia receives less than 200 mm year^{-1}, where the distance from the Andes explains 94% of the spatial variability of the mean annual precipitation (Jobbágy et al. 1995). In the northwest part of the region, annual precipitation decreases almost 7 mm per km, for the first 60 km eastward from the Andes. In this context, Patagonia can be defined as a temperate or cool-temperate region. A characteristic of the temperature pattern is the NW-SE distribution of the isotherms, determined mainly by the presence of the Andes. Mean annual temperature ranges from 12 °C in the northeast part to 3 °C towards the south.

The Chilean Patagonia has a temperate to cold climate, with temperatures that decrease notably from north to south. Thus, on average, southern Patagonia has a temperature of up to 10 °C less than the north. In addition, there is a temperature difference that is divided by the Patagonian Andes of up to 5 °C more to the west of the mountain range than to the east. This is due to the Humboldt Current that carries moisture to the continent, but this is abruptly interrupted a few kilometres inland by the mountains of the mountain range, causing the clouds to precipitate there.

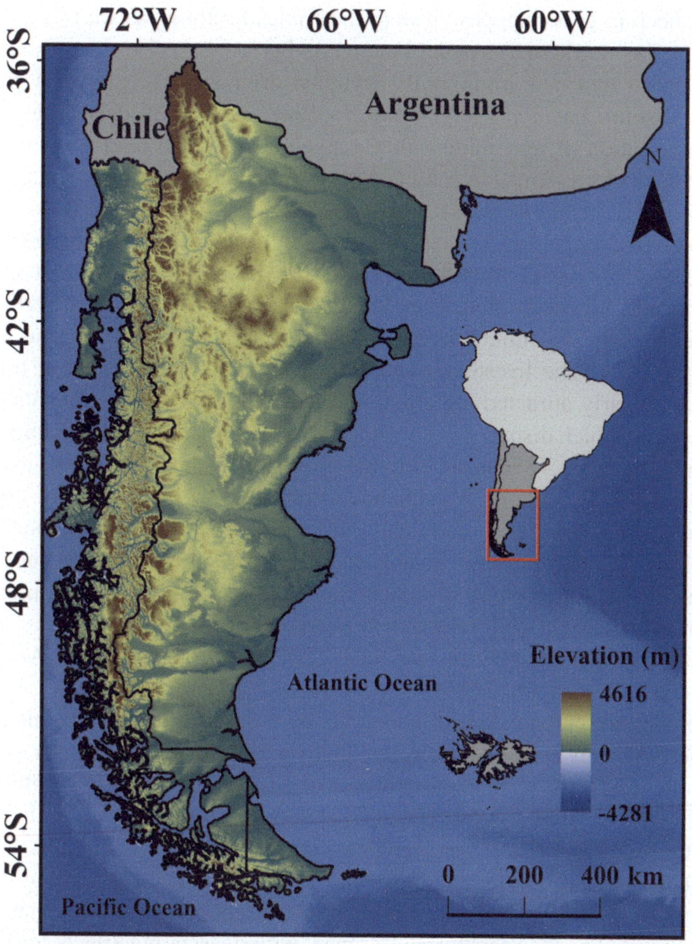

Fig. 1.1 Location of Patagonia region in Argentina and Chile. Elevations are based on data published by Jarvis et al. (2008) and Tozer et al. (2019)

Prior to the colonization of Patagonia, there were different ethnic groups (e.g. mapuches, tehuelches, selknam, yámanas), which were reduced, absorbed or exterminated (Borrero 1997; Martinic 1997). Fences and the concept of private property limited the opportunity for migratory hunting. Martinic (1997) reported that the "cultural exchange was unidirectional, from the colonists to the Indians, which affected the latter's hunting and fishing tools and customs relationship with natural resources, social behaviour, health and their very survival as people". In most cases, survivors became ranch employees. In 1866, a colony of Welsh immigrants had established in the lower valley of the Río Chubut (near the Atlantic coast, Argentina), and some settled in the west as sheep farmers in the Andes foothills (Mainwaring 1983). Later, in 1885, the first sheep farmers arrived and formal colonization started (e.g. mainly Spanish, Scottish, English, German and French) on the most suitable land for sheep in grass steppes close to the Andes foothills, as well as at southern Santa Cruz and northern Tierra del Fuego (Barbería 1995). The success of the first settlers caused that new immigrants occupied drier areas (e.g. steppe grasses in central and south Patagonia Argentina). In 1908, a national legislation prevented liberal distribution of remaining public rangelands and the land was divided into equal-sized blocks assigned to settlers without considering sheep-carrying capacity (Barbería 1995).

In the case of Chile, the first landmark of the formal occupation of Patagonia by the Chilean State was the founding of Fort Bulnes (year 1843), the first settlement of Republican Chile in the area. Then Punta Arenas was founded (year 1848), which came to control the maritime flow between the Atlantic and Pacific oceans, given its strategic position. The livestock expansion in the Malvinas, aimed at trade with England, had early attracted the attention of the successive governments of Chile and Argentina, which disputed the territorial sovereignty of southern Patagonia. "At the beginning of the 1870s, the Chilean Government had already instructed the Governor of the Colonia de Magallanes to promote the settlement of colonizers from Malvinas, offering them free land" (Harambour 2012). In 1876, the Governor of Magallanes travelled to Port Stanley to offer land concessions and buy animals. With this, the definitive phase of Patagonia's colonization began, which until then had been unsuccessful and confined to the precarious Chilean exclave of Punta Arenas, in the Magellan Strait, and to a few constructions on the Argentinean Atlantic coast (Harambour 2015).

The process of colonization of Patagonia by the Argentinean and Chilean states was conceived and executed from the metropolitan cities at Buenos Aires and Santiago. It was assisted by the expansion of British and German surplus capital, favoured both by racial criteria for the allocation of privileges and by the articulation of transnational corruption networks linking national authorities and local European elites. Corruption, and particularly the exercise of power and the influence of connections, articulated commercial and family networks that allowed States to materialize their territorial presence, with devastating effects for the original populations (Harambour 2017).

At present, Patagonia is a sparsely populated region of about three million people in 1,043,076 km^2 area, which determines a density of 1.9 inhabitants km^2. However, this region presents the lower proportion of inhabitants in rural zones (in Argentina only represents 9.9%). According to the Human Development Index (HDI), the Patagonia region in Argentina has a human development considered as very high (HDI = 0.861), with a Gini income coefficient of 0.379 in 2018 (1 = perfect inequality). Chilean Patagonia is also an area of low population density ranging from one person every 10 km^2 (Arturo Prat province) to 0.8 inhabitants per km^2 in the Aysén region (INE 2018).

Chilean Patagonia has a territory of 330,800 km^2 and a population of 2,440,460 inhabitants. The percentage of rural population reaches 27%, while the urban population is mostly located in the capital cities of Temuco, Valdivia, Puerto Montt, Coyhaique and Punta Arenas, which together account for 33% of the total population (INE 2018). In 2019, this territory contributed a near 9% to the national GDP (155,190 million Chilean pesos). In terms of human development, its average HDI reaches to 0.778, compared to the national HDI of 0.847. The distribution of income is relatively uneven (Gini 0.48), while land is distributed more evenly (Gini 0.26) among large properties.

The main economic activities in Patagonia, in both countries, have been mining (e.g. coal, gold, silver), livestock (notably sheep), agriculture (crops and fruit production near the Andes and valleys), fisheries and oil and gas. Sheep farming introduced in the late nineteenth century has been a principal economic activity, while tourism has increased its contribution in the last decades. Traditionally, the region has exploited its hydroelectric potential by building big hydroelectric dams in several rivers. Wind is a significant renewable energy source that is increasing in Patagonia, which could generate a very large amount of electrical energy.

The region has some of the most extensive wilderness areas in the planet (e.g. grasslands forests and shrublands) as well as many low impacted coastal marine ecosystems. Natural grasslands comprise almost 30% of the Americas' total (White et al. 2000), dominating the landscape in a diversity of regions including the Patagonia steppe (Argentina). Grasslands make a significant contribution to food security by providing part of the feed requirements of ruminants used for meat, wool and milk production. Profound changes, affecting key ecosystem functions and ultimately human well-being, have occurred in temperate grasslands. Livestock grazing for over 400 years has reduced soil organic carbon stocks by an estimated 22% and net primary production by 24% (Piñeiro et al. 2006).

In addition, Patagonia's temperate forests are widely recognized for their rich biodiversity and ecosystem services. They hold significant recreational values, are also key for carbon sequestration and storage, and play a pivotal role in water regulation (Armesto 2009; Peri et al. 2012). In Chile, where most of southern temperate forests are found, around 78% of the original forest remain (Luebert and Pliscoff 2006), thanks to large masses of remote forests in the southern part of the country, much of which is within protected areas. However, the unprotected forests have been subjected to different pressures over time and in the present. Among them is their replacement by fast-growing exotic plantations, which generally take the form of short rotation, even-aged monocultures of *Pinus radiata* and *Eucalyptus* species. The expansion of non-native tree plantations has important ecological consequences (Heilmayr et al. 2016). They alter singular natural resources, by reducing the diversity of fauna and flora as compared to native forests and scrublands (Paritsis and Aizen 2008). Non-native trees tend to consume more water than native trees and can be associated with reduced seasonal water provision (Lara et al. 2009).

The Patagonian Shelf Large Marine Ecosystem extends along the Southern Atlantic coast of South America from the Río de la Plata to Tierra del Fuego, covering an area of about 1.2 million km^2 (0.18% is protected). It contains a very rich biodiversity with species of warm, temperate and cold waters that support sea birds, marine mammal and invertebrates, and abundant fishery resources (Muñoz Sevilla and Le Bail 2017). In the Patagonian shelf, in recent years, a series of biological invasions including algae, molluscs, hydroids, bryozoans, ascidians and crustaceans occurred in marine environments because of involuntary transport or voluntary introduction, always with severe consequences not only for local biodiversity but also for local economies (Orensanz et al. 2002; Bigatti and Penchaszadeh 2008). Marine and coastal ecosystems provide food for people, opportunities for recreation and research, maritime transport for trade, renewable energies such as wind, wave and tidal power and a reservoir for fuels such as oil. However, there is increasing evidence of overfishing, toxic pollution, invasive species, nutrient over-enrichment, habitat degradation and biodiversity loss (Sherman and McGovern 2011).

Alike most ecosystems in the planet, Patagonian ecosystems are threatened by climate change and variability. High elevation areas have warmed in the northern Andes (Hofstede et al. 2014). Anthropogenic warming seems to have affected tree growth and increased recruitment above tree-line in some places (Daniels and Veblen 2004; Lutz et al. 2013) and some tree species below tree-line (Feeley et al. 2011).

There is a lack of information regarding the effects of individual and multiple drivers. For example, we know little about the combined effect of climate change and overgrazing and the consequences for pastoral livelihoods that depend on rangelands. Gaitán et al. (2014) found that temperature and the amount and seasonal distribution of precipitation were important controls of vegetation structure in Patagonian rangelands. Greco et al. (2014) reported that suitable areas for meadows (e.g. very productive areas for livestock production) would decrease by 7.8% in 2050 due to changes in climate. Thus, if evapotranspiration and drought stress increase as temperature increases and rainfall decreases in water-limited ecosystems, ranchers might be seriously affected due to a reduction of stocking rate and therefore families' income. In Argentina alone, the number of farmers (mainly family enterprises) exposed to climatic hazards (drought) is approximately 70,000–80,000 that hold 14 to 15 million sheep. Temperate forests will also be affected by climate change. A major drought from 1998 to 1999 coincident with a very hot summer led to extensive dieback in a *Nothofagus* species (Suarez et al. 2004). In the same way, the other dominant *Nothofagus* species presented several periodic droughts that have triggered a forest decline since 1940s (Rodríguez-Catón et al. 2016).

Thus, Patagonia, alike many of the last frontiers of the planet, faces increasing pressures from multiple and interacting drivers. The ES approach can help to understand and foresee the potential impacts of such interactions on ecosystems and people and contribute to a more balanced planning and decision-making regarding conservation, natural resources management and development. In this chapter, we (i) present the conceptual framework for the sustainable management of the natural ecosystems in Patagonia that guides the chapters of the book and (ii) highlight the importance of the ES approach for decision-making.

2 An Ecosystem Service–Based Framework for the Sustainable Management of the Natural Ecosystems of Patagonia

The concept of ecosystem services (ES) emphasizes the multiple connections between ecosystems and human well-being (MEA 2005). The focus on human-nature interactions differentiates the ES concept from narrower environmental management (Baker et al. 2013) and conservation perspectives (e.g. biological conservation). Since its introduction by early studies (Westman 1977; Ehrlich and Ehrlich 1981), research on ES has advanced significantly (Balvanera et al. 2020), and at present, it integrates ecological, economic, cultural and policy aspects. At the same time, the notion of ecosystem services approach (ESA), as an attempt to capture and visualize how natural ecosystem processes provide benefits to human society, has become highly attractive to policy makers by its focus on a broader societal involvement and the use of market instruments (Verburg et al. 2016). However, although the ES concept has been adopted at high-level policy frameworks (e.g. Convention on Biological Diversity, Intergovernmental Platform on Biodiversity and Ecosystem Services, World Bank's Global Partnership for Wealth Accounting, the Valuation of Ecosystem Services and EU Biodiversity Strategy), there is a discrepancy between the considerable conceptual understanding of ES concept in science and the limited practical application thereof (Díaz et al. 2006, 2015; Posner et al. 2016; Lautenbach et al. 2019; Matzek et al. 2019).

Essentially, the ES concept refers to the different goods and benefits that society obtains directly or indirectly from natural ecosystems that contribute to human well-being (Daily 1997; Costanza et al. 1997). Biodiversity conservation at different levels (e.g. genes, species and ecosystems) supports the ecological processes and functions that sustain ES and human well-being (Daily 1997). According to the Millennium Ecosystem Assessment (MEA 2005), ES usually are divided into four categories (Fig. 1.2): (i) provisioning services (e.g. food, timber and medicines), (ii) regulating services (e.g. water regulation, air quality maintenance, pollination, pest control and climate control), (iii) cultural services (e.g. recreation opportunities, sense of place, opportunities for education and identity) and (iv) supporting services (e.g. soil formation, primary productivity, biogeochemistry, nutrient cycling and provisioning of habitat). Provisioning services are likely the best recognized ES, since they contribute directly to human material well-being, unlike supporting services that contribute indirectly to human well-being by maintaining the processes and functions necessary for provisioning, regulating and cultural services (Kremen and Ostfeld 2005). Regulating ES such as water supply and quality or pest and disease control is also greatly recognized (De Groot et al. 2002). In turn, cultural ES are perceived as important for fulfilling basic human and social needs of a broad spectrum of stakeholders in many social-ecological contexts (Martín-López et al. 2012; Oteros-Rozas et al. 2014).

Our conceptual framework for the sustainable management of natural ecosystems in Patagonia starts from a non-dichotomous conception of the relationships

Fig. 1.2 Examples of different ecosystem services in Patagonia: (**a**) timber from *Nothofagus pumilio* forests; (**b**) food from fisheries in the coast of the Atlantic Ocean; (**c**) species habitat (e.g. *Hippocamelus bisulcus*); (**d**) sheep production in natural steppe grasslands; (**e**) recreation opportunities (e.g. Perito Moreno glacier); (**f**) water regulation and provision in mountain landscapes; (**g**) food from fisheries in the Pacific Ocean; (**h**) ornamental non-timber forest products and (**i**) opportunities for education and knowledge

between society and ecosystems. This determines a socio-ecosystem composed of different sub-systems: (i) a biophysical sub-system (ecosystem and biodiversity), where natural ecosystems and processes allow the provision of ES; (ii) a political sub-system ruled by the economic-productive activities, which greatly influence over public policies and (iii) a socio- cultural sub-system that reflects the arrangement and institutional functioning, public policies, the social organization of companies or families that take advantage of natural resources from a particular cultural dynamic (Fig. 1.3).

The multifaceted role of biodiversity supports the delivery of ES (Díaz et al. 2006; De Groot et al. 2010; Perera et al. 2018) in many ways. Genetic diversity of wild species is involved in the quality and production of provisioning services (e.g. food and wood) or contributes to the resistance to pest and climate variations. Species richness enhances provision (e.g. biomass production) and regulating services (e.g. water regulation and soil C sequestration). Finally, the biotic interaction among different species (e.g. predation, parasitism, competition and facilitation) sustains supporting services such as pollination, mycorrhizae interactions, seed dispersal and nitrogen fixing. In the Argentinean Patagonia, Rosas et al. (2020) evaluated the spatial relationships between biodiversity and ES in *Nothofagus* forests, and Chillo et al. (2018) evaluated the direct and indirect effects of biodiversity on

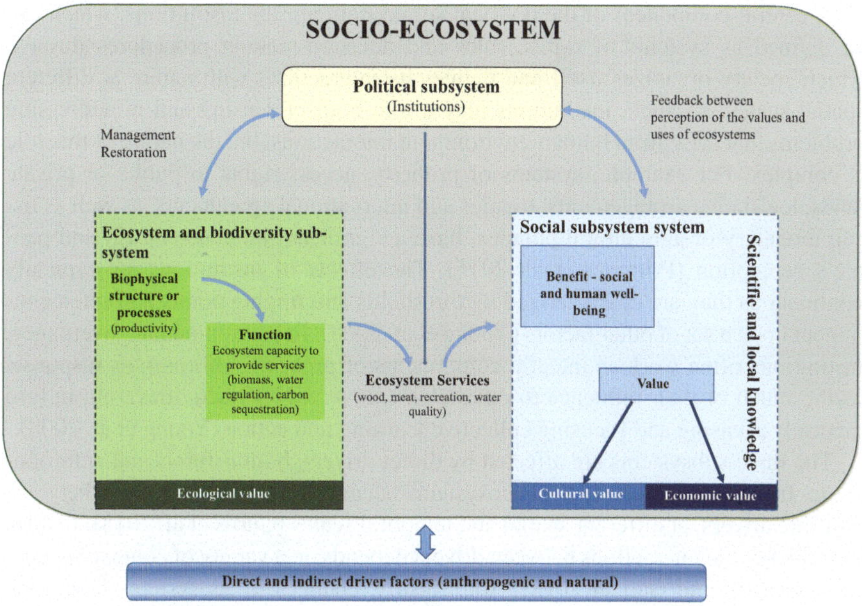

Fig. 1.3 Conceptual framework for sustainable management of natural ecosystems in Patagonia. (Adapted from Braat and De Groot 2012)

the provision of multiple ES in the native mixed forest and how the land-use intensity influenced over these variables. Studying these interactions allows a better understanding of the relationships between biodiversity and ES and of the effectiveness of biodiversity conservation strategies in providing both long-term sustainable biodiversity conservation and provision of ES (Costanza et al. 2007).

Ecosystem services sustain material or non-material benefits, which can be valued by people in different ways. Values in this context can be considered as a measure of the benefit from an ES. Some of those values (e.g. instrumental utilitarian values) can be expressed in monetary terms, whereas others do not, such as cultural or relational values.

Ecosystem services are generated at many ecological scales, and usually in large-scale processes, which may be driven by the sum impact of small-scale processes (Levin 1992). The spatial scales at which ES was supply has strong impacts for decision-making and influence over the valuing of different stakeholders related to ecosystem management. There are spatial scale mismatches; for example, the conservation of biodiversity that often represents a global value, it frequently occurs at the expense of local communities or farmers, whose use of natural resources for food or grazing animals may conflict with conservation actions and values. Since ES values and trade-offs have differential impacts on different contexts and spatial scales, high-level policies (e.g. national or regional) should be designed taking into account requirements for lower level policy implementation (e.g. local) to be sensitive to multiple trade-offs between the different ES (Hauck et al. 2013).

A central component of the political sub-system are the institutions, which can be defined as systems of rights, rules and decision-making procedures through which society organizes itself and defines its interactions with nature at different spatial scales. As such, institutions play a role both in causing and in addressing problems that arise from human-environment interactions, but the nature of this role is complex. For example, systems of property, access rights to public or private lands, legislative arrangements, treaties and international agreements, as well as fiscal, monetary or agricultural policies, have a significant influence on ES and people's perception (Primmer et al. 2015). The effects of institutions are typically nonlinear as they are characterized by thresholds and tipping points and often contingent upon a set of other factors (Young et al. 2008). In environmental governance, institutions often work as integral components of processes of complex response, owing much of their influence to other elements (e.g. NGO activities), but in turn themselves easing and focusing collective learning and action (Young et al. 2008).

The three subsystems are affected by direct drivers, both naturals and anthropogenic. In most cases, changes in ecosystems occur through the interaction between multiple drivers, at different spatial and temporal scales (Curtis et al. 2018). In turn, the effect of the interactions between drivers depends on a variety of context-specific factors (Kolb and Galicia 2018). Additionally, changes in ecosystems feed back drivers in complex ways (MEA 2005; IPBES 2019). For example, altered ecosystems create new opportunities and constraints on land use, induce institutional changes in response to resource degradation and scarcity and lead to social effects such as changes in income inequality (Laterra et al. 2019).

Our conceptual framework highlights direct drivers, but indirect drivers can also be important, such as population growth. Direct drivers can be natural (e.g. volcanic eruptions are very common on the Chilean Patagonia) or anthropogenic. Among the latter, land-use change and climate change are highly relevant. Land-use change is a process by which human activities (e.g. logging, road building and mining) transform the natural landscape, usually emphasizing the functional role of land for economic activities. Land-use changes are often nonlinear and might trigger feedbacks to the system, stress living conditions and threaten vulnerable people. Climate change affects ES (e.g. net primary productivity and water regulation) through extreme events, such as prolonged droughts, cyclones and floods. Long-term sustainability implies that ecosystems and their ES must be resilient to these drivers' effects and variations in the long term (Lindenmayer and Franklin 2002).

3 Importance of Ecosystem Service Assessments for Better Natural Resources and Environmental Decision-Making

Virtually all problems in natural resource and environmental management involve decisions: choices about planning, organizing, directing and controlling that must be made among alternative actions to achieve an objective (Conroy and Peterson 2013). At a very basic level, decision-making, either private or public, is about connecting decisions to objectives, and structured decision-making (Hammond et al. 2015) is just a formalized way of accomplishing that connection. This connection might seem obvious, but in many cases, problems in the management of resources can be or even directly caused, by poor framing of the decision problem.

Ecosystem services–based decision-making, as a new paradigm, means turning the ES recognition into incentives and institutions such as established policy regimes, processes and norms, which will guide wise investments and actions to conserve natural capital, on a large scale (Daily and Matson 2008). Yet, the ways through which ES knowledge is integrated into natural resources and environmental decision-making are complex (Fig. 1.4). The ES approach challenges in many ways the typically top-down, technocratic and linear processes that characterize much policy-making around the world. Firstly, the ES approach calls for a systemic view of environmental problems (Fig. 1.3), whereas at present, problems tend to be addressed in a fragmented manner. Forest management, for example, scarcely takes into account the role of forests in regulating water supply and the management of pastures scarcely recognizes the externalities it produces on watercourses. Secondly, an ES approach is inherently participatory, whereas top-down policies, such as the

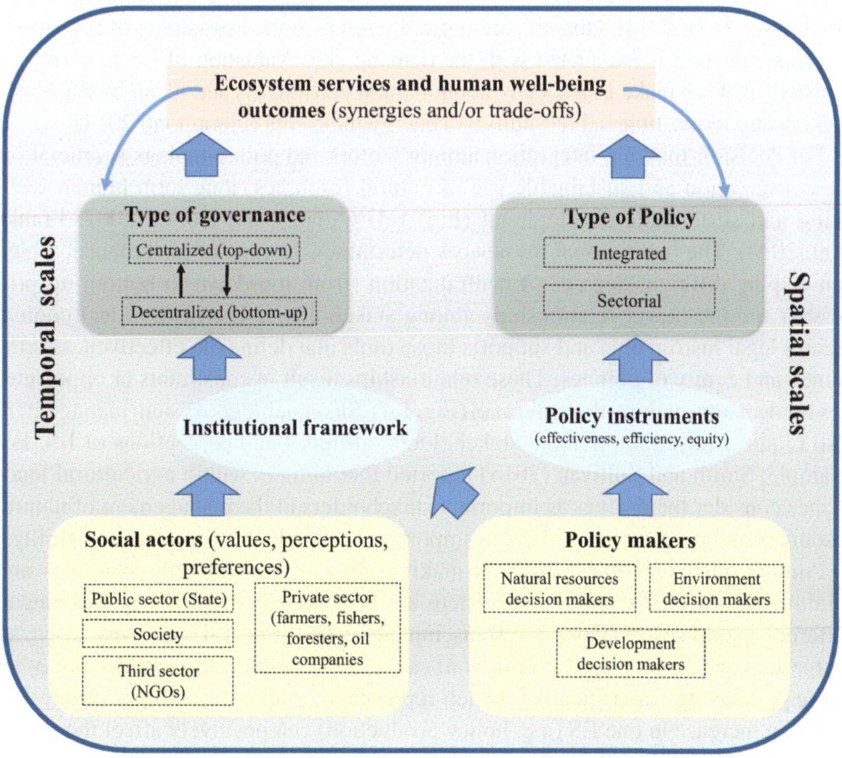

Fig. 1.4 Relationships between social actors and policy-makers for better decision-making that can influence the provision of ecosystem services

centralized creation of public protected areas, do not consider participatory processes, since their conservation objectives are usually detached from human well-being. Thirdly, the use of ES in decision-making calls for the analysis of interactions between the ecological system and the social system; decisions are important not only for the protection of a resource (water, soil, forest) or an ecosystem (wetland, watershed) but also for the well-being of people. Decision-making that adopts the ES is committed to deliver not only conservation outcomes but also well-being outcomes.

In this context, the incorporation of ES in decision-making should involve a process, which identifies key policy issues. During this process, the use of an ES approach is expected to provide new insights, the parts of the system to be considered in relation to these policy issues, the sensitivity of the considered ES to natural or social factors of change, the appropriate methods for ES assessment and the feasibility of implementation in practice (Beaumont et al. 2017). Assessment involves at least three steps: identification of key ES and beneficiaries, spatial representation of ES and beneficiaries, and valuation of ES. The SE assessment aims to reduce the deterioration of biodiversity and ecosystems and improve decision-making through the generation of knowledge about ecosystem functions and their contribution to society and translating this information into measures that can be communicated to decision makers and the community (Abson et al. 2014). Quantification usually refers to the assessment of ES supply, whereas valuation is associated with the demand side. Valuation of ES is useful for decisions that we make as a society about natural ecosystems and it can be expressed in monetary terms, time, labour units and energy indicators (Fioramonti 2014).

For decision-making, integration among sectors and policy-makers is crucial for the conservation and sustainable use of natural resources, long-term human well-being and sustainable development (Fig. 1.4) (Adams and Hutton 2007; Franks et al. 2014). The institutional framework determines the type of governance, which can display different degrees of centralization (from top-down to bottom-up processes) and the power relationships among stakeholders. The institutional context creates legal instruments and supports those tools that define the effectiveness, efficiency and equity of policies. These relationships result in constraints or opportunities for decision-making that affect ES (Fig. 1.4). Sustainable ecosystem management also requires information about stakeholders' attitudes and perceptions of ES. For example, Smith and Sullivan (2014) reported that farmers within agricultural landscapes consider themselves as important stakeholders in the management of natural resources and perceive several ES as important for productivity and sustainability.

Among other outcomes, decision-making aims at strengthening synergies and minimizing trade-offs among ecosystem services and among ES beneficiaries at different spatial scales (Gorg and Rauschmayer 2009) (Fig. 1.4). In many cases, an increase in one ES (e.g. food production) can negatively affect the provision of other ES (e.g. drinking water quality), which represents a trade-off among multiple ES, while an increase in one ES (e.g. honey production) can positively affect the provision of other ES (e.g. fruit production), which can be usually perceived as synergetic.

Stakeholder involvement in decision-making and ES assessment is important in order to understand their values and needs (Menzel and Teng 2009) and, therefore, their expected conducts towards nature. Different stakeholder groups can be expected to have different perspectives on the relative importance of the different ES (Vermeulen and Koziell 2002) and their actions and relative power can lead to very different conservation outcomes. Felipe-Lucia et al. (2015) demonstrated the relevance of power relationships in determining access to ES and its potential impacts on ES flows. According to these authors, identifying and targeting such power relationships is essential for delineating strong environmental management policies while reducing the trade-offs among ES and therefore reducing social inequalities and conflicts.

Integrating the concept of ES into planning and policy-making processes may require vertical policy integration between different levels of government institutions (e.g. international, national, regional and local) as well as horizontal integration across different sectors (e.g. agricultural, urban and tourism) with usually contradictory objectives (Saarikoski et al. 2018). In this context, collaborative approaches (concept of bottom-up governance) are a good option that emphasize the importance of knowledge accumulation, collective learning and sensitivity to changes in the framework of adaptive governance of social-ecological systems (Fraser et al. 2006).

In Patagonia, profound changes are occurring that are affecting key ecosystem functions and ultimately human well-being. This is at least partially due to the fact that most ES go unrecognized and undervalued in economic decisions (production and transaction), government policies and management practices. Historically, markets have largely focused on provisioning services (e.g. timber products, livestock and fish) while neglecting the interdependent roles of regulating services (e.g. erosion and climate control), supporting services (e.g. nutrient cycling) and cultural services (e.g. recreation, local identity and tourism).

In this context, this book aims to reveal the importance of biodiversity and ES to sustain human benefits and well-being and the factors that contribute to their impairment. The chapters in this book cover most of the elements in the conceptual framework presented here (Fig. 1.3) and provide recommendations for a better management and governance of Patagonian ecosystems.

References

Abson DJ, von Wehrden H, Baumgärtner S, Fischer J, Hanspach J, Härdtle W, Heinrichs H, Klein AM, Lang DJ, Martens P, Walmsley D (2014) Ecosystem services as a boundary object for sustainability. Ecol Econ 103:29–37

Adams WM, Hutton J (2007) People, parks and poverty: political ecology and biodiversity conservation. Conserv Soc 5:147–183

Armesto JJ (2009) Annual reports to council Ecological Society of America August 2009. Bull Ecol Soc of America 90(4):360–431

Baker J, Sheate WR, Phillips P, Eales R (2013) Ecosystem services in environmental assessment: help or hindrance? Environ Impact Assess Rev 40:3–13

Balvanera P, Pérez-Harguindeguy N, Perevochtchikova M, Laterra P, Cáceres DM, Langle-Flores A (2020) Ecosystem services research in Latin America 2.0: expanding collaboration across countries, disciplines, and sectors. Ecosyst Ser 42:e101086

Barbería E (1995) Los dueños de la tierra de la Patagonia Austral: 1880-1920. Universidad Federal de la Patagonia Austral, Río Gallegos
Beaumont NJ, Mongruel R, Hooper T (2017) Practical application of the Ecosystem Service Approach (ESA): lessons learned and recommendations for the future. Int J Biodiv Sci Ecosyst Ser Manage 13(3):68–78
Bigatti G, Penchaszadeh PE (2008) Invertebrados del Mar Patagónico, diagnóstico de la problemática actual y potencial de su conservación y manejo. In: Zar MA (ed) Estado de conservación del Mar Patagónico y áreas de influencia. Edición del Foro, Puerto Madryn, pp 105–133
Borrero LA (1997) The origins of ethnographic subsistence patterns in Fuego-Patagonia. In: McEwan C, Borrero LA, Prieto A (eds) Patagonia: natural history, prehistory and ethnography at the uttermost end of the Earth. British Museum Press, London, pp 60–81
Braat LC, de Groot R (2012) The ecosystem services agenda: bridging the worlds of natural science and economics, conservation and development, and public and private policy. Ecosyst Sers 1:4–15
Chillo V, Vázquez DP, Amoroso MM, Bennett EM (2018) Land-use intensity indirectly affects ecosystem services mainly through plant functional identity in a temperate forest. Funct Ecol 32:1390–1399
Conroy MJ, Peterson JT (2013) Decision making in natural resource management: a structured, adaptive approach. Wiley, Washington, DC
Costanza R, Arge R, de Groot R, Farber S, Grasso M, Hannon B, Limburg K, Naeem S, Oneill RV, Paruelo J, Raskin RG, Sutton P, van den Belt M (1997) The value of the world's ecosystem services and natural capital. Nature 387(6630):253–260
Costanza R, Fisher B, Mulder K, Liu S, Christopher T (2007) Biodiversity and ecosystem services: a multi-scale empirical study of the relationship between species richness and net primary production. Ecol Econ 61:478–491
Curtis PG, Slay CM, Harris NL, Tyukavina A, Hansen MC (2018) Classifying drivers of global forest loss. Science 361(6407):1108–1111
Daily GC (1997) Nature's services. Island Press, Washington, DC
Daily GC, Matson PA (2008) Ecosystem services: from theory to implementation. PNAS 105(28):9455–9456
Daniels LD, Veblen TT (2004) Spatio-temporal influences of climate on altitudinal tree-line in northern Patagonia. Ecology 85(5):1284–1296
De Groot RS, Wilson MA, Boumans RM (2002) A typology for the classification, description and valuation of ecosystem functions, goods and services. Ecol Econ 41(3):393–408
De Groot RS, Alkemade R, Braat L, Hein L, Willemen L (2010) Challenges in integrating the concept of ecosystem services and values in landscape planning, management and decision-making. Ecol Complex 7(3):260–272
Díaz S, Fargione J, Chapin FS, Tilman D (2006) Biodiversity loss threatens human well-being. PLoS Biol 4(8):e277
Díaz S, Demissew S, Carabias J, Joly C, Lonsdale M, Ash N, Larigauderie A, Adhikari J, Arico S, Báldi A, Bartuska A, Baste I, Bilgin A, Brondizio E, Chan KMA, Figueroa V, Duraiappah A, Fischer M, Hill R, Koetz T, Leadley P, Lyver P, Mace G, Martin-Lopez B, Okumura M, Pacheco D, Pascual U, Pérez E, Reyers B, Roth E, Saito O, Scholes R, Sharma N, Tallis H, Thaman R, Watson R, Yahara T, Hamid Z, Akosim C, Al-Hafedh Y, Allahverdiyev R, Amankwah E, Asah S, Asfaw Z, Bartus G, Brooks L, Caillaux J, Dalle G, Darnaedi D, Driver A, Erpul G, Escobar-Eyzaguirre P, Failler P, Fouda A, Fu B, Gundimeda H, Hashimoto S, Homer F, Lavorel S, Lichtenstein G, Mala W, Mandivenyi W, Matczak P, Mbizvo C, Mehrdadi M, Metzger JP, Mikissa JB, Moller H, Mooney H, Mumby P, Nagendra H, Nesshover C, Oteng-Yeboah A, Pataki G, Roué M, Rubis J, Schultz M, Smith P, Sumaila R, Takeuchi K, Thomas S, Verma M, Yeo-Chang Y, Zlatanova D (2015) The IPBES conceptual framework: connecting nature and people. Curr Opin Environ Sust 14:1–16
Ehrlich PR, Ehrlich AH (1981) The causes and consequences of the disappearance of species. Q Rev Biol 1:82–85

Feeley KJ, Silman MR, Bush MB, Farfan W, Cabrera KG, Malhi Y, Meir P, Revilla NS, Quisiyupanqui MNR, Saatchi S (2011) Upslope migration of Andean trees. J Biogeogr 38(4):783–791

Felipe-Lucia MR, Martín-López B, Lavorel S, Berraquero-Díaz L, Escalera-Reyes J, Comín FA (2015) Ecosystem services flows: why stakeholders' power relationships matter. PLoS One 10(7):e0132232

Fioramonti L (2014) How numbers rule the world: the use and abuse of statistics in global politics. Zed Books, London

Franks DM, Davis R, Bebbington AJ, Ali SH, Kempa D, Scurrahg M (2014) Conflict translates environmental and social risk into business costs. PNAS 111:7576–7581

Fraser EDG, Dougill AD, Mabee WE, Reed M, McAlpine P (2006) Bottom up and top down: analysis of participatory processes for sustainability indicator identification as a pathway to community empowerment and sustainable environmental management. J Environ Manag 78:114–127

Gaitan JJ, Oliva GE, Bran DE, Maestre FT, Aguiar MR, Jobbagy EG, Buono GG, Ferrante D, Nakamatsu VB, Ciari G, Salomone JM, Massara V (2014) Vegetation structure is as important as climate for explaining ecosystem function across Patagonian rangelands. Ecology 102:1419–1428

Gorg C, Rauschmayer F (2009) Multi-level-governance and the politics of scale-the challenge of the millennium ecosystem assessment. In: Kutting G, Lipschutz R (eds) Environmental governance, power and knowledge in a local-global world. Routledge, New York, pp 81–99

Greco RD, Nielsen CK, Didier KA (2014) Climate change and conservation implications for wet meadows in dry Patagonia. Environ Conserv 41(2):122–131

Hammond JS, Keeney RL, Raiffa H (2015) Smart choices: a practical guide to making better decisions. Harvard Business Review Press, New York

Harambour, A (2012) Borderland sovereignties: postcolonial colonialism and state making in Patagonia, Argentina and Chile, 1840s-1922. Doctoral thesis, State University of New York, Stony Brook. New York, USA

Harambour A (2015) Documentos inéditos Para la historia de Magallanes. Memorias de William blain en Malvinas y Patagonia (c. 1881-1890). Magallania 43(2):223–249

Harambour A (2017) Soberanía y corrupción: La construcción del Estado y la propiedad en Patagonia Austral (Argentina y Chile, 1840-1920) 1. Historia (Santiago) 50(2):555–596

Hauck J, Gorg C, Varjopuro R, Ratamaki O, Jax K (2013) Benefits and limitations of the ecosystem services concept in environmental policy and decision-making: some stakeholder perspectives. Environ Sci Pol 25:13–21

Heilmayr R, Echeverría C, Fuentes R, Lambin EF (2016) A plantation-dominated forest transition in Chile. Appl Geogr 75:71–82

Hofstede R, Calles J, López V, Polanco R, Torres F, Ulloa J, Cerra M (2014) Los Páramos Andinos ¿Qué sabemos? Estado de conocimiento sobre el impacto del cambio climático en el ecosistema páramo. UICN, Quito

Instituto Nacional de Estadísticas (INE) (2018) Síntesis de resultados CENSO 2017. Santiago, Chile

IPBES (2019) In: Díaz S, Settele J, Brondízio ES, Ngo HT, Guèze M, Agard J, Arneth A, Balvanera P, Brauman KA, Butchart SH, Chan KMA, Garibaldi LA, Ichii K, Liu J, Subramanian SM, Midgley GF, Miloslavich P, Molnár Z, Obura D, Pfaff A, Polasky S, Purvis A, Razzaque J, Reyers B, Roy Chowdhury R, Shin YJ, Visseren-Hamakers J, Willis KJ, Zayas CN (eds) Summary for policymakers of the global assessment report on biodiversity and ecosystem services of the Intergovernmental Science-Policy Platform on Biodiversity and Ecosystem Services. IPBES Secretariat, Bonn. 56 pp.

Jarvis A, Guevara E, Reuter HI, Nelson AD (2008) Hole-filled SRTM for the globe: version 4: data grid. CGIAR Consortium for Spatial Information

Jobbágy EG, Paruelo JM, León RJC (1995) Estimación de la precipitación y de su variabilidad inter-anual a partir de información geográfica en el NW de la Patagonia, Argentina. Ecol Austral 5:47–53

Kolb M, Galicia L (2018) Scenarios and story lines: drivers of land use change in southern Mexico. Environ Dev Sust 20(2):681–702
Kremen C, Ostfeld RS (2005) A call to ecologists: measuring, analyzing, and managing ecosystem services. Front Ecol Environ 3:540–548
Lara A, Little C, Urrutia R, McPhee J, Álvarez-Garretón C, Oyarzún C, Soto D, Donoso P, Nahuelhual L, Pino M, Arismendi I (2009) Assessment of ecosystem services as an opportunity for the conservation and management of native forests in Chile. For Ecol Manag 258(4):415–424
Laterra P, Nahuelhual L, Vallejos M, Berrouet L, Pérez EA, Enrico L, Salas D (2019) Linking inequalities and ecosystem services in Latin America. Ecosyst Ser 36:e100875
Lautenbach S, Mupepele AC, Dormann CF, Lee H, Schmidt S, Scholte SS, Seppelt R, van Teeffelen A, Verhagen W, Volk M (2019) Blind spots in ecosystem services research and challenges for implementation. Reg Environ Chang 19:2151–2172
Levin SA (1992) The problem of pattern and scale in ecology. Ecology 73:1943–1967
Lindenmayer D, Franklin J (2002) Conserving forest biodiversity: a comprehensive multiscaled approach. Island Press, Washington, DC
Luebert F, Pliscoff P (2006) Sinopsis bioclimática y vegetacional de Chile. Ed Universitaria, Santiago
Lutz DA, Powell RL, Silman MR (2013) Four decades of Andean timberline migration and implications for biodiversity loss with climate change. PLoS One 8(9):e74496
Mainwaring MJ (1983) From the Falklands to Patagonia. Allison & Busby, London
Martinic BM (1997) The meeting of two cultures. Indians and colonists in the Magellan Region. In: McEwan C, Borrero LA, Prieto A (eds) Patagonia. Natural history, prehistory and ethnography at the uttermost end of the Earth. British Museum Press, London, pp 85–102
Martín-López B, Iniesta-Arandia I, García-Llorente M, Palomo I, Casado-Arzuaga I, García del Amo D, Gómez-Baggethun E, Oteros-Rozas E, Palacios-Agundez I, Willaarts B, González JA, Santos Martín F, Onaindia M, López Santiago C, Montes C (2012) Uncovering ecosystem service bundles through social preferences. PLoS One 7(6):e38970
Matzek V, Wilson KA, Kragt M (2019) Mainstreaming of ecosystem services as a rationale for ecological restoration in Australia. Ecosyst Ser 35:79–86
Menzel S, Teng J (2009) Ecosystem services as a stakeholder-driven concept for conservation science. Conserv Biol 24:907–909
Millennium Ecosystem Assessment (MEA) (2005) Ecosystems and human Well-being: synthesis. World Resources Institute, Washington, DC
Muñoz Sevilla NP, Le Bail M (2017) Latin American and Caribbean regional perspective on ecosystem based management (EBM) of large marine ecosystems goods and services. Environ Dev 22:9–17
Orensanz JM, Schwindt E, Pastorino G, Bortolus A, Casas G, Darrigran G, Elías R, López Gappa JJ, Obenat S, Pascual M, Penchaszadeh P, Píriz ML, Scarabino F, Spivak E, Vallarino EA (2002) No longer the pristine confines of the world ocean: a survey of exotic marine species in the southwestern Atlantic. Biol Inv 4:115–143
Oteros-Rozas E, Martín-López B, González JA, Plieninger T, López CA, Montes C (2014) Sociocultural valuation of ecosystem services in a transhumance social-ecological network. Reg Environ Chan 14:1269–1289
Paritsis J, Aizen MA (2008) Effects of exotic conifer plantations on the biodiversity of understory plants, epigeal beetles and birds in *Nothofagus dombeyi* forests. For Ecol Manag 255(5–6):1575–1583
Perera A, Peterson U, Martínez Pastur G, Iverson L (2018) Ecosystem services from forest landscapes: Broadscale considerations. Springer, Cham
Peri PL, Ladd B, Pepper DA, Bonser SP, Laffan SW, Amelung W (2012) Carbon (δ 13C) and nitrogen (δ 15N) stable isotope composition in plant and soil in southern Patagonia's native forests. Global Chang Biol 18(1):311–321

Peri PL, Caballe G, Hansen NE, Bahamonde HA, Lencinas MV, von Müller AR, Ormaechea S, Gargaglione V, Soler R, Sarasola M, Rusch V, Borrelli L, Fernandez ME, Gyenge J, Tejera LE, Lloyd CE, Martínez Pastur G (2017) Silvopastoral systems in Patagonia, Argentina. In: Gordon AM, Newman SM, Coleman BRW (eds) Temperate agroforestry systems. CAB International, Wallingford, pp 252–273

Piñeiro G, Paruelo JM, Oesterheld M (2006) Potential long term impacts of livestock introduction on carbon and nitrogen cycling in grasslands of southern South America. Global Chang Biol 12(7):1267–1284

Posner SM, McKenzie E, Ricketts TH (2016) Policy impacts of ecosystem services knowledge. PNAS 113(7):1760–1765

Primmer E, Jokinen P, Blicharska M, Barton DN, Bugter R, Potschin M (2015) Governance of ecosystem services: a framework for empirical analysis. Ecosyst Serv 16:158–166

Rodríguez-Catón M, Villalba R, Morales M, Srur A (2016) Influence of droughts on *Nothofagus pumilio* forest decline across northern Patagonia, Argentina. Ecosphere 7(7):e01390

Rosas YM, Peri PL, Bahamonde H, Cellini JM, Barrera MD, Huertas Herrera A, Lencinas MV, Martínez Pastur G (2020) Trade-offs between management and conservation for the provision of ecosystem services in the southern Patagonian forests. In: Stanturf J (ed) Achieving sustainable management of boreal and temperate forests. Burleigh Dodds Series in Agricultural Science, Cambridge. Chapter 6, 33 pp.

Saarikoski H, Primmer E, Saarela S-R, Antunes P, Aszalós R, Baró F, Berry P, Garcia Blanco G, Goméz-Baggethun E, Carvalho L, Dick J, Dunford R, Hanzu M, Harrison PA, Izakovicova Z, Kertész M, Kopperoinen L, Köhler B, Langemeyer J, Lapola D, Liquete C, Luque S, Mederly P, Niemelä J, Palomo I, Martínez Pastur G, Peri PL, Preda E, Priess JA, Santos R, Schleyer C, Turkelboom F, Vadineanu A, Verheyden W, Vikström S, Young J (2018) Institutional challenges in putting ecosystem service knowledge in practice. Ecosyst Ser 29:579–598

Schilling ME, Carlson RW, Tassara A, Conceição RV, Bertotto GW, Vásquez M, Muñoz D, Jalowitzki T, Gervasoni F, Morata D (2017) The origin of Patagonia revealed by Re-Os systematics of mantle xenoliths. Precambrian Res 294:15–32

Sherman K, McGovern G (2011) Toward recovery and sustainability of the World's large marine ecosystems during climate change. IUCN, Gland

Smith HF, Sullivan CA (2014) Ecosystem services within agricultural landscapes-farmers' perceptions. Ecol Econ 98:72–80

Suarez ML, Ghermandi L, Kitzberger T (2004) Factors predisposing episodic drought-induced tree mortality in *Nothofagus*-site, climatic sensitivity and growth trends. Ecology 92(6):954–966

Tozer B, Sandwell DT, Smith WHF, Olson C, Beale JR, Wessel P (2019) Global bathymetry and topography at 15 arc sec: SRTM15+. Earth Space Sci 6(10):1847–1864

Verburg R, Selnes T, Verweij P (2016) Governing ecosystem services: national and local lessons from policy appraisal and implementation. Ecosyst Ser 18:186–197

Vermeulen S, Koziell I (2002) Integrating global and local values: a review of biodiversity assessment. IIED, London

Westman WE (1977) How much are nature's services worth? Science 197(4307):960–964

White R, Murray S, Rohweder M (2000) Pilot analysis of global ecosystems: grassland ecosystems. World Resources Institute, Washington, DC

Young OR, King LA, Schroeder H (2008) Institutions and environmental change: principal findings, applications, and research frontiers. The MIT Press, Cambridge/London, 397 pp

Chapter 2
Assessment of Provisioning Ecosystem Services in Terrestrial Ecosystems of Santa Cruz Province, Argentina

Yamina Micaela Rosas, Pablo L. Peri, and Guillermo Martínez Pastur

Abstract Provisioning ecosystem services play an important role in the development of regional economies. Traditional managements usually intensify the supply of provisioning services, without consideration of other services (e.g. cultural and supporting) and biodiversity. The objective of this chapter was to characterize main provisioning ecosystem services and potential biodiversity in different terrestrial ecosystems (native forests, shrublands and grasslands) of Santa Cruz Province (Southern Patagonia, Argentina) and to identify potential trade-off areas between provisioning ecosystem services and biodiversity conservation values. We found that non-forested areas exhibited higher supply of provisioning ecosystem services and biodiversity values than forested areas, where potential trade-off areas were located in humid steppes and shrublands. Particularly, in *Nothofagus* forests landscape, provisioning ecosystem services and biodiversity increased with forest cover, where *N. antarctica* forests type showed more potential trade-off areas than other *Nothofagus* forests type, while new potential protected areas were located when different forest types were combined (*N. antarctica* and *N. pumilio*). These results can be used by decision-makers to improve management and conservation strategies on private lands.

Keywords Ecosystem services · Biodiversity values · Ecological areas · *Nothofagus* forest types · Trade-offs · Landscape scale

---

Y. M. Rosas (✉)
Centro Austral de Investigaciones Científicas (CADIC), Consejo Nacional de Investigaciones Científicas y Técnicas (CONICET), Ushuaia, Tierra del Fuego, Argentina
e-mail: micaela.rosas@cadic-conicet.gob.ar

P. L. Peri
National Agricultural Technology Institute, Río Gallegos, Argentina

G. Martínez Pastur
Consejo Nacional de Investigaciones Científicas y Técnicas, Centro Austral de Investigaciones Científicas y Técnicas, Ushuaia, Tierra del Fuego, Argentina

P. L. Peri et al. (eds.), *Ecosystem Services in Patagonia*, Natural and Social Sciences of Patagonia, https://doi.org/10.1007/978-3-030-69166-0_2

1 Introduction

Natural ecosystems provide multiple services and goods to people, usually named as ecosystem services (ES) (MEA 2005). There are different ES (provisioning, supporting, regulating and cultural) and are known as provisioning ecosystem services (PrES), the most important ES for societies (Ala-Hulkko et al. 2019). Natural capital includes all natural resources that society uses. ES are provide by biotic organisms or an interaction with abiotic processes (Haines-Young and Potschin 2018). In this context, the last version of ES classification (CICES v.5) includes abiotic ES related to mineral substances that are used as energy sources (e.g. crude fossil fuels). The cascade model for the landscape in Southern Patagonia proposed by Rosas et al. (2019a) linked the forest ecosystem with social systems and identified the potential synergies (positives and negatives) between management and conservation planning. The different terrestial ecosystems of the region determined specific PrES related to biophysical characteristics (e.g. climate, topography and vegetation), and where policy decisions impacted on how these services were obtained through the management strategies implementation (Peri et al. 2016a, b, c; Perera et al. 2018).

Patagonian ecosystems (e.g. steppes and native forests) provide different ES to local people (Laterra et al. 2011). However, when ecosystems are managed only to maximize PrES, many other ES (e.g. regulating or cultural) and biodiversity values are usually undervalued (Thompson et al. 2011; Oñatibia et al. 2015; Martínez Pastur et al. 2017; Perera et al. 2018). Sheep production is one of the most important economic activities in Santa Cruz Province based on extensive grazing and provide PrES as lamb and wool animal. Different studies analysed sheep breeding in the province and determined sheep-carrying capacity (Andrade et al. 2016), trade-off between livestock and biodiversity (Pedrana et al. 2010; Peri et al. 2013; Travaini et al. 2015) and the impact of grazing on soil properties (Peri et al. 2016a). Oil production is also another important economic activity and provides abiotic PrES as crude oil, where the wells establishments, accessibility infrastructure, pipelines and other oil facilities presented different impacts on natural areas by generating habitat fragmentation (Buzzi et al. 2019) and increasing potential desertification processes (Del Valle et al. 1998; Gaitán et al. 2019). Fiori and Zalba (2003) determined that vegetation recovery on seismic lines is extremely poor and facilitates expansion of exotic invasive plants.

In addition, native forest ecosystems provide PrES such as timber wood, fibre or firewood (Gea et al. 2004), food (e.g. fruits, nuts, mushrooms, honey or spices), pharmaceutical plants and other non-woody industrial products (MEA 2005). Studies had been developed to determine timber production of different *Nothofagus* species, especially for *N. pumilio* forests (NP) (Peri et al. 2019a), and to define new silvicultural proposals (Gea et al. 2004; Martínez Pastur et al. 2009, 2019) considering different economics and conservation values. In addition, silvopastoral systems, which combine trees and grasslands or pastures under grazing in the same unit of land, became an economical, ecological and social productive alternative in N. antarctica forests (NA) (Peri and Ormaechea 2013; Peri et al. 2016b), which combine trees and grasslands or pastures under grazing in the same unit of land, became an economical, ecological and social productive alternative in Patagonia. Silvopastoral systems are designed to increase the provision of ES from managed forests, such as livestock (e.g. cattle, goats and sheep) that generates different products (e.g. meat, milk, wool and leather) (Peri et al. 2016b).

In the last years, the interest to understand the relationship between ES supply and biodiversity had increased (Currie 2011; Mace et al. 2012; Maes et al. 2014). Biodiversity had been defined as critical to support ES delivery (Mori et al. 2017) through its role in functional processes (Thompson et al. 2011; Harrison et al. 2014). In fact, some authors suggested that biodiversity itself can be considered as an ES (Mace et al. 2012). In Santa Cruz Province, there are antecedents related to conservation of emblematic species (e.g. *Lama guanicoe*) (Pedrana et al. 2010; Travaini et al. 2015), endangered species (e.g. *Hippocamelus bisulcus*) (Vila et al. 2006; Flueck and Smith-Flueck 2012) and endemic species of darkling beetles (Carrara and Flores 2013) and lizard (Breitman et al. 2014).

Understanding the connections between ES (especially PrES) and biodiversity has been a challenge due to multiple (e.g. ecological, social and scales) perspectives (De Groot et al. 2010; Thompson et al. 2011), mainly in remote areas due to lack of data (Martínez Pastur et al. 2017). Recent methodologies have improved the assessment of species distributions, synergies and trade-offs among ES and biodiversity at different spatiotemporal scales (Raudsepp-Hearne et al. 2010; Cordingley et al. 2016) using scarce available data from field works and remote sensing approaches (Martínez Pastur et al. 2016b).

In the Patagonian region, some studies analysed the impacts of livestock on plants biodiversity (Peri et al. 2013, 2016a, c) and changes on arthropods richness and abundance (Sola et al. 2016; Lescano et al. 2017). In addition, Rosas et al. (2019a) tried to describe the importance of these connections in Southern Patagonian forests, as well as some studies of plant and insect assemblages in non-managed (Peri and Ormaechea 2013; Peri et al. 2019a, b) and harvested *Nothofagus* forests (Gargaglione et al. 2014). During the last years, several studies conducted in Southern Patagonia reported maps of supporting, regulating (Peri et al. 2018, 2019b) and cultural ES (Martínez Pastur et al. 2016a; Rosas et al. 2019a). In addition, potential biodiversity maps (PBM) combining potential habitat suitability maps of different taxa were developed (Martínez Pastur et al. 2016b; Rosas et al. 2018, 2019b, c). PBM, that synthetize the information of several species, became a useful tool to define better management and conservation planning (Rosas et al. 2019b), define the effectiveness of the current protected areas network (Rosas et al. 2018), and identify hotspot areas (Rosas et al. 2019c) and different trade-offs among ES and biodiversity (Martínez Pastur et al. 2017).

In this context at landscape level, the main challenge is to decide the best option of land use management (production and/or conservation) (Carpenter et al. 2009; Raudsepp-Hearne et al. 2010; Cordingley et al. 2016). Mapping methodologies had been used to support policy decisions (De Groot et al. 2010; Maes et al. 2012) by incorporating landscape heterogeneity (Martínez Pastur et al. 2017). Land-use decisions depend on public policies such as the national law no. 26331/07 that defined forest areas under different uses (timber, restoration, conservation). However, the use of these information (e.g. supply of PrES) by public and private policies is scarce (Braat and De Groot 2012). PrES and biodiversity integration analysis may improve the current conservation plans (e.g. identify areas with the highest PBM values), increase the landscape multi-functionality (e.g. combination of sustainable economic activities) or reduce economic costs of companies (Raudsepp-Hearne et al. 2010; Mori et al. 2017).

The objective of this chapter was to analyse the different provisioning ecosystem services (PrES) and potential biodiversity (PBM) in terrestrial ecosystems of Santa Cruz Province (Southern Patagonia, Argentina), with special emphasis on *Nothofagus* forest landscapes. Also, we aimed to identify the (i) potential trade-offs between PrES and MPB outside of the networking protected area, (ii) areas with high MPB and low PrES values to suggest new potential protected areas and (iii) areas with low MPB and high PrES values where conflicts are low and intensification of the management activities is possible.

2 Study Case in Santa Cruz Province

2.1 Study Area

Southern Patagonia includes Santa Cruz Province (Argentina), which covers 243,943 km^2 (Fig. 2.1a) and presents a variety of terrestrial ecosystems dominated by dry steppes in the north and centre; humid steppes and shrublands in the south; and sub-Andean grasslands, *Nothofagus* forests and alpine vegetation occupying a narrow strip near the Andes mountains (Oliva et al. 2004) (Fig. 2.1b). The province presents 7% of the total area under protection (Fasioli and Díaz 2011), while most of the areas are private lands (93%). National parks mainly preserve forests and ice fields close to the mountains in the west (e.g. Perito Moreno National Park), and provincial reserves mainly protect special features in the steppe landscape (e.g. Meseta Espinosa y El Cordón Provincial Reserve) (Fig. 2.1c). Despite this, most of the protected area networks are located near the Andean mountains, where *Nothofagus* forest types are not equally protected (Rosas et al. 2019a). These forests types are distributed from 46° to 52° SL, in a wide range of rainfall, temperature patterns and elevation gradients (Veblen et al. 1996; Peri and Ormaechea 2013; Peri et al. 2019a) (Fig. 2.1d). Detail of *Nothofagus* forests, which names are related to lakes and cities, showed *Nothofagus* forest types distribution (Fig. 2.1d I, II, III, IV and V), where 69% of NP forests (2246 km^2) are protected and mainly distributed in the north and central areas of the province, 82% of mixed evergreen forests (180 km^2) are protected in central areas. NA forests (1699 km^2) prevail in the southern area and only 16% are under protection.

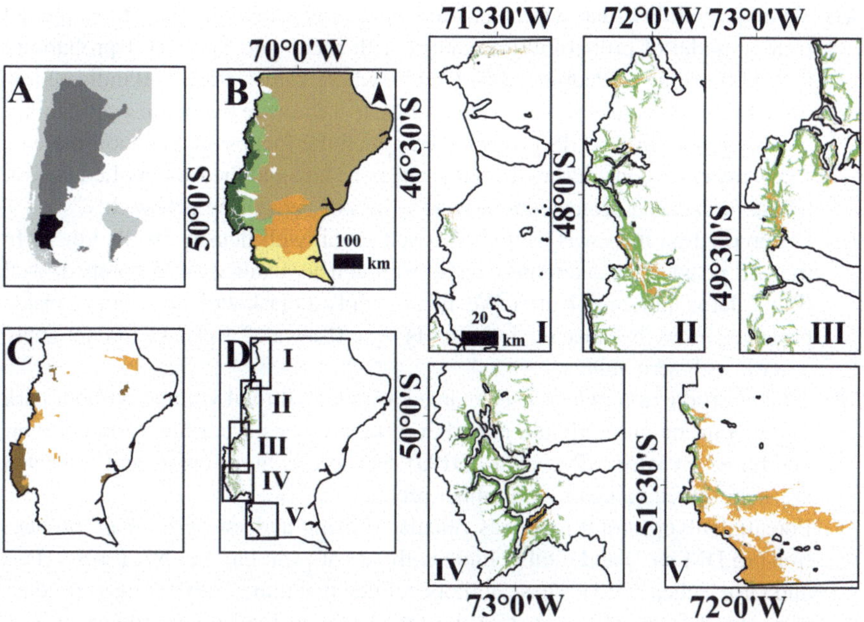

Fig. 2.1 Characterization of the study area: (**a**) location of Argentina (dark grey) and Santa Cruz Province (black); (**b**) main ecological areas (brown = dry steppe, yellow = humid steppe, orange = shrublands, light green = sub-Andean grasslands, dark green = *Nothofagus* forests and alpine vegetation) (modified from Oliva et al. 2004); (**c**) protection areas (orange = provincial reserves, brown = national parks); (**d**) *Nothofagus* forests (light green = *N. pumilio*, orange = *N. antarctica*, dark green = mixed forests) (CIEFAP-MAyDS 2016); more detail of *Nothofagus* forests: (I) Lago Buenos Aires, (II) Lago Pueyrredón, (III) Lago San Martín, (IV) Lago Argentino, (V) Río Turbio

Table 2.1 Proxies and units of provisioning ecosystem services

Type	Division	Proxy	Unit
Provision	Nutrition	Sheep presence probability	Probability of sheep presence km^{-2}
		Potential silvopastoral	Index
	Plants and fibre	Total volume without bark	$m^3 ha^{-1}$
	Oil production	Oil well density	Well km^{-2}

2.2 *Materials and Methods*

2.2.1 Provisioning Ecosystem Services Map

We elaborated one provisioning ecosystem services map (PrESM) considering four proxies based on CITES divisions (MEA 2005; Maes et al. 2012, 2014; Haines-Young and Potschin 2018) (Table 2.1).

Each proxy map was built using a geographical information system (GIS) project and rasterized at 90 × 90 m resolution using the nearest resampling technique:

(i) Sheep probability map was calculated using sheep stocking density estimated from a model of probability of contact with sheep per ranch (0–1 probability km^{-2}) according to Pedrana et al. (2011), where values close to 0 indicate low probability of occurrence and values close to 1 indicate the highest probability of occurrence. In the GIS project, we applied the focal statistics tool to create a new raster by considering the values near 10 km, and then we applied a mask where forests and protected areas had values of zero sheep probability.

(ii) Oil production map was estimated based on oil well density (wells km^{-2}). In the GIS project, we calculated the oil well density using 21,426 points of well (http://datos.minem.gob.ar//). We did not apply a mask with zero values inside protected areas, because oil activities is legal (Law n° 2.185), we did not apply a mask with zero values.

(iii) Timber production map was calculated as potential total volume without bark (TVWB m^3 ha^{-1}) of NP and mixed evergreen forests according to the provincial forest inventory (Peri et al. 2019a). NA forests, non-forests and protected areas presented value of zero timber production.

(iv) Potential silvopastoral map was calculated using understory biomass production (kg DM ha^{-1}) and total volume without bark (m^3 ha^{-1}) of NA forests (Peri and Ormaechea 2013). These authors defined that understory biomass production varied from <500 to >2500 kg DM ha^{-1} and total volume without bark varied from <100 to >200 m^3 ha^{-1}. In the GIS project, we applied the reclassify tool to classify the rasters from 1 to 4. Then, we calculated a potential silvopastoral index considering that 70% of NA forests had livestock and 30% is used to obtain poles wood or firewood (potential silvopastoral index = biomass production × 0.7 + total volume without bark × 0.3). Then, the equation was integrated into the GIS project. We applied a mask where NP and mixed evergreen forests, non-forests and protected areas represent zero potential silvopastoral value.

The four proxy maps were rescaled from 0 to 100 and combined (sum values for each pixel) to obtain the final PrESM. This map was rasterized to present scores that varied from 0 to 100.

2.2.2 Potential Biodiversity Map

We elaborated a potential biodiversity map (PBM), using 119 potential habitat suitability maps of different taxonomic group species (Rosas et al. 2017, 2018, 2019b, c). These maps used a large database: (i) one endangered mammal (*Hippocamelus bisulcus*) in *Nothofagus* forests by using 300 plots from National Park Administration and different studies (Vila et al. 2006); (ii) 47 species of birds by using 5512 plots (Darrieu et al. 2009) and one international web platform of bird collection (https://ebird.org/); (iii) 7 species of lizards by using 250 plots (Cruz et al. 2005; Ibargüengoytía et al. 2010; Fernández et al. 2011; Breitman et al. 2014); (iv) 10 species of darkling beetles by using 310 plots from CEI (Colección Entomológica del Instituto Argentino de Investigaciones de las Zonas Áridas, IADIZA) and (V) 53 species of vascular plants by using 5915 plots from PEBANPA Network (Peri et al. 2016c), native forests provincial inventories and data from FAMA INTA laboratory (Forestal, Agricultura y Manejo del Agua). The database also was complemented with data of the selected species using the Sistema Nacional de Datos Biológicos of Ministerio de Ciencia, Tecnología e Innovación Productiva (www.datosbiologicos.mincyt.gob.ar). Environmental Niche Factor Analysis (ENFA, Hirzel et al. 2002) and Biomapper 4.0 software (Hirzel et al. 2004) were used for species potential habitat suitability mapping based on 41 potential explanatory variables (climate, topography, and other variables related to landscape), which were rasterized at 90 × 90 m resolution using the nearest resampling technique on a GIS project. The GIS methods used here were described in Rosas et al. (2017, 2018, 2019b, c). The maps for each taxonomic group species were combined (average values for each pixel) to obtain four potential biodiversity maps (birds, lizards, darkling beetles and plants) and one potential habitat suitability (mammal). We used a mask of NDVI <0.005 to exclude ice, water or bare soil. The five maps were weighted by a group importance index from 0.5 to 1.0 that combined ENFA index (Hirzel et al. 2002) and endemism of each species. The five weight maps were rescaled from 0 to 100 and then combined (sum values for each pixel) to obtain the final PBM for the province. This map also was rasterized to present scores that varied from 0 to 100.

2.2.3 Landscape Analyses

We calculated the mean of each PrES proxies, PrESM and PBM using a hexagonal binning processes (each hexagon = 250,000 ha) for the full province and for forest landscape matrix (each hexagon = 5000 ha). We analysed the maps considering the influence of the different ecological areas (Oliva et al. 2004) and forest landscape matrix (combination of grasslands and the different forest types) (Peri and Ormaechea 2013; Peri et al. 2019a) by using one-way ANOVAs and Tukey post-hoc test. The hexagonal GIS methods used here were previously described by Rosas et al. (2019c).

Additionally, we analysed the performance of PrESM to detect potential trade-offs with the biodiversity for the total area, forest landscapes and main forest types. For this, we categorized the PBM (low, medium and high) considering equal number of hexagons. For the whole province, the thresholds of potential biodiversity were as follows: low <41%, medium 42–74% and high <75%, and for forest landscape, the thresholds were as follows: (i) G – low <35%, medium 36–47% and high <48%; (ii) G + F – low <52%, medium 53–62% and high <62%; (iii) F – low <67%, medium 68–76% and high <77%. Also, based only on G + F and F hexagons, we classified each one considered the main forest types (NA and NP), according to the most abundant forest type cover inside each one. Finally, we want to identify the (i) potential trade-offs outside the networking protected areas, (ii) potential new protected areas (high MPB and low PrES values) and (iii) potential areas where the economic activities are maximized through intensive management (low MPB and high PrES values). For this, we built a new map crossing PBM (low, medium and high) and PrESM (low and high) categories. These new maps were classified considering equal number of hexagons: (i) for the entire province, the selected thresholds were as follows: PBM – low <41%, medium 42–74% and high <75%; PrESM – low <24% and high >25%; (ii) for forest landscape matrix, the selected thresholds were as follows: PBM – low <41%, medium 42–54% and high <55%; PrESM – low <22% and high >23%.

2.3 *Results and Discussion*

2.3.1 Provisioning Ecosystem Services Map

Sheep presence probability and oil well density proxies in Santa Cruz province occurred in most of the ecological areas (Fig. 2.2), while total volume without bark and potential silvopastoral proxies were specifically from *Nothofagus* forests (Fig. 2.3).

Sheep presence probability map presented values from zero (e.g. natural protected areas) to 1.00 (e.g. best grazing areas) (Fig. 2.2a). The provincial mean value was 0.41 sheep presence probability km^{-2}, where 19% of the area had low values (<0.20), 60% showed values between 0.20 and 0.80 and 14% presented high values (<0.80). Sheep probability values decreased from south to centre where steppes prevailed and from east to west where sub-Andean grasslands dominated (Oliva et al. 2004). ANOVAs showed that sheep probability map presented significant differences among the different ecological areas (Table 2.2), where humid steppe and shrublands had the highest values (0.71 and 0.68, respectively), followed by the dry

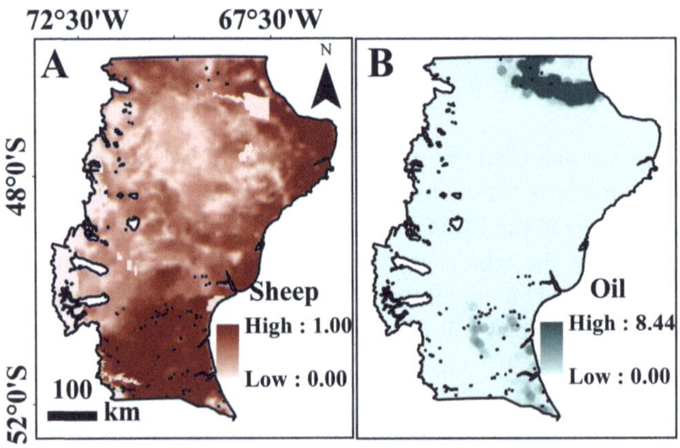

Fig. 2.2 Provisioning ecosystem services of Santa CruzPprovince: (**a**) sheep probability (probability of sheep presence km^{-2}), where dark red represents the highest values and light red the lowest probabilities values and (**b**) oil production (wells km^{-2}), where dark blue represents the highest density values and light blue the lowest density values

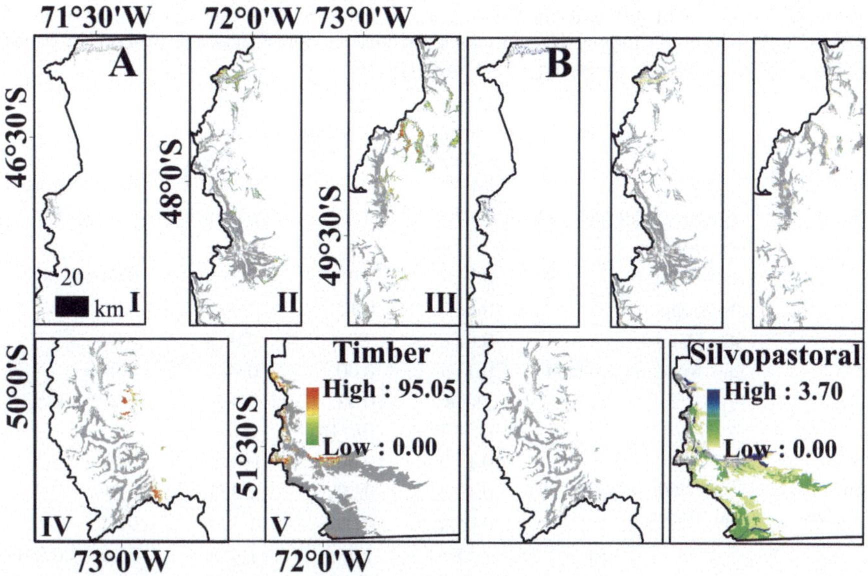

Fig. 2.3 Provisioning ecosystem services from *Nothofagus* forests of Santa Cruz Province: (**a**) timber production of *N. pumilio* and evergreen mixed forest (TVWBm^3 ha^{-1}), red represents the highest volume values and green the lowest volume values and (**b**) potential silvopastoral of *N. antarctica* forest (adimensional), blue represents the highest values and yellow the lowest values. Details of *Nothofagus* forests: (I) Lago Buenos Aires, (II) Lago Pueyrredón, (III) Lago San Martín, (IV) Lago Argentino, (V) Río Turbio. Grey colour indicates *Nothofagus* forests with 0 value

steppe (0.38), and the lowest values were found in sub-Andean grasslands, and *Nothofagus* forests and alpine vegetation (0.21 and 0.07, respectively).

Several studies evaluated the impact of livestock grazing on ecosystem properties (e.g. soil variables and vegetation cover; Peri et al. 2016a) in rangelands, where their extension and economic importance highlight the necessity of sustainable management proposals to supply the demand of an increasing human population (Ala-Hulkko et al. 2019). In Patagonia, extreme climatic condition was used as the most important predictor together with the land-use management in modelling soil carbon concentrations (Peri et al. 2016a). Also, Peri et al. (2013) reported that grass vegetation cover decreased and soil erosion increased due to high livestock stocking rates under continuous grazing in the studied area. In the last 70 years, the degradation of Patagonia steppe had increased due to an inadequate land management (e.g. overgrazing, heterogeneous and large paddocks and continuous grazing) (Del Valle et al. 1998; Gaitán et al. 2019). In addition, sheep presence probability presented significant differences among forest landscape matrix, where the highest values (0.17) were found in the grassland areas (grassland cover >70%). Despite this, when grasslands with forest matrix (G + F) were considered, the combination with NA forests presented the highest values (0.38) as well as when only forest cover (F) was considered. Ecotone areas of NA forests with grasslands have been identified as very important zone for livestock production, where forage species and tree cover

Table 2.2 ANOVAs of different provisioning ecosystem services for terrestrial ecosystems of Santa Cruz Province, considering different ecological areas and forest landscape matrix (grasslands and forests, grasslands and forests types and forest types)

Terrestrial ecosystems			Sheep probability (sheep. km^{-2})	Oil production (wells. km^{-2})	Timber production (TVWB $m^3.ha^{-1}$)	Potential silvopastoral (adimensional)
Ecological areas	Forests and alpine vegetation		0.07 a	0.00	0.60 b	0.08 b
	Humid steppe		0.71 c	0.03	0.21 a	0.04 ab
	Dry steppe		0.38 b	0.12	0.00 a	0.00 a
	Shrublands		0.68 c	0.03	0.05 a	0.00 a
	Sub-Andean grasslands		0.21 a	0.00	0.04 a	0.00 a
	F(p)		28.58 (<0.001)	0.53 (0.716)	9.96 (<0.001)	4.41 (0.002)
Forest landscape matrix	(i) Grasslands and forests	G	0.17 b	0.00	0.38 a	0.02 a
		G + F	0.10 a	0.00	2.29 b	0.09 b
		F	0.06 a	0.00	1.39 b	0.41 c
		F(p)	8.55 (<0.001)	1.73 (0.178)	18.55 (<0.001)	95.99 (<0.001)
	(ii) Grasslands and forest types	G + NP-MIX	0.01 a	–	1.19	0.00 a
		G + NP	0.04 ab	–	2.45	0.00 a
		G + NA-NP	0.12 b	–	3.74	0.12 b
		G + NA	0.38 c	–	0.43	0.49 c
		F(p)	26.83 (<0.001)	–	1.44 (0.237)	67.78 (<0.001)
	(iii) Forest types	NP-MIX	0.00 a	–	0.00	0.00 a
		NP	0.02 a	–	1.57	0.01 a
		NA-NP	0.07 ab	–	3.29	0.14 a
		NA	0.10 b	–	0.38	0.84 b
		F(p)	3.41 (0.023)	–	2.50 (0.068)	63.54 (<0.001)

G grasslands, *F* forests, *NA Nothofagus antarctica*, *NP N. pumilio*, *MIX* mixed evergreen forests
F Fisher test, *(p)* probability. Different letters show differences with Tukey test at $p < 0.05$

increased the habitat qualities for animals (e.g. nutrition properties and shelter for animals) (Peri et al. 2013).

Oil well density map presented values from 0 (minimum density) to 8.44 (maximum density) (Fig. 2.2b), with a mean provincial value of 0.09 wells km^{-2}. The highest values occurred mainly in two areas of the province, one near San Jorge Gulf in the northeast and the other area in the southeast area of the province near Rio Gallegos city, where Producción Petrolera Nacional del Petróleo (ENAP) and Yacimientos Petrolíferos Fiscales (YPF) y TOTAL S.A. are the principal operators. ANOVAs showed that there were not significant differences among different ecological areas ($F = 0.53$; $p = 0.716$) or across the forest landscape matrix. Despite the non-significant differences among ecological areas, there was an increase of oil production values in the dry steppes, where punctual activities (e.g. wells, accessibility and pipelines) presented highest impacts (Del Valle et al. 1998; Gaitán et al. 2019) in protected areas (e.g. Meseta Espinosa y El Cordón Provincial Reserve). This indicated potential trade-off with the conservation of endemic species, where Fiori and Zalba (2003) determined that vegetation recovery in pipelines and oil well areas was extremely poor, being oil the only provisioning ES enable to be conducted inside the protected areas. This creates a potential trade-off with the conservation of endemic species.

Total volume without bark (Fig. 2.3a) varied from 0 (NA forest, open lands and protected areas) to 95.05 (maximum volume without bark), with a mean provincial value of 0.06 TVWB m^3 ha^{-1}. NP and mixed evergreen forests (459 km^2) presented values from 0.01 to 95.05 TVWB $m^3.ha^{-1}$. In Santa Cruz, while 52% of native forests presented low values (<30 TVWB m^3 ha^{-1}), 43% had values between 30 and 60 TVWB m^3 ha^{-1} and only 5% of these forests presented high values (>60 TVWB m^3 ha^{-1}) (Fig. 2.3a). Timber production values increased from north to south. In the north, at the Lago Buenos Aires area (Fig. 2.3aI), all native forests are inside natural reserves, and in the centre areas at Río Chico and Lago San Martin, forests presented values from low to medium (Fig. 2.3aII and III). In the south, Lago Argentino and Río Turbio (Fig. 2.3aIV and V) showed values from medium to high, where the highest values were presented near ecotone areas with the humid steppes. As it was expected, ANOVAs showed significant differences in timber production among ecological areas (Table 2.2), where *Nothofagus* forests presented the highest values (0.60 TVWB m^3 ha^{-1}).

Forest landscape matrix analysis showed that timber production was significantly highest when grasslands were combined with forests (2.29 TVWB m^3 ha^{-1}) or where only forest occurred (1.39 TVWB m^3 ha^{-1}). Furthermore, there was no significant differences among forest types. The use of native forests for timber occurred in the Patagonian region since the European colonization in the late nineteenth century, where harvesting for sawmills and firewood still continues in Tierra del Fuego (Gea et al. 2004; Martínez Pastur et al. 2019). In Santa Cruz Province, there are not operating sawmills (Peri et al. 2019a), and most of the NP (69%) and mixed evergreen (82%) forests are inside the protected areas, where other ecosystem services (e.g. cultural) prevail and mostly define the use of the natural forests (Rosas et al. 2019a).

Potential silvopastoral map (Fig. 2.3b) presents values from 0.00 (e.g. NP and mixed evergreen forests, non-forest and protected areas) to 3.70 (e.g. maximum potential silvopastoral in NA forests), with a provincial mean value of only 0.01. NA forests (1432 km^2) presented values from 1.00 to 3.70, where 93% of the area presented low values (<1.60), 6% had values between 1.70 and 2.30 and only 1% showed high values (>2.40) (Fig. 2.3b). Potential silvopastoral values increased from north to south, where Lago Buenos Aires (Fig. 2.3bI) and Lago Argentino (Fig. 2.3bIV) had the lowest values. Lago Pueyrredón and Lago San Martin (Fig. 2.3bII and III) had medium values at low hillside near valleys and lakes. In Río Turbio (Fig. 2.3bV), the proxy presented the highest values near ecotone areas and lowest values in the extreme south of the province.

ANOVAs showed that potential silvopastoral presented significant differences among the different ecological areas (Table 2.2), where *Nothofagus* forests had the highest values (0.08) followed by humid steppes (0.04). In addition, potential silvopastoral presented significant differences among the forest landscape matrix, where the highest values (0.84) were found in NA forests (forest cover >50%). This is because silvopastoral systems combine trees and grasslands or pastures under grazing in the same unit of land, being an economical, ecological and social productive alternative in Patagonia (Peri et al. 2016b). Peri and Ormaechea (2013) identified that more than 90% of NA forests presented silvopastoral activities. This system provides increasing incomes to ranchers due to the combined production of timber and animals and benefits such as the provision of livestock shelter, enhancement of animal welfare and other beneficial effects on soil conservation (Peri et al. 2016b).

2.3.2 Provisioning Ecosystem Services and Potential Biodiversity Map: Identification of Conservation Areas of Interest

The rescale (0–100) of the four proxy maps allowed us to combine them (sum values for each pixel) and develop the final PrESM (Fig. 2.4a), which presented values from 0 (minimum provisioning ecosystem services) to 100 (maximum provisioning ecosystem services) across the landscape. PrESM increased from north to south and decreased from east to west. Medium to high values occurred near seacoast and humid steppe areas, while the lowest values were located near glaciers and mountain areas. In addition, we combined the 119 potential habitat suitability maps of the different taxonomic group species (Rosas et al. 2017, 2018, 2019b, c) to develop the final PBM (Fig. 2.4b), where values varied from 0 (minimum potential biodiversity) to 100 (maximum potential biodiversity). In general, PBM presented similar pattern as PrESM, with medium to high values obtained from the seacoast to the centre of the province.

ANOVAs showed significant changes in PrESM and PBM across different ecological areas (Table 2.3), where the highest values were found at humid steppes (51.87 and 63.77, respectively) and shrublands (43.01 and 66.66, respectively), while the lowest values in Sub-Andean grasslands (12.20 and 35.32, respectively).

In particular, humid steppes and shrublands showed the sheep breeding proxy as the most important PrES, and PBM presented different plant species that highlighted the importance of these areas (e.g. *Carex* spp. and *Festuca pallescens*). Different studies have been developed to understand the plant biodiversity distribution and their importance on the ecosystem function (Peri et al. 2013; Gaitán et al. 2014) and economic activities (Peri et al. 2013). However, few studies focused on biodiversity related to grazing in these ecosystems (Peri et al. 2016c). Nevertheless, potential trade-offs between forage provision and regulating and supporting services (e.g. carbon and nitrogen stocks) have been observed (Oñatibia et al. 2015; Peri et al. 2016a). In fact, negative consequences (e.g. desertification) (Del Valle et al. 1998; Gaitán et al. 2019; Peri et al. 2016a) due to overgrazing (Peri et al.

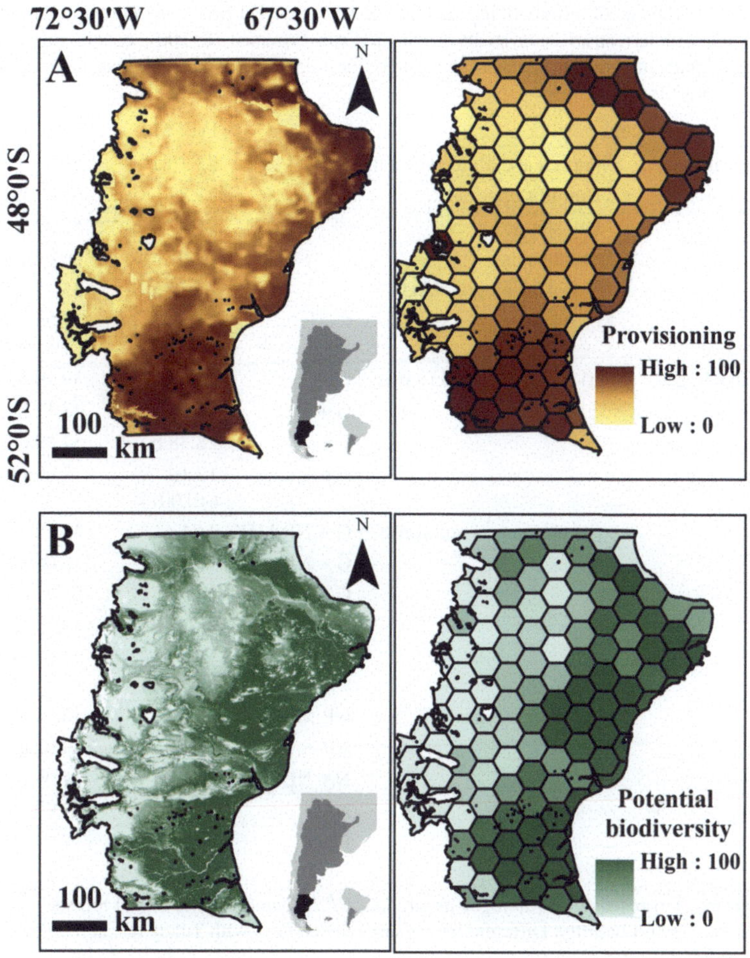

Fig. 2.4 Provisioning ecosystem services and potential biodiversity maps (0–100) of Santa Cruz Province (left) and hexagons of 250,000 ha obtained through the hexagonal binning process (right). (**a**) Provisioning ecosystem services, brown represents the highest values (values close to 100) and yellow the lowest values (values close to 0) and (**b**) potential biodiversity map, dark green represents greater potential biodiversity and light green the lowest potential

2016c), land-use conversion and climate changes (Gaitán et al. 2019) have been reported for the steppe ecosystem. In this context, endemic species with a narrow distribution and high potential habitat suitability values (e.g. *Nyctelia bremi, Liolaemus sarmientoi*) with very specific environmental conditions became an important issue for conservation (Rosas et al. 2018; Rosas et al. 2019b) mainly in areas with a lack of protected areas (e.g. humid steppes). Additionally, bird species presented conflicts with specific economic practices, e.g. shrub removal (e.g. *Junielia tridens*) to increase grasses biomass, which may affect reproductive

Table 2.3 ANOVAs of provisioning ecosystem services (PrESM) and potential biodiversity (PBM) maps in terrestrial ecosystems of Santa Cruz Province (0–100), considering different ecological areas and the forest landscape matrix (grasslands and forests, grasslands and forests types and forest types)

Terrestrial ecosystems			PrESM	PBM
Ecological areas	Forests and alpine vegetation		23.70 a	43.61 a
	Humid steppe		51.87 b	63.77 bc
	Dry steppe		24.49 a	57.39 b
	Shrublands		43.01 b	66.66 c
	Sub-Andean grasslands		12.20 a	35.32 a
	F(p)		12.82 (<0.001)	17.95 (<0.001)
Forest landscape matrix	(i) Grasslands and forests	G	13.74 a	39.70 a
		G + F	17.07 a	48.19 b
		F	28.38 b	54.89 c
		F(p)	12.10 (<0.001)	174.46(<0.001)
	(ii) Grasslands and forest types	G + NP-MIX	4.31 a	45.22 a
		G + NP	10.02 a	46.00 a
		G + NA-NP	24.15 a	52.62 b
		G + NA	46.73 b	49.95 ab
		F(p)	10.41 (<0.001)	9.96(<0.001)
	(iii) Forest types	NP-MIX	0.00 a	46.06 ab
		NP	6.41 a	49.02 a
		NA-NP	20.95 a	54.09 b
		NA	49.05 b	59.90 c
		F(p)	16.68 (<0.001)	17.65 (<0.001)

G grasslands, *F* forests, *NA Nothofagus antarctica*, *NP N. pumilio*, *MIX* mixed forests
F Fisher test, *(p)* probability. Different letters show differences with Tukey test at $p < 0.05$

processes and food for some species (e.g. *Asthenes anthoides, Turdus falcklandii* and *Sturnella loyca*) (Kusch et al. 2016).

Dry steppes presented low values of PrESM (mean of 24.49) and medium values of PBM (mean of 57.39), where both proxies (sheep and oil production) were the most important PrES. These areas occupied more than 60% of the studied province with evident desertification processes (Del Valle et al. 1998) due to the extreme climate conditions and scarce vegetation cover dominated by small shrubs (e.g. *Nassauvia glomerulosa* and *Mulinum spinosum*) and grasses (*Stipa sp.*) (Oliva et al. 2004). Furthermore, oil production greatly affected this area (see Fig. 2.2b) with potential trade-off with the biodiversity. In fact, according to the local regulations, this is the only PrES allowed inside the protected areas (e.g. Mesera Espinosa and El Cordón provincial reserve). These dry steppe areas showed highest values for lizards (e.g. *Liolaemus bribronii, L. fitzingerii, Diplolaemus bibronii* and *Homonota darwinii darwinii*) (Breitman et al. 2014; Rosas et al. 2018), where *H. darwinii darwinii* present the most austral gecko distribution. Also, dry steppes presented medium to high values of potential biodiversity for darkling beetles (Rosas et al. 2019b), with high levels of endemism (Carrara and Flores 2013), e.g., for *Nyctelia fitzroyi* which lived in a narrow area with extreme environmental conditions.

The combination of PBM and PrESM allowed us to locate different areas of conservation interest for Santa Cruz Province (Fig. 2.5a). (i) Potential trade-off areas outside the protected areas (high MPB and high PrES values) decreased from east to west, where protected areas were located (Fig. 2.1c, brown to light green colour). We identified high potential trade-off areas near the seacoast (Fig. 2.5a, brown colour), where the biggest area is located in the centre-west (hexagons = 16), followed by one in the south (hexagons = 11) and the smallest area in the north (hexagons = 2) of the province. Another section with medium potential trade-off areas (medium MPB and high PrESM values) was identified across the province (orange colour). (ii) Potential areas to suggest new protected areas (high MPB and low PrES values) were identified in three little sections (dark green colour): one near Monte Leon National Park in the south, another in the steppe areas (hexagons = 3) and the third near Bosques Petrificados at Jaramillo National Park. Medium potential areas to suggest new protected areas (medium MPB and low PrESM values) were identified mainly in dry steppes areas (green colour) and near big lakes (e.g. Lago Argentino). (iii) Potential areas where conflicts are low and

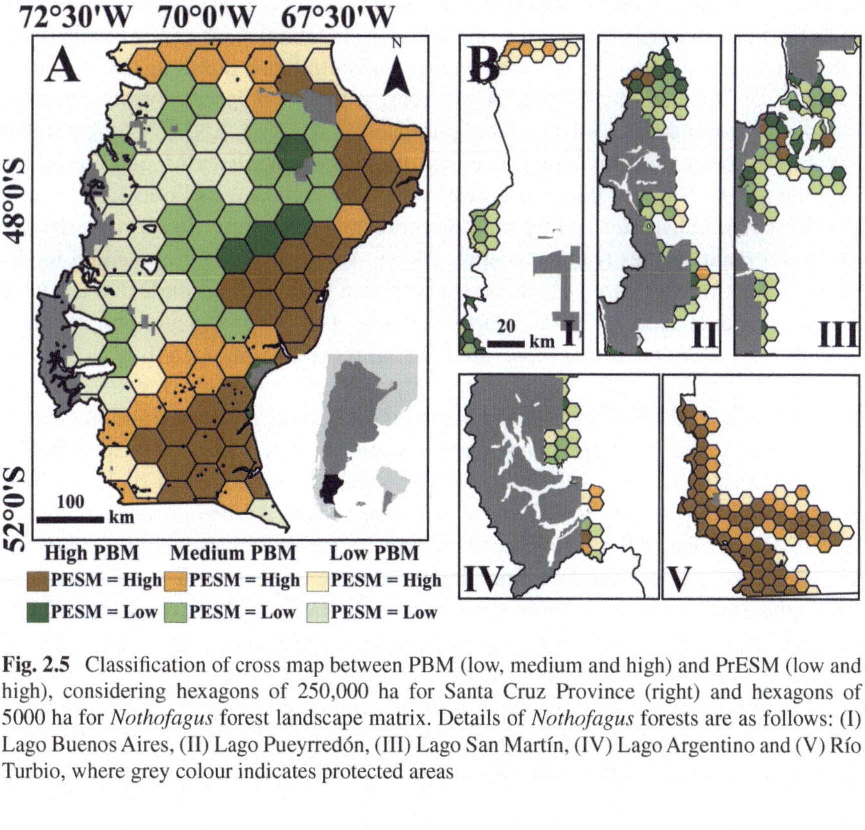

Fig. 2.5 Classification of cross map between PBM (low, medium and high) and PrESM (low and high), considering hexagons of 250,000 ha for Santa Cruz Province (right) and hexagons of 5000 ha for *Nothofagus* forest landscape matrix. Details of *Nothofagus* forests are as follows: (I) Lago Buenos Aires, (II) Lago Pueyrredón, (III) Lago San Martín, (IV) Lago Argentino and (V) Río Turbio, where grey colour indicates protected areas

intensification of the management activities are possible (low MPB and high PrES values) were located in the extreme north and near NA forest in the south (light orange colour) of the province. In fact, ANOVAs showed significant differences of PrESM among the different PBM qualities (low, medium and high) across the province, where PrESM values increased with PBM qualities (Table 2.4).

However, PrESM and PBM showed the lowest values in the west of Santa Cruz Province (see Fig. 2.4 and Table 2.3), where most of the ecosystems are under protection (Fig. 2.1c). Forest landscape matrix ANOVAs showed that the highest values of PrESM and PBM occurred in grasslands with NA forests (forest cover between 30% and 50%) and in areas where NA forests prevail (forest cover >50%) with high sheep probability and potential silvopastoral ESP (Table 2.2). In this context, by crossing PBM and PrESM maps, we located areas of interest for management and conservation planning at landscape level (Fig. 2.5b). (i) Potential trade-off areas decreased from north and south to central-west part of the province (brown to light green colour in Fig. 2.5bV). The biggest area with high potential trade-off (brown colour) is located close to Río Turbio, where NA forests prevail (Fig. 2.1dV). Medium potential trade-offs (orange colour) were identified in Lago Buenos Aires, Lago Argentino and close to Rio Turbio (Fig. 2.5bI, IV, V). (ii) High- and medium-potential areas that suggest new protected areas (dark green and green colour) were located close to Lago Pueyrredón and Lago San Martín (Fig. 2.5bII and III), where few high-potential trade-off areas (brown colour) also occurred due to the existence of different *Nothofagus* forests (NA and NP). (iii) Potential areas where conflicts are low and intensification of the management activities are possible (light orange colour) were located mostly in ecotone areas, where *Nothofagus* forests are combined with grasslands for silvopastoral purposes. In addition, ANOVAs showed that NA forests presented the highest values of PrESM for all PBM qualities (low, medium and high) when the main forest types were considered (Table 2.4).

Forest types characterized by multiple microenvironments allowed the survival of several plant species (Lencinas et al. 2008a; Antos 2009). High potential biodiversity value for understory plants were coincident with studies conducted in Tierra del Fuego (Lencinas et al. 2008a; Martínez Pastur et al. 2016b). In addition, Peri and Ormaechea (2013) identified more shrub and grass species in open than closed NA forests. Our results, identified specific plant species associated with different *Nothofagus* forest types (e.g. *Acaena magellanica, Avenella flexuosa* and *Baccharis magellanica*), and different hotspot areas were identified mainly located in the southernmost part of the province. These hotspot areas presented high-potential habitat suitability values for *Berberis empetrifolia* and *Agrostis capillaris* in NA forests (Rosas et al. 2019c). Silvopastoral management practices increase plant biomass by removing trees and maintaining at the same time more biodiversity values than other proposals (e.g. forest conversion in grasslands) (Peri et al. 2016b). Silvopastoral management also generates positive synergies with biodiversity by enhancing bird, insect and plant richness (Barbier et al. 2008; Peri et al. 2019a). High incoming light levels to understory when canopy trees are removed (thinning practices) provide more energy for plant growth (Antos 2009) and insect richness (Lencinas et al. 2008b). Lantschner and Rusch (2007) also found that birds

Table 2.4 ANOVAs for total provisioning ecosystem services (0–100) and the different ecosystem service types (sheep breeding, oil extraction, timber production and silvopastoral practices) in terrestrial ecosystems of Santa Cruz Province based on different potential biodiversity qualities (low <48%, medium 49–66% and high >100%) for the total province, landscape types (grasslands, grasslands and forests, and forests) and the main forest types

Potential biodiversity			Provisioning	Sheep (sheep.km^{-2})	Oil (wells.km^{-2})	Timber (VTSCm^3.ha^{-1})	Silvopastoral
Total area		Low	18.88 a	26.78 a	1.26	8.03 c	2.03
		Medium	26.63 a	45.86 b	3.24	0.99 a	2.64
		High	38.45 b	69.74 c	5.32	0.00 a	0.00
		F(p)	13.89 (<0.001)	37.35 (<0.001)	0.81 (0.446)	4.31 (0.016)	0.71 (0.495)
Landscapes	G	Low	0.10 a	12.49	–	0.3	0.00 a
		Medium	0.17 b	20.10	–	0.26	0.02 a
		High	0.15 ab	17.20	–	0.58	0.05 b
		F(p)	3.10 (0.051)	2.65 (0.073)	–	2.53(0.081)	15.95 (<0.001)
	G + F	Low	0.06 a	4.52 a	–	0.17 a	0.05 a
		Medium	0.13 a	7.26 ab	–	2.05 ab	0.04 a
		High	0.33 b	16.89 b	–	4.65 b	0.18 b
		F(p)	10.77 (<0.001)	4.31 (0.017)	–	7.41 (0.001)	4.53 (0.014)
	F	Low	0.07 a	3.15	–	0.42	0.09 a
		Medium	0.25 b	7.98	–	1.87	0.32 a
		High	0.54 c	8.42	–	1.96	0.84 b
		F(p)	25.70 (<0.001)	2.11 (0.130)	–	1.27(0.287)	30.48 (<0.001)

(continued)

Table 2.4 (continued)

Potential biodiversity			Provisioning	Sheep (sheep.km^{-2})	Oil (wells.km^{-2})	Timber (VTSCm^3.ha^{-1})	Silvopastoral
Main forest types	NP	Low	0.70 a	0.12 a	–	0.92 a	0.20 a
		Medium	9.69 b	3.55 a	–	12.21 b	1.42 a
		High	40.46 c	14.81 b	–	49.23 c	7.71 b
		F(p)	31.73 (<0.001)	17.20 (<0.001)	–	30.73 (<0.001)	8.23 (0.001)
	NA	Low	39.08	27.21	–	5.36	35.81
		Medium	37.77	19.2	–	7.67	39.58
		High	44.22	13.67	–	6.77	50.95
		F(p)	0.40 (0.673)	1.93 (0.156)	–	0.05 (0.956)	1.12 (0.333)

G grasslands, *F* forests, *NA Nothofagus antarctica*, *NP N. pumilio*
F = Fisher test, *(p)* probability. Different letters show differences with Tukey test at $p < 0.05$

associated of ecotone environments (e.g. forests and grasses areas) moved to managed NA forests and increased the original richness and diversity.

Low values of PrESM (between 24.15 and 20.95) and medium values of PBM (between 52.62 and 54.09) occurred in grasslands with NA-NP (forest cover between 30% and 50%) and NA-NP forest areas (forest cover >50%), where timber production proxy presented the main provision ecosystem service (Table 2.3). There were significant differences of PrESM among different PBM qualities (low, medium and high) when main forest types were considered (Table 2.4). In Santa Cruz Province, plant biodiversity changed through NP forest landscapes (Rosas et al. 2019c). This result was coincident with other studies of NP forests in Tierra del Fuego (Martínez Pastur et al. 2016b); thus values of plant biodiversity increased when different *Nothofagus* forest types were combined (Lencinas et al. 2008a) and insect biodiversity increased in areas with high timber potential (Lencinas et al. 2008b). However, some silvicultural practices (e.g. shelterwood cuts) had negative impacts on insect populations (Spagarino et al. 2001) and increased native and exotic plant species in the harvested areas (Martínez Pastur et al. 2002).

In addition, some mammals such as *Hippocamelus bisulcus* (huemul) can be affected by PrES in NA and NP forested areas (Rosas et al. 2017). Several studies relate the decrease of the huemul habitat to different human impacts (Corti et al. 2013; Briceño et al. 2013) that greatly affected the marginal populations of huemul (e.g. extreme distribution areas in the south and north of the province where silvopastoral activities predominate). Huemul is one of the most vulnerable species with only 350–500 individuals in 50 fragmented subpopulations throughout Patagonia (Díaz and Smith-Flueck 2000), living mainly inside protected areas that represent about 50% of their natural habitat (Vila et al. 2006; Quevedo et al. 2017; Rosas et al. 2017). Therefore, it is necessary to develop new strategies to protect biodiversity outside the protected areas (Mori et al. 2017), where the potential habitat of the huemul is higher (Rosas et al. 2017). Some private initiatives (Ea. Río Condor and Ea. Los Huemules, close to El Chaltén) support this strategy by modifying economic activities inside the ranches, e.g. reducing livestock activity and increasing other activities related to ecotourism. New provincial conservation planning is needed to promote innovative management strategies in productive areas with high-potential habitat suitability values (Smith-Flueck et al. 2011), for example the establishment of corridors and fences to separate cattle and huemul wild populations (Gilbert-Norton et al. 2010; Corti et al. 2011).

The richness of bird species in *Nothofagus* forests were lower in these austral latitudes than in northern hemisphere (Brown et al. 2007; Lencinas et al. 2005); however, most of the species are endemic (e.g. *Agriornis lividus*, *Aphrastura spinicauda* or *Scytalopus magellanicus*). The conservation of forest bird species represented an important challenge for managers, because it is necessary to consider multiple factors such as vegetation structure, connectivity of forest patches with appropriate size and shape to maintain avian diversity, occupancy and turnover rate (Whytock et al. 2018). For this, alternative silvicultural proposals (e.g. variable retention) are necessary to increase the species conservation (Martínez Pastur et al.

2019), where intact patches (e.g. aggregate retention) are combined with single trees in the harvested stands (e.g. dispersed retention). These new proposals maintained some of the original forest structure and micro-environmental conditions in the aggregate patches but increased biodiversity and forest reproduction compared to primary unmanaged forests (Lencinas et al. 2009, 2011; Soler et al. 2016). Variable retention could play a fundamental role for conservation in these forest types, but the influence of retention patterns and the most effective aggregate patch size are still unclear (Martínez Pastur et al. 2019). In this context, the identification of forest areas with potential trade-off between PrES and biodiversity is necessary to develop land-use strategies in the long term (Carpenter et al. 2009; Raudsepp-Hearne et al. 2010; Cordingley et al. 2016).

3 Challenges in the Land-Use Management for Provisioning Ecosystem Services

A key challenge for ecosystem management is to maximize PrES and hold enough biodiversity values across the landscape (Raudsepp-Hearne et al. 2010; Cordingley et al. 2015) to support the society demand (Ala-Hulkko et al. 2019). Some studies showed how human actions improved the delivery of goods (e.g. forage provision) but affected other services (e.g. soil nutrients) or biodiversity (Cardinale 2012; Peri et al. 2016a; Martínez Pastur et al. 2017). According to different studies, it is necessary to protect biodiversity values that are involved in the functional processes and PrES (Thompson et al. 2011; Mace et al. 2012; Maes et al. 2014; Harrison et al. 2014). In the context of spatial and land-use planning, negative and positive interactions had been described between PrES and biodiversity (Cordingley et al. 2016; Turkelboom et al. 2018).

We considered as potential trade-offs those areas with high PrES where intense economic activities affected negatively the provision of another ES and/or the biodiversity conservation. For example, some studies suggested that forest harvesting had potential trade-offs with carbon storage, aesthetic values and habitat quality (Cordingley et al. 2016; Martínez Pastur et al. 2017). In contrast, synergies areas were determined when high levels of PrES occur simultaneously with other services and/or biodiversity. For example, Thompson et al. (2009) reported that 76% of 21 different studies showed a direct relationship between the increase in forest biodiversity and an increase in carbon fixation. These interactions can be managed to reduce costs or to improve the multi-functionality of the managed landscape (Raudsepp-Hearne et al. 2010; Mori et al. 2017). For example, eco-friendly management practices such as silvopastoral systems at landscape level may improve with an integral management of the aesthetic values, protection against soil erosion, increase of long-term understory production and better biodiversity conservation (Peri et al. 2016b).

From results of the present work, we are able to identify those areas whit potential trade-offs and synergies between PrES and biodiversity at regional level and within the forest landscapes. In synthesis, we found the following main aspects:

(i) Humid steppes and shrublands were the most important areas to provided PrES (mainly sheep production), but also presented the major potential biodiversity values where plants, birds and darkling beetles presented medium to high PBM qualities. These areas occupied 19% of the total area of Santa Cruz Province, but less than 3% of these areas were under protection areas. Potential trade-off areas were identified (see Fig. 2.5, brown and orange hexagons), where traditional managements of sheep breeding in private lands (e.g. overgrazing, heterogeneous and large paddocks, and continuous grazing) can affect negatively the biodiversity of plants, birds (Peri et al. 2013, 2016c; Kusch et al. 2016) and darkling beetles (Carrara and Flores 2013). These negative impacts increase degradation processes (Del Valle et al. 1998; Gaitán et al. 2019) and eventually can decrease PrES supply (Peri et al. 2016c).

(ii) Dry steppes provided medium PrES (mainly from sheep breeding and oil production) with medium potential biodiversity values (lizards had the highest PBM quality). These areas occupied 66% of the total province, but scarcely represented the protected areas (less than 3% are protected). We located most of the potential different interaction between PrES and potential biodiversity (see Fig. 2.5), e.g. potential trade-offs (brown and orange hexagons) occurred in specific areas near the seacoast (e.g. livestock) and inside provincial reserves (e.g. due to oil extractive activities). Also, we identified areas that suggest new protected areas (dark green and green hexagons) in the central part of the province.

(iii) Sub-Andean grasslands, native forests and alpine vegetation occupied 17% of the total province and provided less PrES with low potential biodiversity. Forests and alpine vegetation occupied 7% of the Santa Cruz Province; however, these ecosystems are well represented as protected areas (75% protected). Because of this, we considered that these areas had low probability of potential trade-offs.

(iv) Forest landscape matrix presented important values of PrES and potential biodiversity, depending on forest types. Sheep production and silvopastoral systems had the major values when grasslands and NA forests were combined, or when NA forests prevailed. Timber production presented exceptional values when NP forest type was combined with NA or occurred close to grasslands. However, NA forests that represented 41% of the *Nothofagus* forest with exceptional potential biodiversity values were scarcely represented as protected areas (only 16% protected). In contrast, NP and mixed evergreen forests (represented 55% and 4% of the *Nothofagus* forest, respectively) are protected as national parks and provincial reserves (69% and 82%, respectively). We located the biggest area with potential trade-offs in the south, where NA forests prevail (see Fig. 2.5bV, brown and orange hexagons). In this context, areas that suggest new protected areas (dark green and green hexagons) are located in the north where NP is combined with NA forests (see Fig. 2.5bI, II, III). Several potential trade-offs between PrES and biodiversity occurred in NA forests in private lands without any protection and few regulations for conservation.

Landscape analyses allowed us to compare PrES and potential biodiversity for different ecological areas at regional level, as it was also reported in forests landscape at Tierra del Fuego Province, Argentina (Martínez Pastur et al. 2017). The intensification of livestock and forest harvesting without any consideration of other ES and biodiversity can affect the resilience of natural ecosystems (Cardinale 2012; Lindenmayer et al. 2012) as well as biodiversity values (MEA 2005; Mori et al. 2017). The importance of ES and biodiversity conservation incentivises public and private sectors to incorporate these concepts into decision-making (De Groot et al. 2010; Koschke et al. 2012). Recently, scientific and policy agendas on biodiversity have included evaluations of ES by incorporating a monitoring system to determine the effectiveness and progress of implemented public policy (Braat and De Groot 2012; Costanza et al. 2017). For this, it is necessary to consider multiple factors (De Groot et al. 2010), where the characterization and location (e.g. mapping) of ES and biodiversity are necessary to support decision-making at landscape scale (Raudsepp-Hearne et al. 2010; Cordingley et al. 2016; Turkelboom et al. 2018).

At global scale, different advances in economic valuation (De Groot et al. 2012), social perception (Reyers et al. 2013; Quintas-Soriano et al. 2016), conservation planning (Cordingley et al. 2016) and landscape planning (Koschke et al. 2012) for ES maintenance and biodiversity conservation have been developed. In this context, different studies tried to understand these interactions (trade-off and synergies) using different approaches in Argentina, particularly in Patagonia – for example, valuation of PrES from a socio-economic perspective (Laterra et al. 2011), ES provided by different managed ecosystems (Chillo et al. 2018; Rositano et al. 2018) or analyses of several trade-offs in productive ecosystems (e.g. silvopastoral) (Oñatibia et al. 2015; Martínez Pastur et al. 2017; Peri et al. 2016a). However, the challenge to solve trade-offs in the practice still remains. Understanding these relationships and creating maps that link PrES and biodiversity facilitate the connection of main society interests with natural ecosystems (Raudsepp-Hearne et al. 2010; Cordingley et al. 2015).

Our results allowed to (i) obtain empirical information about the provision of different PrES, (ii) define geographical distribution of PrES and potential biodiversity, (iii) identify hot and cold-spot areas, (iv) locate potential trade-off areas between different economic activities and biodiversity conservation values, (v) define areas to suggest new protected areas based on high values of PBM and low values of PrESM and (vi) define areas where the maximization of PrES can reduce ES losses. Moreover, with the identification of these areas of interest, it is possible to promote a balance between management and conservation strategies in private lands and develop new proposals for sustainable management at landscape level (e.g. variable retention harvesting) (Martínez Pastur et al. 2019) and also to contribute in public policies by improving the current management practices on private lands. For example, Law No. 26,331 promotes the use of native forests in a sustainable way to maintain their biodiversity and ecosystem services (Article 5). This law contemplates different uses in conservation categories to manage native forests looking for a balance of different ES provision (Peri and Ormaechea 2013; Peri et al. 2019a). In addition, the "National Plan for the Management of Forests with Integrated Livestock (MBGI)" has defined different guidelines for livestock and forestry activities under the maintenance of the structural and functional components of the native forest and therefore its ecosystem services (Peri et al. 2016b). In this context, our maps can be a powerful tool to develop land use, management and conservation proposals, based on multi-functionality of natural ecosystems (De Groot et al. 2010; Koschke et al. 2012; Maes et al. 2012).

References

Ala-Hulkko T, Kotavaara O, Alahuhta J, Hjort J (2019) Mapping supply and demand of a provisioning ecosystem service across Europe. Ecol Indic 103(7):520–529

Andrade MA, Suárez DD, Peri PL, Borrelli P, Ormaechea SG, Ferrante D, Rivera EH, Sturzenbaun MV (2016) Desarrollo de un modelo para asignación variable de carga animal (MAVC) en Patagonia Sur. Livest Res Rural Dev 28(11):e208

Antos J (2009, Eolss Publishers Co Ltd) Understory plants in temperate forests. In: Owens JN, Gyde Lund H (eds) Forests and forest plants, vol I. Oxford, pp 262–279

Barbier S, Gosselin F, Balandier P (2008) Influence of tree species on understory vegetation diversity and mechanisms involved a critical review for temperate and boreal forests. Forest Ecol Manag 254(1):1–15

Braat LC, De Groot R (2012) The ecosystem services agenda: bridging the worlds of natural science and economics, conservation and development, and public and private policy. Ecosyst Serv 1(1):4–15

Breitman MF, Minoli I, Avila LJ, Medina CD, Sites JW Jr, Morando M (2014) Lagartijas de la provincia de Santa Cruz, Argentina: distribución geográfica, diversidad genética y estado de conservación. Cuad Herpetol 28(2):83–110

Briceño C, Knapp LA, Silva A, Paredes J, Avendaño I, Vargas A, Sotomayor J, Vila AR (2013) Detecting an increase in an endangered huemul *Hippocamelus bisulcus* population following removal of cattle and cessation of poaching in coastal Patagonia, Chile. Oryx 47(2):273–279

Brown C, Anderson CB, Ippi S, Sherriffs M, Charlin R, Mcgehee S, Rozzi R (2007) The autecology of the fío-fío (Elaenia albiceps Lafresnaye and D'Orbigny) in subantarc- tic forests of the Cape Horn biosphere reserve, Chile. An Inst Patagon 35(2):29–40

Buzzi MA, Rueter BL, Ghermandi L, Maldonado F (2019) Influencia de la actividad petrolera y la ganadería ovina en la cubierta del suelo en una región árida y semiárida de la Patagonia Argentina. Geography Res Lett 45(2):661–685

Cardinale BJ (2012) Biodiversity loss and its impact on humanity. Nature 486(7401):59–67

Carpenter SR, Mooney HA, Agard J, Capistrano D, DeFries RS, Díaz S, Dietz T, Duraiappah AK, Oteng-Yeboah A, Pereira HM, Perrings C, Reid WV, Sarukhan J, Scholes RJ, Whyte A (2009) Science for managing ecosystem services: Beyond the Millennium Ecosystem Assessment. PNAS 106(5):1305–1312

Carrara R, Flores GE (2013) Endemic tenebrionids (Coleoptera: Tenebrionidae) from the Patagonian steppe: a preliminary identification of areas of micro-endemism and richness hotspots. Entomol Sci 16(1):100–111

Chillo V, Amoroso MM, Rezzano CA (2018) Silvopastoral use intensity modifies the provision of ecosystem services through changes in diversity in forests of NW Patagonia, Argentina. Ecosistemas 27(3):75–86

CIEFAP-MAyDS (Centro de Investigación y Extensión Forestal Andino Patagónico, AR - Ministerio de Ambiente y Desarrollo Sustentable, AR) (2016) Actualización de la Clasificación de Tipos Forestales y Cobertura del Suelo de la Región Bosque Andino Patagónico. Informe Final

Cordingley JE, Newton AC, Rose RJ, Clarke RT, Bullock JM (2015) Habitat fragmentation intensifies trade-offs between biodiversity and ecosystem services in a heathland ecosystem in southern England. PLoS One 10(6):e0130004

Cordingley JE, Newton AC, Rose R, Clarke R, Bullock J (2016) Can landscape-scale approaches to conservation management resolve biodiversity-ecosystem service trade-offs? J Appl Ecol 53(1):96–105

Costanza R, De Groot R, Braat L, Kubiszewski I, Fioramonti L, Sutton P, Farber S, Grasso M (2017) Twenty years of ecosystem services: how far have we come and how far do we still need to go? Ecosyst Serv 28(1):1–16

Corti P, Shafer ABA, Coltman DW, Festa-Bianchet M (2011) Past bottlenecks and current population fragmentation of endangered huemul deer (*Hippocamelus bisulcus*): implications for preservation of genetic diversity. Conserv Genet 12(1):119–128

Corti P, Saucedo C, Herrera P (2013) Evidence of bovine viral diarrhea, but absence of infectious bovine rhinotracheitis and bovine brucellosis in the endangered huemul deer (*Hippocamelus bisulcus*) in Chilean Patagonia. J Wildl Dis 49(3):744–746

Cruz FB, Kozykariski ML, Perotti MG, Pueta M, Moreno L (2005) Variación diaria de la temperatura corporal en dos especies de lagartos nocturnos (Squamata: Gekkonidae: Homonota) con comentarios sobre el uso de refugios. Cuadernos de Herpetología 18(2):15–22

Currie WS (2011) Units of nature or processes across scales? The ecosystem concept at age 75. New Phytol 190(1):21–34

Darrieu C, Camperi A, Imberti S (2009) Avifauna (Passeriformes) of Santa Cruz province, Patagonia (Argentina): annotated list of species. Rev Mus Argent Cienc Nat 11(1):49–67

De Groot RS, Alkemade R, Braat L, Hein L, Willemen L (2010) Challenges in integrating the concept of ecosystem services and values in landscape planning, management and decision making. Ecol Complex 7(3):260–272

De Groot R, Brander L, Van Der Ploeg S, Costanza R, Bernard F, Braat L, Christie M, Crossman N, Ghermandi A, Hein L, Hussain S, Kumar P, Mcvittie A, Portela R, Rodriguez LC, Ten Brink P, Van Beukeringh P (2012) Global estimates of the value of ecosystems and their services in monetary units. Ecosyst Serv 1(1):50–61

Del Valle HF, Elissalde NO, Gagliardini DA, Milovich J (1998) Status of desertification in the Patagonian region: assessment and mapping from satellite imagery. Arid Land Res Manag 12(2):95–121

Díaz NI, Smith-Flueck JM (2000) The Patagonian huemul: a mysterious deer on the brink of extinction. LOLA, Buenos Aires

Fasioli E, Díaz BG (2011) Cartografía del sistema provincial de áreas protegidas de Santa Cruz (Patagonia Austral, Argentina). Párrafos Geográficos 10(1):174–194

Fernández JB, Smith J Jr, Scolaro A, Ibargüengoytía NR (2011) Performance and thermal sensitivity of the southernmost lizards in the world, *Liolaemus sarmientoi* and *Liolaemus magellanicus*. J Therm Biol 36(1):15–22

Fiori SM, Zalba SM (2003) Potential impacts of petroleum exploration and exploitation on biodiversity in a Patagonian Nature Reserve, Argentina. Biodivers Conserv 12(6):1261–1270

Flueck WT, Smith-Flueck JM (2012) Huemul heresies: beliefs in search of supporting data. 1. Historical and zooarcheological considerations. Anim Prod Sci 52(8):685–693

Gaitán JJ, Bran D, Oliva G, Maestre FT, Aguiar MR, Jobbágy E, Buono G, Ferrante D, Nakamatsu V, Ciari G, Salomone J, Massara V (2014) Plant species richness and shrub cover attenuate drought effects on ecosystem functioning across Patagonian rangelands. Biol Lett 10(10):e20140673

Gaitán JJ, Bran DE, Oliva GE, Stressors PA (2019) Patagonian Desert. Elsevier Inc, Amsterdam

Gargaglione V, Peri PL, Rubio G (2014) Tree-grass interactions for N in *Nothofagus antarctica* silvopastoral systems: evidence of facilitation from trees to underneath grasses. Agrofor Syst 88(5):779–790

Gea G, Martínez Pastur G, Cellini JM, Lencinas MV (2004) Forty years of silvicultural management in southern *Nothofagus pumilio* (Poepp. Et Endl.) Krasser primary forests. Forest Ecol Manag 201(2–3):335–347

Gilbert-Norton L, Wilson R, Stevens JR, Beard KH (2010) A meta-analytic review of corridor effectiveness. Conserv Biol 24(3):660–668

Haines-Young R, Potschin MB (2018) Common international classification of ecosystem services (CICES) V5.1 and guidance on the application of the revised structure. Fabis Consulting Ltd, Nottingham

Harrison PA, Berry PM, Simpson G, Haslett JR, Blicharska M, Bucur M, Dunford R, Egoh B, García-llorente M, Geamănă N, Geertsema W, Lommelen E, Meiresonne L, Turkelboom F (2014) Linkages between biodiversity attributes and ecosystem services: a systematic review. Ecosyst Serv 9:191–203

Hirzel AH, Hausser J, Chessel D, Perrin N (2002) Ecological-niche factor analysis: how to compute habitat- suitability maps without absence data? Ecology 83(7):2027–2036

Hirzel AH, Hausser J, Perrin N (2004) Biomapper 3.1. Division of conservation biology, University of Bern, Bern

Ibargüengoytía NR, Medina SM, Fernández JB, Gutiérrez JA, Tappari F, Scolaro A (2010) Thermal biology of the southernmost lizards in the world: *Liolaemus sarmientoi* and *Liolaemus magellanicus* from Patagonia, Argentina. J Therm Biol 35(1):21–27

Koschke L, Fürst C, Frank S, Makeschin F (2012) A multi-criteria approach for an integrated land-cover-based assessment of ecosystem services provision to support landscape planning. Ecol Indic 21:54–66

Kusch A, Vidal O, Henríquez JM (2016) Remoción de matorrales semi-áridos en Magallanes: Efectos sobre la composición, estructura y rasgos funcionales de los ensambles de aves. An Inst Patagon 44(2):35–48

Lantschner MV, Rusch V (2007) Impacto de diferentes disturbios antrópicos sobre las comunidades de aves de bosques y matorrales de *Nothofagus antarctica* en el NO Patagónico. Ecol Austral 17(1):99–112

Laterra P, Orúe ME, Booman GC (2011) Spatial complexity and ecosystem services in rural landscapes. Agric Ecosyst Environ 154:56–67

Lencinas MV, Martínez Pastur G, Medina M, Busso C (2005) Richness and density of birds in timber *Nothofagus pumilio* forests and their unproductive associated environments. Biodivers Conserv 14(10):2299–2320

Lencinas MV, Martínez Pastur G, Rivero P, Busso C (2008a) Conservation value of timber quality vs. associated non-timber quality stands for understory diversity in *Nothofagus* forests. Biodivers Conserv 17(11):2579–2597

Lencinas MV, Martínez Pastur G, Anderson CB, Busso C (2008b) The value of timber quality forests for insect conservation on Tierra del Fuego Island compared to associated non-timber quality stands. J Insect Conserv 12(5):461–475

Lencinas MV, Martínez Pastur G, Gallo E, Cellini JM (2009) Alternative silvicultural practices with variable retention improve bird conservation in managed South Patagonian forests. For Ecol Manag 258(4):472–480

Lencinas MV, Martínez Pastur G, Gallo E, Cellini JM (2011) Alternative silvicultural practices with variable retention to improve understory plant diversity conservation in southern Patagonian forests. For Ecol Manag 262(7):1236–1250

Lescano MN, Elizalde L, Werenkraut V, Pirk GI, Flores GE (2017) Ant and tenebrionid beetle assemblages in arid lands: their associations with vegetation types in the Patagonian steppe. J Arid Environ 138:51–57

Lindenmayer D, Franklin JF, Lõhmus A, Baker SC, Bauhus J, Beese W, Brodie A, Kiehl B, Kouki J, Martínez Pastur G, Messier C, Neyland M, Palik B, Sverdrup-Thygeson A, Volney J, Wayne A, Gustafsson L (2012) A major shift to the retention approach for forestry can help resolve some global forest sustainability issues. Conserv Lett 5(6):421–431

Mace G, Norris K, Fitter A (2012) Biodiversity and ecosystem services: a multilayered relationship. Trends Ecol Evol 27(1):19–26

Maes J, Egoh B, Willemen L, Liquete C, Vihervaara P, Schägner JP, Grizzetti B, Drakou EG, La Notte A, Zulian G, Bouraoui F, Paracchini ML, Braat L, Bidoglio G (2012) Mapping ecosystem services for policy support and decision making in the European Union. Ecosyst Serv 1(1):31–39

Maes J, Teller A, Erhard M, Murphy E, Murphy P, Paracchini ML, Barredo JI, Grizzetti B, Cardoso A, Somma F, Petersen JE, Meiner A, Gelabert ER, Zal N, Kristensen P, Bastrup-Birk A, Biala K, Romao C, Piroddi C, Egoh B, Florina C, Santos-Martín F, Naruševičius V, Verboven J, Pereira HM, Bengtsson J, Gocheva K, Marta-Pedroso C, Snäll T, Estreguil C, San-Miguel-Ayanz J, Braat L, Grêt-Regamey A, Pérez-Soba M, Degeorges M, Beaufaron C, Lillebø AI, Abdul Malak D, Liquete C, Condé S, Moen J, Ostergard H, Czúcz B, Drakou EG, Zulian G, Lavalle C (2014) Mapping and assessment of ecosystems and their services: indicators for ecosystem assessments under action 5 of the EU biodiversity strategy to 2020. European Union Technical report no. 2014-080, Luxembourg

Martínez Pastur G, Peri PL, Fernández C, Staffieri G, Lencinas MV (2002) Changes in understory species diversity during the *Nothofagus pumilio* forest management cycle. J For Res 7(3):165–174

Martínez Pastur G, Lencinas MV, Cellini JM, Peri PL, Soler R (2009) Timber management with variable retention in *Nothofagus pumilio* forests of Southern Patagonia. For Ecol Manag 258(4):436–443

Martínez Pastur G, Peri PL, Lencinas MV, García Llorente M, Martín López B (2016a) Spatial patterns of cultural ecosystem services provision in Southern Patagonia. Landsc Ecol 31(2):383–399

Martínez Pastur G, Peri PL, Soler Esteban R, Schindler S, Lencinas MV (2016b) Biodiversity potential of *Nothofagus* forests in Tierra del Fuego (Argentina): tools for regional conservation planning. Biodivers Conserv 25(10):1843–1862

Martínez Pastur G, Peri PL, Huertas Herrera A, Schindler S, Díaz-Delgado R, Lencinas MV, Soler R (2017) Linking potential biodiversity and three ecosystem services in silvopastoral managed forest landscapes of Tierra del Fuego, Argentina. Int J Biodivers Sci Ecosyst Serv Manag 13(2):1–11

Martínez Pastur G, Rosas YM, Manríquez MT, Huertas Herrera A, Miller JA, Cellini JM, Barrera MD, Peri PL, Lencinas MV (2019) Knowledge arising from long-term research of variable retention harvesting in Tierra del Fuego: where do we go from here? Ecol Process 8(1):24–40

Millennium Ecosystem Assessment (MEA) (2005) Ecosystems and human wellbeing: current state and trends. Island Press, Washington

Mori AS, Lertzman KP, Gustafsson L (2017) Biodiversity and ecosystem services in forest ecosystems: a research agenda for applied forest ecology. J Appl Ecol 54(1):12–27

Oliva G, Gonzalez L, Ruial P (2004) Áreas Ecológicas. In: Gonzalez L, Rial P (eds) Guía Geográfica Interactiva de Santa Cruz. INTA, Buenos Aires, pp 14–15

Oñatibia GR, Aguiar MR, Semmartin M (2015) Are there any trade-offs between forage provision and the ecosystem service of C and N storage in arid rangelands? Ecol Eng 77:26–32

Pedrana J, Bustamante J, Travaini A, Rodríguez A (2010) Factors influencing guanaco distribution in southern Argentine Patagonia and implications for its sustainable use. Biodivers Conserv 19(12):3499–3512

Pedrana J, Bustamante J, Rodríguez A, Travaini A (2011) Primary productivity and anthropogenic disturbance as determinants of Upland Goose *Chloephaga picta* distribution in southern Patagonia. Ibis 153(3):517–530

Perera A, Peterson U, Martínez Pastur G, Iverson L (2018) Ecosystem services from forest landscapes: Broadscale considerations. Springer, Cham

Peri PL, Ormaechea SG (2013) Relevamiento de los bosques nativos de ñire (*Nothofagus antarctica*) en Santa Cruz: Base para su conservación y manejo. INTA, Buenos Aires

Peri PL, Lencinas MV, Martínez Pastur G, Wardell-Johnson GW, Lasagno R (2013) Diversity patterns in the steppe of Argentinean southern Patagonia: environmental drivers and impact of grazing. In: Morales MB y Traba Diaz J (eds.) Steppe ecosystems: biological diversity, management and restoration. Nova Science Publishers, New York, p 73–96

Peri PL, Ladd B, Lasagno RG, Martínez Pastur G (2016a) The effects of land management (grazing intensity) vs. the effects of topography, soil properties, vegetation type, and climate on soil carbon concentration in Southern Patagonia. J Arid Environ 134:73–78

Peri PL, Hansen NE, Bahamonde HA, Lencinas MV, Von Müller AR, Ormaechea S, Gargaglione V, Soler Esteban R, Tejera L, Lloyd CE, Martínez Pastur G (2016b) Silvopastoral systems under native forest in Patagonia, Argentina. In: Peri PL, Dube F, Varella A (eds) Silvopastoral systems in southern South America, Advances in agroforestry. Springer, Heidelberg, pp 117–168

Peri PL, Lencinas MV, Bousson J, Lasagno R, Soler R, Bahamonde H, Pastur GM (2016c) Biodiversity and ecological long-term plots in Southern Patagonia to support sustainable land management: the case of PEBANPA network. J Nat Conserv 34:51–64

Peri PL, Rosas YM, Ladd B, Toledo S, Lasagno RG, Martínez Pastur G (2018) Modelling soil carbon content in South Patagonia and evaluating changes according to climate, vegetation, desertification and grazing. Sustainability 10(2):e438

Peri PL, Monelos L, Díaz B, Mattenet F, Huertas L, Bahamonde H, Rosas YM, Lencinas MV, Cellini JM, Martínez Pastur G (2019a) Estado y usos de los bosques nativos de lenga, siempreverdes y mixtos en Santa Cruz: Base para su conservación y manejo. INTA, Buenos Aires

Peri PL, Rosas YM, Ladd B, Toledo S, Lasagno RG, Martínez Pastur G (2019b) Modeling soil nitrogen content in South Patagonia across a climate gradient, vegetation type, and grazing. Sustainability 11(9):e2707

Quevedo P, Von Hardenberg A, Pastore H, Álvarez J, Corti P (2017) Predicting the potential distribution of the endangered huemul deer *Hippocamelus bisulcus* in North Patagonia. Oryx 51(2):315–323

Quintas-Soriano C, Martín-López B, Santos-Martín F, Loureiro M, Montes C, Benayas J, García-Llorente M (2016) Ecosystem services values in Spain: a meta-analysis. Environ Sci Pol 55(1):186–195

Raudsepp-Hearne C, Peterson GD, Bennett EM (2010) Ecosystem service bundles for analyzing trade-offs in diverse landscapes. Proc Natl Acad Sci 107(11):5242–5247

Reyers B, Biggs R, Cumming GS, Elmqvist T, Hejnowicz AP, Polasky S (2013) Getting the measure of ecosystem services: a social-ecological approach. Front Ecol Environ 11(5):268–273

Rosas YM, Peri PL, Huertas Herrera A, Pastore H, Martínez Pastur G (2017) Modeling of potential habitat suitability of *Hippocamelus bisulcus*: effectiveness of a protected areas network in Southern Patagonia. Ecol Process 6(1):e28

Rosas YM, Peri PL, Martínez Pastur G (2018) Potential biodiversity map of lizard species in Southern Patagonia: environmental characterization, desertification influence and analyses of protection areas. Amphibia-Reptilia 3(39):289–301

Rosas YM, Peri PL, Bahamonde HA, Cellini JM, Barrera MD, Huertas Herrera A, Lencinas MV, Martínez Pastur G (2019a) Trade-offs between management and conservation for the provision of ecosystem services in the southern Patagonian forests. In: Stanturf J (ed) Achieving sustainable management of boreal and temperate forests. Burleigh Dodds Science Publishing, Cambridge. https://doi.org/10.19103/AS.2019.0057.07

Rosas YM, Peri PL, Carrara R, Flores G, Pedrana J, Martínez Pastur G (2019b) Potential biodiversity map of darkling beetles (Tenebrionidae): environmental characterization, land-uses and analyses of protection areas in Southern Patagonia. J Insect Conserv 23(5–6):885–897

Rosas YM, Peri PL, Lencinas MV, Martínez Pastur G (2019c) Potential biodiversity map of understory plants for *Nothofagus* forests in Southern Patagonia: analyses of landscape, ecological niche and conservation values. Sci Total Environt 682:301–309

Rositano F, Bert FE, Piñeiro G, Ferraro DO (2018) Identifying the factors that determine ecosystem services provision in Pampean agroecosystems (Argentina) using a data-mining approach. Environ Dev 25:3–11

Scholes RJ, Whyte A (2009) Science for managing ecosystem services: Beyond the Millennium Ecosystem Assessment. PNAS 106(5): 1305–1312.

Smith-Flueck JM, Barri J, Ferreyra N, Nuñez A, Tomas N, Guzman J, Jimenez J (2011) Advances in ecology and conservation of *Hippocamelus* species in South America. Anim Prod Sci 51(4):378–383

Sola FJ, Peri PL, Huertas L, Pastur GM, Lencinas MV (2016) Above-ground arthropod community structure and influence of structural-retention management in southern Patagonian scrublands, Argentina. J Insect Conserv 20(6):929–944

Soler R, Schindler S, Lencinas MV, Peri PL, Martínez Pastur G (2016) Why biodiversity increases after variable retention harvesting: a meta-analysis for southern Patagonian forests. For Ecol Manag 369(1):161–169

Spagarino C, Martínez Pastur G, Peri LP (2001) Changes in *Nothofagus pumilio* forest biodiversity during the forest management cycle: 1. Insects Biodivers Conserv 10(12):2077–2092

Thompson I, Mackey B, McNulty S, Mosseler A (2009) Forest resilience, biodiversity, and climate change. A synthesis of the biodiversity/resilience/stability relationship in forest ecosystems. Secretariat of the convention on biological diversity, Montreal. Technical series no. 43, p 67

Thompson ID, Okabe K, Tylianakis JM, Kumar P, Brockerhoff EG, Schellhorn NA, Parrotta JA, Nasi R (2011) Forest biodiversity and the delivery of ecosystem goods and services: translating science into policy. Bioscience 61(12):972–981

Travaini A, Zapata SC, Bustamante J, Pedrana J, Zanón JI, Rodríguez A (2015) Guanaco abundance and monitoring in southern Patagonia: distance sampling reveals substantially greater numbers than previously reported. Zool Stud 54(1):23–35

Turkelboom F, Leone M, Jacobs S, Kelemen E, García-Llorente M, Baró F, Termansen M, Barton DN, Berry P, Stange E, Thoonen M, Kalóczkai Á, Vadineanu A, Castro AJ, Czúcz B, Rockmann C, Wurbs D, Odee D, Preda E, Gómez-Baggethun E, Rush GM, Martínez Pastur G, Palono I, Fick J, Casaer J, van Dijk J, Priess JA, Langemeyer J, Mustajoki J, Kopperoinen L, Baptist MJ, Peri PL, Mukhopadhyay R, Aszalós R, Roy SB, Luque S, Rush V (2018) When we cannot have it all: ecosystem services trade-offs in the context of spatial planning. Ecosyst Serv 29:566–578

Veblen TT, Donoso C, Kitzberger T, Robertus AJ (1996) Ecology of Southern Chilean and Argentinian *Nothofagus* forests. In: Veblen TT, Hill RS, Read J (eds) The ecology and biogeography of nothofagus forests. Yale University Press, Connecticut, pp 293–353

Vila AR, López R, Pastore H, Faúndez R, Serret A (2006) Current distribution and conservation of the huemul (*Hippocamelus bisulcus*) in Argentina and Chile. Mastozool Neotrop 13(2):263–269

Whytock RC, Fuentes-Montemayor E, Watts K, Barbosa De Andrade P, Whytock RT, French P, Macgregor NA, Park KJ (2018) Bird-community responses to habitat creation in a long-term, large-scale natural experiment. Conserv Biol 32(2):345–354

Chapter 3
Grazing Management and Provision of Ecosystem Services in Patagonian Arid Rangelands

Gastón R. Oñatibia

Abstract This chapter discusses how different sheep grazing management of Patagonian arid rangelands impacts on the provision of critical ecosystem services (ES): plant primary productivity (supporting service), forage supply (provisioning service), C and N storage (regulating services), and plant diversity (biodiversity). Particularly, it is evaluated how grazing pressure (land-use intensification), strategic grazing-rests, and herbivore distribution affect these ES, analyzing the synergies and trade-offs among them. Besides, it is assessed how aridity level interacts with grazing management to determine plant attributes linked to the ES of forage supply and plant primary productivity. Moderating the stocking rate, applying strategic grazing-rests according to each ecological site characteristics (mainly the aridity level), and promoting a homogeneous herbivore distribution within each management unit are useful management technologies to maximize the provision of ES, promoting synergies among them. Assessing diverse ES responses and including them in grazing management decisions is critical to promote the sustainability of Patagonian arid rangelands and meet current and future stakeholder needs.

Keywords Biodiversity · C storage · Forage provision · Grazing-rest · Herbivore distribution · Sheep grazing pressure

G. R. Oñatibia (✉)
CONICET-Universidad de Buenos Aires, Instituto de Investigaciones Fisiológicas y Ecológicas Vinculadas a la Agricultura (IFEVA), Buenos Aires, Argentina

Universidad de Buenos Aires, Facultad de Agronomía, Departamento de Recursos Naturales y Ambiente, Cátedra de Ecología, Buenos Aires, Argentina
e-mail: onatibia@agro.uba.ar

P. L. Peri et al. (eds.), *Ecosystem Services in Patagonia*, Natural and Social Sciences of Patagonia, https://doi.org/10.1007/978-3-030-69166-0_3

1 Introduction

Rangeland ecosystems provide multiple ecosystem services (ES) (Sala et al. 2017). In Patagonian rangelands, as in most rangelands around the globe, the main provisioning services are forage supply for domestic herbivores (mainly for wool and meat production) and freshwater supply for drink and irrigation (Sala and Paruelo 1997; Oñatibia et al. 2015; Yahdjian et al. 2015; Sala et al. 2017). The main regulating services are C sequestration and storage, nutrients content and delivery (e.g., N storage), soil erosion control, and soil water storage (Havstad et al. 2007; Oñatibia et al. 2015; Yahdjian et al. 2015; Sala et al. 2017; Zhao et al. 2020). Otherwise, critical supporting services in rangelands are primary productivity, nutrient cycling, soil conservation, and biodiversity conservation (Sala and Paruelo 1997). Though biodiversity has been included in the supporting services according to some classifications, it occupies a separate category due to its key relevance in the sustenance of ecosystems (MEA 2005). As rangelands are largely natural ecosystems, all ES depend in some way on local biodiversity (Havstad et al. 2007). Finally, the most outstanding cultural services provided by rangelands like those in Patagonia are related to human experiences and activities, including traditional lifestyles, spiritual and religious values, aesthetic appreciation, cultural heritage, cultural diversity, scientific knowledge, education, inspiration, recreation, and ecotourism (MEA 2005; Yahdjian et al. 2015; Sala et al. 2017).

The main challenge that rangeland managers face is to sustain the capacity of rangeland ecosystems to support human well-being, which depends on diverse ES. Thus, rangeland management objectives should be based on maintaining the capacity of social-ecological systems to provide goods and services meeting current and future human needs (Briske 2017). This challenge could be even greater given that forecasts indicate that (i) rangeland intrinsic inter-annual variability will be exacerbated as a consequence of climate change (Knapp et al. 2008; Huang et al. 2016; Maestre et al. 2016) and that (ii) grazing pressure is also expected to increase to satisfy the growing demand for meat and dairy products (Thornton 2010). Grazing intensification can produce profound changes in the structure and functioning of ecosystems, affecting biodiversity and ES (Petz et al. 2014; Maestre et al. 2016). This land-use change may also modify the interactions among different ES, and range management decisions should try to take advantage of the knowledge of these interactions to maximize synergies (win-win) and reduce trade-offs (Bennett et al. 2009; Sala et al. 2017). In addition to domestic grazing intensification, most Patagonian rangelands are broadly threatened by an aridity increase as a result of average temperature rise and increases in the frequency and intensity of extreme droughts (Nuñez et al. 2009; Dai 2013). These accelerated changes are unprecedented, making it essential to rapidly address the present state and possible future responses of ES to the processes of land-use intensification and aridity increase (Oñatibia et al. 2020a).

Rangeland ES supply mostly depends on biophysical conditions such as climate, soils and biota, and land-use history (Sala et al. 2017). In Patagonian arid rangelands, ES are mainly controlled by the aridity level, the identity of dominant plant species composing vegetation patches (their population vital rates), and grazing by domestic livestock (mostly sheep) and by native herbivores (Fig. 3.1). Aridity level integrates attributes such as precipitation, temperature, and soil properties, and it directly determines key aspects of rangeland ecosystem structure and functioning, such as the community composition, predominant life-form abundance, and primary production (Milchunas et al. 1988; McNaughton et al. 1989). Domestic grazing pressure (grazing management) is a major control of plant community dynamics, and it may also interact with aridity to determine the structure and functioning of Patagonian rangelands (Fig. 3.1; Oñatibia et al. 2018, 2020a). Since their introduction in the latter nineteenth century, sheep stocks steadily increased during the first 60 years, but they have sharply declined during the last decades as a consequence of generalized resource degradation and the reduction of the forage provision service (Soriano and Movia 1986; Golluscio et al. 1998a; Texeira and Paruelo 2006; Oñatibia et al. 2015, 2020b).

In western semiarid grass steppes of the Subandean District (León et al. 1998; Fig. 3.2a), the primary sign of grazing-induced degradation is the reduction of the dominant grass species cover (*Festuca pallescens*) and the increase of shrubs (León and Aguiar 1985; Aguiar et al. 1996; Oñatibia et al. 2018). These structural changes reduce primary production (supporting service) and herbivore carrying capacity (Aguiar et al. 1996; Verón and Paruelo 2010). In arid grass-shrub steppes of the Occidental District (León et al. 1998; Fig. 3.2b), grazing intensification does not increase shrub abundance, but it reduces preferred grass species biomass, decreasing forage supply (Aguiar and Sala 1998; Cesa and Paruelo 2011; Oñatibia et al.

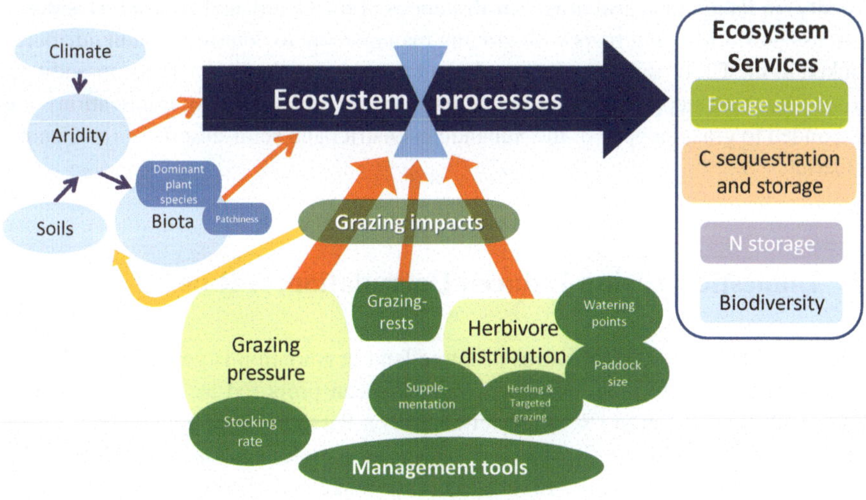

Fig. 3.1 Major controls determining critical ES from Patagonian arid rangelands. The controls of ecosystem processes linked to the ES supply range from ultimate factors, such as the aridity level, to proximate factors, such as vital rates of dominant species, patchiness, and the grazing impacts, which are mediated by the main grazing management tools available for Patagonian rangelands

Fig. 3.2 Paddocks under different grazing pressure management in (**a**) grass steppes of the Subandean District, (**b**) grass-shrub steppes of the Occidental District, and (**c**) semi-deserts of the Central District. The paddock observed to the left of the fence in each photo has been subjected to a higher grazing pressure than the one to the right. (Photo credit: L. Boyero and G. Oñatibia)

2015; Oñatibia and Aguiar 2016). In semi-deserts of the Central District (León et al. 1998; Fig. 3.2c), grazing intensification reduces plant cover by decimating grass species. Even though the dominant dwarf shrub species (*Nassauvia glomerulosa*) can sustain the vegetation patchiness structure under high grazing pressure, the forage provision substantially decreases (Oñatibia et al. 2018, 2020a). As sheep grazing has been proposed as one of the main causes of Patagonian rangeland degradation (Golluscio et al. 1998a), a comprehensive evaluation of the provision of critical ES under different grazing management alternatives (Fig. 3.1), and their interaction with the aridity level, is needed.

In this chapter, I discuss how different sheep grazing management (Fig. 3.1) impacts on plant primary productivity (supporting service), forage supply (provisioning service), C and N storage (regulating services), and plant diversity (biodiversity) in Patagonian arid grass-shrub steppes of the Occidental District. Besides, I discuss how aridity interacts with grazing management to determine plant attributes linked to the ES of forage supply and plant primary productivity. Thus, in addition to grass-shrub steppes of the Occidental District, management implications are expanded to grass steppes of the Subandean District and semi-deserts of the Central District.

2 Domestic Grazing-Induced Degradation: A Rule?

The idea that domestic herbivores promote land degradation is generally accepted. In arid and semiarid rangelands, grazing has been proposed as one of the major causes of desertification (Reynolds et al. 2007). Patagonian rangelands are not an exception (Ares et al. 1990). Words as degradation, deterioration, overgrazing, and desertification are common in grazing and rangeland ecology literature from Patagonia (e.g., Soriano 1956; León and Aguiar 1985; Soriano and Movia 1986; Del Valle et al. 1998; Bertiller et al. 2002; Ares et al. 2003; Bisigato et al. 2005; Verón et al. 2011; Kröpfl et al. 2013; Oñatibia et al. 2020a). These claims are associated with the documented grazing impacts on vegetation and soil, such as undesired changes in species composition, plant cover loss, primary productivity decrease, and soil erosion increase (e.g., Beeskow et al. 1995; Aguiar et al. 1996; Bisigato and Bertiller 1997; Cesa and Paruelo 2011; Palacio et al. 2014; Oñatibia et al. 2018). As a consequence of these negative impacts, it is generally assumed that livestock grazing decreases the global provision of ES. However, depending on management, domestic grazing could promote positive effects on plants, raising productivity, and enhancing forage quality (Posse et al. 2000; Oñatibia and Aguiar 2016, 2019). In this sense, assertions stating that domestic grazing must be removed from Patagonian arid rangelands for conservation goals (Murdoch et al. 2010) appear as requiring at least further critical evaluation.

In many cases, conclusions on grazing-induced degradation come from studies that have some limitations when considering grazing treatments, which would not allow generalizations. A common approach frequently used to evaluate grazing effects is comparing exclosure areas with grazed areas (e.g., Cesa and Paruelo 2011), but studies using this approach usually lack an accurate characterization of grazing management in grazed plots (i.e., stocking rate, grazing system, grazing, and rest seasons), while the relevance of the exclosures' age is dismissed. Another common approach is the use of gradients of grazing pressure inferring from watering point distance (e.g., Bisigato and Bertiller 1997) or indirectly from floristic census ordination (e.g., León and Aguiar 1985). In these cases, the lack of a reference situation (i.e., long-term exclosure) may weaken inferences and conclusions. Besides, this approach frequently relies on a comparison between light and intensive grazing. In this sense, the common inference of grazing-induced degradation, attributed to grazing *per se* when comparing light vs. high grazing pressure, can be considered as a misinterpretation. This may be a semantic point, but strictly, it would be degradation given by grazing intensification instead of degradation by grazing *per se*, since it depends on grazing management. For example, lightly grazed areas may not be degraded, and they may provide a higher level of ES than ungrazed sites (see Sect. 3.1).

An additional experimental issue related to grazing treatments is that contrasting conditions are often deliberately selected in order to record the highest effects (e.g., exclosure vs. intensive grazing or fence contrasts, where there are visually distinguishable differences between both sides). The main limitation is that in this way, the nonlinear effects of the changes in grazing pressure cannot be detected. Otherwise, many times, continuously grazed paddocks (year-round grazing) are also deliberately selected to avoid biases attributed to different grazing management. Therefore, other herbivore management alternatives are left out of the analysis. Aspects as frequency, intensity, and timing of grazing will play important roles in the outcomes of grazing impacts (Golluscio et al. 1998a). The size of the experimental units is another feature frequently not considered. Paddocks or plots are usually small in controlled experiments and very large in gradient studies. These aspects can also strongly influence grazing impacts (Teague et al. 2013; Oñatibia and Aguiar 2018). In short, while it is known that grazing can induce land degradation, it should not be considered as a rule since the sense and size of grazing impacts depend on several grazing management characteristics (Fig. 3.1). Here, it was assessed how different management alternatives can determine the provision of critical ES, through using different complementary approaches, including reference situations, discrete treatments with several levels, and continuous gradients of grazing pressure, to make robust inferences.

3 Grazing Management in Patagonian Rangelands and Ecosystem Services Supply

In Patagonian rangelands, animal husbandry is an important activity in terms of cultural heritage, as domestic livestock production decreases social impacts in comparison with other land uses (e.g., afforestation), for example, reducing the socioeconomic cost of rural to urban migration (Aguiar and Román 2007). Therefore, proper grazing management on natural vegetation is a critical factor for complementarily maximizing provisioning and regulating ES that directly affect human well-being (Oñatibia et al. 2015; Zhao et al. 2020). However, managing arid Patagonian rangelands for livestock production is particularly challenging because of the low average annual precipitation and high year-to-year variability in weather. Besides, managing livestock to achieve goals other than livestock production, such as biodiversity conservation or C sequestration, is even more complex (Sanderson et al. 2020). Improper managing can lead to degraded conditions difficult to revert or even irreversible, as the resilience of these ecosystems is low (Oñatibia et al. 2018). It has been proposed that the main causes of degradation of Patagonian rangelands associated to grazing management are (i) the overestimation of the carrying capacity, leading to overgrazing forage resources with high animal stocking rates; (ii) the inadequate animal distribution in very large and heterogeneous paddocks, which leads to over- and under-graze forage resources at several scales; and (iii) the continuous grazing all the year-round of almost all plant communities, which exacerbates the heterogeneous grazing patterns (Golluscio et al. 1998a). These authors also proposed management technologies to stop and reverse rangeland degradation (see Golluscio et al. 1998a). In this section, the results of the application of these technologies are discussed in light of the empirical evidence obtained during the last decade. Particularly, I discuss how (i) the stocking rate management (grazing pressure), (ii) the implementation of grazing-rest periods, and (iii) the manipulation of the spatial distribution of herbivores affect the ES supply. These management tools are the most widespread technologies that consider the causes of degradation induced by domestic grazing in Patagonia (Fig. 3.1).

3.1 Stocking Rate Management in Grass-Shrub Steppes

Stocking rate, directly related to grazing pressure, is the main grazing management aspect determining grazing impacts on vegetation, soils, animal performance (Hickman et al. 2004; Laca 2009), and, thus, the ES that rangelands provide. It has been proposed that land-use changes that often increase rangeland provisioning ES, as grazing pressure intensification, generally decrease the regulating ES, such as C sequestration and storage, N content, or water holding capacity, essential to maintain the ecosystem integrity (Sala et al. 2017). However, despite having stressed that trade-offs between regulating and provisioning ES are inevitable, some ES may respond similarly to grazing pressure management and specific ecological conditions, creating synergistic situations (Oñatibia et al. 2015).

The grass-shrub steppes of the Patagonian Occidental District exhibit an example of synergies under proper management of the sheep stocking rate (Oñatibia et al. 2015; Oñatibia and Aguiar 2016, 2019). In these steppes, the effect of different sheep stocking rates has been evaluated on (i) the plant aboveground net primary productivity (ANPP), (ii) the forage provision, (iii) C and N storage, and (iv) plant diversity. The supply of these ES has been estimated in ungrazed paddocks (long-term exclosures, >20 years), moderately grazed paddocks (stocking rate: ~0.2 sheep ha^{-1} $year^{-1}$, average during the last 20 years), and intensively grazed paddocks (stocking rate: 0.4–0.5 sheep ha^{-1} $year^{-1}$, average during the last 20 years). The stocking rate in the moderately grazed paddocks could also be considered as light grazing (Oñatibia and Aguiar 2016). All grazed paddocks, where ES were estimated, have been continuously grazed (year-round).

3.1.1 Plant Aboveground Net Primary Productivity (ANPP, Supporting Service)

In the steppes of the Occidental District, ANPP is mainly composed of native perennial grasses and shrubs (Oñatibia and Aguiar 2016). A study during a year with average annual precipitation (Oñatibia and Aguiar 2016) exhibited that total grass ANPP was twice as high in long-term moderately grazed sites than in those ungrazed and intensively grazed sites (Fig. 3.3a). Otherwise, total shrub ANPP did not change as grazing pressure increased (Fig. 3.3a). Among grasses, dominant species exhibited different patterns, which are associated with each species preference degree by sheep (Fig. 3.3b). The dominant preferred species (*Poa ligularis*) had similar ANPP in ungrazed and moderately grazed sites but sharply decreased in intensively grazed sites. The other preferred species (*Bromus pictus*, subordinated) decreased its ANPP with grazing intensification. The dominant species presenting an intermediate preference degree (*Pappostipa speciosa*) exhibited the highest ANPP under moderate grazing conditions. The dominant unpreferred species (*Pappostipa humilis*) was the only species that showed an ANPP increasing trend (not statistically significant)

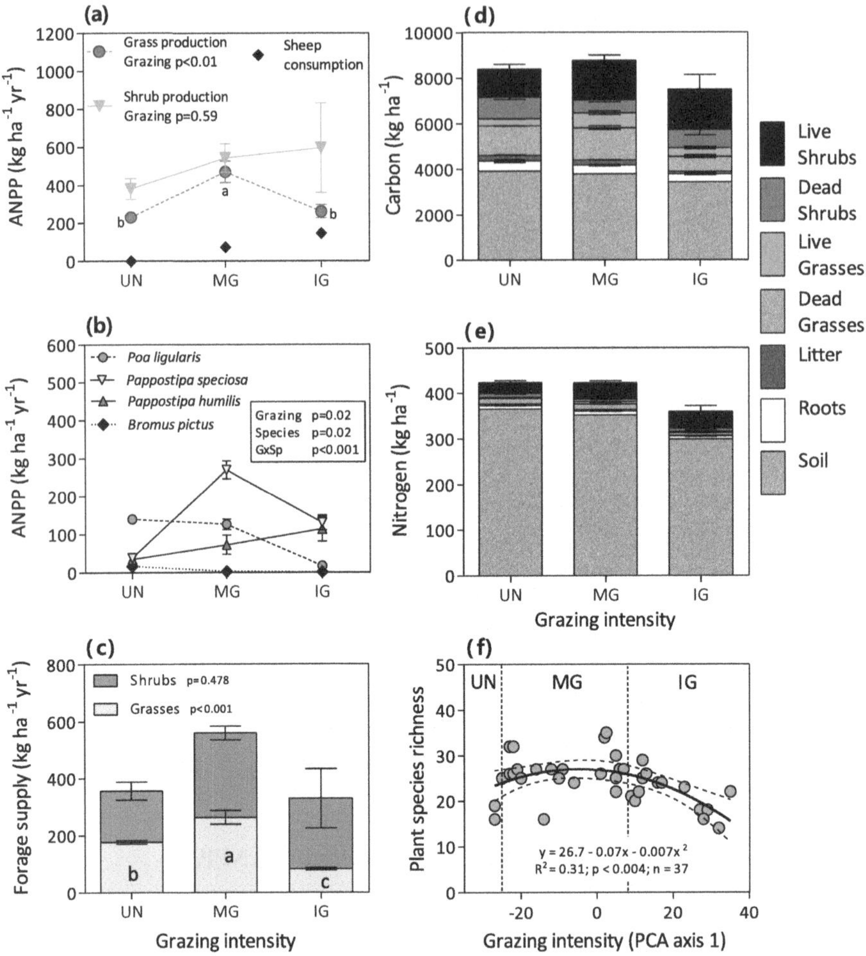

Fig. 3.3 Long-term grazing intensity effects on (**a**) ANPP of grasses and shrubs at life-form level, (**b**) ANPP of dominant native grasses at species level (*Poa ligularis*, *Pappostipa speciosa*, *Pappostipa humilis*, and *Bromus pictus*), (**c**) forage supply by grasses and shrubs, (**d**) carbon stocks, (**e**) nitrogen stocks, and (**f**) plant species richness, in grass-shrub steppes of Occidental District. In panels (**a–e**), the three grazing intensities are ungrazed (UN), moderately grazed (MG; ~0.2 sheep ha^{-1} $year^{-1}$), and intensively grazed (IG; 0.4–0.5 sheep ha^{-1} $year^{-1}$). Symbols and bars correspond to mean values, and vertical lines indicate standard errors. In panel (**a**), different letters indicate significant differences ($p < 0.05$) among grazing intensities within each life-form (Tukey test). In panel (**b**), the table indicates p-values resulting from a two-way ANOVA between species and grazing intensities. Statistical differences among grazing intensities for each grass species are developed in the main text. (Adapted from Oñatibia and Aguiar 2016). In panel (**c**), different letters indicate significant differences ($p < 0.05$) among grazing intensities for each life-form resulting from ANOVA (Tukey test). (Adapted from Oñatibia et al. 2015). In panels (**d**) and (**e**), compartments are aboveground green biomass of grasses and shrubs (green grasses, green shrubs), aboveground standing dead biomass of grasses and shrubs (dead grasses, dead shrubs), litter, belowground biomass of the top 20 cm of soil (roots), and organic matter of the top 5 cm of soil (soil). See the text for pattern description and statistical significances. (Adapted from Oñatibia et al. 2015;

with grazing pressure intensification (Fig. 3.3b). Among shrubs, dominant species did not exhibit ANPP significant changes with grazing pressure intensification (Oñatibia and Aguiar 2016). In other words, moderate grazing promotes an increase of this critical supporting service, mainly increasing grass biomass, without major undesired changes in species composition (Oñatibia and Aguiar 2016). On the contrary, both long-term domestic grazing elimination and grazing pressure intensification reduce ANPP.

3.1.2 Forage Supply (Provisioning Service)

Forage supply is considered the most critical ecosystem service in arid rangelands since human well-being in these ecosystems mostly depends on plant species that provide forage for domestic herbivores (Oñatibia et al. 2020a). This provisioning service is the fraction of aboveground biomass that can be consumed by domestic herbivores, which in Patagonian arid rangelands represents a relatively small fraction of ANPP, strongly determined by species composition (Golluscio et al. 1998a; Oñatibia et al. 2015). In Occidental steppes, among grasses, the forage supply for sheep is mainly determined by the green biomass production of *Poa ligularis* and *Pappostipa speciosa*, while *Bromus pictus* and *Hordeum comosum* contribute a very small fraction, given their low abundance. Other co-dominant grass species like *Pappostipa humilis* and *Pappostipa major* do not provide forage for sheep (Oñatibia et al. 2015; Oñatibia and Aguiar 2019). Among shrubs, twigs and inflorescences of *Adesmia volckmannii* and *Azorella prolifera* (ex *Mulinum spinosum*) also provide forage during later spring and summer (Oñatibia et al. 2015). Besides, since species which can provide forage present different preference degree by sheep, a specific forage aptitude factor is commonly assigned to estimate forage supply by each species (Easdale and Aguiar 2012; Oñatibia et al. 2015).

The study that estimated the impact of sheep stocking rate on the integrated forage biomass provided by all forage species together in Occidental steppes showed that moderate grazing maximized the forage supply provisioning service (Oñatibia et al. 2015). This pattern is mainly explained by the forage grass species, which exhibited the highest forage supply values in moderately grazed sites, intermediate values in ungrazed ones, and minimum values in intensively grazed sites (Fig. 3.3c). Otherwise, stocking rate management did not significantly change the forage supply by shrubs (Fig. 3.3c).

Fig. 3.3 (continued) Golluscio et al. 2009). In panel (**f**), the grazing intensity gradient was inferred from the first eigenvector (axis 1), resulting from a principal component analysis (PCA) of the plant species variance-covariance matrix, which considered 37 field surveys. Vertical dashed lines separate ungrazed (UN), moderately grazed (MG), and intensively grazed (IG) sites. The equation, line (with confidence intervals: dashed lines), R^2, and p-value of the estimated nonlinear regression curve are shown. (Adapted from Perelman et al. 1997)

3.1.3 C and N Storage (Regulating Service)

Estimating the effect of sheep stocking rate on C and N storage indicated that grazing intensification differently impacts on the main reservoirs of Patagonian Occidental steppes (Golluscio et al. 2009; Oñatibia et al. 2015). Carbon storage in green aboveground biomass of grasses was higher in moderately grazed sites than in intensively grazed and ungrazed ones, while C storage in the standing dead biomass of grasses was lower in intensively grazed sites than under the other two conditions (Fig. 3.3d). In contrast, C storage in shrub components (green and standing dead biomass) did not significantly change among grazing conditions. Carbon in the litter was higher in moderately grazed and ungrazed sites than in intensively grazed sites (Fig. 3.3d). Finally, neither the C storage in roots from the top 20 cm of soil nor in the soil organic matter from the top 5 cm were affected by grazing (Fig. 3.3d). A singularity of these Patagonian arid rangelands is the high relative contribution of vegetation to total C storage (Oñatibia et al. 2015). When integrating all the components, total C storage in vegetation and soil was the highest in moderately grazed sites and the lowest in intensively grazed sites (Fig. 3.3d).

Nitrogen storage in aboveground green and standing dead biomass of grasses and shrubs, litter, and roots followed the same patterns as C storage (Fig. 3.3e). In the soil organic matter from the top 5 cm, N storage was higher in moderately grazed and ungrazed sites than in intensively grazed ones (Fig. 3.3e). The relative contribution of soil organic matter to total N storage is more than five times higher than the vegetation contribution. When integrating all the reservoirs, total N storage in vegetation and soil was similar between ungrazed and moderately grazed sites, which exhibited higher N storage than intensively grazed sites.

3.1.4 Plant Diversity (Biodiversity)

Plant diversity is a key component of biodiversity. In Occidental steppes, plant species richness was nonlinearly affected by grazing intensification (unimodal relationship; Perelman et al. 1997; Sala et al. 2017). Intermediate grazing intensities lightly increased the plant richness, while intensive grazing reduced it by producing the local extinction of forage species (Fig. 3.3f). Diversity did not exhibit a significant relationship with this grazing gradient (Perelman et al. 1997). Patterns indicate that biodiversity may be suitably managed through proper grazing pressure management in these steppes.

3.1.5 Ecological Mechanisms: Trade-Offs and Synergies

The ES patterns described in the four previous sections at plant community and ecosystem levels are explained by patterns found at individual plant and population levels (Oñatibia and Aguiar 2016, 2019). Moderate sheep stocking rate maintains the population density of preferred grass species compared to ungrazed sites, while intensive grazing markedly reduces it (Oñatibia and Aguiar 2016, 2019). In turn, moderate grazing pressure increases the green proportion size of grass plants, and, therefore, it raises the forage quantity and accessibility, generating positive effects on the availability of resources (Oñatibia and Aguiar 2016, 2019). This optimization process induced by moderate grazing would stimulate plant productivity mainly by increasing photosynthetic rates in residual tissues, removing older tissues which shade more active younger ones, and accelerating nutrient cycling by feces and urine excretion and physical disturbance (McNaughton 1979; Oñatibia et al. 2015; Oñatibia and Aguiar 2016). In short, at the plant level, productivity would increase by compensatory growth (McNaughton 1983), which would raise the green biomass and revitalize the system, despite the biomass removal by herbivores (Oñatibia et al. 2015). At the population level, the changes in grass species density and plant size distribution driven by sheep grazing also explains the ES increment (Oñatibia and Aguiar 2019). It has been proposed that grazing can generate these positive impacts on vegetation, increasing the global provision of ES, when grazing pressure is light to moderate, the site productivity is low, and the evolutionary history of grazing is long (Milchunas et al. 1988). These conditions can be found in moderately grazed Patagonian steppes (Lauenroth 1998; Oñatibia et al. 2015). Otherwise, ES provided by shrub species were less susceptible, since shrubs behave as a more static component than grasses in the face of grazing pressure changes (Oñatibia et al. 2015; Oñatibia and Aguiar 2016).

Findings at different levels indicate that the studied supporting, provisioning, and regulating ES were generally maximized at a moderate sheep stocking rate (Table 3.1). In relation to domestic herbivore exclusion (no appropriation of

Table 3.1 Changes induced by different sheep stocking rate management in the provision of critical ecosystem services from semiarid steppes of the Patagonian Occidental District

Ecosystem service	Time scale	Stocking rate management: Moderate	Stocking rate management: Intensive	References
ANPP	Long term	+	0	Oñatibia and Aguiar (2016)
Forage supply	Long term	++	–	Oñatibia et al. (2015)
C sequestration and storage	Long term	+	–	Golluscio et al. (2009), Oñatibia et al. (2015)
N storage	Long term	0	– –	Golluscio et al. (2009), Oñatibia et al. (2015)
Plant biodiversity	Long term	0;+	–	Perelman et al. (1997)
Livestock production (meat and wool)	Short term	++	+++	Ares (2007), Oñatibia et al. (2015)
	Long term	++	0;+	
Soil erosion control	Long term	0;+	– –	Biancari et al. (2020)
Water holding capacity (soil storage)	Short/long term	0;+	0;+	Piazza (2016), Pereyra et al. (2017)

Signs indicate increase (+), decrease (–), and no change (0) in the provision level for each ecosystem service, in comparison with the provision level under long-term exclosure conditions. The long-term exclosures of domestic herbivores (>20 years) were considered as a baseline (reference) situation provisioning services) or stocking rate intensification, moderate grazing maximizes the ANPP, the forage supply, the C sequestration, and the plant species richness (Table 3.1), reducing trade-offs among these ES. Indeed, these ES were positively correlated, indicating synergistic relationships. This synergy occurs because moderate grazing promotes grasses productivity (and thus C sequestration), keeping the abundance of forage species high (Oñatibia et al. 2015; Oñatibia and Aguiar 2016). Because of this absence of trade-offs, the long-term exclusion of domestic herbivores is not the best option to increase forage supply and C storage. Although it has been frequently suggested to enhance the provision of critical ES, it is an inappropriate management practice in Patagonian arid steppes. On the other hand, intensive grazing reduces most ES (Table 3.1). These changes are explained by a substantial reduction of preferred species density and biomass or, even, their local extinction. Thus, grazing intensification decreases the forage natural capital which is the basis of the sheep industry in Patagonia (Ares 2007), also affecting other critical ecosystems functions, such as C sequestration and nutrient cycling. Then, the objective of achieving greater secondary productivity (e.g., meet or wool) per area unit in the short term by intensifying the grazing pressure would not be sustainable, since this management reduces the provisioning ecosystem service of forage supply, diminishing future secondary production (Ares 2007; Oñatibia et al. 2015; Table 3.1). This temporal trade-off emerges by not considering the temporal dimension of the appropriation of the same provisioning service (Rodríguez et al. 2006) because increasing the forage supply appropriation in the short term decreases the long-term forage supply.

Assessing diverse ES responses to stocking rate management opens a new perspective for analyzing rural socio-ecosystems in Patagonian rangelands. Overall, domestic grazing at moderate stocking rates is the best option for land use in Occidental grass-shrub steppes, from the perspective of most ecological and socio-economic aspects (Aguiar and Román 2007). Regarding ES not directly studied, such as water holding capacity, soil erosion control, or stability of primary production, it is also expected that moderate stocking rates can maintain high level of provision (Table 3.1), since these ES are generally correlated with soil C storage, soil cover, and plant species richness, respectively (Maestre et al. 2012, 2016; Sala et al. 2017). This knowledge is particularly relevant because the stocking rate is the main management aspect that producers (ranchers) may modify.

3.1.6 Stocking Rate Adjustment (Grazing Pressure Management)

In most rangeland ecosystems grazed by domestic herbivores, the allocation of the number of animals to an area is determined by the amount of forage produced, generally estimated from aboveground productivity or plant cover (Holechek 1988). However, in arid rangelands, a frequent mistake in this decision is not to consider either the different plant species producing biomass (different preference degree by domestic herbivores) or how accessible the forage is (e.g., forage produced by many plants of preferred species is inaccessible to sheep due to the amount of standing dead material protecting them). Neglecting these aspects may lead to inaccurate stocking rates, promoting undesirable outcomes in species composition, and primary and secondary productivity (e.g., Briske et al. 2008).

The stocking rate of moderately grazed sites, where ES were generally maximized (see previous sections), was recommended based on the estimation of the carrying capacity using the pastoral value method for each paddock. The method is based on the relative frequency of all species composing the community, weighed by their specific quality index, which depends on the sheep preference (see Golluscio et al. 1998a; Massara and Buono 2020 for methodological details). Since field measurements are needed to estimate species abundance, a population approach that estimates the green proportion of individual plants can also be applied, achieving a better estimation of forage accessibility (Oñatibia and Aguiar 2019; Oñatibia et al. 2020a). As a result of this management strategy, which is relatively conservative to estimate the carrying capacity, the average proportion of ANPP consumed by sheep under moderately grazed conditions is around 10% of the ANPP. Thus, grazing pressure as the ratio between forage consumption (kg dry matter ha^{-1} $year^{-1}$) and average ANPP (kg dry matter ha^{-1} $year^{-1}$) integrated over a year is around 0.1 (Oñatibia et al. 2018, 2020a). This proportion is close to that found in ecosystems grazed by native herbivores, and it is relatively lower than the theoretical average consumption by domestic herbivores in ecosystems with similar ANPP (Golluscio et al. 1998a; Oñatibia and Aguiar 2016).

The method to estimate the pastoral value at the paddock level requires a great field sampling effort to be representative. However, it can be complemented to spatially explicit estimations of ANPP, using remote sensing data (from satellite or unmanned aerial vehicles), which also allow classifying communities to enhance the efficiency in the field sampling (Paruelo et al. 2004; Irisarri et al. 2012; Easdale et al. 2019). In this way, values of ANPP of each community can be weighted by the forage availability and accessibility proportions, achieving spatially explicit maps of available and accessible forage. Otherwise, since these estimations can be performed annually, the method also allows adjusting the stocking rate to the interannual fluctuations in resource availability (e.g., associated with rainfall or temperature). This aspect is particularly relevant given the increasing occurrence of drought events or prolonged periods of water deficit, which require an adaptive management approach to pursuit preservation as well as maintain viable productive rangelands (Meffe et al. 2012; Allen et al. 2017; Derner et al. 2017). In summary, moderating the stocking rate with this method would allow maintaining sustainable animal production in the long term, preserving the natural forage capital and the provision of critical ES (Oñatibia et al. 2015).

3.2 *Grazing Pressure Management in Grass Steppes and Semi-deserts (Aridity Influence)*

In the grass steppes of the Subandean District and the semi-deserts of the Central District, the effect of sheep grazing pressure on plant cover, vegetation patchiness, and the structure of grass species populations has also been assessed (Oñatibia et al. 2018, 2020a). These vegetation attributes are strongly related to ecosystem functioning and critical ES, such as ANPP, C sequestration, and forage supply (Aguiar and Sala 1999; Flombaum and Sala 2007; Oñatibia and Aguiar 2019). In the Subandean steppes, findings indicated that the dominant forage species (*Festuca pallescens*) presents a unimodal response to increasing grazing pressure, both in its plant cover and population density. This species increases as grazing pressure raises from ungrazed to moderate and decreases as pressure is intensified from moderate to intensive (Fig. 3.4a and b). These results suggest that in the grass steppes, moderating the grazing pressure is an adequate tool to increase forage productivity (Oñatibia et al. 2020a). However, in the Central semi-deserts, grass populations are substantially reduced, even under moderate grazing (Fig. 3.4c). Because grasses are the main forage source for sheep, the ecosystem service of forage provision markedly decreases with grazing under these high aridity conditions (Oñatibia et al. 2020a).

The studied grass steppes of the Subandean District, grass-shrub steppes of the Occidental District, and semi-deserts of the Central District present low, intermediate, and high aridity level, respectively, since they are located across an increasing aridity gradient, which allows estimating the aridity influence (see Oñatibia et al. 2020a for a detailed description of sites). Findings on grazing impacts across this gradient indicate that grazing pressure interacts with aridity to determine ecosystem functioning and ES (Oñatibia et al. 2018, 2020a). Therefore, different grazing pressure must be adjusted depending on the aridity level. Results suggest that grazing pressure should be maintained between 0.2 and 0.3 in grass steppes, between 0.1 and 0.2 in grass-shrub steppes, and it should be around ($\leq$) 0.1 in semi-deserts to sustain the ES provision, mainly the forage supply (Oñatibia et al. 2020a). Patterns found also indicate that an increase in the aridity of Patagonian rangelands as a consequence of climate change (Nuñez et al. 2009; Dai 2013) may require a further reduction of the domestic grazer's pressure, especially in high aridity sites, where plants' tolerance does not compensate the effect of moderate grazing. Finally, it is also recommended to decrease grazing pressure during extended droughts periods to uncouple the biotic stress caused by herbivores from the drought-induced abiotic stress (Oñatibia et al. 2020a).

Fig. 3.4 (continued) The equations, lines (with confidence intervals: *dashed lines*), R^2, and p-values of the estimated nonlinear regressions curves are shown in each panel. Grazing pressure in each site was estimated as the ratio between sheep forage consumption (estimated from the stocking rate) and ANPP, weighed by a fecal index, which relates the fecal pellet density in the site, with the average of fecal pellets density of all grazed sites. (Adapted from Oñatibia et al. 2018, 2020a)

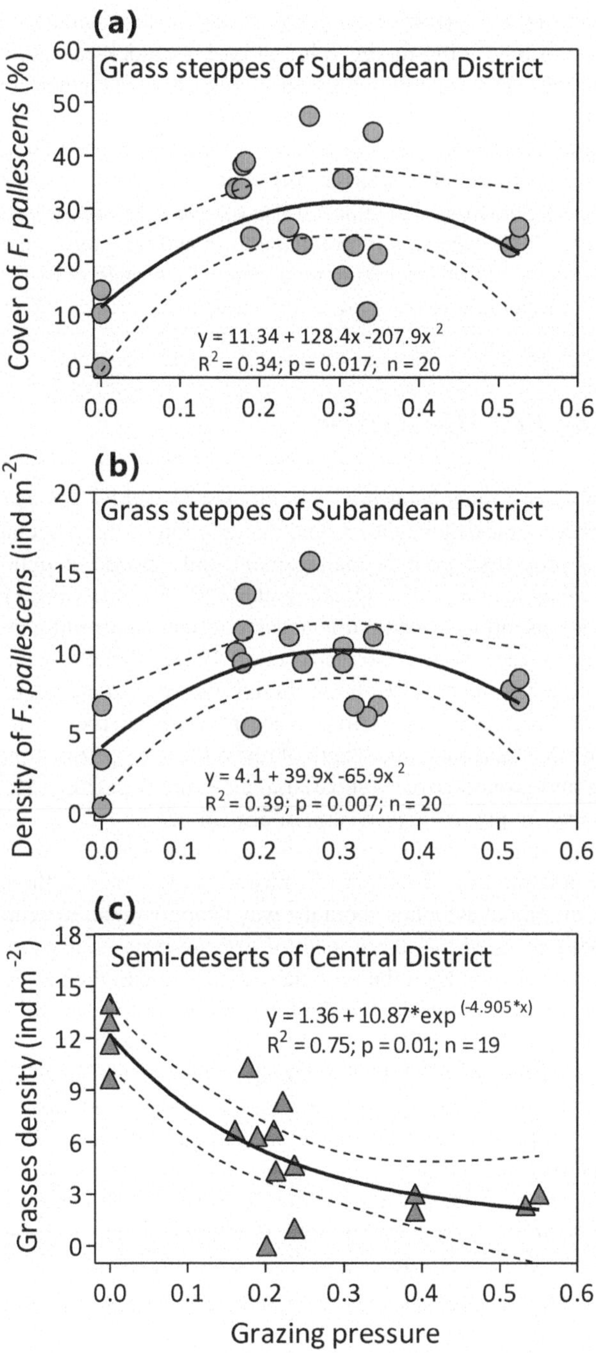

Fig. 3.4 Nonlinear regressions between grazing pressure and (**a**) plant cover (%) and (**b**) population density (individuals m^{-2}) of *Festuca pallescens* in the grass steppes of the Subandean District and (**c**) the density of grasses (individuals m^{-2}) in the semi-deserts of the Central District.

In general terms, across the aridity gradient, adequate management of the grazing pressure according to the forage value of each paddock can improve, on average, the ES supply. Nonetheless, the grazing pressure management by itself is not enough to reduce the high defoliation of preferred species (Oñatibia et al. 2020b). The over-grazing of some areas, which may occur even under light stocking rates, and the under-grazing of other areas at different scales are considered as signs of forage resource degradation. This uneven use pattern may compromise the critical ecosystem service of forage supply in the medium and long term. Sections 3.3 and 3.4 address the impact of other grazing management alternatives that complementarily with light to moderate grazing pressure can increase critical ES provision.

3.3 Grazing-Rest Management

Most Patagonian arid rangelands are continuously grazed by domestic herbivores. Continuous grazing may induce a gradual degradation of the rangeland condition, decreasing ES provision, even in management units grazed at light or moderate stocking rate (Teague et al. 2013; Oñatibia et al. 2020b). It has been proposed that grazing-rests are useful to mitigate this negative effect, controlling defoliation and re-growth of preferred species (Golluscio et al. 1998a; Paruelo et al. 2008; Distel 2013; Oñatibia and Aguiar 2019). Even though the scientific evidence supporting the effectiveness of grazing-rest periods is controversial (Briske et al. 2008), recent studies indicate that increasing the length of rest relative to grazing time increase the plant biomass and ground cover compared to continuous grazing (McDonald et al. 2019), which may improve the provision of critical ES.

In Patagonian steppes, grazing-rests have been proposed more than 20 years ago to reverse degradation processes (e.g., Golluscio et al. 1998a), although in general terms without empirical evidence about the way to apply them. Recently, it has been empirically demonstrated that the impact of grazing-rests on the growth of dominant forage grasses depends on the regional aridity context, the current rainfall, the identity of plant species, and the rest length (Oñatibia 2017; Oñatibia and Aguiar 2019; Oñatibia et al. 2020a). Studies were performed to estimate the effects of different seasonal grazing-rests (at the beginning, at the end, and all along the growing season) on plant performance of dominant forage species (and thus the critical service of forage supply), compared to continuously grazed plants. They were replicated in grass steppes, grass-shrub steppes, and semi-deserts (located across an aridity gradient), to evaluate how the regional aridity level mediates the outputs. The species common to the three communities, *Poa ligularis* (key to forage supply), is functional to evaluate the aridity level effect on plant responses to grazing-rest, by isolating the species phylogenetic component that co-varies with regional aridity (Oñatibia 2017; Oñatibia et al. 2020a).

In general, grazing-rests enhance the growth and reproduction of grasses, especially in low aridity sites, during wet years (in drier sites), and when the rest is throughout the growing season (Oñatibia 2017; Oñatibia and Aguiar 2019; Oñatibia

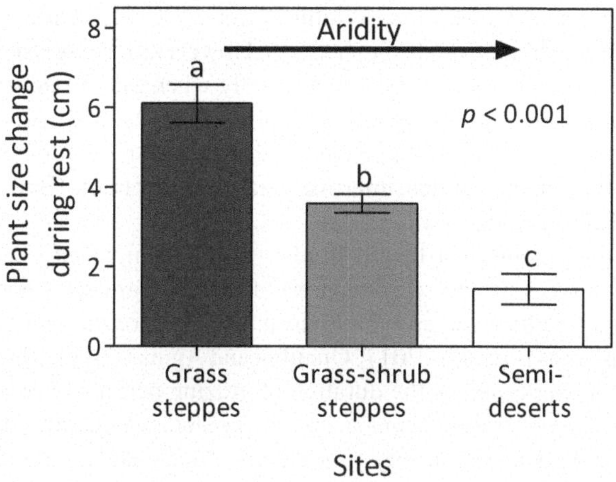

Fig. 3.5 Plant growth during a growing season resting of *Poa ligularis* individuals, estimated through the changes in total plant size, in three sites located across a regional aridity gradient: grass steppes of the Subandean District, grass-shrub steppes of the Occidental District, and semi-deserts of the Central District. Bars correspond to mean values, and vertical lines indicate SE. Different letters indicate significant differences among sites, resulting from a GLM analysis. (Adapted from Oñatibia et al. 2020a)

et al. 2020a). The plant growth after defoliation during a grazing-rest (recovery) of *Poa ligularis*' plants was the highest in the grass steppes, where plant size recovery was twice as high as in the grass-shrub steppes, while it quadrupled the values found in the semi-deserts (Fig. 3.5). High aridity (mainly low water availability) generally limits positive vegetation responses (Bailey and Brown 2011), and rest periods allow the recovery of defoliated plants only when conditions to grow are adequate (Wallace et al. 1984; Müller et al. 2007). Otherwise, when responses of other dominant species were compared with *Poa ligularis* in each site, patterns indicated that species adapted to high aridity level respond to a lesser extent to rests, due to traits that give them drought resistance generally restrict growth rates (Oñatibia 2017).

In grass steppes of the Subandean District, responses of the two dominant grass species (*Festuca pallescens* and *Poa ligularis*) indicate that grazing-rests during the growing season are effective to improve the forage supply and preserve the dominant native grasses under most current rainfall conditions (Oñatibia 2017). In grass-shrub steppes of the Occidental District, the effectiveness of grazing-rest is lower than in grass steppes, increasing forage grass species growth (*Poa ligularis* and *Pappostipa speciosa*), compared with continuous year-round grazing, mainly during growing seasons with rainfall above the average (Oñatibia 2017; Oñatibia and Aguiar 2019). Finally, in semi-deserts, the impact of seasonal grazing-rests is the lowest (Oñatibia 2017; Oñatibia et al. 2020a). Only one forage species (*Poa ligularis*), during an extraordinary wet year, presented higher growth during rest than under continuous grazing (Oñatibia 2017). These patterns confirm that planning strategic grazing-rests would be effective to increase forage supply in low-aridity sites, and/or under wet conditions (Golluscio et al. 1998a; Paruelo et al. 2008; Teague et al. 2013; Oñatibia and Aguiar 2019). However, in the driest environments, grazing-rest management would be effective only when rainfall does not markedly restrict growth (i.e., extremely wet years). In this sense, the ES provision in the driest environments should be managed mainly through the regulation of the stocking rate, together with practices that improve grazing distribution (Bailey and Brown 2011; see the following section).

Regarding the grazing-rest length, studies in Patagonian rangelands show that the performance (growth and reproduction) of plants is greater when the grazing-rest is applied throughout the growing season, and it is considerably lower in mid-season rest situations (Oñatibia 2017; Oñatibia and Aguiar 2019). The length of rest periods can be as important as the duration of grazing periods to control plant performance and ES provision, being a critical aspect, which must be considered according to ecological-site aridity (Laca 2009; McDonald et al. 2019; Oñatibia et al. 2020a). Positive responses would be achieved if rest duration and water availability are enough to recover biomass lost by defoliation (Müller et al. 2007; Teague et al. 2013; Oñatibia and Aguiar 2019). Thus, the greater the aridity (or the water/nutrient deficit), the greater the length to recover defoliated plants vigor (Oñatibia et al. 2020a). However, long-term rests (exclusion of herbivores for several years) should be avoided, since the positive effect on ES disappears if the duration of the rest is too long (Oñatibia et al. 2015; Oñatibia and Aguiar 2016; Sun et al. 2020; Sect. 3.1).

Some aspects must be taken into account to apply strategic grazing-rest management. It must be considered that commonly applied grazing-rest management methods in arid rangelands often have a high failure rate because they generally do not allow flexibility in livestock movement to adjust grazing pressure to variation in forage quality and quantity (Fynn 2012). Besides, rigid approaches do not avoid the same grazing pattern (e.g., grazing of the same forage resources at the same moment, every year). Repeated grazing patterns can promote rangeland degradation processes, even under schemes that include grazing-rests. Otherwise, as arid ecosystems are subject to high climatic uncertainty (e.g., high temporal and spatial variability in rainfall and net primary productivity), the proper design must be flexible, framed in an adaptive management framework with specific productive objectives (Golluscio et al. 1998a; Distel 2013; Oñatibia 2017). In Patagonian rangeland ecosystems, although the impact of grazing-rest on ES can be small and it decreases with aridity, a flexible adjustment of grazing-rests, according to the aridity level, the current rainfall, and the dominant forage species responses, has the potential to benefit the provision of ES in the long term (Oñatibia 2017; Oñatibia and Aguiar 2019; Oñatibia et al. 2020a).

Another aspect to consider to implement this type of management is that ranches must have several paddocks and/or grazing areas where animals can be located (with higher densities) during rests in the growing season of the steppes (Golluscio et al. 1998a). In Patagonian ranches, summer paddocks ("veranadas") or meadows are common and could be used for this purpose. Meadows are much more humid and productive than steppes (Buono et al. 2010) and, therefore, can be used on a rotating basis, as they would require shorter recovery periods and much less surface area to support the same number of animals. Finally, it must be considered that resting a paddock implies reductions in the number of animals at the ranch scale (if the economic condition allows it) or to raise the stocking rates in other paddocks (if ranchers want to keep the animal number constant). All these aspects highlight that ecological solutions to increase the ES provision are, in many cases, limited by economic or social constraints (Oñatibia et al. 2020b).

3.4 Management of the Spatial Distribution of Herbivores and the Heterogeneous Grazing Impacts

Domestic herbivores distribution is heterogeneous at different spatial scales as a consequence of selectivity, being grazing pressure heavy in some localized areas while other areas receive light or no utilization. In arid and semiarid Patagonian rangelands, the inadequate animal distribution has been identified as one of the major causes of forage resource degradation and ES decline (Golluscio et al. 1998a; Golluscio et al. 2005; Oñatibia and Aguiar 2018). The size of paddocks and the distribution of watering points play a key role in determining the heterogeneous impacts of sheep on vegetation attributes closely related to the provision of critical ES (Oñatibia and Aguiar 2018).

In grass-shrub steppes of the Occidental District, it was assessed how paddock size mediates the heterogeneity of continuous sheep grazing effects on vegetation, at a constant stocking rate. Three small paddocks (ca. 110 ha) were compared with three large (ca. 1100 ha) ones, all dominated by the same plant community, containing a single watering point and presenting a similar shape (Oñatibia and Aguiar 2018). Findings show that relationships between vegetation variables and distance from the watering point are in most cases asymptotic exponential, although responses strongly differ between small and large paddocks (Fig. 3.6a and b). In small paddocks, plant cover and density of grass forage species reach a plateau at a shorter distance from the watering point (~200 m) than in large paddocks (~2000 m), where the changes in these vegetation variables are larger and more gradual (Fig. 3.6a and b). In short, vegetation and ES supply spatial heterogeneity throughout the paddock are lower in small than in large paddocks (Oñatibia and Aguiar 2018). Sheep feces density shows an inverse pattern (Fig. 3.6c), indicating that these responses are associated with sheep distribution.

Decreasing paddock size and providing more watering points may counteract the undesired effects of uneven grazing (Barnes et al. 2008; Laca 2009; Bailey and Brown 2011). In Patagonian steppes, these practices are useful tools to improve sheep distribution and their medium- and long-term effects, decreasing the heterogeneity of ES provision between the most impacted and the most avoided areas (Oñatibia and Aguiar 2018). As the paddock size increases, forage utilization and grazing impact on ES provision within a paddock are more heterogeneous. This

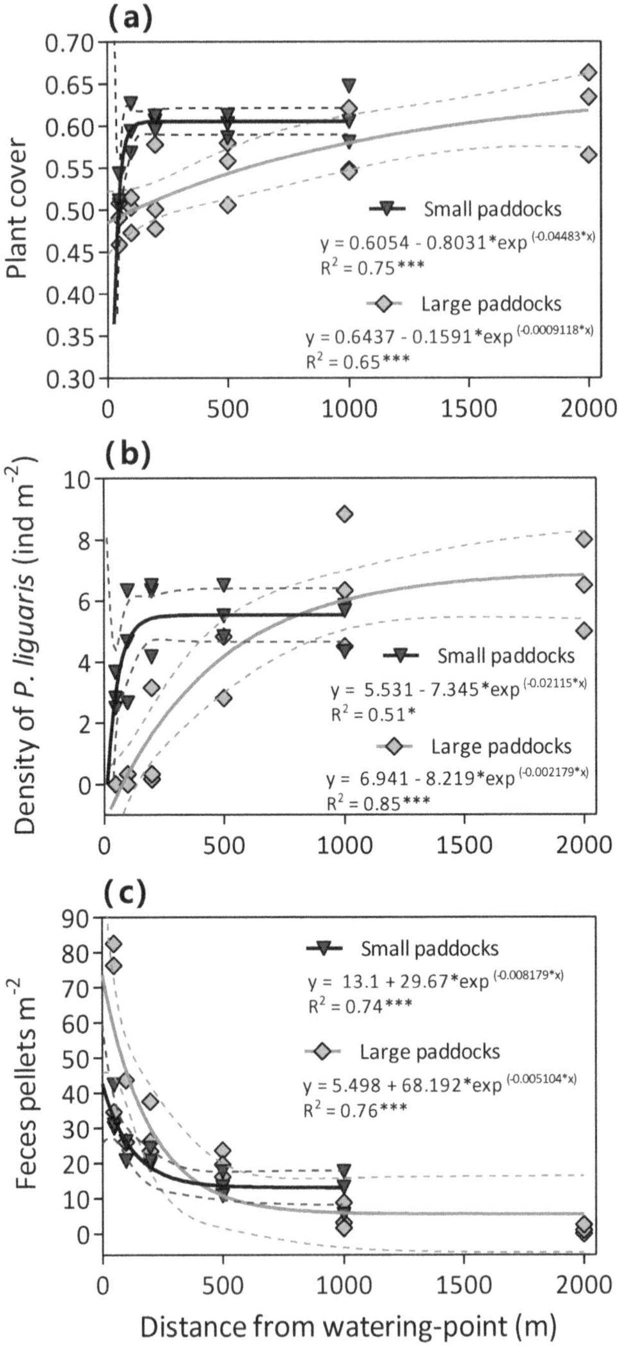

Fig. 3.6 (**a**) Total plant cover, (**b**) specific density of *Poa ligularis* (individuals m^{-2}), and (**c**) sheep feces density (pellets m^{-2}) as function of distance from watering points in small paddocks (green triangles) and large paddocks (red rhombus). Equations, determination coefficients (R^2), and lines (with confidence intervals: dashed lines) representing the selected models (best fit, nonlinear least squares) for each paddock size are shown (small paddocks, dark green line; large paddocks, dark red line). Asterisks next to determination coefficients (R^2) indicate p-values of significant nonlinear models: (*) between 0.05 and 0.01; (**) between 0.01 and 0.001; (***) < 0.001

occurs because the highly selected patches near watering points receive higher grazing pressure, resulting from a greater animal concentration, since the larger the paddock, the higher the number of animals that move together (they generally constitute herds). Otherwise, distant patches are avoided because of the difficulty for herbivores to explore the entire area of large paddocks (Hunt et al. 2007). Contrastingly, as paddock size decreases, the extent of the inaccessible area for herbivores is smaller (increasing forage availability, as herbivores can explore the whole area), while degradation of preferred patches is lower. As a general effect, the carrying capacity in smaller paddocks would increase because the availability of forage is not limited by the unequal distribution of herbivores (Teague et al. 2013; Oñatibia and Aguiar 2018). Thereby, a more homogeneous moderate utilization can be achieved, maximizing the benefits of moderate grazing on ES (Oñatibia et al. 2015; Oñatibia and Aguiar 2016; Oñatibia and Aguiar 2018; Sect. 3.1). Besides, as paddock size is reduced, the relationship between the average animal stocking rate and animal performance in interaction with vegetation is more predictable (Laca 2009), improving the forage utilization efficiency and increasing the effectiveness of management tools to control ES provision (Oñatibia and Aguiar 2018).

In arid Patagonian rangelands, most paddocks are large and include landscape-scale heterogeneity (Golluscio et al. 1998a; Ormaechea and Peri 2015; Oñatibia and Aguiar 2018; Ormaechea et al. 2019). This complexity highlights the relevance of delimiting homogeneous areas when reducing paddocks' size to efficiently promote a more even grazing impact on ES within each management unit (Bailey 2004). This delimitation also allows achieving a heterogeneous use at the landscape level, according to each management unit idiosyncrasy (Fuhlendorf and Engle 2001). The differential use of management units may increase landscape-level use heterogeneity, which may be desired according to productive and conservation goals based on the provision of multiple ES (Fuhlendorf and Engle 2001; Briske et al. 2020).

In addition to paddock size and watering points distribution, strategic supplementation, herding, and targeted grazing can promote a more homogeneous use within a management unit, reducing heterogeneity among plants of different species and even within a single species (Golluscio et al. 1998b; Bailey 2004; Derner et al. 2017; Bailey et al. 2019; Oñatibia et al. 2020b). For example, choosing sites to provide dietary supplements, shade, or shelter from the wind and low temperatures can help to attract herbivores to underutilized areas. Increasing the proportion of defoliated plants of species presenting a low preference degree is challenging since the management of grazing pressure by itself is not enough to achieve this objective (Oñatibia et al. 2020b). Strategic supplementation (e.g., urea, phosphorus, minerals), frequently used to overcome the seasonal deficiencies in forage quality, can be also useful to enhance the use of low preferred species. For example, it has been shown that the consumption of *P. speciosa* (species with low N content) increases when the diet is strategically supplemented with urea (Golluscio et al. 1998b). A combination of smaller paddocks, with watering point creation and strategic supplementation (and also targeted grazing; see Bailey et al. 2019), may lead animals to consume a greater variety of plants, including this kind of species. Such combination of management practices, with an adequate stocking rate allocation and strategic grazing-rest periods, has the potential to achieve a homogeneous moderate use in each management unit and a heterogeneous use of landscapes, increasing the provision of critical ES at both levels.

4 Public Policies to Improve the Provision of Ecosystem Services from Patagonian Arid Rangelands

Most Patagonian arid rangelands are located on lands of private ownership. Thus, decision-making is in private hands (ranchers), who base their management decisions mainly on the short-term provision of the ES of livestock production and forage supply. Moreover, livestock ranches are predominantly made up of small producers (<3000 sheep), who present little capacity to adopt most management technologies proposed here. This chapter discusses the ecological consequences of the different management practices, but the economic dimension is a critical aspect that needs to be considered. In arid Patagonian steppes, the costs of fencing, watering point creation, supplementation, animal movements, changing animal number, and monitoring can only be afforded by large ranches (~15,000 sheep), while medium and small ranches (~ 4000 and ~1000 sheep, respectively) need market instruments that promote investment to encourage management changes (Aguiar and Román 2007). This context makes the role of the state particularly relevant in the technology adoption and support measures for producers to apply resource management strategies. Public policies must promote technology adoption to increase the long-term provision of multiple ES, discouraging ranchers to increase the stocking rate (i.e., to increase short-term livestock provision), since this management practice undermines most critical ES, including long-term livestock production by reducing forage supply (see Table 3.1).

Some recommendations to policy-makers emerge from both the ecological results shown in this chapter and the socioeconomic context that ranchers face. On the one hand, to encourage stocking rate management that promotes a high provision of diverse ES, a certification system of good practices can be implemented, which, in turn, adds value to the products produced in this way. This proposal includes assistance from the state to capture niche markets for certified products. Besides, small ranchers who certify a global ES increase could receive economic incentives. On the other hand, subsidies, tax reductions (100% refund of the investment cost for small ranches), and affordable credit lines to large ranches can be adequate tools to motivate the proposed management practices that require financial investment. Finally, state assistance for the extension of both management practices and available policies, the facilitation of certifications, and the accompaniment to carry out frequent monitoring of critical ES provision through subsidies and human resources are key aspects to implement these public policies. Sustainable production in Patagonian rangelands is a complex issue, and management practices in these ecosystems need a comprehensive analysis, which, besides the perspective focused on the ecological subsystem addressed here, should be complemented with socioeconomic and cultural analyses (Chap. 1).

5 Concluding Remarks

Management of Patagonian arid rangelands is particularly challenging since these ecosystems are undergoing accelerated ecological and social changes, such as climate change and increasing demand for high-quality and safe animal products together with multiple ES. This chapter provides evidence supporting that ES can be maximized through moderating the grazing pressure (stocking rate), applying strategic grazing-rests according to each ecological site characteristics, and promoting a homogeneous herbivore distribution within each management unit. Such combination of grazing techniques enables domestic herbivores to function as positive ecosystem engineers, maintaining rangelands integrity and animal husbandry (Derner et al. 2017). Thus, applying these management technologies can promote ecological and socioeconomic sustainability of the Patagonian arid region, through the optimal provision of ES that meets the needs of most stakeholders, while improving the well-being of livestock producers.

Due to the complexity and dynamics of Patagonian rangelands, management technologies and specific strategies proposed here should not be implemented as fixed recipes, since there is not a unified model of management for these rangelands. Rather, the challenge is how to best manage the rangeland social-ecological systems to provide optimal combinations of ES according to each specific context (Rasch et al. 2016; Briske et al. 2020). In this sense, the proposed management initiatives must be framed within an adaptive management framework (see Meffe et al. 2012; Allen et al. 2017), which provides flexibility when making natural resource management decisions. In other words, the expected results of each management decision should be addressed as predictions to be tested and evaluated in light of the specific objectives. Thus, adaptive grazing management, with monitoring-informed decision-making, allows correcting the management in the face of climatic anomalies or changes in the socioeconomic context. This will concurrently reduce negative environmental impacts, for example, associated with improper stocking rate during droughts, a critical aspect under the current climate change scenarios (Derner et al. 2017). Monitoring to detect trends in the most relevant attributes for management (such as those evaluated in this chapter) and developing early warning systems and predictive temporal models (to know in advance how critical ES provision will change) can help to better design flexible management strategies, reducing uncertainties related to management impacts (Allen et al. 2017; Derner et al. 2017; Sala et al. 2017).

This chapter focuses on the consequences of sheep management in Patagonian steppes, leaving other herbivores and meadows out of the analysis. On the one hand, the inclusion of other domestic herbivores with different requirements and behavior (e.g., cows, horses) can increase the efficiency in the use of forage resources and the provision of ES. On the other hand, while meadows cover a scarce surface, they are particularly relevant for livestock production due to their high productivity (León et al. 1998; Golluscio et al. 1998a). The sustainable intensification of meadows through management technologies such as irrigation, water flow control, inter-sowing, fertilization, and rotational grazing systems with electric fences may increase the forage use efficiency, reducing the salinization and erosion processes induced by domestic herbivores and increasing the global ES provision of Patagonian ranches.

In conclusion, grazing management objectives in Patagonian rangelands, which have generally focused on forage provision as being the main source of animal husbandry, should incorporate diverse ES in the decision-making process (e.g., forage supply, C sequestration, biodiversity conservation, nutrient cycling) to promote long-term rangeland sustainability. Proper adaptive grazing management has the potential to simultaneously maximize the whole provision of ES, creating synergies among provisioning, regulating, and supporting ES (Oñatibia et al. 2015). Incorporating critical ES and their related interactions into range management objectives is needed to improve the multiple benefits from Patagonian arid rangelands and meet the current and future stakeholder needs.

References

Aguiar MR, Román ME (2007) Restoring forage grass populations in arid Patagonian rangeland. In: Restoring natural capital: science, business, and practice. Island Press, Washington, pp 112–121

Aguiar MR, Sala OE (1998) Interactions among grasses, shrubs, and herbivores in Patagonian grass-shrub steppes. Ecol Austral 8:201–210

Aguiar MR, Sala OE (1999) Patch structure, dynamics and implications for the functioning of arid ecosystems. Trends Ecol Evol 14:273–277

Aguiar MR, Paruelo JM, Sala OE, Lauenroth WK (1996) Ecosystem responses to changes in plant functional type composition: an example from the Patagonian steppe. J Veg Sci 7:381–390

Allen CR, Angeler DG, Fontaine JJ, Garmestani AS, Hart NM, Pope KL, Twidwell D (2017) Adaptive management of rangeland systems. In: Rangeland systems. Springer, Cham, pp 373–394

Ares J (2007) Systems valuing of natural capital and investment in extensive pastoral systems: lessons from the Patagonian case. Ecol Econ 62:162–173

Ares J, Beeskow AM, Bertiller M, Rostagno M, Irrisarri M, Anchorena et al (1990) Structural and dynamics characteristics of overgrazed lands of northern Patagonia, Argentina. In: Managed grasslands. Elsevier, Amsterdam, pp 149–175

Ares J, Del Valle H, Bisigato A (2003) Detection of process-related changes in plant patterns at extended spatial scales during early dryland desertification. Glob Chang Biol 9:1643–1659

Bailey DW (2004) Management strategies for optimal grazing distribution and use of arid rangelands. J Anim Sci 82:147–153

Bailey DW, Brown JR (2011) Rotational grazing systems and livestock grazing behavior in shrub-dominated semi-arid and arid rangelands. Rangel Ecol Manag 64:1–9

Bailey DW, Mosley JC, Estell RE, Cibils AF, Horney M, Hendrickson JR et al (2019) Synthesis paper: targeted livestock grazing: prescription for healthy rangelands. Rangel Ecol Manag 72:865–877

Barnes MK, Norton BE, Maeno M, Malechek JC (2008) Paddock size and stocking density affect spatial heterogeneity of grazing. Rangel Ecol Manag 6:380–388

Beeskow AM, Ellisalde NO, Rostagno CM (1995) Ecosystem changes associated with grazing intensity on the Punta Ninfas rangelands of Patagonia, Argentina. J Range Manag 48:517–522

Bennett EM, Peterson GD, Gordon LJ (2009) Understanding relationships among multiple ecosystem services. Ecol Lett 12:1394–1404

Bertiller MB, Ares JO, Bisigato AJ (2002) Multiscale indicators of land degradation in the Patagonian Monte, Argentina. Environ Manag 30:0704–0715

Biancari L, Aguiar MR, Cipriotti PA (2020) Grazing impact on structure and dynamics of bare soil areas in a Patagonian grass-shrub steppe. J Arid Environ 179:104197

Bisigato AJ, Bertiller MB (1997) Grazing effects on patchy dryland vegetation in northern Patagonia. J Arid Environ 36:639–653

Bisigato AJ, Bertiller MB, Ares JO, Pazos GE (2005) Effect of grazing on plant patterns in arid ecosystems of Patagonian Monte. Ecography 28:561–572

Briske DD (2017) Rangeland systems: processes, management and challenges. Springer Nature, Cham, p 661

Briske DD, Derner JD, Brown JR, Fuhlendorf SD, Teague WR, Havstad KM et al (2008) Rotational grazing on rangelands: reconciliation of perception and experimental evidence. Rangel Ecol Manag 61:3–17

Briske DD, Coppock DL, Illius AW, Fuhlendorf SD (2020) Strategies for global rangeland stewardship: assessment through the lens of the equilibrium–non-equilibrium debate. J Appl Ecol 57:1056–1067

Buono G, Oesterheld M, Nakamatsu V, Paruelo JM (2010) Spatial and temporal variation of primary production of Patagonian wet meadows. J Arid Environ 74:1257–1261

Cesa A, Paruelo JM (2011) Changes in vegetation structure induced by domestic grazing in Patagonia (Southern Argentina). J Arid Environ 75:1129–1135

Dai A (2013) Increasing drought under global warming in observations and models. Nat Clim Chang 3:52–58

Del Valle HF, Elissalde NO, Gagliardini DA, Milovich J (1998) Status of desertification in the Patagonian region: assessment and mapping from satellite imagery. Arid Land Res Manag 12:95–121

Derner JD, Hunt L, Euclides Filho K, Ritten J, Capper J, Han G (2017) Livestock production systems. In: Rangeland systems. Springer, Cham, pp 347–372

Distel R (2013) Manejo del pastoreo en pastizales de zonas áridas y semiáridas. Revista Argentina de Producción Animal 33:53–64

Easdale MH, Aguiar MR (2012) Regional forage production assessment in arid and semi-arid rangelands–a step towards social–ecological analysis. J Arid Environ 83:35–44

Easdale MH, Umaña F, Raffo F, Fariña C, Bruzzone O (2019) Evaluación de pastizales patagónicos con imágenes de satélites y de vehículos aéreos no tripulados. Ecol Austral 29:306–314

Flombaum P, Sala OE (2007) A non-destructive and rapid method to estimate biomass and aboveground net primary production in arid environments. J Arid Environ 69:352–358

Fuhlendorf SD, Engle DM (2001) Restoring heterogeneity on rangelands: ecosystem management based on evolutionary grazing patterns: we propose a paradigm that enhances heterogeneity instead of homogeneity to promote biological diversity and wildlife habitat on rangelands grazed by livestock. Bioscience 51:625–632

Fynn RW (2012) Functional resource heterogeneity increases livestock and rangeland productivity. Rangel Ecol Manag 65:319–329

Golluscio RA, Deregibus VA, Paruelo JM (1998a) Sustainability and range management in the Patagonian steppes. Ecol Austral 8:265–284

Golluscio RA, Paruelo JM, Mercau JL, Deregibus VA (1998b) Urea supplementation effects on the utilization of low-quality forage and lamb production in Patagonian rangelands. Grass Forage Sci 53:47–56

Golluscio RA, Pérez JA, Paruelo JM, Ghersa CM (2005) Spatial heterogeneity at different grain sizes in grazed versus ungrazed sites of the Patagonian steppe. Ecoscience 12:103–109
Golluscio RA, Austin AT, Martínez GCG, Gonzalez-Polo M, Sala OE, Jackson RB (2009) Sheep grazing decreases organic carbon and nitrogen pools in the Patagonian steppe: combination of direct and indirect effects. Ecosystems 12:686–697
Havstad KM, Peters DP, Skaggs R, Brown J, Bestelmeyer B, Fredrickson E et al (2007) Ecological services to and from rangelands of the United States. Ecol Econ 64:261–268
Hickman KR, Hartnett DC, Cochran RC, Owensby CE (2004) Grazing management effects on plant species diversity in tallgrass prairie. J Range Manag 57:58–65
Holechek JL (1988) An approach for setting the stocking rate. Rangelands 10:10–14
Huang J, Yu H, Guan X, Wang G, Guo R (2016) Accelerated dryland expansion under climate change. Nat Clim Chang 6:166–171
Hunt LP, Petty S, Cowley R, Fisher A, Ash AJ, MacDonald N (2007) Factors affecting the management of cattle grazing distribution in northern Australia: preliminary observations on the effect of paddock size and water points. Rangel J 29:169–179
Irisarri JGN, Oesterheld M, Paruelo JM, Texeira MA (2012) Patterns and controls of aboveground net primary production in meadows of Patagonia. A remote sensing approach. J Veg Sci 23:114–126
Knapp AK, Beier C, Briske DD, Classen AT, Luo Y, Reichstein M et al (2008) Consequences of more extreme precipitation regimes for terrestrial ecosystems. Bioscience 58:811–821
Kröpfl AI, Cecchi GA, Villasuso NM, Distel RA (2013) Degradation and recovery processes in semi-arid patchy rangelands of northern Patagonia, Argentina. Land Degrad Dev 24:393–399
Laca EA (2009) New approaches and tools for grazing management. Rangel Ecol Manag 62:407–417
Lauenroth WK (1998) Guanacos, spiny shrubs and the evolutionary history of grazing in the Patagonian steppe. Ecol Austral 8:211–215
León RJC, Aguiar MR (1985) El deterioro por uso pasturil en estepas herbáceas patagónicas. Phytocoenologia 13:181–196
León RJC, Bran D, Collantes M, Paruelo JM, Soriano A (1998) Grandes unidades de vegetación de la Patagonia extra andina. Ecol Austral 8:125–144
Maestre FT, Quero JL, Gotelli NJ, Escudero A, Ochoa V, Delgado-Baquerizo M et al (2012) Plant species richness and ecosystem multifunctionality in global drylands. Science 335(6065):214–218
Maestre FT, Eldridge DJ, Soliveres S, Kéfi S, Delgado-Baquerizo M, Bowker MA et al (2016) Structure and functioning of dryland ecosystems in a changing world. Annu Rev Ecol Evol Syst 47:215–237
Massara V, Buono G (2020) Métodos de Evaluación de Pastizales en Patagonia Sur. INTA ediciones, p 288
McDonald SE, Lawrence R, Kendall L, Rader R (2019) Ecological, biophysical and production effects of incorporating rest into grazing regimes: a global meta-analysis. J Appl Ecol 56:2723–2731
McNaughton SJ (1979) Grazing as an optimization process: grass-ungulate relationships in the Serengeti. Am Nat 113:691–703
McNaughton SJ (1983) Compensatory plant growth as a response to herbivory. Oikos 40:329–336
McNaughton SJ, Oesterheld M, Frank DA, Williams KJ (1989) Ecosystem-level patterns of primary productivity and herbivory in terrestrial habitats. Nature 341:142–144
Meffe G, Nielsen L, Knight RL, Schenborn D (2012) Ecosystem management: adaptive, community-based conservation. Island Press, Washington, pp 313
Milchunas DG, Sala OE, Lauenroth W (1988) A generalized model of the effects of grazing by large herbivores on grassland community structure. Am Nat 132:87–106
Millennium Ecosystem Assessment (2005) Ecosystems and human well-being: synthesis. Island Press, Washington, DC
Müller B, Frank K, Wissel C (2007) Relevance of rest periods in non-equilibrium rangeland systems–a modelling analysis. Agric Syst 92:295–317

Murdoch W, Ranganathan J, Polasky S, Regetz J (2010) Using return on investment to maximize conservation effectiveness in Argentine grasslands. PNAS 107:20855–20862

Nuñez MN, Solman SA, Cabré MF (2009) Regional climate change experiments over southern South America. II: climate change scenarios in the late twenty-first century. Clim Dyn 32:1081–1095

Oñatibia GR (2017) Presión de pastoreo y selectividad en estepas áridas y semi-áridas: retroalimentación planta-herbívoro a distintos niveles (PhD thesis). University of Buenos Aires, pp 188

Oñatibia GR, Aguiar MR (2016) Continuous moderate grazing management promotes biomass production in Patagonian arid rangelands. J Arid Environ 125:73–79

Oñatibia GR, Aguiar MR (2018) Paddock size mediates the heterogeneity of grazing impacts on vegetation. Rangel Ecol Manag 71:470–480

Oñatibia GR, Aguiar MR (2019) Grasses and grazers in arid rangelands: impact of sheep management on forage and non-forage grass populations. J Environ Manag 235:42–50

Oñatibia GR, Aguiar MR, Semmartin M (2015) Are there any trade-offs between forage provision and the ecosystem service of C and N storage in arid rangelands? Ecol Eng 77:26–32

Oñatibia GR, Boyero L, Aguiar MR (2018) Regional productivity mediates the effects of grazing disturbance on plant cover and patch-size distribution in arid and semi-arid communities. Oikos 127:1205–1215

Oñatibia GR, Amengual G, Boyero L, Aguiar MR (2020a) Aridity exacerbates grazing-induced rangeland degradation: a population approach for dominant grasses. J Appl Ecol 57:1999–2009

Oñatibia GR, Golluscio RA, Texeira M, Paruelo JM (2020b) Controls of forage selective defoliation by sheep in arid rangelands. Ecosphere 11(11):e03285

Ormaechea SG, Peri PL (2015) Landscape heterogeneity influences on sheep habits under extensive grazing management in Southern Patagonia. Livest Res Rural Dev 27(6):105

Ormaechea SG, Peri PL, Cipriotti PA, Distel RA (2019) El cuadro de pastoreo en los sistemas extensivos de Patagonia Sur. Percepción y manejo de la heterogeneidad. Ecol Austral 29:174–184

Palacio RG, Bisigato AJ, Bouza PJ (2014) Soil erosion in three grazed plant communities in northeastern Patagonia. Land Degrad Dev 25:594–603

Paruelo JM, Golluscio RA, Guerschman JP, Cesa A, Jouve VV, Garbulsky MF (2004) Regional scale relationships between ecosystem structure and functioning: the case of the Patagonian steppes. Glob Ecol Biogeogr 13:385–395

Paruelo JM, Pütz S, Weber G, Bertiller M, Golluscio RA, Aguiar MR, Wiegand T (2008) Long-term dynamics of a semiarid grass steppe under stochastic climate and different grazing regimes: a simulation analysis. J Arid Environ 72:2211–2231

Perelman SB, León RJC, Bussacca JP (1997) Floristic changes related to grazing intensity in a Patagonian shrub steppe. Ecography 20:400–406

Pereyra DA, Bucci SJ, Arias NS, Ciano N, Cristiano PM, Goldstein G, Scholz FG (2017) Grazing increases evapotranspiration without the cost of lowering soil water storages in arid ecosystems. Ecohydrology 10:e1850

Petz K, Alkemade R, Bakkenes M, Schulp CJ, van der Velde M, Leemans R (2014) Mapping and modelling trade-offs and synergies between grazing intensity and ecosystem services in rangelands using global-scale datasets and models. Glob Environ Chang 29:223–234

Piazza MV (2016) Impacto de los herbívoros domésticos sobre el reciclado de C y N: cambios mediados por la vegetación y el ambiente. (PhD thesis). University of Buenos Aires, pp169

Posse G, Anchorena J, Collantes MB (2000) Spatial micro-patterns in the steppe of Tierra del Fuego induced by sheep grazing. J Veg Sci 11:43–50

Rasch S, Heckelei T, Oomen RJ (2016) Reorganizing resource use in a communal livestock production socio-ecological system in South Africa. Land Use Policy 52:221–231

Reynolds JF, Smith DMS, Lambin EF, Turner BL, Mortimore M, Batterbury SP et al (2007) Global desertification: building a science for dryland development. Science 316:847–851

Rodríguez JP, Beard TD Jr, Bennett EM, Cumming GS, Cork SJ, Agard J et al (2006) Trade-offs across space, time, and ecosystem services. Ecol Soc 11:28

Sala OE, Paruelo JM (1997) Ecosystem services in grasslands. Nature's services: societal dependence on natural ecosystems. Island Press, Covelo, pp 237–251
Sala OE, Yahdjian L, Havstad K, Aguiar MR (2017) Rangeland ecosystem services: Nature's supply and humans' demand. In: Rangeland Systems. Springer, Cham, pp 467–489
Sanderson JS, Beutler C, Brown JR, Burke I, Chapman T, Conant RT et al (2020) Cattle, conservation, and carbon in the western Great Plains. J Soil Water Conserv 75:5A–12A
Soriano A (1956) Aspectos ecológicos y pasturiles de la vegetación patagónica relacionados con su estado y capacidad de recuperación. Revista de Investigaciones Agrícolas 10:349–372
Soriano A, Movia CP (1986) Erosión y desertización en la Patagonia. Interciencia 11:77–83
Sun J, Fu B, Kemp D, Zhao W, Liu G, Han G et al (2020) Reconsidering the efficiency of grazing exclusion using fences on the Tibetan Plateau. Sci Bull 65:1405–1414
Teague R, Provenza F, Kreuter U, Steffens T, Barnes M (2013) Multi-paddock grazing on rangelands: why the perceptual dichotomy between research results and rancher experience? J Environ Manag 128:699–717
Texeira M, Paruelo JM (2006) Demography, population dynamics and sustainability of the Patagonian sheep flocks. Agric Syst 87:123–146
Thornton PK (2010) Livestock production: recent trends, future prospects. Philos Trans R Soc, B 365:2853–2867
Verón SR, Paruelo JM (2010) Desertification alters the response of vegetation to changes in precipitation. J Appl Ecol 47:1233–1241
Verón SR, Paruelo JM, Oesterheld M (2011) Grazing-induced losses of biodiversity affect the transpiration of an arid ecosystem. Oecologia 165:501–510
Wallace LL, McNaughton SJ, Coughenour MB (1984) Compensatory photosynthetic responses of three African graminoids to different fertilization, watering and clipping regimes. Bot Gaz 145:151–156
Yahdjian L, Sala OE, Havstad KM (2015) Rangeland ecosystem services: shifting focus from supply to reconciling supply and demand. Front Ecol Environ 13:44–51
Zhao Y, Liu Z, Wu J (2020) Grassland ecosystem services: a systematic review of research advances and future directions. Landsc Ecol 35:793–814

Chapter 4
Synergies and Trade-Offs Among Ecosystem Services and Biodiversity in Different Forest Types Inside and Off-Reserve in Tierra del Fuego, Argentina

Josela Carrasco, Yamina Micaela Rosas, María Vanessa Lencinas, Andrés Bortoluzzi, Pablo L. Peri, and Guillermo Martínez Pastur

Abstract Land use planning is mainly based on monetary values of provisioning ecosystem services (ES). However, many other non-monetary ES and biodiversity provide values for human well-being, and it should be included in the decision-making. The objective of this chapter was to characterize different ES (provisioning, cultural, supporting, regulating) and potential biodiversity in different forest types in Tierra del Fuego Province, Argentina. We map and extract information for provision of ES and biodiversity and compare them through univariate and multivariate methods. We found that each forest type showed different potential biodiversity that determines the need of specific conservation and management strategies. Forest types presented different types and levels of provision of the studied ES, where several synergies and trade-offs were observed according to the current economic activities. Beside this, ES and potential biodiversity of the forests are not equally represented in the currently protected natural reserve network, compared to the values at landscape level. These outputs can be used to improve the current land use planning and the effectiveness of conservation at landscape level.

J. Carrasco · Y. M. Rosas · M. V. Lencinas
Laboratorio de Recursos Agroforestales, Centro Austral de Investigaciones Científicas (CADIC), Consejo Nacional de Investigaciones Científicas y Técnicas (CONICET), Ushuaia, Tierra del Fuego, Argentina

A. Bortoluzzi
Facultad de Ciencias Agrarias, Universidad Nacional del Litoral (UNL), Esperanza, Santa Fe, Argentina

P. L. Peri
National Agricultural Technology Institute, Río Gallegos, Argentina

G. Martínez Pastur (✉)
Consejo Nacional de Investigaciones Científicas y Técnicas, Centro Austral de Investigaciones Científicas y Técnicas, Ushuaia, Tierra del Fuego, Argentina
e-mail: gpastur@conicet.gov.ar

P. L. Peri et al. (eds.), *Ecosystem Services in Patagonia*, Natural and Social Sciences of Patagonia, https://doi.org/10.1007/978-3-030-69166-0_4

Keywords Ecosystem services · Biodiversity conservation · Forest landscapes · Natural reserve network · Native forest law in Argentina · Synergies and trade-offs

1 Introduction

The concept of ecosystem services (ES) refers to the goods and benefits that the society obtains from the natural ecosystems (Daily 1997). Society obtains material goods and services from ecosystems (e.g., timber products, food, clean water), as well as other economic (e.g., market value, salaries) and non-economic values for the well-being of individuals and communities. However, when ecosystem management goals include only a limited set of goods and services, such as timber, other ES may be overlooked and therefore undervalued (Perera et al. 2018), leading to trade-offs among the potential users. In this chapter, we recognize three groups of ES from the natural forests (MEA 2005): (i) provisioning ES, such as timber, fiber or firewood (Gea Izquierdo et al. 2004), food production (e.g., livestock), and other non-woody industrial products (Peri et al. 2016a; Quintas-Soriano et al. 2016); (ii) cultural ES, related to aesthetic, beauty, recreation and ecotourism, artistic inspiration, and a sense of place for local communities (de Groot et al. 2002; Martínez Pastur et al. 2016a); and (iii) regulating and supporting ES, which contribute to climate stability by removing greenhouse gases and other pollutants and increasing soil retention and water regulation (Panagos et al. 2015; Quintas-Soriano et al. 2016) or contribute to provision of habitat for wild plants and animals (de Groot et al. 2002).

Biodiversity is a critical factor for supplying ES (Mori et al. 2017). In fact, biodiversity itself can be considered as one ecosystem service, e.g., when it is the basis of nature-based tourism or diseases regulation (Mace et al. 2012); however, this approach is still controversial. Biodiversity in native forests regulates ecosystem functions, productivity, and provision of several ES. There are antecedents that describe the importance of biodiversity in Southern Patagonia, including provision of different materials, genetic diversity, biotic interactions with economic importance (e.g., pollination, mycorrhizal fungi, and nitrogen-fixing), and ecological processes that regulate the dynamics of these ecosystems (e.g., Gea Izquierdo et al. 2004; Birch et al. 2010; Gargaglione et al. 2014; Martínez Pastur et al. 2017; Peri et al. 2016b, 2017; Hewitt et al. 2018; Rosas et al. 2019a; Goldenberg et al. 2020).

Recent methodologies have improved the assessment of synergies and trade-offs among ES and biodiversity at different spatiotemporal scales (Raudsepp-Hearne et al. 2010; Martínez Pastur et al. 2016a, b, 2017; Peri et al. 2017; Rosas et al. 2019a, 2020). These studies allowed defining better planning of land use, including the effectiveness of natural reserve networks (Rosas et al. 2019b). The design of natural reserve networks is mainly attributable to their location (e.g., remote areas or border areas between countries) or scenic values (e.g., majestic landscapes or unique features like glaciers), but there is an increasing concern to add more areas with unique potential biodiversity. In this context, suitable reserve networks should take into consideration (i) different forest types; (ii) unique potential biodiversity that these areas can host; (iii) provision of different ES, as well as the consideration of potential synergies and trade-offs among these services and biodiversity that the reserves host; and (iv) the human impacts or the economic activities within the reserve network (Rosas et al. 2019a, 2020).

Southern Patagonia in general, and Tierra del Fuego in particular, is composed by different ecosystem types with extreme environmental conditions, from arid steppes to dense temperate forests (Moore 1983) with varied levels of human impacts (e.g., livestock grazing, desertification) that can greatly affect local biodiversity (Rosas et al. 2019a, b, 2020). Areas with low human influence often provide ecosystem values for the preservation of endemic natural features, species, and biodiversity (Inostroza et al. 2016; Rosas et al. 2020). However, this Patagonian ecosystem is vulnerable to human disturbance and climate change (Lencinas et al. 2011; Lindenmayer et al. 2012; Peri et al. 2019). The objective of this chapter was to characterize different ES (provisioning, cultural, supporting, regulating) and potential biodiversity in different forest types (inside and off-reserves) in Tierra del Fuego, Argentina, and determine potential synergies and trade-offs among them. Additionally, we aimed to answer the following questions: (i) Do forest types present different potential biodiversity to justify different conservation or management strategies (understory plants were used as proxies)? (ii) Do forest types present different provision of ES that can be used to propose differential conservation or management strategies (eight services were used as proxies)? (iii) What is the distribution of the best forests (ES and potential biodiversity) in the current natural protected reserve network? (iv) How can current land use planning effectiveness be enhanced?

2 Study Area

The study was carried out in the southernmost area of Patagonia (52.7–55.1° SL, 63.7–68.5° WL) that included the Argentinean sector of the Fuegian Archipelago (21.3 thousand km^{-2}) (Fig. 4.1a) that belongs to Tierra del Fuego Province, Antártida e Isla del Atlántico Sur. In the archipelago, the Andes Mountains runs from west to east and define the relief and climate of the region and receive the influence of the Antarctica and both oceans (Pacific and Atlantic). A rainfall and temperature gradients from north to south define the vegetation, with grasslands in the north and forests in the south (Allué et al. 2010; Kreps et al. 2012). This province has a population density of 6.0 inhabitants km^{-2} mainly located in two cities (97.5% of the total population), Río Grande close to sheep ranching and oil extraction areas and Ushuaia close to major tourism areas. National parks and provincial reserves mainly preserved native forests (Fig. 4.1c). Two traditional provisioning ES prevailed in Tierra del Fuego forests, timber harvesting and livestock grazing (Peri et al. 2016a; Martínez Pastur et al. 2019). The use of forest resources has increased in terms of saw-timber industry since colonization, but decreased the extractions for firewood.

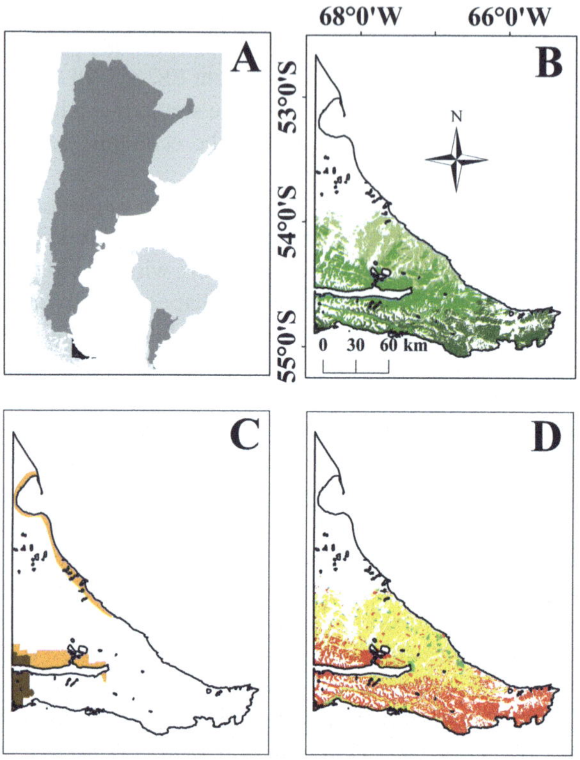

Fig. 4.1 Location of the study area indicating Argentina in South America and the Isla Grande of Tierra del Fuego in the southernmost portion of Patagonia (**a**) and characterization of the landscape, including: (**b**) main forest types (*Nothofagus antarctica* = pale green, *N. pumilio* = green, mixed evergreen = dark green) based on Collado (2001), (**c**) natural reserves (brown = national park, orange = provincial reserves), and (**d**) forest classifications according ordination planning of the National Law 26,331/07 (red = high, yellow = medium, green = low value)

Firewod was the main source of energy, but loss importance with the introduction of natural gas in the 1970s (Gea Izquierdo et al. 2004). *Nothofagus pumilio* forests are economically and in extension the most important (Martínez Pastur et al. 2002, 2016a), while *N. antarctica* is one of the main deciduous native tree used for silvopastoral systems (Peri et al. 2016a, 2017). There are three main forest types in the region (Fig. 4.1b): the north dominated by *N. antarctica*, central and southern areas dominated by *N. pumilio*, and mixed evergreen forests occupying middle altitude in mountain valleys, as well as lakes shores and Beagle channel (Collado 2001; Allué et al. 2010; Mestre et al. 2017). *N. antarctica* forests enable the diversification of farm products by sustaining sheep and cattle production, which provides income from meat, wool, and a range of wood products including poles, firewood, and timber for rural construction purposes. In addition, silvopastoral system in the region provides other ES such as water regulation, biodiversity conservation, soil and water quality, carbon sequestration, recreation, and cultural identity. These forests are growing in a mosaic with different structure and floristic composition as a result of livestock grazing and silvicultural management, in interaction with natural (e.g., drought) and anthropogenic (fires, introduction of plant species) factors (Peri and Ormaechea 2013; Martínez Pastur et al. 2017, 2019). Beside this, forestlands were also used for tourism, and this activity increases during the last decades (Martínez Pastur et al. 2016a). These forests occurred in private and public lands (near 50% each), and different natural reserves (national parks and provincial reserves) exist (Fig. 4.1c). *Nothofagus* forests presented unique cultural ES, e.g., (i) aesthetic values, which included unique natural landscapes (e.g., Tierra del Fuego National Park) (Fig. 4.1c); (ii) existence values, which included unique species of flora and fauna (e.g., *Campephilus magellanicus*, magellanic woodpecker); (iii) local identity represented by heritage, folklore, traditions, art, and local workers (e.g., ranching, forestry, artisanal fishing, mining, and oil extraction); and (iv) recreational activities such as winter sports, hiking, trekking, climbing, riding, camping, kayaking, and sport fishing.

The National Law 26,331/07 of Argentina defined the minimum environmental protection budgets for enrichment, restoration, conservation, harvesting, and sustainable management of the native forests, as well as the environmental services that forests provide to society (MAyDS 2017). The objectives of the law were to (i) promote the conservation through the spatial planning of the native forestlands (OTBN) and regulate the expansion of the agricultural frontier; (ii) implement regulations and controls to decrease the forest loss, promoting the maintenance of the native forest cover over time; (iii) improve and maintain the ecological and cultural values that benefit the society; and (iv) encourage the activities of enrichment, conservation, restoration, and sustainable management of native forests (Martínez Pastur et al. 2020). The OTBN was defined by the local forest authority in each province, based on the forest area distribution, the connectivity, the connection with existing protected network areas, the biological values, the conservation status, the timber potential, the watershed conservation, and the indigenous and local community uses. The OTBN defined three categories for native forests (Fig. 4.1d): (i) red category representing forest areas of high conservation value; (ii) yellow category with forests with medium conservation value that may be subject to different sustainable use, tourism, or scientific research; and (iii) green category where low conservation forests value can be partial or completely transformed into other land uses (agriculture).

The proposed study included several modeling and statistical analyses for the comparison of (i) different forest types (NP = *Nothofagus pumilio*, NA = *N. antarctica*, MIX = mixed broadleaved and evergreen forests), (ii) natural reserves categories (NP = national park, PR = provincial natural reserves), and (iii) OTBN classifications (red = high value, yellow = medium value, green = low value) according to the National Law 26,331/07.

3 Provision of Different Ecosystem Services at Landscape Level

For this study we selected several proxies for the different ES: (i) cultural including aesthetic, existence, local identity, and recreational values; (ii) regulating through the net primary productivity and supporting characterized by the habitat quality; and (iii) provisioning characterized through timber (e.g., wood for sawmills) and silvopastoral values (e.g., livestock capacity and wood for different uses as lumber) (based on Martínez Pastur et al. 2016a, 2017, 2019).

Four maps were modeled for cultural ES, three related to human interactions (e.g., physical or intellectual) with biotic systems, ecosystems, and landscapes and other map related to spiritual values with the biota systems, ecosystems, and landscapes. For these maps, we worked with the methodology and data obtained by Martínez Pastur et al. (2016a) for Tierra del Fuego, based on Casalegno et al. (2013) and Wood et al. (2013). This methodology used geo-referenced digital photos from public databases on the web, evaluating the social and biophysical importance of the cultural ES in the landscape through the quantification of number of digital images that local people and visitors published. Using published raw data and results of Martínez Pastur et al. (2016a), we applied the kernel density tool into a GIS, which allows calculating the density of photos around each cell of a raster. The four proxies were considered: (i) aesthetic values were related to the characteristics of living systems that allowed for aesthetic experiences, (ii) existence values were related to the characteristics of living systems that generate value feelings for the society, (iii) local identity values were related to characteristics of living systems that are sound in terms of local culture and heritage, and (iv) recreation values were related to elements of nature used for human entertainment. Each map was rescaled from 0 to 100 using a linear scale by a function tool into ArcMap 10.0 software (ESRI 2011).

Cultural services in the forested areas (Fig. 4.2) were mainly influenced by accessibility and the landscape characteristics (e.g., mountains and forests). Areas with low accesibility (e.g., Península Mitre) presented low provision of ES, because people cannot reach to the benefits and goods that they provide. Hotspots occurred due to specific values, e.g., shipwrecks, penguin colonies, national parks, or natural reserves, while coldspots are relate to the lack of routes or homogeneous flat landscapes.

Regulating services was described using the net primary productivity (NPP) as proxy, between 2000 and 2009 ($g.m^{-2}.year^{-1}$), based on the methodology and information provided by Zhao and Running (2010). Into a GIS project, the final map had a resolution of 90 × 90 m. Supporting services were characterized by the human footprint index (HFI) as a proxy related to habitat quality (Rosas et al. 2020). In a GIS project, the habitat was calculated as the inverse value of the human footprint map (1-HFI), obtaining the natural habitat proxy map in a grid of 90 × 90 m. Both maps were rescaled from 0 to 100 linear scale by the same tool as was described before.

NPP increased from north to south and from dry grasslands to humid forests and increased to the west where peatlands are intermingled with mixed evergreen

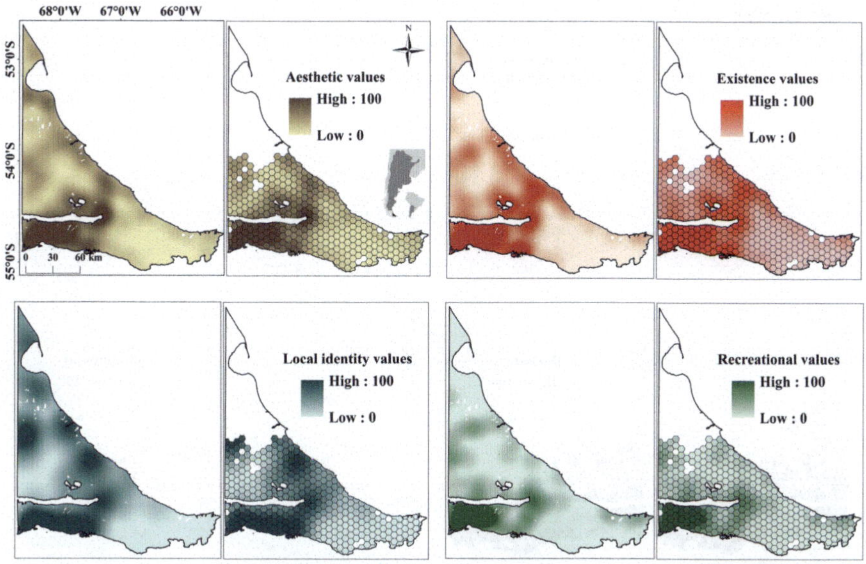

Fig. 4.2 Cultural ecosystem services (aesthetic, existence, local identity, and recreational values) based on Martínez Pastur et al. (2016a). Each figure included the model outputs (left) and hexagons used for further analyses (right) (hexagon = 5000 ha)

forests (Fig. 4.3). Beside this, values were greater in valleys and decreased with the altitude near tree line and alpine grasslands. Habitat quality decreased with human infrastructure, mainly related to accessibility through routes and paths. The highest values are in southern and west areas of the province, and lowest values were related to those areas with ranching and oil extraction activities.

Provisioning services included timber material of *N. pumilio* and mixed forests (material-fiber division of CICES) (Haines-Young and Potschin 2018). In a GIS project, we classified these forest types according to (i) timber forests and non-timber forests (including those forests with legal restrictions of environmental protection or high conservation values) and (ii) harvested forests before 2000 and between 2000 and 2019. Then, we used the normalized difference vegetation index (NDVI) (Remer et al. 2008) to characterize the harvesting intensity (e.g., remnant canopy) and the vigor of the forest stands (e.g., regeneration status). Crossing both shapes and grids, we extracted and rescaled each pixel using a linear scale by a function tool into ArcMap 10.0 software (ESRI 2011) and classified them in 0 = non-timber, 0.1–0.2 = timber forests recently harvested (2000–2019), and 0.2–1.0 timber forests without harvesting or harvested before 2000. The final map (90 × 90 m) was rescaled from 0 to 100 linear scale by the same tool as was described before.

Provisioning services also included the silvopastoral potential values of *N. antarctica* forests, belonging to the nutrition division of CICES (Haines-Young and Potschin 2018). For this, we defined a silvopastoral index (SI) based on the potential understory biomass (BIO = <1000, 1000–3000, >3000 kg $DW.ha^{-1}$) and tree height

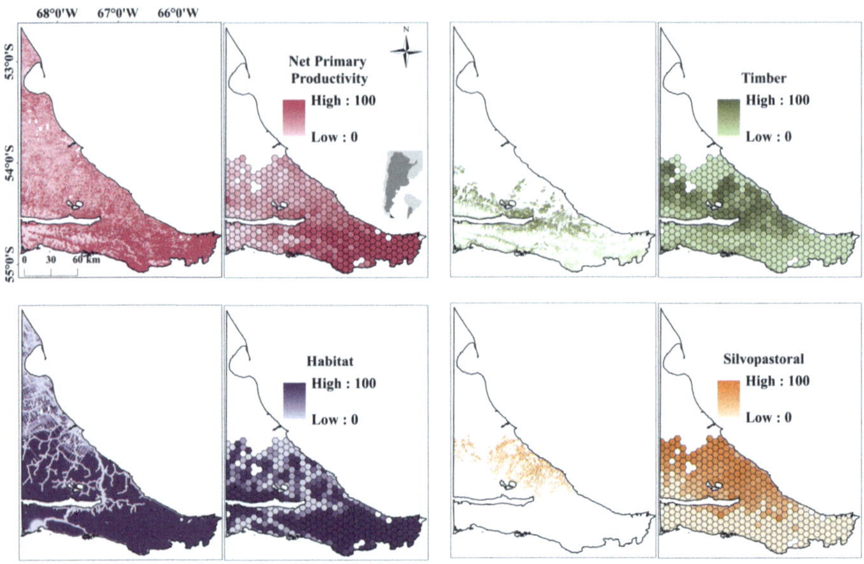

Fig. 4.3 Regulating (net primary productivity), supporting (habitat), and provisioning (timber and silvopastoral values) ecosystem services based on Martínez Pastur et al. (2017, 2019). Each figure included the model outputs (left) and hexagons used for further analyses (right) (hexagon = 5000 ha)

Table 4.1 Silvopastoral index defined by the understory biomass (BIO = <1000, 1000–3000, >3000 kg DW.ha^{-1}) and tree height (*H* = <7, 7–9, >9 m) of the *Nothofagus antarctica* forests (Peri 2009)

BIO	*H*	SI
<1000	<7	1.0
1000–3000	<7	1.3
>3000	<7	1.6
<1000	7–9	1.7
1000–3000	7–9	2.0
>3000	7–9	2.3
<1000	>9	2.4
1000–3000	>9	2.7
>3000	>9	3.0

(H = <7, 7–9, >9 m) (Peri 2009). In a GIS project, we generated a new map using a reclassify tool into ArcMap 10.0 software (ESRI 2011) from 1.0 to 3.0 considering the different understory biomass and tree height. Then, the silvopastoral index was defined considering a differential importance of *N. antarctica* forests, 70% for livestock (understory biomass rescaled from 0 to 100) and 30% for timber values as wood or firewood use (site quality of the stand defined as tree height rescaled from 0 to 100) (Table 4.1). The final map (90 × 90 m) was rescaled from 0 to 100 linear scale by the same tool as was described before.

The timber potential was closely related to the distribution of *N. pumilio* forests close to Tolhuin city in the center area of the Island (Fig. 4.3), where most of saw-mills are located. Silvopastoral uses are directly related to *N. antarctica* forests, but some cattle breeding occurred in the *N. pumilio* forests close to Fagnano Lake and eastern marginal areas of Tolhuin city.

The different ES presented in Figs. 4.2 and 4.3 were then combined to obtain maps for the three ES types (Fig. 4.4). All the rasterized maps (0–100 values) were combined into ArcMap 10.0 software (ESRI 2011), and the resulting average values were rasterized again to obtain final values of each ES type from zero (lower provision of the service) to 100 (maximum provision of the service). This combination map maintains the same patterns but highlights some areas where the different ES were overlapped; e.g., in the map that combine cultural services, higher values are shown around the national routes and the higher mountains.

4 Potential Biodiversity for *Nothofagus* Forested Landscapes

We employed the map of potential biodiversity (MPB) developed by Martínez Pastur et al. (2016b) for Tierra del Fuego (Fig. 4.4) for this variable. This map used a large database of understory plants of *Nothofagus* forests (535 plots) from PEBANPA Network (Peri et al. 2016b). Environmental Niche Factor Analysis

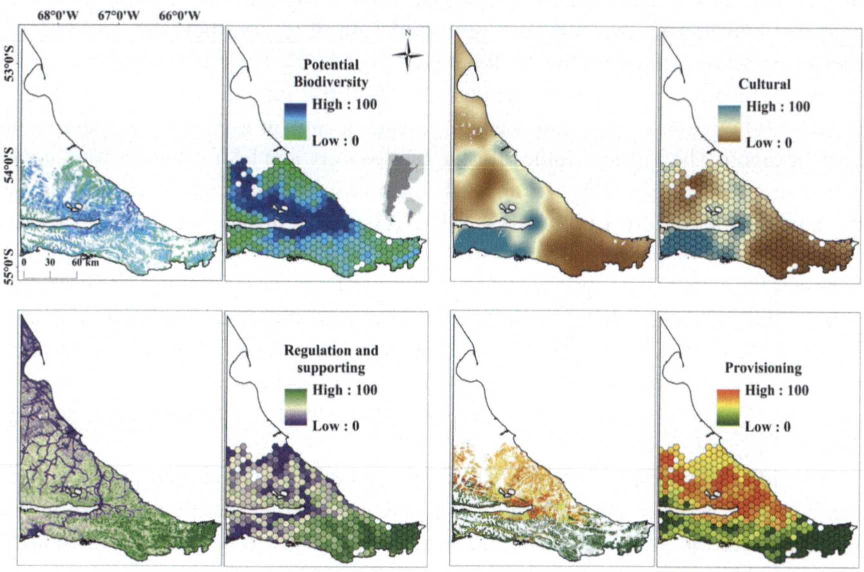

Fig. 4.4 Potential biodiversity (Martínez Pastur et al. 2016b) and combination maps for cultural, regulating, supporting, and provisioning ecosystem services. Each figure included the model outputs (left) and hexagons used for further analyses (right) (hexagon = 5000 ha)

(Hirzel et al. 2002) was used to map habitat suitability for each understory plant species based on 41 potential explanatory variables (climate, topography, and other variables related to landscape), which were rasterized at 90 × 90 m resolution using the nearest resampling technique on ArcMap 10.0 software (ESRI 2011) in Biomapper 4.0 software (Hirzel and Le Lay 2008). The habitat suitability maps for the understory species were combined (average values for each pixel) to obtain a MPB for each forest type (NP, NA, and MIX) at provincial level. These maps were rasterized to present scores that varied from 0 to 100 (average values of potential habitat suitability for all the studied species). For further comparisons, we classified the MPB values in low, medium, or high, where each forest-type class contains an equal quantity of pixels.

Potential biodiversity was higher in the ecotone areas of the main forest types (central area of the island near the Fagnano Lake and Tolhuin city) than in drier forests located in the upper mountains and in large peatlands. The occurrence of high biodiversity areas in more diverse landscapes was previously reported for Patagonia (e.g., Rosas et al. 2019a, b), where many species live in optimum and marginal environments across the ecotone areas (Riesch et al. 2018).

5 Provision of Ecosystem Services in *Nothofagus* Forested Landscapes and Protected Natural Areas

In this section, the distribution of potential biodiversity and ES of the forests was analyzed considering the currently protected natural reserve network and special protection areas defined by the National Law 26,331/07. For this analysis, we used a hexagonal binning processes (hexagon = 5000 ha) into ArcMap 10.0 software (ESRI 2011) to extract the values where forests occupy at least 10% of the area of each hexagon. The average values for the biodiversity and ES for the forested areas were presented in Figs. 4.2, 4.3, and 4.4. Each hexagon was classified considering the most abundant forest type (MIX, NP, NA), the occurrence of natural reserves (national parks or provincial reserves), or the different protection degrees of the National Law 26,331/07 (green and yellow, or red which defined the conservation values priorities). For these classifications, we used the zonal statistics tool into ArcMap 10.0 software (ESRI 2011), allowing to extract the mean values of each hexagon ($n = 363$) for potential biodiversity and ES and different analysis of variance (ANOVA) (Tables 4.2, 4.3, 4.4, and 4.5). In all tests, means were compared by Tukey test ($p < 0.05$).

The different forest types presented significant differences in their potential biodiversity (NP > NA > MIX), with greater potential outside the reserves than in the formal protection areas (Table 4.2). Because *N. pumilio* forests share understory species richness between the other two forest types (*N. antarctica* in the north and mixed evergreen forests in the south), these showed the highest potential biodiversity (Martínez Pastur et al. 2016b; Rosas et al. 2019b). Most of the natural reserves

Table 4.2 Multiple ANOVAs for the different forest types (FT) and their inclusion on natural reserves (NR, national parks and provincial reserves) analyzing potential biodiversity (BIO) and cultural ecosystem services (CULT) discriminated in their components: aesthetic values (AEST), existence values (EXIS), local identity (LI), and recreation values (RV)

Treatment	Level	BIO	CULT	AEST	EXIS	LI	RV
A: FT	MIX	43.6a	34.0b	32.4b	35.9b	40.1b	27.8c
	NP	51.4b	23.4a	24.3a	24.2a	27.1a	17.9b
	NA	45.9a	20.3a	18.7a	22.1a	35.7ab	4.5a
	F (*p*)	*15.77 (<0.001)*	*6.53 (0.002)*	*6.18 (0.002)*	*6.24 (0.002)*	*5.29 (0.005)*	*20.19 (<0.001)*
B: NR	Yes	44.0a	35.7b	35.6b	38.7b	43.3b	25.1b
	No	49.9b	16.1a	14.7a	16.0a	25.3a	8.4a
	F (*p*)	*18.86 (<0.001)*	*47.05 (<0.001)*	*56.54 (<0.001)*	*55.97 (<0.001)*	*21.76 (<0.001)*	*39.34 (<0.001)*
A × B	*F* (*p*)	*3.67 (0.027)*	*20.68 (<0.001)*	*15.87 (<0.001)*	*17.48 (<0.001)*	*17.12 (<0.001)*	*22.17 (<0.001)*

MIX mixed broadleaved and evergreen *Nothofagus* forests, *NP N. pumilio*, *NA N. antarctica*. Numbers varied between 0 (minimum provision) and 100 (maximum provision)
F Fisher test, *(p)* probability. Different letters showed differences with Tukey test at $p < 0.05$

Table 4.3 Multiple ANOVAs for the different forest types (FT) and their inclusion on natural reserves (NR, national parks and provincial reserves) analyzing supporting and regulating ecosystem services (S&R) discriminated in their components, primary productivity net (PPN) and habitat quality (HAB), and the provisioning ecosystem services (PROV) discriminated in their components, potential timber (TIM) and silvopastoral values (SILVO)

Treatment	Level	S&R	PPN	HAB	PROV	TIM	SILVO
A: FT	MIX	61.6b	41.0	82.0b	5.0a	8.8a	1.2a
	NP	62.5b	37.7	87.3b	40.6b	39.6c	41.6b
	NA	50.9a	34.1	67.7a	49.6c	23.6b	75.6b
	F (*p*)	*19.72 (<0.001)*	*2.96 (0.053)*	*22.68 (<0.001)*	*46.73 (<0.001)*	*23.85 (<0.001)*	*80.34 (<0.001)*
B: NR	Yes	50.8a	30.9a	70.7a	29.4	20.9	37.9
	No	65.8b	44.2b	87.3b	34.0	27.0	41.0
	F (*p*)	*72.20 (<0.001)*	*42.15 (<0.001)*	*36.52 (<0.001)*	*1.71 (0.192)*	*2.24 (0.136)*	*0.54 (0.461)*
A × B	*F* (*p*)	*19.37 (<0.001)*	*19.85 (<0.001)*	*13.11 (<0.001)*	*0.78 (0.460)*	*2.02 (0.135)*	*0.11 (0.896)*

MIX mixed broadleaved and evergreen *Nothofagus* forests, *NP N. pumilio*, *NA N. antarctica*. Numbers varied between 0 (minimum provision) and 100 (maximum provision)
F Fisher test, *(p)* probability. Different letters showed differences with Tukey test at $p < 0.05$

were not located following its potential biodiversity. These were created considering other characteristics, e.g., large inhabited areas close to international borderline (Tierra del Fuego National Park) or inaccessible for most users (Corazón de la Isla provincial reserve). This aspect of natural reserve networks location criteria was also pointed for the continental Patagonia (Rosas et al. 2017). Similar situation was identified for the OTBN process of the National Law 26,331/07, where low and

Table 4.4 Simple ANOVAs for the different protection degrees of the National Law 26,331/07 (NL) analyzing potential biodiversity (BIO) and cultural ecosystem services (CULT) discriminated in their components: aesthetic values (AEST), existence values (EXIS), local identity (LI), and recreation values (RV)

Treatment	Level	BIO	CULT	AEST	EXIST	LI	RV
NL	G-Y	51.9b	17.7	17.5	19.1	30.5b	8.0a
	R	45.8a	18.8	17.7	18.4	20.6a	14.1b
	F *(p)*	*51.65* *(<0.001)*	*0.31* *(0.580)*	*0.01* *(0.925)*	*0.13* *(0.723)*	*15.49* *(<0.001)*	*11.05* *(0.001)*

G-Y green and yellow categories which indicated low and medium conservation values and *R* red which indicated high conservation priorities. Numbers varied between 0 (minimum provision) and 100 (maximum provision)
F Fisher test, *(p)* probability. Different letters showed differences with Tukey test at $p < 0.05$

Table 4.5 Simple ANOVAs for the different protection degrees of the National Law 26,331/07 (NL) analyzing supporting and regulating ecosystem services (S&R) discriminated in their components, primary productivity net (PPN) and habitat quality (HAB), and the provisioning ecosystem services (PROV) discriminated in their components, potential timber (TIM) and silvopastoral values (SILVO)

Treatment	Level	S&R	PPN	HAB	PROV	TIM	SILVO
NL	G-Y	59.3a	38.8a	79.5a	50.9b	39.8b	62.1b
	R	71.3b	47.7b	94.8b	10.1a	10.4a	9.7a
	F *(p)*	*84.85* *(<0.001)*	*30.65* *(<0.001)*	*78.69* *(<0.001)*	*364.89* *(<0.001)*	*135.04* *(<0.001)*	*341.33* *(<0.001)*

G-Y green and yellow categories which indicated low and medium conservation values and *R* red which indicated high conservation priorities. Numbers varied between 0 (minimum provision) and 100 (maximum provision)
F Fisher test, *(p)* probability. Different letters showed differences with Tukey test at $p < 0.05$

medium protection levels (green and yellow categories) presented more potential biodiversity than the high protection level areas (red category). This contradicts the national law criteria because the provincial government defined the priority based on the current forest uses to avoid conflicts with the industry (e.g., ranching and forestry) and for their inaccessibility (e.g., Península Mitre) rather than conservation values. In the land planning (Fig. 4.1d), we can clearly see that the highest protection status was concentrated in one forest type (MIX) where most of the provisioning services occurred (Fig. 4.3). This contradiction was previously reported by Martínez Pastur et al. (2020).

Cultural ES also presented significant differences among forest types and natural reserves (Table 4.2). Higher values were found in mixed broadleaved and evergreen forests than in pure forests (NP > NA). Cultural ES were not related to the potential biodiversity. Because plant richness in these austral forests are quite low (Lencinas et al. 2008), there is a trend to valorize other values (e.g., landscapes, water bodies, or shipwrecks) (Martínez Pastur et al. 2016a). It was not surprising that cultural services are more appreciated by the people outside the natural reserves (average values and even for each subtype) due to the lack of access of most of the protected areas. Furthermore, there are no differences for the average cultural ecosystem service provision between categories of the OTBN, neither for aesthetic or existence values. However, while local identity was higher in green and yellow areas due to traditional farming (e.g., gauchos), recreation was higher in red areas due to hiking, skiing, and climbing activities in mountain areas (Martínez Pastur et al. 2020).

The supporting and regulating ES indicators were greater in southern and humid forests (NP and MIX) compared to northern dry forests (NA) which are intermingled with grasslands close to the ecotone (Table 4.3). These differences among forests were mainly due to habitat quality than primary productivity net. However, these ES were significantly highest outside the formal natural reserves (Table 4.3). In addition, the indicators of supporting and regulating ES were greater in red areas of OTBN (high conservation values) than in the lower-quality categories (green and yellow) (Table 4.4). This is because most of the inaccessible forests (high habitat quality due to low human impacts) and humid forests with peatlands (high PPN) were included in the red category.

Provisioning ES differed among the forest types in the landscape (Table 4.3). *Nothofagus antarctica* forests had the highest provision of ES explained by the silvopastoral activities, followed by *N. pumilio* forests which presented the highest timber potential. Mixed forest showed the lowest provisioning values. As it was expected, the provisioning ES were greater outside the reserves due to the same reasons that were pointed out for other ES. Another reason is the unbalanced representation of forest types in reserves, e.g., *N. antarctica* that present the highest values were scarce in the reserves, while mixed forests were one of the most represented in the provincial reserves.

Finally, the OTBN under the framework of the National Law 26,331/07 did not equally protect the values of the different forest types (Table 4.5). The red category had the lowest provisioning ES (timber and silvopastoral). This is because OTBN defines high conservation value forests as those located in inaccessible, remote areas, or upper basins, avoiding conflicts with stakeholders (e.g., ranchers and sawmills) in areas with timber forests. This is evident when the OTBN landscape pattern was compared with forest type distribution (Fig. 4.1).

6 Synergies and Potential Trade-Offs Among the Ecosystem Services and Biodiversity in *Nothofagus* Forested Landscapes

In the previous section, we showed that the provision of the ES and the potential biodiversity greatly varied across the forested landscapes and relationships detected according to the different characteristics of the main tree species, e.g., *N. antarctica* closely related to silvopastoral, *N. pumilio* linked to timber values, and *N. betuloides* with supporting and regulating services. However, these univariate analyses did not represent the synergies and potential trade-offs among different ES. For this, a multivariate statistical analysis was performed, including: (i) a principal component analysis (PCA) with a matrix of potential biodiversity and provision of ES for 363 hexagons delimited in Tierra del Fuego, exploring the relationships among forest types, natural reserves, and forest classifications of OTBN; and (ii) multi-response permutation procedures (MRPP) with Bray-Curtis distance testing differences in composition of potential biodiversity, cultural, supporting and regulating, and provisioning ES in the 363 hexagons classified by forest types, natural reserves, and OTBN (high values, medium-low values). In the PCA, significance of eigenvalues was evaluated using Monte Carlo permutation test (999 randomizations). The MRPP was evaluated using T, p-value, and A (McCune et al. 2002). The statistic T describes the separation between groups (the more negative is T, the stronger the separation) and has an associated p-value determined by numerical integration of the Pearson type III distribution. A is the chance-corrected within-group agreement, which describes the within-group homogeneity compared to the random expectation ($A = 1$ when all items are identical within groups, $A = 0$ when heterogeneity within groups equals expectation by chance, and $A < 0$ if there is less agreement within groups than expected by chance). Then, pairwise groupings were tested to determine potential significance in the differences (Zimmerman et al. 1985). We used the software PC-ORD (McCune and Mefford 1999) to conduct PCA and MRPP analyses.

In the PCA (Fig. 4.5), we selected the two first components, which retained 80% of the cumulative variance. The Axis 1 (46% of the explained variance, $p < 0.001$) was positively associated with provisioning > biodiversity > cultural and negatively with supporting and regulating ES. Axis 2 (34% of explained variance, $p < 0.001$) was positively associated with biodiversity > provisioning > supporting and regulating ES and negatively with cultural ES. Provisioning ES were related to potential biodiversity, as it was pointed by several authors (Quijas et al. 2010; Duru et al. 2015; Soliveres et al. 2016). Biodiversity regulates ecosystem processes and determines delivery of several ES, where the adequate combination of biotic and abiotic components must occur at any particular place and time (de Groot et al. 2010; Mace et al. 2012). Martínez Pastur et al. (2017) analyzed the silvopastoral values in Tierra del Fuego, and determined that plant species richness (as surrogate for potential biodiversity) was inversely linked with high provisioning ES values, due to high cattle stocking rate occurred in the less valuable *N. antarctica* forest. While forests with complete crown cover intermixed with other forested lands (close forest-forest ecotone) were areas with high potential biodiversity, the open forests near sea shores (close forest-steppe ecotone) are the best zones for livestock production (Lencinas et al. 2008; Peri et al. 2016a). Cordingley et al. (2016) also found that potential biodiversity and provisioning ES were related to ecosystem characteristics in England (e.g., size of heathland patches). On the other hand, the analysis showed that cultural ES, as well as supporting and regulating, were not fully related to provisioning or potential biodiversity. This was mainly because most of the provisioning services occurred in non-touristic places (Martínez Pastur et al. 2016a, 2017).

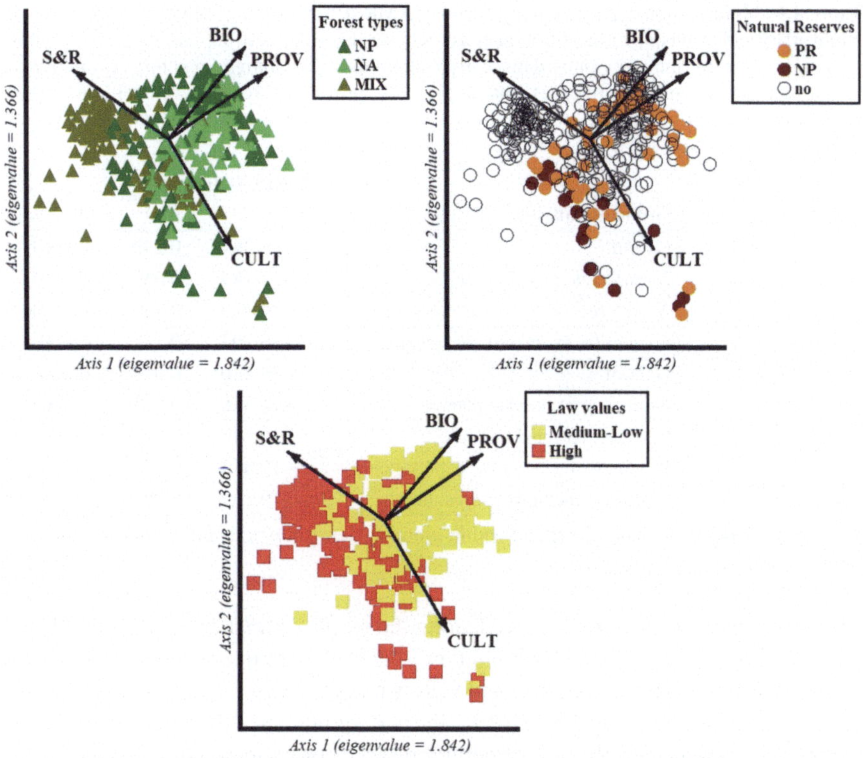

Fig. 4.5 Principal component analysis (PCA) of potential biodiversity (BIO) and cultural (CULT), supporting and regulating (S&R), and provisioning (PROV) ecosystem services. Each point corresponded to a hexagon classified according to the different forest types (NP = *Nothofagus pumilio*, NA = *N. antarctica*, MIX = mixed evergreen), natural reserves (PR = provincial reserves, NP = national park, no = none), and forest classifications according to the ordination planning of the National Law 26,331/07 (medium-low values, high values)

Contrary to our findings, there are studies that described positive relations between biodiversity and cultural ES (Hough 2014; Sandifer et al. 2015).

Mixed evergreen and *N. antarctica* forests were clearly differentiated along the Axis 1. While mixed forest was more related with highest values of supporting and regulation ES, *N. antarctica* forest linked with provisioning ES. On the other hand, *N. pumilio* forests occupied an intermediate position between the other forest types, with lower values of supporting and regulating and high provisioning ES values (Fig. 4.5). MRPP detected significant differences ($p < 0.001$) among the three forest types for the multivariate analyses (Table 4.6). Similarly, significant differences were found among the groups conformed by national park, provincial reserve, and none reserve hexagons, as well as between medium-low and high value hexagons categorized with the National Law 26,331/07 (Fig. 4.5). National park values were opposite to potential biodiversity and provisioning ES. While the highest values of cultural ES were exclusively associated with natural provincial reserves, those plots

Table 4.6 Multi-response permutation procedure (MRPP) comparing differences for potential biodiversity and provision of the different ecosystem services (cultural, supporting and regulating, provisioning) according to groups defined by forest types (*Nothofagus pumilio*, *N. antarctica*, mixed evergreen), natural reserves (national park, provincial reserves, none), and forest classifications according to the ordination planning of the National Law 26,331/07 (high values, medium-low values)

		MRPP statistics		
Factor	Group comparison	*T*	*A*	*p*
Forest types	Overall	−106.360	0.249	<0.001
	N. antarctica vs. *N. pumilio*	−27.144	0.057	<0.001
	N. antarctica vs. mixed evergreen	−116.159	0.349	<0.001
	N. pumilio vs. mixed evergreen	−80.903	0.187	<0.001
Natural reserves	Overall	−24.440	0.058	<0.001
	None vs. provincial reserves	−14.338	0.025	<0.001
	None vs. national park	−19.649	0.039	<0.001
	Provincial reserves vs. national park	−19.983	0.172	<0.001
Law values	Overall = medium-low vs. high	−119.160	0.197	<0.001

T statistic of MRPP, *A* chance-corrected within-group agreement, *p* probability associated with *T*

not placed in reserves were dispersed along the whole ordination space. Finally, the red category according to the National Law 26,331/07 was mainly related to cultural and supporting and regulating ES, and yellow category was related to potential biodiversity and provisioning ES. MRPP detected significant differences ($p < 0.001$) among the existence or absence of natural reserves or national parks for the multivariate analyses (Table 4.6).

These analyses highlighted synergies and potential trade-offs among the different ES and biodiversity. (i) Firstly, there was a trade-off between provisioning ES and potential biodiversity. Management proposals (e.g., shelterwood cuts and thinning for silvopastoral purposes) associated to provisioning ES generate great changes in the structure and function of these forests, both in the timber and silvopastoral activities (e.g., Peri et al. 2017; Martínez Pastur et al. 2019). These management proposals can lead to local extinction of several species as it was described in these forests for plants, birds, mammals, spiders, insects, and other taxa (e.g., fungi, mosses, spiders) (Lencinas et al. 2011, 2015; Argañaraz et al. 2020). (ii) There was a synergy between national parks in Tierra del Fuego and cultural ES related to the presence of two forest types (e.g., mixed evergreen and *N. pumilio* forested landscapes). In addition, while provincial reserves mainly included *N. pumilio* forests, *N. antarctica* forests were left aside of the formal protection areas. The OTBN does not protect all ES types and biodiversity and clearly avoided the trade-offs with provisioning services (Martínez Pastur et al. 2020).

7 Recommendations for Management and Conservation Planning: The Potential Use of Ecosystem Services and Biodiversity Concepts in *Nothofagus* Forested Landscapes

The intensification of agroforestry and timber management has already shown that optimization of one or some ES (e.g., provisioning) is likely to reduce diversity and system stability (Cardinale et al. 2012; Lindenmayer et al. 2012), as well as biodiversity (MEA 2005). Recently, science and policy agendas on biodiversity moved to include ES assessments and recognized the crucial task of monitoring ES to determine the effectiveness of policy frameworks (Geijzendorffer and Roche 2013; Liquete et al. 2016).

Several studies showed how biodiversity attributes affected provision of ES by impacting the underlying natural processes (Díaz et al. 2006; Harrison et al. 2014). There are studies conducted at species level, group of species, or at broader scale by analyzing the biodiversity impact on a single or multiple ES (Diaz et al. 2007; Luck et al. 2009; Poirazidis et al. 2011; Harrison et al. 2014). However, in Tierra del Fuego, studies that relate biodiversity and ecosystem service have only been performed at stand or ranch level (e.g., Lencinas et al. 2008, 2011, 2015) and few at landscape level considering different ES (Martínez Pastur et al. 2017; Peri et al. 2017). The relationship between species and ES is often complex (Mace et al. 2012).

For one species, the recommendations for management and conservation are quite simpler, e.g., the provisioning services of *N. pumilio* forest providing timber material. However, in the long term, we must include those species that allowed the survival of the forests (e.g., mycorrhizae) (Hewitt et al. 2018). In this context, species are the key in the supply of ES because biodiversity is the main factor where ecosystems sustain their functions with an evident consequence on human economy (Luck et al. 2003; Deliège and Neuteleers 2015; Martínez Pastur et al. 2017; Peri et al. 2017).

The results of this chapter provide information for management or conservation planning based on the potential use of ES and potential biodiversity for the *Nothofagus* forested landscapes in Tierra del Fuego. The management can incorporate recommendations for cultural services (Martínez Pastur et al. 2016a). Most of the cultural services were related to tourism and recreation activities inside natural reserves demonstrating synergies between conservation and management. In addition, ranches activities were identified as local identity values, and then, this occurrence also increases the cultural values of these areas under management. However, timber harvesting generated disservices and potential trade-offs with cultural ES, because tourism disagrees with the impact produced by the harvesting in the natural environments (Ahtikoski et al. 2011). In this sense, the maps generated in this chapter show the potential areas with trade-offs between these activities. Therefore, in these areas, it is highly recommended to implement another silvicultural practices more related to close-to-nature harvesting (e.g., Brunet et al. 2010; Brang et al. 2014).

We demonstrated that provisioning ES are related to silvopastoral and timber activities. Livestock production (80% cattle and 20% sheep) is the main annual income for farmers. The adaptive management plan in silvopastoral systems may adjust the stocking rate to forage availability (net primary production) and include the protection of tree regeneration against animal (domestic or native) overbrowsing (Peri et al. 2016a). Silvopastoral proposals maintained most of the original grassland diversity and the native herbivore populations (e.g., *Lama guanicoe*) (Soler et al. 2013). In this context, the challenge for sustainable silvopastoral production is to consider other ES into management plans, especially regulating and supporting. Timber production mainly in primary *N. pumilio* forest occurred in a short time (Gea Izquierdo et al. 2004; Martínez Pastur et al. 2007). These human impacts can be compared with large natural impacts (e.g., windthrow); however, these events eventually occurred and produced mosaic patterns in the landscape (Rebertus et al. 1997). Harvesting tends to homogenize the forest structure, increasing some services (e.g., C fixation) but decreasing other multiple ES, as well as biodiversity conservation (e.g., Martínez Pastur et al. 2019). In this context, the challenge for sustainable timber forestry is to apply technical solutions that achieve an equilibrium among ecological, economic, and social values (Lindenmayer et al. 2012), e.g., variable retention proposals reduced some trade-offs (conservation values increased in the harvested stands).

Several authors suggested that maintaining high levels of biodiversity in managed ecosystems had a positive effect on provisioning ES (e.g., plant diversity in agro-forest ecosystems according to Quijas et al. 2010 and Duru et al. 2015). In addition, maintaining heterogeneous landscapes provides high spatiotemporal biodiversity from small patches to the landscape level, increasing the resilience and stability of ecological processes in changing environments or after any kind of impact (e.g., Lencinas et al. 2011; Tscharntke et al. 2012; Geijzendorffer and Roche 2013).

Understanding the processes behind forest ES provision, as well as their trade-offs with biodiversity conservation, is a useful tool to support spatial planning and land management (Poirazidis et al. 2011; Carvalho-Santos et al. 2015). Conservation strategies must change from land sparing (e.g., creation of national parks and provincial reserves) to land sharing (e.g., increasing the conservation values in the managed landscapes), where OTBN of the National Law 26,331/07 is an excellent planning opportunity. However, it is necessary to define better implementation of this tool by combining main policy objectives with information landscape ES and potential biodiversity (Rosas et al. 2019a, 2020).

Improved decision-making in land management requires empirical information of the impact of landscape-scale conservation management approaches on trade-offs between biodiversity and ES (de Groot et al. 2010; Birch et al. 2010; Newton et al. 2012; Hodder et al. 2014; Cordingley et al. 2016). Productive system proposals (tourism, silvopastoral, timber) in Tierra del Fuego generated few changes in biodiversity at landscape level by maintaining the original forest types, the native fauna (e.g., large ungulates as *Lama guanicoe*), and most of the original understory plants (Soler et al. 2013; Peri et al. 2016a). Our results highlight the needs of policies to connect biodiversity conservation with ES maintenance and enhance its supply. Further studies should address the inclusion of more ES to develop new management alternatives. Incorporating traditional conservation strategies for species and habitat protection within the broader context of social-ecological systems and ES delivery can lead to added benefits for biodiversity through closer integration of conservation and productive policies (Martínez Pastur et al. 2017, 2020; Peri et al. 2017). For developing these new management practices, it is essential to quantify the synergies and trade-offs between biodiversity and ES provision that should be incorporated into decision support tools to reach better spatial planning.

Acknowledgments This research is part of the degree thesis and research scholar of Josela Carrasco at the Universidad Nacional del Litoral (UNL). We also want to thank the financial support of the project "Sinergias y conflictos entre las actividades económicas y los socio-ecosistemas de Tierra del Fuego: Mantenimiento de la productividad y los servicios ecosistémicos en el largo plazo" (CONICET, PUE 2016, Res 3334/16) and the project "Modeling endangered species forest habitats, and updating forest land use plans in Argentina in support of the UN sustainable development goals" (NRA ROSES 2018, NASA 18-SLSCVC18-0019, USA).

References

Ahtikoski A, Tuulentie S, Hallikainen V, Nivala V, Vatanen E, Tyrväinen L, Salminen H (2011) Potential trade-offs between nature-based tourism and forestry: a case study in Northern Finland. Forests 2:894–912

Allué C, Arranz JA, Bava J, Beneitez JM, Collado L, García López JM (2010) Caracterización y cartografía fitoclimáticas del bosque nativo subantártico en la Isla Grande de Tierra del Fuego (Patagonia, Argentina). For Syst 19:189–207

Argañaraz CI, Martínez Pastur G, Ramírez M, Grismado C, Lencinas MV (2020) Ground-dwelling spiders and understory vascular plants on Fuegian austral forests: community responses to variable retention management and their association to natural ecosystems. For Ecol Manag 474:e118375

Birch J, Newton AC, Alvarez Aquino C, Cantarello E, Echeverria C, Kitzberger T, Schiappacasse I, Tejedor Garavito N (2010) Cost-effectiveness of dryland forest restoration evaluated by spatial analysis of ecosystem services. PNAS 107:21925–21930

Brang P, Spathelf P, Larsen JB, Bauhus J, Boncčìna A, Chauvin C, Drössler L, García-Güemes C, Heiri C, Kerr G, Lexer MJ, Mason B, Mohren F, Mühlethaler U, Nocentini S, Svoboda M (2014) Suitability of close-to-nature silviculture for adapting temperate European forests to climate change. Forestry 87(4):492–503

Brunet J, Fritz Ö, Richnau G (2010) Biodiversity in European beech forests: a review with recommendations for sustainable forest management. Ecol Bull 53:77–94

Cardinale BJ, Duffy JE, Gonzalez A, Hooper DU, Perrings C, Venail P, Narwani A, Mace GM, Tilman D, Wardle DA, Kinzig AP, Daily GC, Loreau M, Grace JB, Larigauderie A, Srivastava DS, Naeem S (2012) Biodiversity loss and its impact on humanity. Nature 486:59–67

Carvalho-Santos C, Sousa-Silva R, Goncalves J, Honrado JP (2015) Ecosystem services and biodiversity conservation under forestation scenarios: options to improve management in the Vez watershed, NW Portugal. Reg Environ Chang 16:1557–1570

Casalegno S, Inger R, DeSilvey C, Gaston KJ (2013) Spatial covariance between aesthetic value and other ecosystem services. PLoS One 8(6):e68437

Collado L (2001) Tierra del Fuego forest: analysis of their stratification through satellite images for the forest province inventory. Multequina 10:1–16

Cordingley JE, Newton AC, Rose R, Clarke R, Bullock J (2016) Can landscape-scale approaches to conservation management resolve biodiversity-ecosystem service trade-offs? J Appl Ecol 53:96–105

Daily GC (1997) Nature's services. Island Press, Washington, DC

de Groot RS, Wilson MA, Boumans RM (2002) A typology for the classification, description and valuation of ecosystem functions, goods and services. Ecol Econ 41(3):393–408

de Groot RS, Alkemade R, Braat L, Hein L, Willemen L (2010) Challenges in integrating the concept of ecosystem services and values in landscape planning, management and decision making. Ecol Complexity 7(3):260–272

Deliège G, Neuteleers S (2015) Should biodiversity be useful? Scope and limits of ecosystem services as an argument for biodiversity conservation. Environ Val 24:165–182

Díaz S, Fargione J, Chapin FS, Tilman D (2006) Biodiversity loss threatens human well-being. PLoS Biol 4:e277

Diaz S, Lavorel S, de Bello F, Quetier F, Grigulis K, Robson TM (2007) Incorporating plant functional diversity effects in ecosystem service assessments. PNAS 104:20684–20689

Duru M, Therond O, Martin G, Martin-Clouaire R, Magne MA, Justes E, Journet EP, Aubertot JN, Savary S, Bergez J, Sarthou JP (2015) How to implement biodiversity based agriculture to enhance ecosystem services: a review. Agron Sustain Dev 35:1259–1281

ESRI (2011) ArcGIS Desktop: Release 10. Environmental Systems Research Institute Inc., Redlands

Gargaglione V, Peri PL, Rubio G (2014) Tree-grass interactions for N in *Nothofagus antarctica* silvopastoral systems: evidence of facilitation from trees to underneath grasses. Agrofor Syst 88:779–790

Gea Izquierdo G, Martínez Pastur G, Cellini JM, Lencinas MV (2004) Forty years of silvicultural management in southern *Nothofagus pumilio* primary forests. For Ecol Manag 201(2–3):335–347

Geijzendorffer IR, Roche P (2013) Can biodiversity monitoring schemes provide indicators for ecosystem services? Ecol Indic 33:148–157

Goldenberg MG, Oddi FJ, Gowda JH, Garibaldi L (2020) Effects of firewood harvesting intensity on biodiversity and ecosystem services in shrublands of northern Patagonia. For Ecosyst 7:e47

Haines-Young R, Potschin MB (2018) Common international classification of ecosystem services (CICES) V5.1 and guidance on the application of the revised structure. Available from www.cices.eu

Harrison PA, Berry PM, Simpson G, Haslett JR, Blicharska M, Bucur M, Dunford R, Egoh B, Garcia-Llorente M, Geamănă N, Geertsemah W, Lommelen E, Meiresonne L, Turkelboom F (2014) Linkages between biodiversity attributes and ecosystem services: a systematic review. Ecosyst Serv 9:191–203

Hewitt R, Taylor L, Hollingsworth T, Anderson CB, Martínez Pastur G (2018) Variable retention harvesting influences belowground plant-fungal interactions of *Nothofagus pumilio* seedlings in forests of southern Patagonia. PeerJ 6:e-5008

Hirzel AH, Le Lay G (2008) Habitat suitability modelling and niche theory. J Appl Ecol 45:1372–1381

Hirzel AH, Hausser J, Chessel D, Perrin N (2002) Ecological-niche factor analysis: how to compute habitat- suitability maps without absence data? Ecology 83(7):2027–2036

Hodder KH, Newton AC, Cantarello E, Perrella L (2014) Does landscape-scale conservation management enhance the provision of ecosystem services? Int J Biodiv Sci Ecosyst Ser Manage 10:71–83

Hough RL (2014) Biodiversity and human health: evidence for causality? Biodivers Conserv 23:267–288

Inostroza L, Zasada I, König HJ (2016) Last of the wild revisited: assessing spatial patterns of human impact on landscapes in Southern Patagonia, Chile. Reg Environ Chan 16(7):2071–2085

Kreps G, Martínez Pastur G, Peri PL (2012) Cambio climático en Patagonia Sur: Escenarios futuros en el manejo de los recursos naturales. Ed INTA, Buenos Aires

Lencinas MV, Martínez Pastur G, Rivero P, Busso C (2008) Conservation value of timber quality versus associated non-timber quality stands for understory diversity in *Nothofagus* forests. Biodivers Conserv 17:2579–2597

Lencinas MV, Martínez Pastur G, Gallo E, Cellini JM (2011) Alternative silvicultural practices with variable retention to improve understory plant diversity conservation in southern Patagonian forests. For Ecol Manag 262:1236–1250

Lencinas MV, Kreps G, Soler R, Peri PL, Porta A, Ramírez M, Martínez Pastur G (2015) *Neochelanops michaelseni* (Pseudoscorpions: Chernetidae) as potential bioindicator in managed and unmanaged *Nothofagus* forests of Tierra del Fuego. J Arach 43:406–412

Lindenmayer D, Franklin JF, Lõhmus A, Baker S, Bauhus J, Beese W, Brodie A, Kiehl B, Kouki J, Martínez Pastur G, Messier C, Neyland M, Palik B, Sverdrup-Thygeson A, Volney J, Wayne A, Gustafsson L (2012) A major shift to the retention approach for forestry can help resolve some global forest sustainability issues. Conserv Let 5:421–431

Liquete C, Cid N, Lanzanova D, Grizzetti B, Reynaud A (2016) Perspectives on the link between ecosystem services and biodiversity: the assessment of the nursery function. Ecol Indic 63:249–257

Luck GW, Daily GC, Ehrlich PR (2003) Population diversity and ecosystem services. Trends Ecol Evol 18:331–336

Luck GW, Harrington R, Harrison PA, Kremen C, Berry PM, Bugter R, Dawson TP, de Bello F, Diaz S, Feld CK, Haslett JR, Hering D, Kontogianni A, Lavorel S, Rounsevell M, Samways MJ, Sandin L, Settele J, Sykes MT, van den Hove S, Vandewalle M, Zobel M (2009) Quantifying the contribution of organisms to the provision of ecosystem services. Bioscience 59:223–235

Mace GM, Norris K, Fitter A (2012) Biodiversity and ecosystem services: a multilayered relationship. Trends Ecol Evol 27(1):19–26

Martínez Pastur G, Peri PL, Fernández MC, Staffieri G, Lencinas MV (2002) Changes in understory species diversity during the *Nothofagus pumilio* forest management cycle. J For Res 7(3):165–174

Martínez Pastur G, Lencinas MV, Peri PL, Moretto A, Cellini JM, Mormeneo I, Vukasovic R (2007) Harvesting adaptation to biodiversity conservation in sawmill industry: technology innovation and monitoring program. J Tech Manage Innov 2(3):58–70

Martínez Pastur G, Peri PL, Lencinas MV, García Llorente M, Martín López B (2016a) Spatial patterns of cultural ecosystem services provision in Southern Patagonia. Landsc Ecol 31(2):383–399

Martínez Pastur G, Peri PL, Soler R, Schindler S, Lencinas MV (2016b) Biodiversity potential of *Nothofagus* forests in Tierra del Fuego (Argentina): tool proposal for regional conservation planning. Biodivers Conserv 25(10):1843–1862

Martínez Pastur G, Peri PL, Huertas Herrera A, Schindler S, Díaz Delgado R, Lencinas MV, Soler R (2017) Linking potential biodiversity and three ecosystem services in silvopastoral managed forests landscapes of Tierra del Fuego, Argentina. Int J Biodiv Sci Ecosyst Ser Manage 13(2):1–11

Martínez Pastur G, Rosas YM, Toro Manríquez M, Huertas Herrera A, Miller JA, Cellini JM, Barrera MD, Peri PL, Lencinas MV (2019) Knowledge arising from long-term research of variable retention harvesting in Tierra del Fuego: where do we go from here? Ecol Proc 8:e24

Martínez Pastur G, Schlichter T, Matteucci SD, Gowda JH, Huertas Herrera A, Toro Manríquez M, Lencinas MV, Cellini JM, Peri PL (2020) Synergies and trade-offs of national conservation policy and agro-forestry management over forest loss in Argentina during the last decade. In: Lorenzo C (ed) Latin America in times of global environmental change, The Latin American studies book series. Springer, Cham, pp 135–155

McCune B, Mefford MJ (1999) PC-ORD: multivariate analysis of ecological data; Version 4 for Windows (user's guide). MjM software design, Corvallis

McCune B, Grace JB, Urban DL (2002) Analysis of ecological communities, vol 28. MjM Software Design, Corvallis

Mestre L, Toro Manríquez M, Soler R, Huertas Herrera A, Martínez Pastur G, Lencinas MV (2017) The influence of canopy-layer composition on understory plant diversity in southern temperate forests. For Ecosyst 4:6

Millennium Ecosystem Assessment (MEA) (2005) Ecosystems and human wellbeing: current state and trends. Island Press, Washington, DC

Ministerio de Ambiente y Desarrollo Sustentable (MAyDS) (2017) Informe de estado de implementación 2010–2016 de la Ley N°26.331 de Presupuestos Mínimos de Protección Ambiental de los Bosques Nativos. MAyDS, Buenos Aires, Argentina

Moore DM (1983) Flora of Tierra del Fuego. Anthony Nelson, United Kingdom - Missouri Botanical Garden, USA. 396 pp.

Mori AS, Lertzman KP, Gustafsson L (2017) Biodiversity and ecosystem services in forest ecosystems: a research agenda for applied forest ecology. J Appl Ecol 54(1):12–27

Newton AC, Hodder K, Cantarello E, Perrella L, Birch JC, Robins J, Douglas S, Moody C, Cordingley J (2012) Cost-benefit analysis of ecological networks assessed through spatial analysis of ecosystem services. J Appl Ecol 49:571–580

Panagos P, Borrelli P, Poesen J, Ballabio C, Lugato E, Meusburger K, Montanarella L, Alewell C (2015) The new assessment of soil loss by water erosion in Europe. Environ. Sci Policy 54:438–447

Perera A, Peterson U, Martínez Pastur G, Iverson L (2018) Ecosystem services from forest landscapes: Broadscale considerations. Springer, Cham

Peri PL (2009) Relevamiento de los bosques nativos de ñire (*Nothofagus antarctica*) de Tierra del Fuego. Ed INTA, Buenos Aires

Peri PL, Ormaechea S (2013) Relevamiento de los bosques nativos de ñire (*Nothofagus antarctica*) en Santa Cruz: base para su conservación y manejo. INTA, Buenos Aires

Peri PL, Hansen N, Bahamonde HA, Lencinas MV, Von Müller AR, Ormaechea S, Gargaglione V, Soler R, Tejera L, Lloyd CE, Martínez Pastur G (2016a) Silvopastoral systems under native forest in Patagonia, Argentina. In: Peri PL, Dube F, Varella A (eds) Silvopastoral systems in southern South America, Advances in agroforestry 11, chapter 6. Springer, Bern, pp 117–168

Peri PL, Lencinas MV, Bousson J, Lasagno R, Soler R, Bahamonde H, Martínez Pastur G (2016b) Biodiversity and ecological long-term plots in Southern Patagonia to support sustainable land management: the case of PEBANPA Network. J Nat Conserv 34:51–64

Peri PL, López D, Rusch V, Rusch G, Rosas YM, Martínez Pastur G (2017) State and transition model approach in native forests of Southern Patagonia (Argentina): linking ecosystemic services, thresholds and resilience. Int J Biodiv Sci Ecosyst Ser Manage 13(2):105–118

Peri PL, Rosas YM, Ladd B, Toledo S, Lasagno RG, Martínez Pastur G (2019) Modelling soil nitrogen content in South Patagonia across a climate gradient, vegetation type, and grazing. Sust 11:e2707

Poirazidis K, Schindler S, Kati V, Martinis A, Kalivas D, Kasimiadis D, Wrbka T, Papageorgiou AC (2011) Conservation of biodiversity in managed forests: developing an adaptive decision support system. In: Li C, Lafortezza R, Chen J (eds) Landscape ecology and forest management: challenges and solutions in a changing globe. Higher Education Press, Springer, New York, pp 380–399

Quijas S, Schmid B, Balvanera P (2010) Plant diversity enhances provision of ecosystem services: a new synthesis. Basic Appl Ecol 11:582–593

Quintas-Soriano C, Martín-López B, Santos-Martín F, Loureiro M, Montes C, Benayas J, García-Llorente M (2016) Ecosystem services values in Spain: a meta-analysis. Environ Sci Pol 55(1):186–195

Raudsepp-Hearne C, Peterson GD, Bennett EM (2010) Ecosystem service bundles for analyzing tradeoffs in diverse landscapes. PNAS 107(11):5242–5247

Rebertus AJ, Kitzberger T, Veblen TT, Roovers LM (1997) Blowdown history and landscape patterns in the Andes of Tierra del Fuego, Argentina. Ecology 78(3):678–692

Remer LA, Kleidman RG, Levy RC, Kaufman KJ, Tanré D, Mattoo S, Martins JV, Ichoku C, Koren I, Yu H, Holben BN (2008) Global aerosol climatology from the MODIS satellite sensors. J Geophys Res 113:D14S07

Riesch R, Plath M, Bierbach D (2018) Ecology and evolution along environmental gradients. Curr Zool 64(2):193–196

Rosas YM, Peri PL, Huertas Herrera A, Pastore H, Martínez Pastur G (2017) Modeling of potential habitat suitability of *Hippocamelus bisulcus*: effectiveness of a protected areas network in Southern Patagonia. Ecol Proc 6(1):28–42

Rosas YM, Peri PL, Martínez Pastur G (2018) Potential biodiversity map of lizard species in Southern Patagonia: environmental characterization, desertification influence and analyses of protection areas. Amphib Reptil 39(3):289–301

Rosas YM, Peri PL, Bahamonde HA, Cellini JM, Barrera MD, Huertas Herrera A, Lencinas MV, Martínez Pastur G (2019a) Trade-offs between management and conservation for the provision of ecosystem services in the southern Patagonian forests. In: Stanturf J (ed) Achieving sustainable management of boreal and temperate forests. Chapter 6. Burleigh Dodds Science Publishing, Cambridge, UK, 33 pp

Rosas YM, Peri PL, Lencinas MV, Martínez Pastur G (2019b) Potential biodiversity map of understory plants for *Nothofagus* forests in Southern Patagonia: analyses of landscape, ecological niche and conservation values. Sci Tot Environ 682:301–309

Rosas YM, Peri PL, Carrasco J, Lencinas MV, Pidgeon AM, Politi N, Martinuzzi S, Martínez Pastur G (2020) Improving the design of natural reserves networking in *Nothofagus* forests of Southern Patagonia using maps of potential biodiversity, human footprint maps, and the priorization of the spatial conservation values. In: Shit PK, Reza H, Das P, Sankar Bhunia G (eds) Spatial modeling in forest resources management: rural livelihood and sustainable development. Springer-Nature: Environmental Science and Engineering, Heidelberg

Sandifer PA, Sutton-Grier A, Ward B (2015) Exploring connections among nature, biodiversity, ecosystem services, and human health and well-being: opportunities to enhance health and biodiversity conservation. Ecosyst Ser 12:1–15

Soler R, Martínez Pastur G, Lencinas MV, Borrelli L (2013) Seasonal diet of *Lama guanicoe* (Camelidae: Artiodactyla) in a heterogeneous landscape of South Patagonia. Bosque 34(2):129–141

Soliveres S, Manning P, Prati D, Gossner M, Alt F, Arndt H, Baumgartner V, Binkenstein J, Birkhofer K, Blaser S et al (2016) Locally rare species influence grassland ecosystem multifunctionality. Phil Trans R Soc B 371:20150269

Tscharntke T, Tylianakis JM, Rand TA, Didham RK, Fahrig L, Batary P, Bengtsson J, Clough Y, Crist TO, Dormann CF, Ewers RM, Fründ J, Holt RD, Holzschuh A, Klein AM, Kleijn D, Kremen C, Landis DA, Laurance W, Lindenmayer D, Scherber C, Sodhi N, Steffan-Dewenter I, Thies C, van der Putten WH, Westphal C (2012) Landscape moderation of biodiversity patterns and processes: eight hypotheses. Biol Rev 87:661–685

Wood SA, Guerry AD, Silver JM, Lacayo M (2013) Using social media to quantify nature-based tourism and recreation. Sci Rep 3:e2976

Zhao M, Running SW (2010) Drought-induced reduction in global terrestrial net primary production from 2000 through 2009. Science 329:940–943

Zimmerman GM, Goetz H, Mielke PW Jr (1985) Use of an improved statistical method for group comparisons to study effects of prairie fire. Ecology 66(2):606–611

Chapter 5
Shrubland Management in Northwestern Patagonia: An Evaluation of Its Short-Term Effects on Multiple Ecosystem Services

Matías G. Goldenberg, Facundo J. Oddi, Juan H. Gowda, and Lucas A. Garibaldi

Abstract Sustainable management of natural resources is one of the greatest challenges that humanity is facing today. Part of the global energy demand is supplied by woody biomass whose production needs to ensure the maintenance of structures and functions that forests provide through biodiversity and ecosystem services. The general objective of this chapter is to review the effects of shrubland management for woody fuel production on biodiversity and different ecosystem services (ES) in northwestern Patagonia shrublands. This revision is based on results obtained from a field experiment evaluating a harvesting intensity gradient in three contrasting site qualities. Regrowth biomass production increases with harvesting intensity, and increments are higher at the high- and medium-quality sites. Consequently, the highest extraction intensity is the selected alternative for firewood production as economic metrics indicate. Wood energy density of shrubland species is higher than that observed in other woody species commonly used as fuels, highlighting the potential for energy supply in these ecosystems. Effect of harvesting on biodiversity varies with site quality: it is negative at the low-quality site and positive at the high-quality site, and biodiversity is maximized at intermediate harvesting intensity at the medium-quality site. Regarding the fire protection ES, the effects are general: firewood harvesting reduces both fuel continuity and live fuel moisture content at all

M. G. Goldenberg (✉) · F. J. Oddi · L. A. Garibaldi
Universidad Nacional de Río Negro. Instituto de Investigaciones en Recursos Naturales, Agroecología y Desarrollo Rural, San Carlos de Bariloche, Río Negro, Argentina

Consejo Nacional de Investigaciones Científicas y Técnicas. Instituto de Investigaciones en Recursos Naturales, Agroecología y Desarrollo Rural,
San Carlos de Bariloche, Río Negro, Argentina
e-mail: mgoldenberg@unrn.edu.ar

J. H. Gowda
Laboratorio Ecotono, INIBIOMA, CONICET-Universidad Nacional del Comahue,
San Carlos de Bariloche, Río Negro, Argentina

P. L. Peri et al. (eds.), *Ecosystem Services in Patagonia*, Natural and Social Sciences of Patagonia, https://doi.org/10.1007/978-3-030-69166-0_5

three sites. Litter cover, associated with soil formation ES, also decreases with harvesting intensity in all the sites, but the size effect depends on both site quality and time since intervention. When integrating the multiple ES, it is concluded that northwestern Patagonia shrublands could provide bioenergy ensuring biodiversity and ES, but the optimal level of intervention will depend on site conditions and decision-maker's preferences.

Keywords Silviculture · Bioenergy · Sustainability

1 Shrublands as Bioenergy and Ecosystem Services Suppliers

Sustainable use of renewable energy sources is essential to face current global change challenges. While the provisioning ecosystem service (ES) of biomass for energy reduces the dependency on nonrenewable energy sources, and thus on greenhouse gas emissions, it also generates pressure on forest resources. Today the concept of sustainable production also implies the maintenance of biodiversity and other multiple ES while designing natural resources management plans.

Shrublands have suitable properties for energy production, and biomass derived from these ecosystems is gaining importance (González-González et al. 2017). Indeed, this forest type has the potential to complement biomass derived from fast-growing species (Karp and Shield 2014). Forest management has historically focused on favoring the provisioning ES, but the response of biodiversity and other multiple ES to forest management has being less studied (Steffan-Dewenter et al. 2007; Bennett et al. 2009; Carpenter et al. 2009). Although the topic of ES is popular in ecology research, experimental tests of relationships between natural resource management and multiple ES are still necessary (de Groot et al. 2010). Forest management activities by promoting provisioning ES could have, for example, negative effects on supporting ES such as soil formation (de Groot et al. 2010; Cimon-Morin et al. 2013; Biber et al. 2015). This trade-off generates negative environmental impacts that are not incorporated when decision-making is driven by provisioning ES alone. On the other hand, there could be win-win scenarios, considering that some forest interventions may, for example, enhance understory diversity (Battles et al. 2001) and fire control (Goldenberg et al. 2020b).

2 Origin, Composition, and Dynamics of Shrublands of Northwestern Patagonia

Human activities drive shrubland dynamics in northwestern Patagonia. An important expansion process has occurred associated with the colonization of the region by European settlers at the end of the nineteenth century (Willis 1914; Veblen et al. 2003). Ever since, shrublands are showing a slow retraction (Gowda et al. 2012).

Because of their human-related origin, these fire-prone communities are largely associated with valley bottoms, roads, and urban centers (Gowda 2013). Replacement of shrublands by coniferous afforestation has also occurred in a smaller magnitude in northern Patagonia during the past century (Laclau 1997).

Currently, its distribution covers approximately 130,000 hectares in the province of Río Negro (CIEFAP y MAyDS 2016; Oddi et al. 2020). It is mainly located in the sub-Antarctic phytogeography province (semi-deciduous forest district) defined by Cabrera (1976), over a strip of rugged topography on both sides of the Andes Mountains (latitudes 37° to 55° S) (Reque et al. 2007). These shrublands are often dominated by ñire (*Nothofagus antarctica* (Forst.) Oerst.) but also form mixed communities with other species such as radal (*Lomatia hirsuta* (Lam.) Diels ex J.F. Macbr.), laura (*Schinus patagonicus* (Phil.) I. M. Johnst. ex Cabrera), retamo (*Diostea juncea* (Gillies ex Hook.) Miers) (Gyenge et al. 2009), caña colihue (*Chusquea culeou* (Desv.)), maitén (*Maytenus boaria* (Mol.)), palo piche (*Fabiana imbricata* Ruiz et Pavón), notro (*Embothrium coccineum* J. R. Forst. et G. Forst.), and chacay (*Discaria chacaye* (G. Don) Tortosa) (Fig. 5.1). Both shrubland types (pure ñire and mixed) are distributed mainly in valley bottoms and low slopes up to 1200 m a.s.l. (Reque et al. 2007) (Fig. 5.1).

Conservation of forest ecosystem services and sustainable management are set up under the current legal framework (National Law 26.331 and Provincial Law 4.552 of Río Negro). Native shrublands are one of the forest types with the highest plant diversity in the region (Speziale et al. 2010), and nowadays it is no longer legally possible to replace them by exotic species across large part of their distribution. These shrublands are interesting forest systems to study, for their productive potential, which has been little explored (Reque et al. 2007; Grosfeld et al. 2019). Most of the woody species that compose north Patagonia shrublands are heliophiles and resprouters (Rusch et al. 2017) tolerant to fire, the most recurrent disturbance in the region (Kitzberger and Veblen 1999). Resprouting individuals exhibit fast initial growth after such disturbance (Tiribelli et al. 2018), and their wood is locally used for energy (firewood). In absence of fire and cattle, shrublands are displaced by longer-lived, obligate seed dispersers such as ciprés de la cordillera (*Austrocedrus chilensis* (D. Don) Pic.Serm. & Bizzarri) or coihue (*Nothofagus dombeyi* (Mirb.) Oerst.) (Kitzberger and Veblen 1999). Nonetheless, this dynamic is sometimes affected by fire associated with the higher flammability of these communities (Blackhall et al. 2012, 2017) representing an enhanced fire risk to people and goods (de Torres et al. 2012).

3 Current Shrubland Land Use

Regulations that prevent these communities to be burned for pastures have reduced their livestock potential, leading to a reduction of small livestock (sheep). Consequently, livestock activity is extensive and with very low profitability. However, firewood is managed according to the needs of producers, favoring in

Fig. 5.1 Landscapes dominated by shrublands in Río Negro province, Patagonia, Argentina. Upper photography shows low slopes with mixed shrublands. Lower photography shows a valley bottom with pure ñire shrublands

some cases the generation of clearings. Because shrublands are close to routes and urban centers, their management is feasible under the regulatory framework (Gowda 2013), and biomass production for energy could be an interesting alternative. Their species have good regeneration capacity (Raffaele and Veblen 1998; Veblen et al. 2003) and high initial growth (Tiribelli et al. 2018), which would allow a quick recovery after the intervention. Nevertheless, no silvicultural plans have been developed in the region to manage shrublands for energy purpose.

4 Experimental Setup for Firewood Production in Shrublands Without Livestock

As named above, firewood harvesting has traditionally been combined with extensive livestock keeping, so the response of the vegetation to harvesting methods and intensity may be directly affected by changes in livestock feeding behavior and habitat choice. In 2013, a long-term experiment was established in northwestern Patagonia (Rio Negro province, Argentina) to evaluate the effects of shrubland management on multiple ecosystem services, among them, the regional potential for bioenergy production. Three sites covering a productivity gradient and thus differing in their apparent forestry quality were selected (thereafter, high-, medium-, and low-quality sites), and eight 31.5 m × 45.0 m permanent plots protected from cattle were established on each site (total plots = 24). Between 2013 and 2014, in each site, six plots were harvested in strips of increasing width (1.5, 2.5, and 3.5 m), resulting in two plots with 30% shrub cover removal, two plots 50% removal, and two plots 70% removal, leaving the remaining two plots as controls (Goldenberg et al. 2020b).

This field experiment was designed to explore trade-offs between the intensity of use of a resource under a market-based approach and the provision of other ES with non-market values (Chap. 1). Hence, 4 and 5 years after interventions, different indicators were measured as proxies of multiple ecosystem services (Table 5.1) to evaluate their response to harvesting intensity across different site qualities. In this chapter, we review the effects that were found and discuss the synergies and trade-offs between biomass extraction, bioenergy production, diversity, and ES supply.

Table 5.1 ES considered in this chapter. Only some indicators per ES class are used as an example

ES classification	ES measure	Indicator	Unit
Provisioning ES	Fuelwood (energy)	Net present value Biomass volume	US$ ha^{-1} m^3 ha^{-1} $year^{-1}$
Biodiversity	Taxonomic and functional diversity	Shannon index	H´
Regulating ES	Fire protection	Fuel amount	m^2 ha^{-1}
Supporting ES	Soil formation	Leaf litter cover	%

5 Provisioning ES and Economic Valuation

Economic valuation is an important component of the sustainability analysis of forest management projects. The economic feasibility of these projects is usually quantified through financial indicators, in which the discount rate plays a fundamental role (Price 1997). One of the most widely used financial indicators of forest investment is the net present value (NPV) (Goldenberg et al. 2018). This indicator, as expected, increases with harvesting intensity, and, at high levels of firewood extraction, it is even higher than those from alternatives requiring higher investments that return higher final values (mixed management of firewood/plantation of ciprés de la cordillera with timber purpose) when the discount rate is higher than 2.9% (Goldenberg et al. 2018). The main reason for this fact is the resprouting dynamics of the shrubby species that cuts out initial investment costs (contrary to afforestation alternatives, shrublands are already established in the beginning of the project and resprout following harvesting). These advantages were also observed in systems such as secondary forests of *Drimys winteri* in the Cordillera de la Costa (Chile). There, the profitability of a thinning intensity range becomes negative when administration costs are incorporated in the analysis (Navarro Cárcamo et al. 2010). This highlights the importance of native forest management applying schemes that ensure their natural regeneration in the medium and long term (Goldenberg et al. 2018).

The regeneration capacity of shrublands is evident when evaluating how canopy opening from harvesting affects the regrowth capacity of shrubland species. At plant level, the volume increment of sprouts responds nonlinearly and positively to harvest intensity (Table 5.2 and Fig. 5.2) (Goldenberg et al. 2020b). The strongest responses are observed at the medium- and high-quality sites in radal and retamo, the species with the highest growth rates. Ñire, the shrubland species with the widest distribution in northern Patagonia, had an intermediate response (Table 5.2). As other early succession species used for bioenergy production, in northern Patagonia, shrublands show high rates of juvenile growth with morphological adjustments as resource is released (Willebrand and Ledin 1993; Bond et al. 2001). This is part of a general functional mechanism of sprouting species, which have different strategies for resource exploitation (Neke et al. 2006; Forrester et al. 2013).

Table 5.2 Description of annual coppice biomass production. Mean values of plant annual volume growth (cm^3 $plant^{-1}$ $year^{-1}$) for each species in each site and two harvesting intensities and energy density (ED) mean value of each species

	High		Medium		Low		
	Harvesting intensity (%)						
Species	30	70	30	70	30	70	ED (GJ m^{-3})
Ñire	83.3	342.9	137.1	343.5	100.1	142.0	9.35
Laura	78.2	282.5	59.4	217.28	–	–	8.93
Radal	38.8	893.3	165.5	1164.4	–	–	8.46
Retamo	195.7	686.2	85.6	897.0	–	–	8.84
Notro	153.2	137.9	26.12	300.4	–	–	7.86

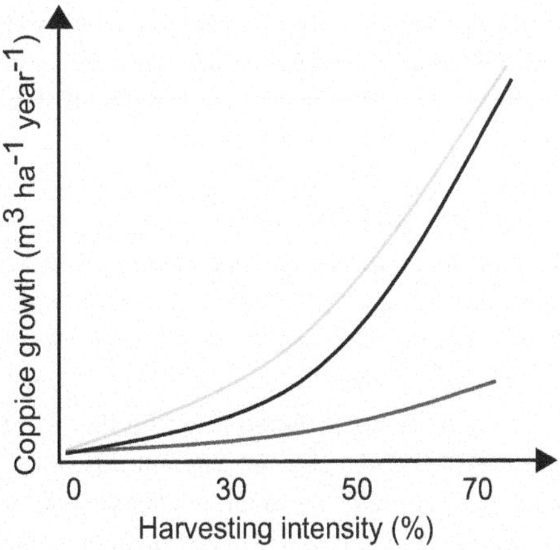

Fig. 5.2 Mean annual increment of volume coppice biomass under a harvesting intensity gradient of northern Patagonia shrublands. Light grey, high-quality site; black, medium-quality site; dark grey, low-quality site

Coppice biomass production at sand level also increases nonlinearly as harvest intensity increases, especially in hillside sites considered in this chapter as medium- and high-quality sites (Fig. 5.2, Goldenberg et al. 2020b). More harvesting results in more firewood volume for sale, but it also favors the growth of shrubs; thereby, harvesting cycles could be shortened when extraction biomass is higher. From a productive perspective, shrublands respond to harvesting in such a way that favors the execution of more intensive interventions (Goldenberg et al. 2020a).

During the firsts 4 years, annual increment in total biomass (e.g., that obtained from the extraction of 100% basal area) was about 3000 kg ha^{-1} $yr.^{-1}$ in the high- and medium-quality sites. The productivity of these stands would be close to that found in the region (Gyenge et al. 2008) with differences probably explained by both the asymptotic growth patterns of these species (Landesmann et al. 2016; Tiribelli et al. 2018) and stand composition. While useful for comparison, this extrapolation should be used with caution because no clear-cuts were explored in the experiment (Goldenberg et al. 2020b). When designing harvesting cycles, however, it should consider that subsequent interventions will be carried out on remnant vegetation; therefore, the entire stand becomes a mosaic with strips in different growth stages. More studies are needed to determine the optimal time interval between harvests across different management scenarios.

Differences between site qualities are also important in the coppice growth response. In low-quality sites, where environmental conditions are limiting for biomass growth (lower temperatures during winter, higher during summer, and lower precipitation during the growing period), coppice growth response to harvesting is low. Therefore, intense interventions under these conditions for coppicing systems would not be feasible (Goldenberg et al. 2020b). Probably under these conditions, other management alternatives such as silvopastoral use could be more appropriate as has being explored in southern Patagonia ñire forests (Peri et al. 2016).

Bioenergy potential depends not only on firewood production (discussed above) but also on the amount of energy in woody biomass. Calorific value quantifies the energy contained within woody mass and wood density the mass contained in wood volume, and both determine energy density. The last one is an important parameter because energy transport and storage costs, and thus the distribution and commercialization of bioenergy, depend on it. Energy density of the shrublands species is higher than other species frequently used for energy (Goldenberg et al. 2020a). Interestingly, the highest energy density is observed in ñire, the dominant species of the northwestern Patagonia shrublands. The woody energy characteristics, productivity, and response to harvesting observed in radal, retamo, and ñire (Table 5.2, Goldenberg et al. 2020a; Goldenberg et al. 2020b) denote that management favoring these species over others could be proposed as a strategy for producing high-quality firewood and other wood fuels.

In conclusion, when analyzing the provisioning ES of biomass for energy linking economic, productive, and firewood characteristics, the advantages of high-intensity use management are clear. As harvesting intensity increases, more biomass product is obtained at the start of the project benefiting the net present value, especially at high discount rates. Furthermore, increasing the extraction of biomass potentially could shorten the time between silvicultural interventions with profitability benefits, in particular, in high- and medium-quality sites. However, using only these indicators for management decision-making can lead to such a pressure of use that can cause resource deterioration (Lara et al. 2013).

6 Potential Trade-Offs Between Firewood Production, Biodiversity, and Other ES

6.1 Biodiversity

Forest interventions, such as biomass harvesting, change site environmental conditions at the stand level. Canopy opening increases both temperatures (and their fluctuations) and wind speed, reducing relative humidity (Trentini et al. 2017). These management-induced environmental changes can modify plant diversity according to site conditions (Fig. 5.3) (Goldenberg et al. 2020b). At the high-quality site, diversity increases with harvesting intensity, probably because when disturbances are more severe, the competitive exclusion is counteracted by the dominant woody species. In fact, high-intensity harvesting at this site is unlikely to create overly stressful conditions for plants because of the natural conditions of this site (high soil moisture levels, attenuated extreme temperatures in summer). The medium-quality site shows a unimodal pattern as that proposed by Connell (1978), which was also observed in other shrublands of the region (Rusch et al. 2005; Peri et al. 2016). At the low-quality site, on the contrary, diversity decreases with harvest intensity because extreme conditions leading to high plant mortality are achieved at low levels of disturbance (Huston 2014) (Fig. 5.3). Indeed, interactions between site quality and harvesting intensity are observed when plant diversity is evaluated with the Shannon taxonomic or functional diversity index, suggesting these shrublands to comply the dynamic equilibrium model proposed by Huston (2014) (Chillo et al. 2020; Goldenberg et al. 2020c).

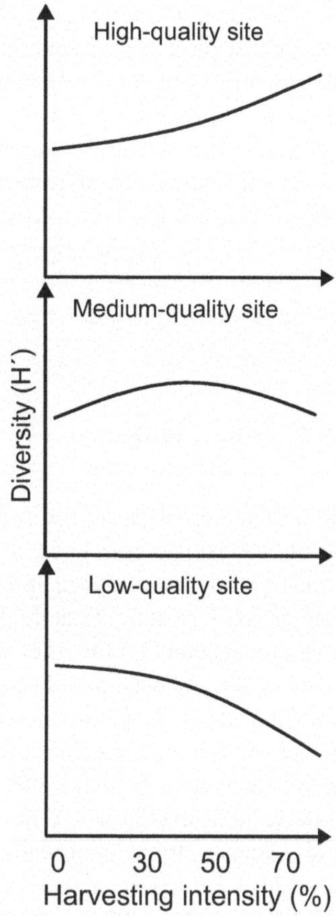

Fig. 5.3 Harvesting intensity effect on Shannon index for three sites

Functional diversity responds with a similar pattern than taxonomic diversity to the disturbance caused by harvesting (Chillo et al. 2020). While harvesting intensity interacts with site quality to account for differences in functional diversity resilience (e.g., the curves of diversity vary with site), site quality has been found as an important factor influencing resilience in taxonomic diversity. In high- and medium-quality sites, under lower and higher harvesting intensities, functional diversity resembles a non-disturbed community more than that found at intermediate harvesting intensities. In the low-quality site, however, the response to lower and higher harvesting intensities presents opposite patterns, and sites with higher harvesting intensities are the ones that resemble the most to non-disturbed communities. The most productive sites (high quality) may not experiment negative transitions from mixed shrubland to open grasslands-woodland as diversity peaked at the beginning and at the end of the experiment, and the traits that characterize a shrubland resembled the control treatment in 4 years, when high levels of biomass extraction were applied. But caution should be taken when determining harvesting levels to apply in unproductive shrublands, as the response pattern is the opposite (Chillo et al. 2020)

The low-quality site seems to be the most diverse, as suggested when unmanaged or control plots are compared (the maximum diversity is observed at this site). Low-quality sites typically have poor soil fertility (Schoenholtz et al. 2000). High nitrogen availability has been shown to reduce species richness mainly through competition mechanisms (light becomes limiting as growth increases, and plants shift from below- to aboveground competition) (Tilman 1987). It has also been shown that low phosphorous availability is associated with high diversity (Blanck et al. 2012). Consequently, both the highest taxonomic and functional shrubland diversities are observed at the low-quality site (Chillo et al. 2020) where soil fertility could be one of the main drivers explaining this pattern (Goldenberg et al. 2020c).

6.2 Fire Protection ES

Shrublands are the most fire-prone communities in northwestern Patagonia (Mermoz et al. 2005). Biomass harvesting is an effective practice to reduce fuel load and, consequently, slowing fire spread and diminishing fire severity (Regos et al. 2016). Therefore, flammability at the landscape level may decrease through shrubland biomass management (Gowda et al. 2019). The fact that biomass harvesting can be also used for fire management becomes this practice especially important in a changing climate, where temperatures are expected to rise (Halofsky et al. 2017). Shrublands occupy urban interface zones (Gowda 2013; Ghermandi et al. 2016); therefore, biomass management could also mitigate or prevent fire damage to human health, safety, and livelihoods (Haines-Young and Potschin 2018; Sil et al. 2019). Nevertheless, forest interventions change stands environmental conditions that may lead to a reduction in live fuel moisture content in the remnant forest structure (Pollet and Omi 2002), leading to a higher flammability of the harvested stand (Cornelissen et al. 2003).

As observed in the shrubland management experiment, relative area basal, a proxy of horizontal fuel continuity, decreases with increasing harvest intensity (Goldenberg et al. 2020c). Since fire is a contagious process (in a fire, each fuel particle is an ignition source for the surrounding fuel particles) (Peterson 2002), disrupting fuel continuity slows down or even stops the spread of fires. The moisture content of fine fuels shows a common pattern in all the sites, and it also decreases with harvesting. In addition, leaf moisture content is lower in ñire (considered the control species because it is present in the three sites) than that observed in mixed samples containing leafs of all species (Goldenberg et al. 2020c). In this sense, the low-quality site is formed by pure ñire and high shrub cover due to extreme stem density of small dimensions, indicating that shrublands under these conditions are highly fire-prone environments, as showed in previous works (Blackhall et al. 2017; Tiribelli et al. 2018). Hence, fire control trough management could be highly important in low-quality sites increasing the relevance of the trade-off with other ES.

More studies are required to determine the optimal harvesting intensity that minimizes fire risk. For this, it is necessary to evaluate how the continuity and moisture of live fuels modify fire activity (occurrence, size, severity) in these shrubland communities. Beyond this, the effect size of harvesting intensity on fuel is much smaller than on fuel continuity. This suggests fuel load to be more important than moisture (Goldenberg et al. 2020c). Another interesting aspect is that the effect of biomass harvesting on fire hazard would last a few years because shrubland species recover quickly (Table 5.2 and Fig. 5.2) producing large bulks of fine fuels. In fact, this chapter has shown that as increasing biomass, harvesting intensity (potentially diminishing fuel continuity) increases the production of fine fuel in the short term (potentiality increasing flammability). This trade-off on fire activity shows that the management of these communities requires complex solutions (Madrigal et al. 2016).

6.3 Soil Formation ES

Partial harvesting or thinning reduces stand density affecting litter production and soil cover (Harrington and Edwards 1999; Huebschmann et al. 1999; Roig et al. 2005; Jandl et al. 2007). This could affect soil properties and have negative long-term consequences, since the accumulation of leaves is one of the main carbon inputs in forest soils and a key component in the soil formation process (Jandl et al. 2007). Leaf litter is one of the main detritus cover type in this community and is mainly coming from woody species (de Paz et al. 2013). Harvesting intensity decreases leaf litter cover in the short term due to the decrease in stand density (Goldenberg et al. 2020c). Therefore, the soil formation ES could be compromised. Nonetheless, leaf litter accumulation decreases with harvesting intensity, but the magnitude is diminished with time since harvesting. Being that almost all the woody species show vigorous resprout under high harvesting intensities (Goldenberg et al. 2020b), leaf litter cover increases as the sprouts grow and sprouts are bigger under high harvesting intensities. Probably, as woody species regrowth and the detritus input increases, the leaf litter cover would be reestablished.

When comparing the three different site qualities, the high-quality site has the highest levels of litter cover while the low-quality site had the lowest. In the low-quality site, regrowth is not as vigorous as in the medium- and high-quality sites (Goldenberg et al. 2020b). Considering all the discussed parameters, the low-quality site would be the most fragile to the high levels of harvesting intensity in terms of soil formation ES.

7 Integrating Multiple Dimensions

This chapter focused on the short-term changes produced by biomass harvesting in biodiversity and provisioning, regulating, and supporting ES. It was shown that increases in harvesting intensity trigger growth rates benefiting the economic returns of the intervention (higher net presents income and potentially shorter rotations).

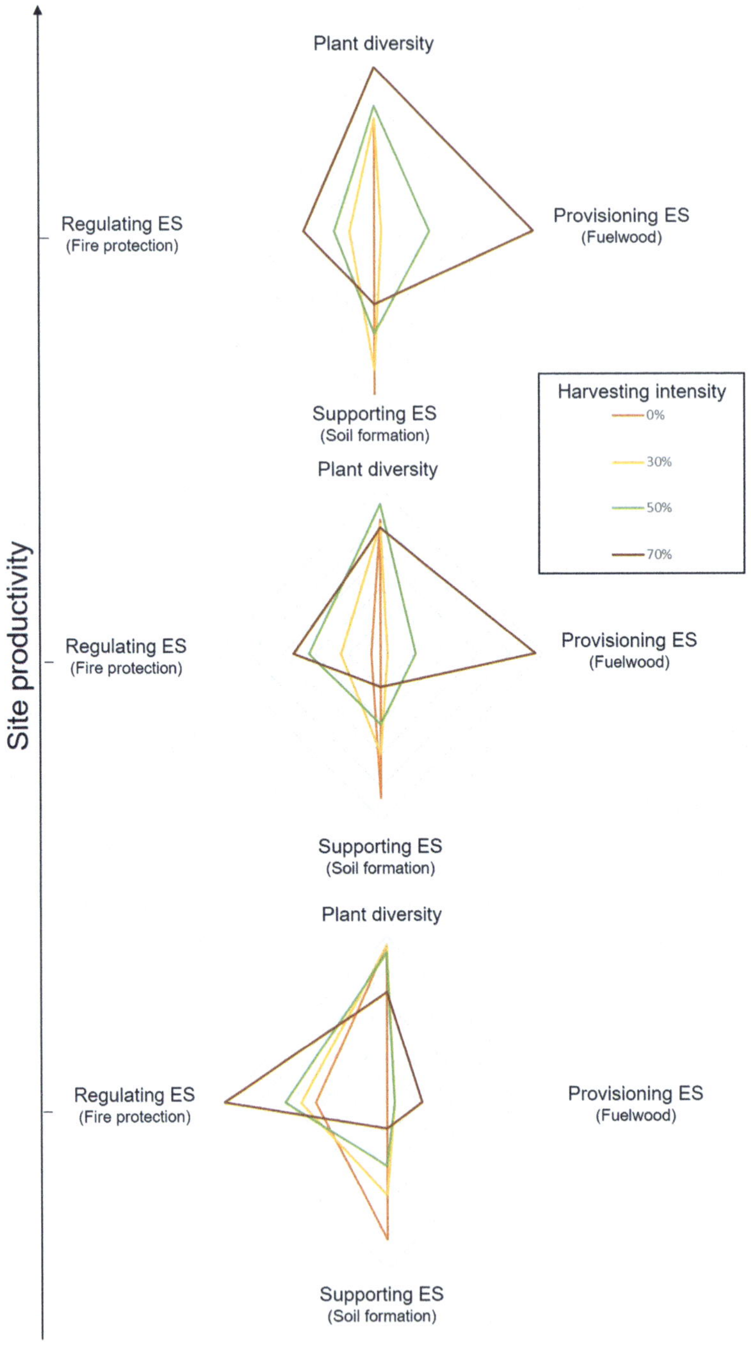

Fig. 5.4 Conceptual model of how this ecosystem could behave. Specifically, this figure shows some indicators representing different ES and trade-offs in response to harvest intensity and site productivity (site quality in this chapter). The maximum relative values correspond to the maximum measured value in all the experiment. All indicators are expressed in a-dimensional values (0–1) and in terms of "more is better" for ES provision

However, when other ES are included in the analysis, the optimal level of intervention seems to depend on site quality and thus productivity (Fig. 5.4). Hence, trade-offs between provisioning ES and supporting ES are revealed by integrating the indicators analyzed in this chapter (Fig. 5.4). While this pattern is observed in all the sites, trade-offs tend to be weaker as site productivity increases. For example, harvesting intensity is more important to fire protection ES in the lowest productive site, but this has also a stronger negative effect on both the supporting ES of soil formation and diversity. Also, coppice biomass growth is low in this site; therefore, provisioning ES of firewood is also compromised.

8 Guidelines and Recommendations

Historically, northwestern Patagonia shrublands have being used for livestock production. However, through this chapter, it has been showed that northern Patagonian shrublands can be managed for bioenergy (firewood or other wood energy formats), maintaining biodiversity and the provision of other ES. The complementation between biomass management with the current land use in these ecosystems should be studied in the future. The following recommendations can be bring for government actors, producers, and/or professionals based on the importance of the ecosystem services of Patagonia's natural resources for better decision-making in the sustainability of natural ecosystems: (i) Bioenergy use of native shrublands can be feasible both to produce energy and to provide biodiversity and other ES; (ii) bioenergy could complement the plantation of native species of high timber value, so this could be especially important to developed silvicultural systems that provide both energy and wood; and (iii) it remains necessary to give social value and to determine thresholds of the different indicators associated with different ES to achieve sustainable forest management plans.

References

Battles JJ, Shlisky A, Barrett RH, Heald RC, Allen-Diaz BH (2001) The effects of forest management on plant species diversity in a Sierran conifer forest. For Ecol Manag 146: 211–222.

Bennett EM, Peterson GD, Gordon L J (2009) Understanding relationships among multiple ecosystem services. Ecol Letters 12:1394–1404

Biber P, Borges JG, Moshammer R, Barreiro S, Botequim B, Brodrechtová Y, Brukas V, Chirici G, Cordero-Debets R, Corrigan E, Eriksson LO, Favero M, Galev E, Garcia-Gonzalo J, Hengeveld G, Kavaliauskas M, Marchetti M, Marques S, Mozgeris G, Navrátil R, Nieuwenhuis M, Orazio C, Paligorov I, Pettenella D, Sedmák R, Smreček R, Stanislovaitis A, Tomé M, Trubins R., Tuček J, Vizzarri M, Wallin I, Pretzsch H, Sallnäs O (2015) How Sensitive Are ecosystem services in European forest landscapes to silvicultural treatment? Forests 6:1666–1695

Blackhall M, Raffaele E, Veblen TT (2012) Is foliar flammability of woody species related to time since fire and herbivory in Northwest Patagonia, Argentina? J Veg Sci 23:931–941

Blackhall M, Raffaele E, Paritsis J, Tiribelli F, Morales JM, Kitzberger T, Gowda JH, Veblen TT (2017) Effects of biological legacies and herbivory on fuels and flammability traits: a long-term experimental study of alternative stable states. J Ecol 105:1309–1322

Blanck Y, Gowda J, Mårtensson L, Sandberg J, Fransson A (2012) Plant species richness in a natural Argentinian matorral shrub-land correlates negatively with levels of plant phosphorus. Plant Soil 345:11–21

Bond WJ, Midgley JJ, Bond WJ, Midgley JJ (2001) Ecology of sprouting in woody plants: the persistence niche. Trends Ecol Evol 16:45–51

Cabrera A (1976) Regiones Fitogeográficas Argentinas. In: Enciclopedia Argentina de Agricultura y Jardinería. ACME, Buenos Aires, Argentina

Carpenter SR, Mooney HA, Agard J, Capistrano DS, DeFries R, Díaz S, Dietz T, Duraiappah AK, Oteng-Yeboah A, Pereira HM, Perrings C, V. Reid W, Sarukhan J, Scholes RJ, Whyte A (2009) Science for managing ecosystem services: Beyond the Millennium Ecosystem Assessment. PNAS 106:1305–1312

Chillo V, Goldenberg M, Perez-Mendez N, Garibaldi L (2020) Diversity, functionality, and resilience under increasing harvesting intensities in woodlands of northern Patagonia. For Ecol Manag 474:18349

CIEFAP, MAyDS (2016) Actualización de la Clasificación de Tipos Forestales y Cobertura del Suelo de la Región Bosque Andino Patagónico. Informe Final, CIEFAP

Cimon-Morin JÔ, Darveau M, Poulin M (2013) Fostering synergies between ecosystem services and biodiversity in conservation planning: A review. Biol Cons 166:144–154

Connell JH (1978) Diversity in tropical rain forests and coral reefs. Science 199(4335):1302–1310

Cornelissen JHC, Lavorel S, Garnier E, Díaz S, Buchmann N, Gurvich DE, Reich PB, ter Steege H, Morgan HD, van der Heijden MGA, Pausas JG, Poorter H (2003) A handbook of protocols for standardised and easy measurement of plant functional traits worldwide. Aust J Bot 51:335–380

de Groot RS, Alkemade JRM, Braat L, Hein LG, Willemen L (2010) Challenges in integrating the concept of ecosystem services and values in landscape planning, management and decision making. Ecol Complex 7:260–272

de Paz M, Gobbi ME, Raffaele E (2013) Mantillo de las especies leñosas de matorrales del no de la patagonia: Abundancia, composición, estructura y heterogeneidad. Bol Soc Argent Bot 48:525–541

de Torres CM, Biscayart C, Ghermandi L (2012) Wildland-urban interface fires and socioeconomic conditions: a case study of a northwestern Patagonia City. Environ Manag 49:876

Forrester DI, Collopy JJ, Beadle CL, Baker TG (2013) Effect of thinning, pruning and nitrogen fertiliser application on light interception and light-use efficiency in a Young Eucalyptus nitens plantation. For Ecol Manag 288:21–30

Ghermandi L, Beletzky NA, de Torres Curth MI, Oddi FJ (2016) From leaves to landscape: a multiscale approach to assess fire hazard in wildland urban interface areas. J Environ Manag 183:925–937

González-González BD, Sixto H, Alberdi I, Esteban L, Guerrero S, Pasalodos M, Vázquez A, Cañellas I (2017) Estimation of shrub biomass availability along two geographical transects in the Iberian Peninsula for energy purposes. Biomass Bioenergy 105:211–218

Goldenberg MG, Gowda JH, Casas C, Garibaldi LA (2018) Efecto de la tasa de descuento sobre la priorización de alternativas de manejo del matorral Norpatagónico argentino. Bosque 39(2):217–226

Goldenberg MG, Cardoso YA, Oddi FJ, Garibaldi LA (2020a) Fuelwood energy characteristics and biomass equations of the dominant species of northern Patagonia shrublands. South For 82(1):56–64

Goldenberg MG, Oddi FJ, Amoroso MM, Garibaldi LA (2020b) Effects of harvesting intensity and site conditions on biomass production of northern Patagonia shrublands. Eur J For Res. https://doi.org/10.1007/s10342-020-01292-6

Goldenberg MG, Oddi FJ, Gowda JH, Garibaldi LA (2020c) Effects of firewood harvesting intensity on biodiversity and ecosystem services in shrublands of northern Patagonia. For Ecosyst 7:47

Gowda JH, Kitzberger T, Premoli AC (2012) Landscape responses to a century of land use along the northern Patagonian forest-steppe transition. Plant Ecol 213:259–272

Gowda JH (2013) ¿Qué protegemos y qué deberíamos proteger con la ley de bosques? Aportes para la primera revisión del ordenamiento territorial de los bosques de Río Negro. Rev Asoc Argent Ecol Paisajes 4:46–62

Gowda JH, Tiribelli F, Mermoz M, Kitzberger T, Morales JM (2019) Fragmentation modulates the response of dichotomous landscapes to fire and seed dispersal. Ecol Model 392:22–30

Grosfeld J, Chauchard L, Gowda JH (2019) Debates: ¿Podemos manejar sustentablemente el bosque nativo de Patagonia Norte? Ecol Austral:156–163

Gyenge JE, Fernández ME, Sarasola M, Schlichter T (2008) Testing a hypothesis of the relationship between productivity and water use efficiency in Patagonian forests with native and exotic species. For Ecol Manag 255:3281–3287

Gyenge JE, Fernández ME, Sarasola M, de Urquiza M, Schlichter T (2009) Ecuaciones para la estimación de biomasa aérea y volumen de fuste de algunas especies leñosas nativas en el valle del río Foyel, NO de la Patagonia Argentina. Bosque 30:95–101

Haines-Young R, Potschin MB (2018) Common international classification of ecosystem services (CICES) V5.1 and guidance on the application of the revised structure. Available from www.cices.eu

Halofsky JS, Halofsky JE, Hemstrom MA, Morzillo AT, Zhou X, Donato DC (2017) Divergent trends in ecosystem services under different climate-management futures in a fire-prone forest landscape. Clim Chang 142:83–95

Harrington TB, Edwards MB (1999) Understory vegetation, resource availability, and litterfall responses to pine thinning and woody vegetation control in longleaf pine plantations. Can J For Res 29:1055–1064

Huebschmann MM, Lynch TB, Wittwer RF (1999) Needle litterfall prediction models for even-aged natural shortleaf pine (Pinus echinata Mill.) stands. For Ecol Manag 117:179–186

Huston MA (2014) Disturbance, productivity, and species diversity: empiricism versus logic in ecological theory. Ecology 95:2382–2396

Jandl R, Lindner M, Vesterdal L, Bauwens B, Baritz R, Hagedorn F, Johnson DW, Minkkinen K, Byrne KA (2007) How strongly can forest management influence soil carbon sequestration? Geoderma 137:253–268

Karp A, Shield I (2014) Bioenergy from plants and the sustainable yield challenge. New Phytologist 179:15–32

Kitzberger T, Veblen TT (1999) Fire-induced changes in northern Patagonian landscapes. Landsc Ecol 14:1–15

Laclau P (1997) Los ecosistemas forestales y el hombre en el sur de Chile y Argentina. Boletín Técnico Fundación Vida Silvestre (34)

Landesmann JB, Gowda JH, Kitzberger T (2016) Temporal shifts in the interaction between woody resprouters and an obligate seeder tree during a post-fire succession in Patagonia. J Veg Sci 1-11

Lara A, Reyes R, Urrutia R (2013) Bosques nativos. en Centro de Análisis de Políticas Públicas, Universidad de Chile eds. Informe País: Estado del Medio Ambiente en Chile 2013. Nova Graf, Santiago, Chile

Madrigal J, Fernández-Miguela I, Hernando C, Guijarro M, Vega-Nieva DJ, Tolosana E (2016) Does forest biomass harvesting for energy reduce fire hazard in the Mediterranean basin? A case study in the Caroig Massif (Eastern Spain). Eur J For Res 136:13–26

Mermoz M, Kitzberger T, Veblen TT (2005) Landscape influences on occurrence and spread of wildfires in Patagonian forests and shrublands. Ecology 86:2705–2715

Navarro Cárcamo C, Herrera MA, Aranda FD, Donoso PJ (2010) Evaluación de la rentabilidad del manejo en bosques secundarios de canelo (Drimys winteri) en la Cordillera de la Costa de Valdivia, Chile. Bosque 31:209–218

Neke KS, Owen-smith N, Witkowski ETF (2006) Comparative resprouting response of Savanna woody plant species following harvesting: the value of persistence. For Ecol Manag 232:114–123

Oddi FJ, Goldenberg M, Cardozo Y, Garibaldi LA (2020) Manejo de matorrales y bioenergía. Rev Asoc Argent Ecol Paisajes 9(1):58–61

Peri PL, Bahamonde HA, Lencinas MV, Gargaglione V, Soler R, Ormaechea S, Marínez Pastur G (2016) A review of silvopastoral systems in native forests of Nothofagus antarctica in southern Patagonia, Argentina. Agrofor Sys 90(6):933–960

Peterson GD (2002) Contagious disturbance, ecological memory, and the emergence of landscape pattern. Ecosystems 5:329–338

Pollet J, Omi PN (2002) Effect of thinning and prescribed burning on crown fire severity in ponderosa pine forests. Int J Wildland Fire 11:1–10

Price C (1997) Twenty-five years of forestry cost-benefit analysis in Britain. Forestry 70:171–189

Raffaele E, Veblen TT (1998) Facilitation by nurse shrubs of resprouting behavior in a post-fire shrubland in northern Patagonia, Argentina. J Veg Sci 9:693–698

Regos A, Aquilué N, López I, Codina M, Retana J, Brotons L (2016) Synergies between forest biomass extraction for bioenergy and fire suppression in mediterranean ecosystems: insights from a storyline-and-simulation approach. Ecosystems 19:786–802

Reque JA, Sarasola M, Elena M (2007) Caracterización silvícola de ñirantales del norte de la Patagonia para la gestión forestal sostenible. Bosque 28:33–45

Roig S, del Río M, Cañellas I, Monteroet G (2005) Litter fall in Mediterranean Pinus pinaster Ait. stands under different thinning regimes. For Ecol Manag 206:179–190

Rusch V, Sarasola M, Schlichter T (2005) Indicadores de Biodiversidad en Bosques Nothofagus. IDIA XXI Forestales. INTA 5(8):8–14

Rusch VE, Rusch GM, Goijman AP, Varela SA, Claps L (2017) Ecosystem services to support environmental and socially sustainable decision-making. Ecol Austral 27:162–176

Schoenholtz SH, Van Miegroet H, Burger JA (2000) A review of chemical and physical properties as indicators of forest soil quality: challenges and opportunities. For Ecol Manag 138:335–356

Sil A, Fernandes PM, Rodrigues AP, Alonso JM, Honrado JP, Perera A, Azevedo J (2019) Farmland abandonment decreases the fire regulation capacity and the fire protection ecosystem service in mountain landscapes. Ecoyst Serv 2212-0416

Speziale KL, Ruggiero A, Ezcurra C (2010) Plant species richness-environment relationships across the Subantarctic-Patagonian transition zone. J Biogeogr 37:449–464

Steffan-Dewenter I, Kesslerc M, Barkmannc J, Bos MM, Buchori D, Erasmi S, Faust H, Gerold G, Glenk K, Gradstein SR, Guhardja E, Harteveld M, Hertel D, Hohn P, Kappas M, Kohler S, Leuschner C, Maertens M, Marggraf R, Migge-Kleian S, Mogea J, Pitopang R, Schaefer M, Schwarze S, Sporn SG, Steingrebe A, Tjitrosoedirdjo SS, Tjitrosoedirdjo S, Twele A, Weber R, Woltmann L, Zeller M, Tscharntke T (2007) Tradeoffs between income, biodiversity, and ecosystem functioning during tropical rainforest conversion and agroforestry intensification. PNAS 104:4973–4978

Tilman D (1987) Secondary succession and the pattern of plant dominance along experimental nitrogen gradients. Ecol Monogr 57:189–214

Tiribelli F, Kitzberger T, Manuel J (2018) Changes in vegetation structure and fuel characteristics along post -fire succession promote alternative stable states and positive fire -vegetation feedbacks. J Veg Sci 1-11

Trentini CP, Campanello PI, Villagra M, Ritter LJ, Ares A, Goldstein G (2017) Thinning of loblolly pine plantations in subtropical Argentina: impact on microclimate and understory vegetation. For Eco Manag 384:236–247

Veblen TT, Kitzberger T, Raffaele E, Lorenz DC (2003) Fire history and vegetation changes in northern Patagonia, Argentina. In: Veblen TT, Baker WL, Montenegro G, Swetnam TW (eds) Fire and climatic change in temperate ecosystems of the Western Americas, Ecological studies (analysis and synthesis). Springer, New York

Willebrand EVA, Ledin S (1993) Willow coppice systems in short rotation forestry: effects of plant spacing, rotation length and clonal composition on biomass production. Biomass Bioenergy 4:323–331

Willis B (1914) El Norte de la Patagonia. Dirección de Parques Nacionales, Buenos Aires, Argentina

Chapter 6
Silvopastoral Systems in Northern Argentine-Chilean Andean Patagonia: Ecosystem Services Provision in a Complex Territory

Verónica Chillo, Ana H. Ladio, Jaime Salinas Sanhueza, Rosina Soler, Daniela F. Arpigiani, Carlos A. Rezzano, Andrea G. Cardozo, Pablo L. Peri, and Mariano M. Amoroso

Abstract Silvopastoral systems (SPS) are sustainable production systems, characterized by a great biodiversity and multifunctionality compared with other livestock production systems. Northern Argentine-Chilean Andean Patagonia is a complex socio-ecological system where the provision and perception of ecosystem services

V. Chillo (✉) · D. F. Arpigiani · M. M. Amoroso
Universidad Nacional de Río Negro, Instituto de Investigaciones en Recursos Naturales, Agroecología y Desarrollo Rural, Río Negro, Argentina

Consejo Nacional de Investigaciones Científicas y Técnicas (CONICET), Instituto de Investigaciones en Recursos Naturales, Agroecología y Desarrollo Rural, Río Negro, Argentina

A. H. Ladio
Grupo de Etnobiologia, INIBIOMA. CONICET y Universidad Nacional del Comahue, Bariloche, Argentina

J. Salinas Sanhueza
Instituto Forestal (INFOR), sede Patagonia, Coyhaique, Región de Aysén, Chile

R. Soler
Centro Austral de Investigaciones Científicas (CADIC) – CONICET, Ushuaia, Argentina

C. A. Rezzano
Universidad Nacional de Río Negro, Instituto de Investigaciones en Recursos Naturales, Agroecología y Desarrollo Rural, Río Negro, Argentina

A. G. Cardozo
Agencia de Extensión Rural El Bolsón, Instituto Nacional de Tecnología Agropecuaria (INTA), IFAB, INTA-CONICET, El Bolsón, Río Negro, Argentina

P. L. Peri
National Agricultural Technology Institute, Río Gallegos, Argentina

P. L. Peri et al. (eds.), *Ecosystem Services in Patagonia*, Natural and Social Sciences of Patagonia, https://doi.org/10.1007/978-3-030-69166-0_6

(ES) vary depending on multiple contexts. We aim to characterize multiple ES and the associated benefits from SPS under different contexts and key factors. We propose a conceptual model of the socio-ecosystem that considers socio-historical contexts, forest ecosystems, cultural contexts, relational values, and anthropogenic aspects. We provide a deep review of published information on these factors and how these are related to the provision and perception of ES. We also consider natural and anthropogenic drivers to assist government and institutions. Finally, we discuss the main knowledge gaps that need to be addressed to attain a sustainable management. There is a need for a multidisciplinary and regional approach that can serve as a new interpretive framework for managers and decision-makers. In particular, the inclusion of relational values with nature and the visibility of anthropocentric factors can be useful tools for local development.

Keywords Livestock · Native forest · Socio-ecosystem · Sustainable management

1 Introduction

Silvopastoral systems (SPS) are sustainable production systems, characterized by a great biodiversity and multifunctionality compared with other livestock production systems (e.g., conversion of native forests to forage). SPS intentionally integrate grasslands or pasture forage with livestock in a complex system where the interactions between components are actively managed to ensure sustainability (Salinas et al. 2017). The modern concept of sustainability implies ecological and economic viability, as well as social acceptability (Roseland 2000; Matthies et al. 2017). In this context, SPS aim to increase the provision of ecosystem services (ES) (e.g., forest, non-forest products, livestock), with the minimum impact on regulation and support services, and maintain the cultural services associated with forests.

The Andean zone of north Patagonia contains great ecological and socio-cultural diversity, encompassing the province of Río Negro in Argentina and the region of Aysén in Chile. This territory with common environmental and socio-cultural characteristics has led to the development of SPS as a similar production option. However, the actual use of SPS (predominantly pastoral) is still far from an integral management of the system. In most cases, management only considers the livestock component, while the medium- and long-term planning of pasture and forest resources has not been fully integrated.

In this region, livestock grazing in native forests is an historical practice. Prior to "criollos" settlement by the end of the nineteenth century, native communities based their productive activities on livestock, with a strong transhumance component to both sides of the Andes Mountains (Cardozo 2014; Rusch and Varela 2019). The first permanent settlers maintained an economic structure based on livestock, but they practiced forest clearing and fires to increase agricultural and livestock sectors in the region, as well as the use and commercialization of large amounts of native forest resources (Kitzberger and Veblen 1999; Sotomayor and Barros 2016; Rusch and Varela 2019). Besides the strong social and economic transformation in the territory, livestock production continues to be an important activity due to local traditions and rural home incomes (Bondel 2008; Guitart Fité 2008; Cardozo 2014). At different scales, livestock still remains as the main production being a main factor in the transformation of the natural landscape (Easdale 2007).

In order to attain sustainable development through the provision of multiple ES, it is necessary to consider the territorial complexity of the socio-ecosystem of SPS, where multiple factors interact at local and regional scale. For this, we propose a conceptual framework based on a socio-ecosystem to integrate the territorial complexity of Andean North Patagonia. This framework includes socio-historical contexts, forest ecosystems, and cultural contexts, and it considers relational values and anthropogenic aspects, two key factors that have not been integrated into previous socio-ecological analyses. First, we provide a review of the published information on these factors at the local scale including natural and anthropogenic disturbance drivers to assist governments and institutions. Then, we review the backgrounds related to the provision and perception of ES in SPS. Finally, we integrate current knowledge about ES within the proposed socio-ecosystem framework to discuss main knowledge gaps and needed management practices.

2 Silvopastoral Systems in Andean North Patagonia of Argentina and Chile

In the Andean forests of the Río Negro Province, Argentina, Cardozo (2014) identified and characterized two main typologies of livestock-based producers and different socio-productive strategies, grouped into "small" and "medium" categories. Both types carry out mixed livestock, mainly cattle with sheep and goats. The workforce is family-run, with temporary salaried employees. Small producers (up to 1000 ha) practice intensive continuous grazing (year-round) with winter food supplementation. The cattle herd is small (<10 breeding cows), and the average total stocking rate is 5.6 livestock units (LU)/ ha. A significant percentage of income is extra-property (>50%). Medium producers (up to 8000 ha) use a great diversity of environments and forest types, generally in an altitude gradient, which allows extensive grazing and seasonal management of livestock (winter-summer). The cattle herd is larger (>40 breeding cows) with mean total stocking rate of 1.2 LU/ha (Cardozo 2014). A large part of their income is from production (>50%), complementing the livestock activity with summer agrotourism (e.g., hiking, horseback riding, camping, shelter, rafting, fishing).

Medium-sized producers' farm is covered with native forests in more than 70% of the area (Cardozo 2014), traditionally used for both livestock activities and forest extraction (silvopastoral use). Livestock remains in the valleys and at middle slopes with a dominance of mixed forest of ciprés de la cordillera (*Austrocedrus chilensis*) and coihue (*Nothofagus dombeyi*) during autumn, winter, and spring (winter areas "invernadas"). In the summer, they move animals to highland forests (summer areas "veranada") dominated by lenga (*Nothofagus pumilio*) and ñire (*N. antarctica*) (Cardozo 2014; Arias Sepúlveda and Chillo 2017; Chillo et al. 2018; Amoroso et al. 2018). In this type of farms, it is common to find gaps in small areas to create grazing areas and corrals (pastures or "pampas"). In the forest, timber and firewood extraction is usually carried out to open the canopy to obtain wood products and increase of understory forage availability. In these open canopy areas, cattle spend more time and more intense use than in closed forest, creating heterogeneity in silvopastoral use in the same farm (Chillo et al. 2018, Amoroso et al. 2018).

In general, the complexity of this anthropogenic disturbance is associated with intensity gradients given by the heterogeneity in the landscape (exposure and slope, presence of streams, forest type) and management strategies (opening of pampas, presence of paddocks, and handling areas). Consequently, this results in a landscape's matrix of great spatial heterogeneity that includes areas with low-medium intense use and areas degraded by overexploitation (logging and overgrazing). This pattern has been recorded both in summer grazing areas of lenga forest (Quinteros et al. 2012, Quinteros 2018) and in mixed evergreen forests during winter grazing (Chillo et al. 2018, Amoroso et al. 2018) (Fig. 6.1).

In the north Patagonian forests of the Aysén Region (Chile), the heterogeneity of socio-productive environments and the geography determines the differentiation of producers, depending on farm size and availability of capital and technology. In this sentence, there are three types of exploitations: small producers (a peasant-type farm with an agricultural area of 12 basic irrigated hectares and the income mainly is from farming or forestry) and medium and large farmers with an agricultural area that provides significant commercial returns and benefits (PASO 2000). The

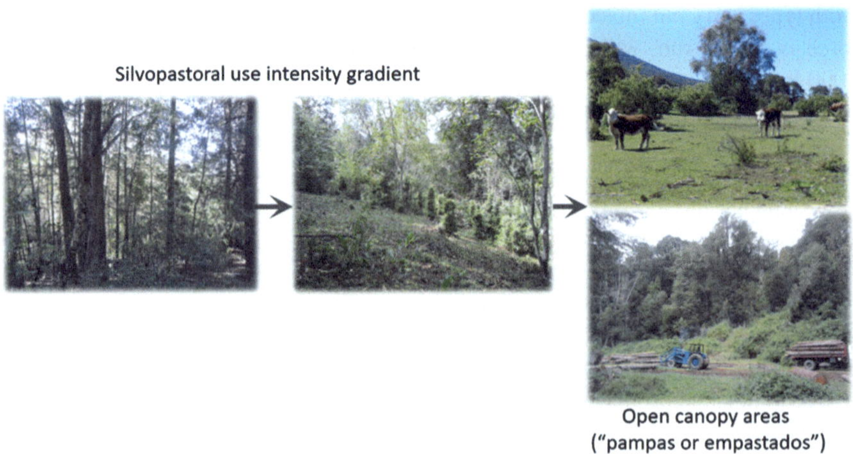

Fig. 6.1 Typical silvopastoral use gradient in winter areas of mixed ciprés-coihue forests at Río Negro Province, Argentina

productive activity in the Aysén region supports around 2285 cattle farms in 1,461,475 ha, mainly slaughtered in other regions. INE (2007) indicates that beef cattle reach 143,919 heads, mainly Coyhaique Province which accounts for 50% (FIA 2016). Of the total number of cattle, 60% is produced for small farmers in paddocks with meadows and forests.

In the region, there are two clearly defined cattle production systems, breeding and fattening, both of which are dominated by the Overo Colorado and Hereford bovine breeds. The bovine breeding system is developed by small and medium producers and is mainly sell directly on the farms or indirectly through local cattle fairs. This income is complemented by other farm activities such as the firewood, horticulture, and non-wood forest products. The fattening system is carried out mainly by large producers (more than 1000 breeding cows) (IICA et al. 2006). In the last 20 years, the average carrying capacity of the region has remained at around 0.15 EAU/ha (equivalent animal unit), which implies that the region can support a maximum of 189,735 EAU under current conditions.

The territory of the Aysén Region represents a recent history of colonization (Ortega and Brüning 2004), which has transformed large extensions of deciduous lenga and ñire forests into rangelands and scattered fragments of these native forests (Veblen et al. 1996, Sanchez et al. 2011). Thus, cattle ranching constitutes the basis of deep-rooted traditions in the local population since the beginning of colonization (Ortega and Brüning 2004). The use of the forest by livestock producers is associated with partial cutting for firewood (energy use), creating canopy openings that increase understory biomass for cattle grazing (Fig. 6.2). The occupation of the forest by cattle is not continuous and permanent, but it is accentuated in dry seasons with very hot summers or winters with very cold temperatures, where the use of the forest provides comfort to the cattle. Thus, the system of livestock uses lenga forests in summer and ñire forests in winter.

Fig. 6.2 Illustration of a typical silvopastoral system in ñire forest of the region of Aysén, Chile

3 Conceptual Framework of the Socio-ecosystem

According to Pascual et al. (2017), better life quality of local inhabitants would be achieved by linking nature material and intangible goods for people. Moreover, this idea should be expanded by considering different visions of nature from users or stakeholders. For example, native people values nature for human well-being (Rozzi 2012). In order to understand the ways of obtaining life quality in interaction with nature, it is first necessary to distinguish that different societies have different perceptions and knowledge (Pascual et al. 2017). In this context, we define "relational values" as the degrees of people responsibility in actions for perpetuation of ecological cycles. These different human-nature interactions are multidimensional and depend on multiple factors in a temporal trajectory such as social, economic, historical, and environmental.

The case of SPS in the Andean northern Patagonia in Chile and Argentina experienced through the history economic-social and environmental events that determined its current state. To achieve higher life quality, people have developed different management strategies according to their different perceptions of "quality." Therefore, different conditioning contexts determine different ways of valuing and using native forests.

Here we present a conceptual framework for better understanding of the socio-ecosystem of SPS in Andean North Patagonia of Chile and Argentina (Fig. 6.3). In this heuristic model, multidimensional human-nature interactions are considered, including different contexts that arise from such complex socio-ecosystem. The existing forest systems and the socio-historical context are structuring elements in SPS interacting with the cultural context (Fig. 6.3). This creates different relational values of interaction with forests. This occurs because different social actors have different conceptions of environmental uses with contrasting possibilities to implement the vision in SPS.

The socio-historical context represents those external social and political factors that have directly affected the region over time, such as the creation of provincial/regional states, land colonization, and policies assigned to the use of land through different laws and public institutions. This set of decisions and policies has shaped the evolution of the SPS in Argentine and Chilean Patagonia over the last 300 years. For example, the new societies established at expenses of original people were a result of social and political decisions that determined a particular trajectory and interactions with nature.

The forest ecosystems represent the environmental context (ecosystem and biodiversity subsystem in Fig. 1.3, Chap. 1), characterized by the different forest types where the SPS are developed. The dominant plant community formation is the humid, deciduous, or evergreen temperate forest with different species composition according to altitude, latitude, and aspect. Together with a consistent variation in climate attributes, geomorphology characteristics, substrate, and topography, there is a variation in both vegetation and fauna species (Donoso 1993).

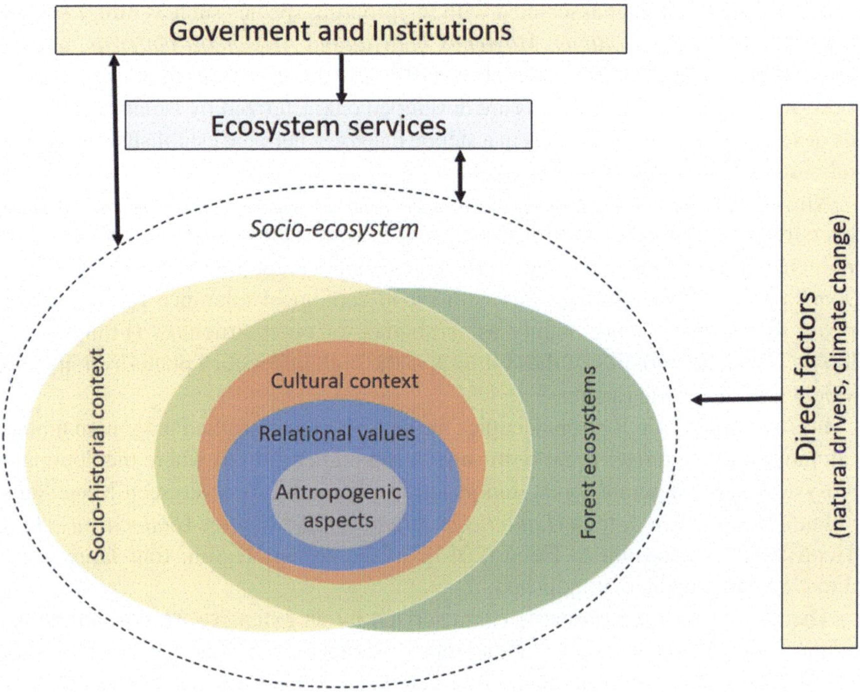

Fig. 6.3 Proposed conceptual framework of the socio-ecosystem (modified from Fig. 1.3) for the analysis of silvopastoral systems in Andean North Patagonia

In the north of the Argentine Patagonia, the western and altitudinal limit of the temperate forest is occupied by the lenga deciduous forest. This strip of forests protects a large part of the headwaters, being mainly monospecific, although mixed forests in humid environments are also common associated mainly with other species of the genus *Nothofagus* (*N. dombeyi*, *N. obliqua*, *N. alpina*) (González et al. 2006; Lara et al. 2014). These forests are dense, with large individuals (25–30 m tall) that gradually lose height with altitude until they blend with the coihue forest. Even in places where legna forest reach the highest productivity, these forests support a relatively low understory richness, except when it is associated with grasslands, peatlands, wetlands, and ñire forests. This evergreen species forms pure forests with tall and old individuals in humid extreme environments (3000–2500 mm/year) (Suarez and Kitzberger 2010) and gradually lose height and productivity when rainfall decrease (2000–1400 mm/year), where mixed forests with ciprés de la cordillera (*A. chilensis*) occurs. The mixed evergreen forests of coihue and ciprés are characterized by closed and dense canopies forming even-aged and stratified stands (Veblen et al. 1992a, b, Dezzotti 1996). The proportion of ciprés in these mixed forests is inversely proportional to soil moisture availability and altitude. As precipitation decreases to the east, ciprés forests dominates the mesic sites (900–1600 mm/year) forming pure and dense stands. In colder and wetter sites in the mountain, ciprés becomes a thicket associated with re-sprouting species such as ñire, *Lomatia hirsuta*, *Schinus patagonicus*, *Maytenus boaria*, and *Aristotelia chilensis*, among others (Donoso et al. 2006; Lara et al. 2014). At the driest distribution of ciprés (600–900 mm/year), open forests are developed characterized by isolated individuals or semi-closed to open stands in a steppe matrix, where the establishment of new individuals is sporadic (Villalba y Veblen 1997).

Ñire forests grow across the precipitation gradient and in a wide environmental diversity including valley bottoms, steep slopes with shallow soils, floodplain environments, and from the tree line to the forest-steppe ecotone (Veblen et al. 1996, Donoso et al. 2006). This species with broad ecological tolerance forms pure or mixed forests and occurs mainly as scrublands or open structures (Frangi et al. 2004). The characteristics of these forests is the high understory plant diversity and widely used for cattle raising.

In Chile, the Aysén Region occupies the eastern sectors of the Patagonian mountain range and comprises areas with valleys and higher rainfall where the dominant ecosystem is a temperate forest dominated by lenga and the evergreen Magellan's coihue (*N. betuloides*), tepa (*Laurelia phillipiana*), and mañíos (*Podocarpus* spp.) (Hepp 2019). According to Hepp (2014), in the Aysén Region, four large agro-climatic zones can be distinguished.

The coastal or insular zone is characterized by an extensive islands and archipelagos network located in the Patagonian channels and fjords and covered with evergreen forest mainly dominated by Magellan's coihue, mañíos, canelo (*Drimys winteri*), and other species. The predominant climate is temperate humid littoral with precipitations that exceed 3000 mm/year, reaching 7000 mm/year in some sectors. The islands have shallow and rich in organic matter soils on rocky material, not suitable for agricultural or forestry productive development. The humid agro-climatic zone is located in the continental western slope of the Andes mountain range. The predominant climate is the coastal humid temperate, characterized by average rainfall of 2640 mm/year. The predominant vegetation is the evergreen forest of Magellan's coihue, tepa, and *Podocarpus* spp., with other species in the understory such as canelo, notro (*Embothrium coccineum*), and myrtle (*Luma apiculata*).

The Intermediate zone, located on the eastern aspect of the Andes mountain range, is characterized by valleys that emerge towards the east of side of the region. This area has various climates, depending mainly on the altitude. At high altitudes, the boreal Andean climate predominates, while in the low areas, the intermediate humid temperate climate occurs with an average annual rainfall of 1150 mm/year. The predominant natural vegetation is the *Nothofagus* deciduous forest, especially lenga and ñire in certain more humid sectors and in transition towards steppe areas, respectively. This area was the most affected by large forest fires during the colonization period, particularly the beginnings of the twentieth century.

Finally, the steppe zone is associated with the steppe biome with tussock grasslands, in areas bordering Argentina. At high sites, the boreal Andean climate shares transitional characteristics towards colder conditions, with dominance of ñire forest.

However, the predominant climate is cold climate with an average annual precipitation 588 mm/year.

The regional cultural context is diverse and multicultural, determined from the ancestral existence of native peoples, the "criollos," and European immigration of the late nineteenth and early twentieth century, and the most recent immigration resulting in the current urban centers. The Andean North Patagonia has been inhabited by different communities of native peoples such as Mapuche, Tehuelche, or Mapuche-Tehuelche. The original territory of the Mapuche people occupied areas in southern Chile and Argentina, from the Pacific to the Atlantic oceans. These societies traveled long distances for the exchange of goods by synchronizing the cycles of nature in the complementary use of different environments (Ladio and Molares 2014, 2017). Currently, they live in subsistence conditions in marginal areas and correspond to the typology of small producers. In their subsistence systems, the role of the family and community action are key organizational aspects for the management of the SPS. For example, in the province of Río Negro, about 70% of livestock-forestry producers are considered small (small land size with low number of animals) and depend on family working force, with little contract labor (Cardozo 2014). In this context, the lack of young people working in family farms is recognized as the main problem for management and technological innovation (Cardozo 2014).

The main community organization for indigenous, artisan associations, and/or farmer's cooperatives to sale products (wool, timber, artefacts) in rural markets is essential to overcome the limiting conditions of development. In these communities, there are several ways of collaboration such as the exchange of goods (firewood for animals) and joint work in the construction of corrals and fences and animals marking. The community organization within the cultural context guarantees the governance within the SSP.

In Fig. 6.3, the relational model, based on Muradian and Pascual (2018), represents the set of society's preferences, principles, and virtues that explain the degree of responsibility towards nature. In case of the SPS of Andean North Patagonia from Chile and Argentina, we can distinguish the following: (1) domination, when degradation of nature is intentional; (2) exploitation, when nature is only considered an element of consumption and services; (3) custody, when it is protected for its intrinsic value; (4) management, when practices are carried out allowing the perpetuation of ecological cycles and cultural life systems simultaneously; (5) the ritualized exchange; and (6) devotion, when nature is established based on religiosity and mutual integration. In the Andean-Patagonian forests, these typologies can be described by their socio-environmental history. The characterization of the main actors interacting with forests from these points of view is essential for the understanding of the logic applied to management and the knowledge and values that are put into practice.

In the case of Mapuche-Tehuelche communities, the logic under which they use the landscape corresponds to a relational model linked to management, devotion, and ritualized exchange, where mutual care has prevailed between people and forest (Rozzi 2012; Castillo and Ladio 2017). But also other forms of interaction have been developed in the territory. The colonization of the Andean-Patagonian forests by Spanish and criollos began in the Chilean territory in the mid-nineteenth century. After more than 250 years of war between the Spanish and the Araucana nation, the advance on forest occurs by following a relational model of domination. With the consolidation of the Argentine state at the end of the nineteenth century, the occupation and use of the region were established under the relational model of exploitation of natural systems. A production model based on extensive cattle ranching on natural grasslands in the hands of few landowners occurred on lands usurped from native people. As a consequence, the notion of private property was established, followed by large-scale deforestations and fires that reduced the forest cover, mainly ciprés and coihue (Ladio and Molares 2014). This determined the displacement of small and medium farmers to the driest and less productive areas in the region. In both countries, since the twentieth century, the main drivers of use of the landscape correspond to logging of native and exotic forests, tourism, and livestock under an exploitation model based on market utilities.

Finally, with the creation of national parks (i.e., Lanin, Alerces, and Nahuel Huapi) in the middle of the twentieth century, a custody model was added to the Andean northern Patagonian based on SPS, which included logics of forest care directly linked to the valuation of ecosystem services and diversity. In the last 20 years, the parks have incorporated the figure of co-management with native communities, accounting for the valuation of the cultural context in the maintenance of biodiversity. Consequently, the different anthropogenic disturbances in SPS of north Andean-Patagonia arise from the link between the multiple contexts proposed for this socio-ecosystem.

Finally, the value and perception of different ES also vary depending on anthropogenic aspects. For example, local ecological knowledge (LEK) of farmers assist them to better manage nature, with greater governance and resilience (management of water, soil, biota, animal health) (Cardoso et al. 2015; Castillo and Ladio 2017). It is important the integration of LEK with scientific knowledge through mechanisms of social learning and interchange (Ladio 2017), as well as a link with state institutions of extension and technology transfer (Cardozo 2014). Also, it needs an adequate articulation between public and private institutions to obtain their welfare objectives (Ladio 2017; Castillo et al. 2020), as well as access to communication (telephones, internet, and roads), health, and education to rural inhabitants. Another important aspect is the access to a complementary use of landscapes for the collection of non-wood forest products (food and medicines) (Molares and Ladio 2012; Morales et al. 2017, 2018).

4 External Drivers of Change

Governments and institutions appear transversely within the socio-ecosystem model. In the SPS of north Andean-Patagonia, we identified mainly state institutions in their multiple levels of action. In Río Negro, there are two main laws regulating the forest use: Provincial Forestry Law No. 757 (1972) that establishes the regulatory framework for the defense, improvement, and use of forests and Provincial Law of Native Forests No. 4552 (2010) that establishes the complementary norms for the conservation and sustainable use of native forests under the National Law of Minimum Budgets No. 26331. Law 26,331 promotes the conservation of native forests through land planning, sustainable management, and land-use change regulations. Native forests have been classified according to three conservation categories: red (high forests conservation value for ancestral uses, gathering of non-timber forest products, scientific research, "respectful" tourism, conservation plans, ecological restoration), yellow (medium forests conservation value for sustainable productive activities and tourism under management and conservation plans), and green (low forest conservation value where land-use change is allowed).

At the national level, during the 2010–2014 period, more than 70% of the budget has been destined to SPS plans in yellow areas. Furthermore, articulation of public policies for silvopastoral development has been developed in a joint institutional agreement between the Ministry of Agriculture, Livestock, and Fisheries and the Ministry of Environment and Sustainable Development (SAyDS) and the National Agricultural Institute (INTA). This general agreement named "National Forest Management with Integrated Livestock (MBGI)" aims mainly to (i) contribute sustainable use of native forests as a tool of development and according to sustainability criteria and minimum standards established by Law No. 26,331 and (ii) strengthen the provinces by promoting capacity building for implementing MBGI plans and (ii) establish a monitoring system. To expand silvopastoral land-use systems and farmer adoption, a multiagency, interdisciplinary, and participatory strategy is required.

In Chile, there are mainly two state regulations related to incentives for the establishment and management of agroforestry systems. The System of Incentives for the Agro-environmental Sustainability of Degraded Soils (SIRSD-S) aims to recover the productive potential of degraded agricultural soils. The Institute of Agricultural Development (INDAP) and the Agricultural Livestock Service (SAG) are the institutions of the Ministry of Agriculture responsible of the administration of this financing instrument to different farmers according to their production size. In relation to agroforestry systems, the financing instrument incentives to establish trees cover under SPS use for soil conservation.

The objectives of Law N° 20,283 of Recovery of the Native Forest and Forest Development are the protection, recovery, and improvement of native forests, in order to ensure forest sustainability and environmental policy. It was promulgated in 2008 and contains regulatory and promotional aspects for the management and recovery of the Chilean native forest. The National Forestry Corporation (CONAF) is the institution responsible for its implementation. Unfortunately, Law N° 20,283 does not consider financing the implementation of SPS in native forest due to the lack of information on natural regeneration. However, there is the Native Forest Research Fund (FIBN) that finances competitive research projects to incorporate improvements to Law 20,283 including the topic forests and livestock.

Finally, there are several direct drivers that affect the operation and structure of SPS (Fig. 6.3). The study and description of the main forest disturbances in northern Patagonia (39–41° S) have been reported previously (González et al. 2014; Lara et al. 2014; Srur et al. 2020). The main natural disturbances in these forests are fire, volcanism, landslides, snow avalanches, wind, droughts, forest decay, insect's population explosions, and bamboo blooms (González et al. 2014). Fire has been used as an instrument of agricultural management in humid forests or hunting in the ecotone between the forest and the steppe by the native communities, being a driver of land-use change (Willis 1914; Kitzberger and Veblen 1999) and, recently, urban development. The most important anthropic disturbance factors in the region are deliberate fires, overgrazing (mainly cattle and sheep), invasion of exotic plant and animal species, and timber overexploitation (Veblen et al. 1992b; Relva and Veblen 1998: Rovere 2008; Rovere et al. 2014). The increase of urbanization in Patagonia (Lantschner et al. 2008) is another disturbance factor that modifies ecosystems and introduces numerous invasive exotic species (Rovere et al. 2014). Moreover, there is an increase demand of pristine land by wealthy foreigners with the logic of private reserves or extensive livestock production. This concentration in few landowners generates a process of local emigration to urban areas with a decrease in their life quality.

5 Ecosystem Services in Silvopastoral Systems of Andean North Patagonia

The Andean-Patagonian forests sustain a wide variety of ecosystem services (ES) including provisioning, regulating and supporting, and cultural. In general, SPS in ñire forests have been the most studied, probably because they represent the main forest type where livestock farming takes place. In northern Patagonia in particular, the high use intensity of ñire forests has triggered degradation processes due to overuse of livestock and wood harvesting (Rusch and Varela 2019). ES and the associated benefits in SPS with different use intensities occurring in forest types of northwestern Patagonia are shown in Table 6.1.

Harvestable wood and meat production constitute the main provisioning services in SPS. Lenga and ciprés de la cordillera forests represent the main timber resources in the northern region of the Andean-Patagonian forests. In the case of ñire forests, firewood, poles, and rods (maintenance of fences, construction of paddocks and small buildings) are the most important products. The use of firewood has also been strongly linked to the regional development and the establishment of rural and urban communities. For example, the average biomass firewood consumption (for heating, cooking, fuel) in a Patagonian rural community is 12,000 kg/year/home (1479 kg/year/person), with significantly higher values during winter (63 kg/day/home) than in summer (18.5 kg/day/home) (Morales et al. 2018).

Table 6.1 Identification of main ecosystem services (ES) deriving from SPS in Andean North Patagonia, based on Common International Classification of Ecosystem Services (CICES) Version 5.1

Specific ES	Division	Simple descriptor (proxy)	Service	Goods and benefits	Reference
Provisioning	Biomass	Materials from wild plants used for energy	Volume of harvested wood	Processed timber Fuel wood Fencing construction	Morales et al. (2018)
	Biomass	Livestock grazing outdoors and/or raised in housing	Increase in weight or numbers of cattle herd per year	Meat produced at abattoir, eggs, milk sold on farm or in shops, wood	Rusch et al. (2019); Salinas et al. (2017)
	Biomass	Food from wild plants (leaves, flowers, fruits, roots, bulbs, stems, exudates, barks and seeds)	Harvestable volume of wild plants and fungi	Berries and fungi as food. Fruits for the production of jam, infusions. Fiber for dyes. Leaves for the production of infusion, resins	Morales and Ladio (2009); Salinas (2019); Rusch et al. (2017); Chillo et al. (2018); Chamorro and Ladio (2020)
	Water	Drinking water from surface, ground, and belowground	Volume and characteristics of water from rivers, creeks, and aquifers	Potable water, mineral water, reduced energy costs (irrigation)	
Regulating and supporting	Regulation of physical, chemical, biological conditions	Controlling or preventing soil loss	The capacity of vegetation to prevent or reduce the incidence of soil erosion	Reduction of damage and associated costs of sediment input to water courses	Chillo et al. (2018); Gyenge et al. (2019)
	Regulation of physical, chemical, biological conditions	Regulating the flows of water in our environment	The capacity of vegetation to retain water and release it slowly	Watershed protection	Gyenge et al. (2019); Quinteros (2018)
	Regulation of physical, chemical, biological conditions	Ensuring the organic matter in our soils is maintained	Decomposition of plant litter	Maintenance of soil formation and quality	Arias Sepúlveda and Chillo (2017); Chillo et al. (2018); Goldenberg et al. (2020)
	Regulation of physical, chemical, biological conditions	Providing habitats for wild plants and animals that can be useful to us	Important nursery habitats (forest patches, shrublands)	Sustainable populations of useful or iconic species that contribute to a service in another ecosystem	Blackhall et al. (2008); Chillo et al. (2018b); Quinteros (2018); Gyenge et al. (2019)

(continued)

Table 6.1 (continued)

Specific ES	Division	Simple descriptor (proxy)	Service	Goods and benefits	Reference
Cultural	Direct, in situ	Using the environment for sport and recreation, sports, etc.	Ecological qualities of woodland that make it attractive to hiker; private gardens	Recreation, fitness; de-stressing or mental health; nature-based recreation	
	Direct, in situ	Watching plants and animals where they live; using nature to distress	Mix of species in a woodland of interest to wildlife watchers	Agro-tourism, rock painting	
	Indirect, remote	Using nature as a national or local emblem	Spiritual and religious values. Places where religious ceremonies and various aboriginal traditions take place	Social cohesion, cultural icon	Ladio and Morales (2017)
	Indirect, remote	The things in nature that we want future generations to enjoy or use	Endangered species (huemul, pudú pudú) or habitats (Alerce forests)	Moral well-being. Sense of belonging	

Ñire, as a main source of heat energy for the region, has been gradually replaced by fossil energy. However, in some rural and peri-urban communities, ñire still represents the most important and valuable energy resource. According to surveys in rural and urban households, ñire is also one of the species with the highest commercial circulation at the regional level (Arre et al. 2015; Morales et al. 2017). In Chile, the livestock use opened forest as a result of partial cutting (Fig. 6.2) that increases forage availability (Sanchez et al. 2011; Salinas 2016). In addition, the ñire wood is used in a variety of household utensils such as baskets and containers, furniture, and instruments. These types of products are particularly important for small landowners, artisans, and indigenous communities (Salinas and Acuña 2017).

Livestock for meat production represents the most important use of the forest in the entire region in terms of area. Livestock (sheep and cattle) production in the Andean area of northern Patagonia benefit from ecosystems contribution by providing forage biomass (Rusch et al. 2019). The ñire forests are the most used forest type for livestock production in terms of stocking rates due to high forage biomass. In non-managed ñire forests located in the transition with the Patagonian steppe, native plants productivity is between 200 and 350 kg/DM/ha, and under SSP management, understory productions increase up to 2500–3300 kg/DM/ha (Salinas et al. 2017). However, there are various types that farmers use in seasonal movements of livestock depending on the availability of forage during the year (Cardozo 2014). The lenga forests are used for summer grazing and the middle-slope mixed ciprés-coihue forests during winter. In ñire and lenga forests, livestock grazing combined with firewood extraction affect natural regeneration by transforming the forest into shrubby grasslands (Quinteros et al. 2013; Quinteros 2018; Vila and Borrelli 2011; Rusch et al. 2017). In mixed ciprés-coihue forests, this negative effect on the natural regeneration of the canopy is not clear, mainly because during the summer cattle are moved to other environments (Amoroso et al. 2018).

Wild plants for nutrition, materials, or medicine have been the basis of the food and health systems of the local populations in the region. More than 500 species with important uses had been recognized, including leaves, flowers, fruits, roots, bulbs, stems, exudates, barks, and seeds (Tacón Clavaín 2004; Molares and Ladio 2009; Mattenet et al. 2015; Salinas 2019). Forests also provide plants for dyeing and aromatic use, as well as forages and edible fungi relevant to the material and spiritual culture of local communities. There have also been examples of cultivating seeds, cuttings, or transplanting useful native plants for domestication (Ladio and Molares 2017). However, there are no official records or legislation regarding to non-wood forest products. It is interesting to mention that grazing can favor the development and growth of edible fruit bushes, edible fungi (e.g., *Morchella* spp.), and/or ferns used as cut foliage (Rusch et al. 2017). Thus, sites under silvopastoral use may present high plants diversity with known cultural value and uses (Chillo et al. 2018). Furthermore, in ñire forests of Chile, 40 species are used for medicinal purposes, 36 species for food production, 16 species as ornamental, and 8 species for dyes (Salinas 2019). From these, michay (*Berberis microphylla*) and maqui (*Aristotelia chilensis*) are distinguished especially by their potential due to their antioxidant properties (Arena et al. 2017; Chamorro and Ladio 2020).

Regarding the regulating and supporting services, soil quality in ñire forests severely diminished due to water and wind erosion and soil compaction in overused paddocks. Overgrazing decreases the nitrogen content in the system due to plant cover decrease and the loss of endo- and ectomycorrhizae mainly in radal (*Lomatia hirsuta*) and notro (*Embothrium coccineum*) trees. Also, the decrease in vegetation cover due to overgrazing decreases litter quantity and quality (Gyenge et al. 2019). Conversely, decomposition rate is favored in mixed ciprés-coihue forests with intermediate intensities of silvopastoral use compared with closed forests (Chillo et al. 2018). This is mainly because canopy openings allow the development of a heliophytic understory plant community, characterized by herbaceous species with membranous leaves, higher N content, and low specific leaf area (Arias Sepúlveda and Chillo 2017).

The regulation service provided by vegetation in mountain areas is affected by silvopastoral use since tree removal modifies soil stability through changes in litter input (quality and quantity) for the formation of organic matter (Gyenge et al. 2019; Chillo et al. 2018; Goldenberg et al. 2020). In forests with different site productivity, it has been shown that the increase in the intensity of firewood extraction modifies understory plant taxonomic and functional diversity vegetation cover by increasing the soil exposure to wind and drought. Because more exposed soils are prone to compaction and moisture loss, intensity management interventions should be adjusted to avoid degradation processes. In other forests of Patagonia, Bahamonde et al. (2015) highlighted that one of the main benefits of SPS is a greater resilience of the soil to the loss of nutrients compared to grasslands. In Chile, SPS have also been recognized in the improvement of goods and services associated with soil nutrition and microbial activity (Alfaro et al. 2018; Ortiz et al. 2020).

Regarding the regulation of water flow, overgrazed ñire forests reduced soil water retention capacity and increases surface runoff due to trampling and compaction that decreases soil porosity (Gyenge et al. 2019). In the areas where cattle are concentrated, water quality in lenga forests located at the head of the watershed is affected by increasing suspended sediments by 72% and phosphorus by 50% (Quinteros 2018).

Habitat provision by SPS for wild plants and animals can be considered as a proxy of biodiversity conservation. Grazing promotes the increases abundance of opportunistic herbaceous species (native and exotic) and reduces richness and abundance of palatable plants that are not tolerant to herbivory and/or increased light (Blackhall et al. 2008; Chillo et al. 2018b; Quinteros 2018; Gyenge et al. 2019). Canopy openings that increase light availability for the understory not always imply a decrease in biodiversity but rather changes in the botanical composition associated with different intensities of use (Chillo et al. 2020). In mixed evergreen forests, intermediate intensities of use increased specific and functional diversity, and this change in the composition of species generates changes in the biophysical provision of ES (Chillo et al. 2018, 2018b).

On the other hand, there is a knowledge gap regarding the role of SPS (synergies or trade-off) in supporting habitat for specialist forest species. For example, some authors described as indirect effect of livestock use the predation of dogs on native fauna, mainly on endangered species such as the iconic huemul (*Hippocamelus bisulcus*) (Silva et al. 2011; Muñoz-Santibañez and Muñoz 2016). In ñire forests of Southern Patagonia, the increase in coarse woody debris in the soil from harvesting provided habitat for small rodents and insects or even promotes trophic interactions between microbial fauna and insects (Gonzalez-Polo et al. 2013). Additionally, open forests with livestock use benefited the group of granivorous and migratory birds, unlike the group of birds that nest in cavities (Benitez et al. 2019). In the case of plants, the provision of habitat would be favorable for those species that tolerate intermediate levels of radiation (Chillo et al. 2018). However, considering the key role of biodiversity conservation to support the processes and functions that sustain ES (Daily 1997), we need to improve the understanding of the relationships between biodiversity and ES.

Finally, there are cultural services such as the areas for recreation, leisure, outdoor sports, local and international tourism, aesthetic value, and enjoyment. It also refers to those landscapes that are fundamental for the maintenance of the local cultural identity, for example, places for religious ceremonies of original people.

The Patagonian indigenous peoples conceive forests as their home as an inseparable part of their cosmology (Ladio and Molares 2017). Recently, there has been an increase of agro-tourism in private lands the properties (mainly camping and rafting spots), but still remains unclear the local and regional impact. In this context, local ecological knowledge (LEK) assist producers to better nature management (management of water, soil, biota, animal health), governance, and resilience (Cardoso et al. 2015; Castillo and Ladio 2017).

6 Main Local Drivers and Knowledge Gaps

The introduction and further expansion of livestock have been related to the occupation of the territory through appropriation models and based on the supply of natural resources like timber forest products. However, both components (forest and livestock) of SSP have been mainly managed separately. It is therefore necessary to work on a systemic vision of SPS management by considering productive and social objectives, relational values, and the interaction between components (forests, livestock, and forages). For this, the socio-ecosystem model proposed here for SPS may be useful for managers and decision-makers, since it considers the different socio-historical, cultural, and environmental contexts. An important aspect that makes the resulting relational model even more complex is the presence of aboriginal communities and large landowners in National Park areas. Because the socio-historical and cultural contexts are changing, it is necessary to analyze the interaction dynamics according to new legislation and the tensions between different models. The comprehensive analysis of the SPS from multiple social-ecological contexts allowed us to identify the main knowledge gaps for the comprehensive management of multiple ES in this complex territory. One of the knowledge gaps detected is the need to recognize and understand the specific relational models to each forest type in SPS of the northern Argentine-Chilean Patagonia. Design new SPS from an integral vision is needed to integrate forest and livestock production with the integrity of multifunctional ecosystems and the well-being of the producer and associated communities.

There is lack of information related to synergies and trade-offs between provision, regulation, support, and cultural ES for the different forest types under SPS use. Also, our knowledge regarding provision ES (firewood, timber, and non-wood products) in SSP had been reported isolated. We need further research to determine grasslands evaluation for stocking rate adjustment and animal supplementation for reducing the browsing effect on regeneration. Still we need to address the following questions: How interventions on the socio-ecosystem can be conducted to achieve favorable changes in SPS management? How forest resources can be managed to obtain structural diversity in the landscape that allows a greater provision of firewood and forage but in turn contributes to biodiversity and forest regeneration? What anthropogenic aspects can be linked to multiple ES management objectives in SPS?

Research has focused on supporting and regulation ES on forest types (mainly in ñire forests) at stand level, without a watershed or landscape view. Thus, information related to water regulation, biodiversity, erosion prevention, and soil fertility at the watershed level are needed. For example, riparian and wetland environments provide multiple ES (provision, regulation, habitat) within these socio-ecosystems and are poorly represented in our current knowledge and research. While traditional livestock farming deteriorated river beds and streams, there are few actions in conjunction with rural residents for river corridors protection or soil recovery through restoration. There is also a knowledge gap regarding habitat provision for specialist species under different SPS intensities at a landscape level. This information is needed for the understanding of the conservation values of SPS compared to more intensive production systems.

Finally, we need to detect and know how regulation and supporting ES are modified by different cultural contexts and relational values. Key questions to be explored are: Does similar socio-historical and cultural contexts determine similar relational values? How different anthropogenic contexts modify the generation, perception, and provision – both local and regional – of regulation and supporting ES? What is the relative importance of the different socio-ecosystem contexts in the generation, perception, and provision of ES in SPS?

7 Recommendations for Decision-Making Based on the Sustainability of Silvopastoral Systems

The consideration of local actors in the decision-making and management processes of the territory is of great importance due to its socio-ecological complexity. For example, we have shown that relational values and anthropogenic aspects are key drivers in the production and provision of ES. Therefore, it is important to achieve a participatory technical-scientific process with local communities for planning. For example, Del Castillo et al. (2019) reported that only 39% of social actors included the knowledge of non-experts from local communities in the context of ecosystem services assessment. Thus, it is essential to include rural inhabitants in the co-generation of information together with the academia, since knowledge come from different sources and perceptions. This may contribute to address specific problems with integrated information on these complex socio-ecosystems.

On the other hand, historically, this region is under a strong conservation policy. For example, with the creation of the National Parks in Argentina (Nahuel Huapi in 1934 and later Lanin, Alerces, and Lago Puelo) and provincial and municipal protected areas, more than 35% of the territory is under different degrees of protection and conservation regime. This protection implies a limitation for local people (that inhabited the land before the creation of the reserves) in SPS implementation, as the grazing areas for livestock had been reduced (no seasonal movements are allowed) and impacted their life quality. In north Andean Patagonia, the paradigm of land sparing implies that land-use intensification occurs outside protected areas with strong competition between urbanization, SPS, and tourism activities. We argue that considering SPS in a complex territorial matrix as proposed in our model implies that proper and sustainable management of SPS can lead to multifunctional landscapes (Soler and Chillo 2018), where generation of provision ES does not generate trade-offs with support and regulating and cultural services at the landscape scale. The agroecological paradigm for the territorial planning is more appropriate for sustainable development in this region (Perfecto and Vandermeer 2012).

In this context, the principles and guidelines for Forest Management with Integrated Livestock (MBGI) contemplate an adaptive management strategy based on a comprehensive vision of the environment by balancing the productive potential of livestock and forestry with the capacity and integrity of forests and the preservation of their services and maintaining and improving the well-being of producers and associated communities. To achieve this policy, there is an urgent need to generate a baseline at the regional level regarding the conservation condition, productive forestry and livestock activities with respective economic analysis, and characterization of local actors involved in decision-making for the use of resources. This territorial baseline based on the description of the current physical, biotic, social, and economic environment and the development of criteria and indicators to define reference states and transition thresholds will allow the elaboration of clear guidelines for livestock and forestry sustainable management in the region.

Bibliography

Alfaro M, Dube F, Zagal E (2018) Soil quality indicators in an Andisol under different tree covers in disturbed Nothofagus forests. Chilean J Agric Res 78:106–116

Amoroso MM, Chillo V, Alcala V et al (2018) Efecto del manejo silvopastoril sobre la estructura y dinámica poblacional de bosques mixtos de ciprés de la cordillera (*Austrocedrus chilensis*) y coihue (*Nothofagus dombeyi*). Ecosistemas 27:33–40

Arena ME, Postemsky PD, Curvetto NR (2017) Changes in the phenolic compounds and antioxidant capacity of *Berberis microphylla* G. Forst. berries in relation to light intensity and fertilization. Sci Hortic 218:63–71

Arias Sepúlveda JE, Chillo V (2017) Cambios en la diversidad funcional del sotobosque y la tasa de descomposición frente a diferentes intensidades de uso silvopastoril en el noroeste de la Patagonia, Argentina. Ecol Austral 27:29–38

Arre J, Molares S, Ladio AH, Kutschker K (2015) Etnobotánica de las plantas leñateras y su circuito comercial en una ciudad de la Patagonia Argentina. Gaia Scientia 9:41–48

Bahamonde H, Peri PL, Martínez Pastur G, Monelos L (2015) Litterfall and nutrients return in *Nothofagus antarctica* forests growing in a site quality gradient with different management uses in Southern Patagonia. Eur J For Res 134:113–124

Benitez J, Barrera MD, Blazina AP, Lencinas MV (2019) Cambios en la diversidad funcional de la comunidad de aves en bosques de *Nothofagus antarctica* de Tierra del Fuego bajo uso silvopastoril. En: Actas del VI Congreso Nacional de Conservación de la Biodiversidad, La Rioja

Blackhall M, Raffaele E, Veblen TT (2008) Cattle affect early post-fire regeneration in a *Nothofagus dombeyi–Austrocedrus chilensis* mixed forest in northern Patagonia, Argentina. Biol Conserv 141:2251–2261

Bondel CS (2008) Transformaciones territoriales y análisis geográfico en ámbitos patagónicos de montaña. La Comarca Andina del Paralelo 42. Tesis doctoral en geografía. Facultad de Humanidades y Ciencias de la Educación. Universidad Nacional de La Plata. La Plata, Argentina

Cardoso MB, Ladio A, Dutrus S, Lozada M (2015) Preference and calorific value of fuelwood species in rural populations in northwestern Patagonia. Biomass Bioenergy 81:514–520

Cardozo AG (2014) Estrategias socio-productivas de establecimientos ganaderos del sudoeste de la provincia de Río Negro, Argentina. Tesis de maestría en Desarrollo Rural. Escuela Para Graduados, Facultad de Agronomía, Universidad de Buenos Aires, Argentina

Cardozo AG (2019) Heterogeneidad socio-productiva de las prácticas de los productores ganaderos en el SO de Río Negro. IV Congreso Internacional Agroforestal Patagónico. 22–26 Abril 2019. Ushuaia, Argentina

Castillo L, Ladio AH (2017) Mammals and birds as ethno-indicators of change: their importance to livestock farmers in Arid Patagonia (Argentina). Environ Dev Sustain. https://doi.org/10.1007/s10668-017-9983-z

Castillo L, Rostagno M, Ladio AH (2020) Ethnoindicators of environmental change: local knowledge used for rangeland management among smallholders of Patagonia. Rangel Ecol Manage 73:594–606

Chamorro M, Ladio AH (2020) Native and exotic plants with edible fleshy fruits utilized in Patagonia, and their role as sources of local functional foods. BMC Complement Altern Med 20:1–16

Chillo V, Amoroso MM, Rezzano C (2018b) La intensidad en el uso silvopastoril modifica la provisión de servicios ecosistémicos através de cambios en la diversidad en bosques del noroeste de la Patagonia Argentina. Ecosistemas 27:75–86

Chillo V, Goldemberg M, Pérez-Méndez N, Garibaldi L (2020) Diversity, functionality, and resilience under increasing harvesting intensities in woodlands of northern Patagonia. For Ecol Manag 474:118349

Chillo V, Vázquez DP, Amoroso MM, Bennett EM (2018) Land-use intensity indirectly affects ecosystem services mainly through plant functional identity in a temperate forest. Funct Ecol 32:1390–1399

Daily GC (1997) Nature's services. Island Press, Washington, DC

Del Castillo D, Di Pasquo FM, Busan TE et al (2019) What role do social actors play in the context of ecosystem services? A review in areas of ecology and conservation biology. Sustentabilidade em Debate 10:116–131

Dezzotti A (1996) *Austrocedrus chilensis* and *Nothofagus dombeyi* stand development during secondary succession, in northwestern Patagonia, Argentina. For Ecol Manag 89:125–137

Donoso C (1993) Bosques templados de Chile y Argentina. Variación, estructura y dinámica. Ecología Forestal, Editorial Universitaria, Santiago de Chile, p 484

Donoso C, Steinke L, Premoli A (2006) *Nothofagus antarctica* (G. Forster) Oerst. In: Donoso Zegers C (ed) Las especies arbóreas de los bosques templados de Chile y Argentina. Autoecología. Marisa Cuneo Ediciones, Valdivia, Chile, pp 401–410

Easdale M (2007) Los sistemas agropecuarios en los valles cordilleranos de Patagonia Norte y su posible evolución. Cuadernos de Desarrollo Rural 58:11–35

FIA (2016) Agenda de Innovación Agraria Territorial, Región de Aysén. Santiago, Chile, p 129

Frangi JL, Barrera MD, Puig de Fábregas J et al (2004) Ecología de los bosques de Tierra del Fuego. In: Arturi MF, Frangi JL, Goya JF (eds) Ecología y manejo de bosques nativos de Argentina. Editorial Universidad Nacional de La Plata, La Plata, pp 1–88

Goldenberg MG, Oddi FJ, Gowda JH, Garibaldi LA (2020) Effects of firewood harvesting intensity on biodiversity and ecosystem services in shrublands of northern Patagonia. For Ecosyst 7:47

González M, Donoso Zegers C, Ovalle P, Martínez Pastur GJ (2006) *Nothofagus pumilio* (Poepp. et Endl) Krasser – lenga, roble blanco, leñar, roble de Tierra del Fuego – Familia: Fagaceae. In: Donoso Zegers C (ed) Las Especies arbóreas de los Bosques Templados de Chile y Argentina: Autoecología

González ME, Amoroso MM, Lara A et al (2014) Ecología de disturbios y su influencia en los ecosistemas forestales templados de Chile y Argentina. In: Donoso C, González ME, Lara A, Donoso P (eds) Ecología Forestal: Bases para el Manejo Sustentable de los Bosques Nativos. Marisa Cuneo Editores, Valdivia

Gonzalez-Polo M, Fernández-Souto A, Austin AT (2013) Coarse woody debris stimulates soil enzymatic activity and litter decomposition in an old-growth temperate forest of Patagonia, Argentina. Ecosystems 16:1025–1038

Guitart Fité E (2008) Caracterización de la ganadería bovina en Patagonia Sur. INTA Esquel. Carpeta Técnica, Economía N° 9, Abril 2008. EEA INTA Esquel

Gyenge J, Rusch V, Weigandt M, Varela S (2019) Capitulo IV. Aspectos ambientales pp 101–125. In: Rusch VE, Varela SA (Comp) 2019. Bases para el manejo de bosques nativos con ganadería en Patagonia Norte. Parte I. Ediciones INTA, 160pp

Hepp C (2014) Caracterización agroclimática de la región de Aysén. In: Hepp C, Stolpe NB (Eds) C.aracterización y propiedades de los suelos de la Patagonia occidental (Aysén). Instituto de Investigaciones Agropecuarias, Centro de Investigación INIA Tamel Aike, Chile. pp. 15–34

Hepp C (ed) (2019) Sistemas de producción de bovinos de carne en la Patagonia húmeda. Instituto de Investigaciones Agropecuarias, Centro de Investigación INIA Tamel Aike, Chile, 134 pp

IICA, ODEPA, INDAP (2006) Pequeña agricultura en Chile: rasgos socioproductivos, institucionalidad y clasificación territorial para la innovación. Apey A, García BA (Eds). Santiago, Chile, 148 pp

INE (2007) Censo Agropecuario y Forestal 2007, resultados por comuna. Disponible en Internet: http://www.ine.cl/canales/chile_estadistico/censos_agropecuarios/censo_agropecuario_07_comunas.php/

Kitzberger T, Veblen TT (1999) Fire-induced changes in northern Patagonian landscapes. Landsc Ecol 14:1–15

Ladio AH (2017) Ethnobiology and research on global environmental change: what distinctive contribution can we make? Ethnobiol Conserv 6. doi: https://doi.org/10.15451/ec2017076.718

Ladio AH, Molares S (2014) El paisaje patagónico y su gente. In: Raffaele E, de Torres CM et al (eds) Ecología e Historia Natural de la Patagonia. Fundación de Historia Natural Félix de Azara, Buenos Aires, pp 205–223

Ladio AH, Molares S (2017) Etnoconservacionismo y prácticas locales en Patagonia: avances y perspectivas. In: Casas A, Torres-Guevara J, Parra F (eds) Domesticación en el Continente Americano. Historia y perspectivas del manejo de recursos genéticos en el Nuevo Mundo. Universidad Agraria La Molina. IIES, Lima, Perú, pp 649–672

Lantschner MV, Rusch V, Peyrou C (2008) Bird assemblages in pine plantations replacing native ecosystems in NW Patagonia. Biodivers Conserv 17:969–989

Lara A, Amoroso MM, Bannister J et al (2014) Sucesión y Dinámica de Bosques Templados en Chile. In: Donoso C, González M, Lara A, Donoso P (eds) Ecología Forestal: Bases para un manejo sustentable de los Bosques Nativos. Marisa Cuneo Editores, Valdivia

Mattenet F, Goyheneix M, Peri PL (2015) Tintes Naturales de Plantas Nativas: Colores de la Patagonia. Ediciones INTA. Secretaría de Agricultura Familia-Universidad Nacional de la Patagonia Austral, Buenos Aires, p 64

Matthies AL, Stamm I, Hirvilammi T, Närhi K (2017) Ecosocial innovations and their capacity to integrate ecological, economic and social sustainability transition. Sustainability 11:2107. https://doi.org/10.3390/su11072107

Molares S, Ladio AH (2009) Ethnobotanical review of the Mapuche medicinal flora: use patterns on a regional scale. J Ethnopharmacol 122:251–260

Molares S, Ladio AH (2012) Mapuche perceptions and conservation of Andean Nothofagus forests and their medicinal plants: a case study from a rural community in Patagonia, Argentina. Biodivers Conserv 21:1079–1093

Morales D, Molares S, Ladio AH (2017) Fire resource management in different landscapes in NW Patagonia. Front Ecol Evol 5:111

Morales D, Molares S, Ladio AH (2018) Seasonal variation in the consumption of biomass fuel in a rural community of arid Patagonia, Argentina. Ethnobiol Conserv 8:2. https://doi.org/10.15451/ec2019-01-8.02-1-23

Muñoz-Santibañez P, Muñoz AE (2016) Conflictos entre fauna silvestre y agricultura en Chile. Agronomia y Forestal 53:10–17
Muradian R, Pascual U (2018) A typology of elementary forms of human-nature relations: a contribution to the valuation debate. Curr Opin Environ Sustain 35:8–14
Ortega H, Brüning A (2004) Aysén. Panorama histórico y cultural de la XI región. Ediciones LOM, Chile. Disponible en http://www.aisenpanorama.cl/libro.htm (07/09/2015)
Ortiz J, Dube F, Neira P et al (2020) Soil quality changes within a (*Nothofagus obliqua*) forest under silvopastoral management in the Andes Mountain Range, South Central Chile. Sustainability 12:6815
Pascual U, Balvanera P, Díaz S et al (2017) Valuing nature's contributions to people: the IPBES approach. Curr Opin Environ Sustain 26–27:7–16
Perfecto I, Vandermeer J (2012) Separación o integración para la conservación de biodiversidad: la ideología detrás del debate "land-sharing" frente a "land-sparing". Ecosistemas 21:180–191
Quinteros CP, Feijóo MS, Arias N et al (2013) Dieta de verano de bovinos pastoreando en bosques de lenga (*Nothofagus pumilio*) y mallines de Chubut, Argentina. Revista de la Facultad de Ciencias Agrarias 45:285–292
Quinteros CP, López Bernal P, Gobbi M, Bava J (2012) Distance to flood meadows as a predictor of use of *Nothofagus pumilio* forest by livestock and resulting impact, in Patagonia, Argentina. Agrofor Syst 84:261–272
Quinteros P (2018) Propiedades ecosistémicas de bosques de *Nothofagus pumilio* afectadas por diferente intensidad de uso ganadero en Chubut, Patagonia Argentina. Revista Ecosistemas 27:24–32
Relva MA, Veblen TT (1998) Impacts of large herbivores on *Austrocedrus chilensis* forests in Patagonia, Argentina. For Ecol Manag 108:27–40
Roseland M (2000) Sustainable community development: integrating environmental, economic, and social objectives. Prog Plan 54:73–132
Rovere A (2008) Ensayo de restauración con *Austrocedrus chilensis* (Cupressaceae) en Patagonia, Argentina. In: González-Espinosa M, Rey Benayas JM, Ramírez-Marcial N (eds) Restauración de bosques en América Latina. MundiPrensa, Mexico City, pp 6–21
Rovere AE, Martínez Pastur G, Anderson C, et al (2014) Ecorregión: Bosques Patagónicos. II Simposio de Restauración Ecológica en Argentina: enfoques y prioridades. Universidd Maimónides, pp. 32
Rozzi R (2012) Biocultural ethics: recovering the vital links between the inhabitants, their habits, and habitats. Environmental Ethics 34:27–50
Rusch V, López DR, Cavallero L, Rusch GM, Garibaldi LA, Grosfeld JE, Peri PL (2017) Modelo de Estados y Transiciones de los ñirantales del NO de la Patagonia como herramienta para el uso silvopastoril sustentable. Ecol Austral 27:266–278
Rusch VE, Fariña C, Borrelli L, Cardozo AG (2019) Capítulo III. Los componentes forrajeros, sotobosque. Páginas 70–100. In: Rusch VE, Varela SA (Comp) 2019. Bases para el manejo de bosques nativos con ganadería en Patagonia Norte. Parte I. Ediciones INTA, 160pp
Rusch VE, Varela SA (Comp) (2019) Bases para el manejo de bosques nativos con ganadería en Patagonia Norte. Parte I. Ediciones INTA, 160pp
Salinas J (2016) Experiencia de manejo silvopastoral en dos renovales coetáneos de *Nothofagus antarctica* (G.Forst.) Oerst. (Ñirre) en la Región de Aysén. In: Barros S (ed) Revista Ciencia e Investigación Forestal 22(1). Instituto Forestal, Chile
Salinas J (2019) Investigación de PFNM de los bosques de Ñirre (*Nothofagus antarctica*) de la zona sur austral de Chile. Informe técnico Instituto Forestal, Chile, p 50
Salinas J, Acuña B (2017) Protección Individual de Tocones; Una Estrategia para la Continuidad del Bosque de Ñirre (*Nothofagus antarctica*). Instituto Forestal, Chile. Documento de Divulgación N° 42., p 16
Salinas J, Peri P, Hepp C, Acuña B (2017) Sistemas Silvopastorales en Bosques de Ñirre (*Nothofagus antarctica* (G.Forst.) Oerst.) de la Región de Aysén. Instituto Forestal, Chile. Documento de Divulgación N° 43, p 60

Sanchez L, del Pozo A, Acosta B et al (2011) Hacia un sistema de uso silvopastoral con árbol nativo. In: Barros S (ed) Revista Ciencia e Investigación Forestal 17(2). Instituto Forestal, Chile, pp 175–187

Silva C, Repetto F, Droguett D, et al (2011) Actas de Taller: Hacia un plan para la conservación del Huemul *Hippocamelus bisulcus* (Molina, 1782) en la zona austral de Chile. Nov. 11-12/2010, Punta Arenas, Chile

Soler R, Chillo V (2018) Sinergias y antagonismos entre manejo agroforestal y conservación en paisajes multi-funcionales en Latinoamérica. Ecosistemas 27:1–3

Sotomayor A, Barros S (eds) (2016) Los Sistemas Agroforestales en Chile. Instituto Forestal, Chile, p 458

Srur A, Amoroso MM, Mundo I et al (2020) Forest dynamics in the Argentinean Patagonian Andes. Lessons learned from dendroecology. In: Pompa-Garcia M, Camarero JJ (eds) Latin American Dendroecology: combining tree-ring sciences and ecology in a mega diverse territory. Springer

Suarez ML, Kitzberger TT (2010) Differential effects of climate variability on forest dynamics along a precipitation gradient in northern Patagonia. J Ecol 98:1023–1034

Tacón Clavaín A (2004) Manual de productos forestales no madereros. CIPMA, Valdivia, Chile, p 64

Veblen TT, Kitzberger T, Lara A (1992a) Disturbance and forest dynamics along a transect from Andean rain forest to Patagonian shrubland. J Veg Sci 3:507–520

Veblen TT, Mermoz M, Martin C et al (1992b) Ecological impacts of introduced animals in Nahuel Huapi National Park, Argentina. Cons Biol 6:71–83

Veblen TT, Kitzberger T, Rebertus A (1996) Perturbaciones y dinámica de regeneración en bosques andinos del Sur de Chile y Argentina. In: Armesto JJ, Villagrán C, Arroyo MK (eds) Ecología de los bosques nativos de Chile. Editorial Universitaria, Santiago de Chile, pp 169–197

Vila A, Borrelli L (2011) Cattle in the Patagonian forests: feeding ecology in Los Alerces National Reserve. For Ecol Manag 261:1306–1314

Villalba R, Veblen TT (1997) Regional patterns of tree population age structure in northern Patagonia: climatic and disturbance influences. J Ecol 85:113–124

Willis B (1914) El norte de la Patagonia. Comisión de estudios hidrológicos, Buenos Aires

Chapter 7
Ecosystem Services Values of the Northwestern Patagonian Natural Grasslands

Luciana Ghermandi and Sofía L. Gonzalez

Abstract Achieving sustainable development is an urgent challenge for nations that not only must ensure the maintenance of the productivity of natural resources but also must know the environmental, economic, and social costs that their exploitation entails. The scarce knowledge and inadequate policies have diminished the capacity of ecosystems to sustain productivity and benefits that they provide, putting at serious risk the life quality of the human beings that inhabit them. The scientific advance and the population awareness have made the use of northwestern Patagonian natural grasslands for agricultural, forest exotic plantations, livestock, mining, and real estate more problematic without evaluating the effects, often irreversible that accompany them. We are aware of the debate around the term ecosystem services, which we believe helps to make visible the complex environmental problems and should always take into account human needs. This chapter refers to the main services provided by the northwestern Patagonia natural grasslands to discuss some of their ecological, economic, and social dimensions. Based on our knowledge of 30 years of this ecosystem, we want to contribute with elements for future generations who will be heirs of this land and its benefits, elements that, at the same time, are of interest to those responsible for planning the territory.

Keywords Aesthetic values · Biomass production · Desertification · Overgrazing · Soil erosion

---

L. Ghermandi (✉) · S. L. Gonzalez
Instituto Nacional de Investigaciones en Biodiversidad y Medioambiente (INIBIOMA), Universidad Nacional del Comahue (UNCOMA), Consejo Nacional de Investigaciones Científicas y Técnicas (CONICET), San Carlos de Bariloche, Río Negro, Argentina
e-mail: sgonzalez@comahue-conicet.gob.ar

P. L. Peri et al. (eds.), *Ecosystem Services in Patagonia*, Natural and Social Sciences of Patagonia, https://doi.org/10.1007/978-3-030-69166-0_7

1 The Ecological-Socioeconomic or the Food Value of the Natural Ecosystems

At present, it is impossible to separate the economic from the social and ecological concepts, although each discipline can see its focus as a priority (Lankoski and Thiemb 2020). According to the classification of the ecosystem services (ES) (see Chap. 1), the presence and maintenance of biodiversity can be considered as a support service, while food production is considered as a provisioning ES. In the northwest of Patagonia (Río Negro Province), there are the most productive ranches in the region because they are located in a sector of the west-east precipitation gradient that is still favorable to the production of biomass to feed cows and sheep (Easdale and Madariaga 2009; Easdale and Aguiar 2012). The abovementioned provisioning and supporting ES overlap; this means that they must found a balance between them, a balance that allows people who live in the region do business in it or make political decisions to be in agreement with the management decisions that also must be sustained over time (Herrero et al. 2015).

One of the concepts of ecology most abused by politicians is the sustainability of an ecosystem, landscape, and region. "[…] It is in the hands of humanity to make development sustainable, that is, to ensure that it meets the needs of the present without compromising the ability of future generations to meet their own" (UN 1987), and in 2015, the organization of the United Nations transmitted a final document entitled "Transforming our world: the 2030 agenda for sustainable development" in which it was declared: "We are determined to liberate humanity from the tyranny of poverty and deprivation, and to heal and protect our planet. We are determined to take the bold and transformative steps that are urgently needed to set the world back on the path of sustainability and resilience" (UN 2015). There is no doubt that land use often involves the sacrifice of some ES in favor of others. A classic example is the monoculture over large areas, where soybean is grown; it is very difficult for the forest to return (Andrade-Díaz et al. 2019). In that case, the change in land use means a rather brutal sacrifice of various ES provided by the natural forests in the past. Even regarding the need to produce an enormous amount of soybean, the discussion is open. Reboratti (2010) has a critical vision: "Between 1980 and 2005 soybean production in Argentina grew over 15 million hectares, turning soybean and its derivatives into the main exportation good of the country. This expansion had many environmental, economic and social impacts that gave rise to a heated dispute."

In the grasslands of northwestern Patagonia, the principal land use is the extensive sheep grazing. The situation seems less dramatic in terms of loss of ES. However, extensive livestock farming without clear and precise indications on what stocking rate can be maintained over time has led to desertification, soil erosion, replacement of palatable species by other less preferred by cattle, loss of biodiversity, and invasion of exotic species (Gaitán et al. 2014; Franzese and Ghermandi 2014; Bidinost et al. 2020; Valenta et al. 2020).

These ecosystems are subject to natural disturbances such as fires caused by lightning storms and volcanic ash accumulation due to the presence of one of the most active chains of volcanoes in the world (in Chile, with prevailing winds from the west) (Ghermandi et al. 2004, 2013, 2015; Oddi and Ghermandi 2016). Livestock production must be considered in this framework and in an arid climatic context. A large part of Patagonia is called desert, e.g., this represents a great challenge if we want to maintain the greatest possible ecological, economic, and social value of the region (Gaitán et al. 2019).

As much as the historical vocation of livestock production is invoked, this vocation is just over a hundred years old (León and Aguiar 1985). The precariousness of land titles was a drag on regional development because of the lack of security of permanence discouraged all long-term investment, as well as the adoption of conservation management practices. On the contrary, the highest possible immediate gain was the goal, that is, overgrazing, which is the germ of desertification (Coronato 2015).

We are aware of the history of use of these lands, the resistance to change, the little presence of government plans sustained over time, and the climatic restrictions. However, we think that a detailed and participatory analysis of the possibilities of comprehensive development of northwestern Patagonia would be useful to carry out changes in view of a better future for the inhabitants of this isolated and semi-abandoned region (Nuñez et al. 2020). The decision is fundamentally political and is related to a desire for change from an extractive model to another that added value to the products that the earth generously grants us. It is not a question of agreeing or not with mining or afforestation. The problem to be solved is how to do good mining and how to cut and work in situ the pine trunks that now grow uncontrollably in Patagonia, responding to a logic of subsidies and not accompanied by the production of only timber products (Bava et al. 2015). Afforestation has a value if the silvicultural work is done and if the wood is transformed into boards and furniture, which enter in the commercial chain. If pines from plantations remain in the ecosystem, they are eventually burned in a fire as occurred in the San Ramón ranch, for example, in 1999 (Defossé and Dentoni 1999; Paritsis et al. 2018). In this context, the reaction of the population cannot be expected to be positive, being, in addition, that uncontrolled afforestation is foci of pine invasions (Franzese et al. 2017). When deciding to intervene in nature, there is no doubt that many management decisions have to be made. Nature alone cannot be expected to regulate situations that are artificially created by humans, supposedly for society's benefit.

2 Ecosystem Services Categories (MA, TEEB, and CICES) for Natural Grasslands at Northern Patagonia

Achieving sustainable development is an urgent challenge for nations that not only must ensure the maintenance of the productivity of natural resources but also must know the environmental, economic, and social costs that their exploitation entails. The scarce knowledge and inadequate policies have diminished the capacity of ecosystems to sustain productivity and benefits they provide, putting at serious risk the life quality of the human beings that inhabit them. The scientific advance and the population awareness have made the use of northwestern Patagonia grasslands for agricultural, forestry, livestock, mining, and real estate more problematic without evaluating the effects, often irreversible, that accompany them. We are aware of the debate around the term "ecosystem service" which, however, we believe helps to make visible the complex environmental problems, which should always take into account human needs. This chapter refers to the principal services provided by the northwestern Patagonia grasslands to discuss some of their ecological, economic, and social dimensions. We will present the aesthetic as cultural service, food as provisioning service, and soil as supporting service based on MA classification (Table 7.1) and the major threats that contribute to their degradation.

In Table 7.1, we show the classifications of ecosystem services categories used in the Millennium Ecosystem Assessment (MA), Economics of Ecosystems and Biodiversity (TEEB), and Common International Classification of Ecosystem Services (CICES). MA provides a classification that is globally recognized and used in sub-global assessments; TEEB provides an updated classification, based on the MA, which is used in ongoing national TEEB studies across Europe; and CICES provides a hierarchical system, building on the MA and TEEB classifications but tailored to accounting.

3 The Aesthetic Values of the Natural Grasslands

What do we experience when we walk through a Patagonian landscape? It probably depends on whether we see it as beautiful or "all the same" as European tourists, accustomed to more changing landscapes (van Zanten et al. 2014), tend to comment. Beauty is a good that escapes definitions, and, to overcome this difficulty, we speak of "tastes" (Dronova 2019). *De gustibus non est disputandum* (there's no accounting for taste) say a famous Latin proverb. However, all human beings love some things: symmetry, series, variations, tones, certain colors, and certain smells that indicate health or good condition of something (Evans et al. 2012). We like to see the horizon, the sky, and the clouds (not everyone, it is true, because there are agoraphobics). We like the transparent water that runs in the rivers and the calmer one in the lakes. We love the sea. The steppe distresses us because of its austere climate, with sunny summers, no trees in sight, and snowy icy winters. However, anyone who has seen the embroidery of frost on the vegetation in the Patagonian steppe knows that anguish is not the only thing that winter brings us (Hansson and Norberg 2009).

The scenic beauty can be a source of money, due to its tourist exploitation, but this is a secondary thing. The efforts of nations to preserve landscapes are also due to the desire to preserve their aesthetic value, not just the economy (Cooper et al. 2016). The economy, in general, is the cause of the destruction and degradation of the ecosystems, making the landscapes uniform (e.g., see the fields cultivated with soybeans in Argentina, if you have doubts about the latter; Fig. 7.1).

Table 7.1 Ecosystem services categories used in Millennium Ecosystem Assessment (MA), the Economics of Ecosystems and Biodiversity (TEEB), and Common International Classification of Ecosystem Services (CICES)

MA categories	TEEB categories	CICES categories
Food (fodder) (Provisioning service)	Food	Biomass (nutrition)[a] Biomass (materials from plants, algae, and animals for agricultural use)
Fresh water	Water	Water (for drinking purposes) (nutrition) Water (for non-drinking purposes) (materials)
Fiber, timber	Raw materials	Biomass (fibers and other materials from plants, algae, and animals for direct use and processing)
Genetic resources	Genetic resources	Biomass (genetic materials from all biota)
Biochemicals	Medicinal resources	Biomass (fibers and other materials from plants, algae, and animals for direct use and processing)
Ornamental resources	Ornamental resources	Biomass (fibers and other materials from plants, algae, and animals for direct use and processing) Biomass-based energy sources Mechanical energy (animal based)
Air quality regulation	Air quality regulation	(Mediation of) gaseous/air flows
Water purification and water treatment	Waste treatment (water purification)	Mediation (of waste, toxics, and other nuisances) by biota Mediation (of waste, toxics, and other nuisances) by ecosystems
Water regulation	Regulation of water flows/ moderation of extreme events	(Mediation of) liquid flows
Erosion regulation	Erosion prevention	(Mediation of) mass flows[a]
Climate regulation	Climate regulation	Atmospheric composition and climate regulation
Soil formation (supporting service)	Maintenance of soil fertility	Soil formation and composition[a]
Pollination	Pollination	Life cycle maintenance, habitat, and gene pool protection
Pest regulation Disease regulation	Biological control	Pest and disease control
Primary production Nutrient cycling (supporting services)	Maintenance of life cycles of migratory species (incl. nursery service) Maintenance of genetic diversity (especially in gene pool protection)	Life cycle maintenance, habitat, and gene pool protection Soil formation and composition (Maintenance of) water conditions Life cycle maintenance, habitat, and gene pool protection
Spiritual and religious values	Spiritual experience	Spiritual and/or emblematic

(continued)

Table 7.1 (continued)

MA categories	TEEB categories	CICES categories
Aesthetic values (cultural service)	Aesthetic information	Intellectual and representational interactions[a]
Cultural diversity	Inspiration for culture, art, and design	Intellectual and representational interactions Spiritual and/or emblematic
Recreation and ecotourism	Recreation and tourism	Physical and experiential interactions[a]
	Information for cognitive development	Intellectual and representational interactions Other cultural outputs (existence, bequest)

[a]Indicated the ecosystem services which were addressed in this chapter (source: Biodiversity Information System for Europe)

Fig. 7.1 (**a**) Farm in Salta Province deforested due to soybean cultivation (Source: Greenpeace). (**b**) Landscape of northwestern Patagonia (Río Negro Province) (Source: L. Ghermandi)

We could fully exploit Patagonia obtaining many provision services (e.g., oil, minerals, cows, sheep, and timber from exotic forest plantations), and we could use small land portions for intensive crops of fruits and vegetables under irrigation and protected from frost. In this way, the region would give us all the possible ecosystem services. We could even do things well, and not induce desertification processes, controlling the stocking rate and leaving the degraded grasslands at rest so that they recover and continue to give us infinitely what we want. However, many things tell us that this is not the way for a sustainable land use. We have to consider landscape beauty, and we cannot find it only in memories or on cinema, TV, PC, and smartphone screens. The aesthetic value of landscapes contributes to human well-being, but studies linking biodiversity and ecosystem services and the relationship among the aesthetic perception of landscapes, ecological value, and biodiversity are been scarcely analyzed. Society needs of this approach to change our culture of nature and promote more comprehensive conservation policies (Tribot et al. 2018).

4 The Grasslands ES Associated to Soils, Drivers, and Processes

Soils play a crucial role in ecosystem functioning and supplies important ES for human survival as provisioning services (e.g., food, fiber, energy, biodiversity), regulation services (e.g., decomposition and cycling of organic matter, nutrients, water infiltration, gas exchange), and supporting services as the provision of habitat (Pulleman et al. 2012; Lal et al. 2013; Baer and Birgé 2018; FAO 2020). Grassland ES associated with soils are the control of erosion and soil formation (Bengtsson et al. 2019) (Table 7.1). Control of erosion is a regulating ES that protects the soil from wind and water erosion (Fu et al. 2011) and is frequently estimated by soil conservation amount taking into account the actual and potential soil erosion of the grasslands (Fu et al. 2011). Soil formation is a supporting ES, and the maintenance of soil fertility is the priority (Zhao et al. 2020).

Several human activities are drivers that degrade grasslands, being overgrazing one of the most significant (Geist and Lambin 2004). Heavy stocking rates can affect the control of soil erosion ES by altering the structural conditions of the soils and reducing the cover vegetation increasing bare soil. Reduction of vegetation cover by overgrazing promotes soil erosion that enhances soil nutrient (C, N, and P) loss from topsoil layers, decreasing soil fertility and affecting other ES as climate regulation (Li et al. 2007; Peri et al. 2016). For example, the capacity of very severe grasslands to store in soil C, N, and P was 474%, 198%, and 56% less than lightly degraded grassland (Li et al. 2009).

Soil erosive processes are one of the most threats to soil ES. According to recent estimates, land area globally affected by wind erosion encompasses 549 M ha, 54.0% of which are severely affected (Lal 2003). In Argentina, the arid and semiarid regions cover 75% of the territory having these areas high risk of soil degradation. In the period 1950–2000, it was showed that 33.1% of the areas of the country presented a potential rate of wind erosion of more than 150 ton ha^{-1} $yr.^{-1}$ (Colazo et al. 2019). Most of the areas affected by wind erosion corresponded to Patagonia region. In Rio Negro Province, 61.1% of the territory is affected by wind erosion, being 18.5% slight, 70.7% moderate, and 10.8% severely affected (Table 7.2). Slight to moderate wind erosion levels prevail in the NW Patagonia (Bran et al. 2015). Erosive processes are favored by the intense winds (mean velocity of 30 km h^{-1}), which were more frequent during spring-summer (Godagnone and Bran 2009). In this region, the scarce presence of torrential summer rains dismisses the importance of water erosion. However, during the last years, the frequency of storms in summer-autumn was increased producing relevant erosive processes (Becker et al. 2012).

A great extension of Patagonia region is affected by desertification, a serious process of land degradation. In the 1990s, various national institutions elaborated the first map of desertification in Patagonia (del Valle et al. 1998). Of the whole area studied, 73.5 million ha (93.6%) had some signs of desertification, being 60% of the territory included in moderate to severe and severe categories. The Patagonian provinces that were included in these categories were in order: Santa Cruz (38%),

Table 7.2 Area of Río Negro Province affected by different levels of wind and water erosion (in hectares and percentage)

	Area	
Type and level of erosion	ha	%
Wind erosion		
Slight	2,300,016	11.33
Moderate	8,766,421	43.18
Severe	1,344,680	6.62
Water erosion		
Slight	462,126	2.28
Moderate	4,336,516	21.36
Severe	2,138,792	10.54
Total	19,348,551	95.31

Total area of the province is 20,301,200 ha. The total area in the table did not included water bodies (modified from Bran et al. 2015)

Neuquén (37%), Chubut (31%), and Río Negro (26%). Although currently there are no regulations related to control of soil erosion, existing programs evaluate land degradation processes and wind erosion in pilot sites (Land Degradation Assessment in Drylands, FAO 2020; National Observatory of Land Degradation and Desertification, ONDTyD; Ministry of Environment and Sustainable Development 2020; Monitoring of Arid and Semiarid Regions, MARAS, INTA 2020; Ecology and Biodiversity plots of natural environments in Southern Patagonia, PEBANPA, Peri et al. 2016).

In 2008, a team of researchers of Argentina installed 350 monitors in all Patagonia (MARAS), in which long-term soil and vegetation indicators are monitored. This system uses a standardized methodology with a common database, which allows comparing different rangelands in Patagonia. The main purpose is to obtain an environmental base of reference to monitor the trend of the desertification process and detect early warnings that aid management decisions to mitigate or reverse the problem (Oliva et al. 2019). We consider of utmost importance to maintain the financing of desertification monitoring projects and to investigate areas where these processes were not studied (e.g., large sectors of northwestern Patagonia).

Desertification is defined by the United Nations Convention to Combat Desertification (UNCCD 1994) as "the process of degradation of arid, semi-arid and dry sub-humid areas resulting from several factors, including climatic variations and human activities," which also cause socioeconomic problems suffered by the rural population established in arid Patagonia (Mazzoni and Vázquez 2010). However, only recently the politicians have put the emphasis on the social aspect focusing on rural employment and the permanence of the population in the territory. The consequences of desertification on the environment, production, and society are currently considered in an integrative form (Coronato 2015).

The main drivers of erosion in Patagonia are low precipitation and overgrazing by sheep. Oil and mining activities, wood extraction, and fires are other causes of soil loss (Coppa 2004; Bran et al. 2015). The more diffuse economic activity in

Patagonia is the extensive domestic livestock breeding (León and Aguiar 1985). For example, in Rio Negro more than 90% of the total land is used for livestock production (Bran et al. 2015). Domestic herbivores were introduced in Patagonia near 1880 and displaced the only large native herbivore, e.g., guanaco (*Lama guanicoe*) adapted to the xeric conditions of rangelands. Guanacos produce low impact on the vegetation and soil. They consume less food because they digest dry grasses and recycle nitrogen better and cut the grasses instead of pulling it up like the sheep (Baldi 2012). The replacement of mobile herbivores by fenced herbivores, the lack of knowledge about the structure and functioning of the grasslands, and the implementation of traditional management from other countries and the economic expectation of the society contributed to the overgrazing of grasslands (Coronato 2015). In the 1950 decade, Patagonia had the largest number of sheep, but the fall of the international wool market due to the increasing use of synthetic fiber along the increasing degradation of grasslands led to a production slowdown. The sheep number decreased from 22 million heads in the 1950 decade to 14 million in the 1980 decade. Many lands were abandoned, and many people migrated from rural communities to urban centers (Mazzoni and Vázquez 2010). The desertification is the result of an inadequate livestock system in an arid and fragile ecosystem (Coronato 2015). Overgrazing reduces the vegetation cover as main provisioning service (e.g., forage biomass), increasing the surface area of bare soil resulting in greater exposure of the soil to erosive processes. Soil erosion alters regulating services (e.g., carbon and nitrogen sequestration and storage), decreasing the capacity of ecosystems to support production (Golluscio et al. 1998; Paruelo and Aguiar 2003; Oñatibia et al. 2015). In that sense, vegetation prevents soil erosion and maintains ecosystem services mediated by soil (Guerra et al. 2016).

The consequences of soil erosion include the alteration of carbon and nitrogen cycles inducing carbon emissions to the atmosphere (Lal et al. 2013). Soil fertility is reduced as was demonstrated in several studies carried out in Chubut Province where soil lost organic matter, nitrogen, and carbon (Carrera et al. 2005; Chartier et al. 2011; Chartier et al. 2013). In degraded areas, soil erosion produces changes in the physical properties of soil increasing water runoff and decreasing water infiltration and retention (Parizek et al. 2002; Chartier and Rostagno 2006). Total water infiltration was reduced from 28.5 mm to 42.9 mm and was associated with grass cover decrease and bare soil increase (Chartier et al. 2011). The water losses by runoff have important effects on vegetation dynamics because they limit the grass seedling recruitment favoring the dominance of shrubs (Chartier et al. 2013). In Santa Cruz Province, overgrazing decreases C stock being the main disturbance that influences ecosystem C levels (between 130 and 50 mg C ha^{-1}, Peri 2011). Peri (2011) found slightly higher total C content in the low grazing intensity areas compared with non-grazed grasslands (130 vs. 120 mg C ha^{-1}) attributed to the C immobilization in the litter. Loss of C was accentuated by soil losses caused by the strong winds.

In northwest Patagonia grasslands, there are few studies about the consequences of soil erosion linked to overgrazing (Gaitán et al. 2009; Bran and Velasco 2017). In the mesic extreme of the west-east precipitation gradient (580 mm yr.$^{-1}$), the areas more affected by erosive processes are wetlands, where the runoff increases topsoil loss (Bran et al. 2006; Bran and Velasco 2017). In the arid extreme of the gradient (260 mm yr.$^{-1}$), soil function indices (e.g., stability, infiltration, and nutrient cycling) had been used as indicators of the degree of soil degradation (Gaitán et al. 2009). In these environments, plant patches constitute "fertility islands" where organic matter, microbial activity, and nutrients are concentrated (Rostagno and del Valle 1988). Grazing pressure increases the distance among plant patches and, consequently, decreases the capacity to retain the resources affecting the soil quality and degrading the grassland (Gaitán et al. 2009). Overgrazing also indirectly affected soil organic carbon reservoir through the increase of bare soil (from 33.8 ton ha^{-1} in enclosure to 7.9 ton ha^{-1} in degraded grasslands).

The MARAS monitors installed in pilot sites allow the identification of post-disturbance (volcanic ash deposition) processes and their impacts on soil and vegetation. In a pilot site (Jacobacci town, Rio Negro Province), the presence of micro-dunes was accentuated by the remobilization of volcanic ash after the Cordon Caulle eruption of 2011 (Ghermandi et al. 2015; Bran and Velasco 2017; Fig. 7.2a). In addition, the bare soil increases changed the micro-meteorological conditions, causing high soil temperatures in summer and cryoturbation in winter favoring erosive processes (Fig. 7.2b).

5 Proposals for Land Use Planning and Grasslands Management in Northwest Patagonia

We consider that to encourage the regulation and sustainable management of ES in northwest Patagonia, there is a need to implement environmental governance considering the provision of the different ES. Governance of ES emerges from multi-level local, regional, national, and international decision-making involving multiple actors (e.g., organizations, institutions, scientific community, stakeholders) with different interests in resource management (Primmer et al. 2015). For that, first is fundamental to identify ES, their threats, and information gaps. Second, there must

Fig. 7.2 (**a**) Micro-dunes generated by wind erosion and (**b**) mounds affected by cryoturbation in Jacobacci town NW Patagonia. (Source: Bran and Velasco 2017, ONDTyD)

be interdisciplinary collaboration between environmental and social scientists and economists investigating how ecosystems function and how resources can be transformed into services with social and economic value. Third, it is necessary to elaborate adequate proposals and regulations of land use and management to control, reverse, or mitigate the ES degradation. Fourth, consider the awareness and education about them and, finally, include compliance and supervision of different regulations (Fig. 7.3).

We presented and analyzed few ES of northwest Patagonian grasslands, but many others must be considered. We believe that it is fundamental to recognize and maintain cultural services as aesthetic value through avoiding uses that uniform the landscape as extensive crops or afforestation. Since several decades ago, several national and provincial programs promote and finance exotic conifer plantations without assessing the effects they may have on the soil and vegetation (Paruelo 2015; Simberloff et al. 2010). We propose demand the control of existing forestations to prevent the advance of species exotics on the landscape by supervision and penalties.

The maintenance of control of soil erosion and soil formation ES of grasslands could be achieved by large-scale and long-term monitoring programs and networks. For example, the aim of MARAS program is to monitor land degradation (Oliva et al. 2019). The usefulness of this large-scale monitoring is to validate the sustainability of livestock production in the arid lands. Another example is PEBANPA network that monitors biodiversity, vegetation cover, and soil properties (e.g., soil carbon) in long-term plots in southern Patagonia to support sustainable land management (Peri et al. 2016). Amount of soil carbon helped to predict the presence of

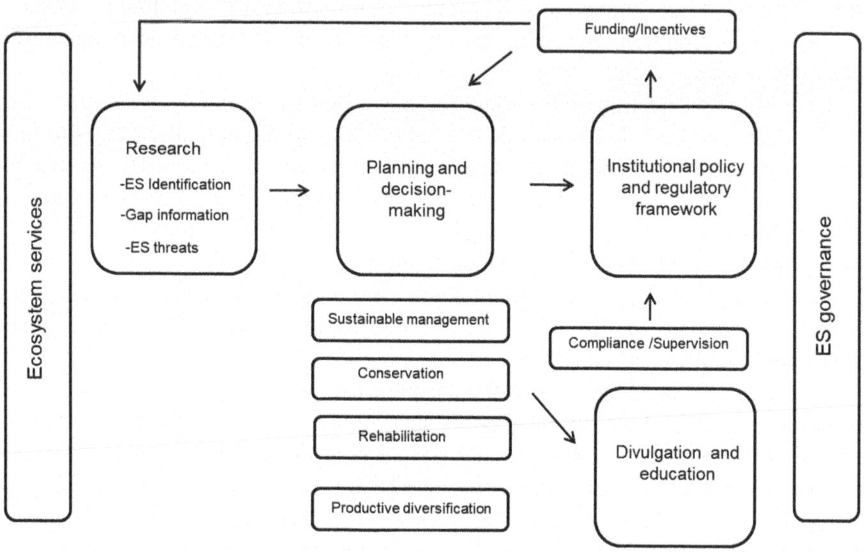

Fig. 7.3 General framework for analyzing the different actors involved in the management of the ecosystem services

threatened species and could be used to prioritize sites in conservation planning (Peri et al. 2019). We propose more financial support for existing programs for appropriated sustainable management of grasslands and conservation of their ES.

On the other hand, there are no national regulations for the protection of soil and control of erosion. However, the attempts to promote an international soil treaty within the framework of existing conventions (e.g., Río Convention) also have so far not gained political impulse (Bodle and Stockhaus 2020). Additionally, soils per se, like water, are not considered as one ES, but it is related to its formation, fertility, and composition (see Table 7.1), even though soil supplies and supported other fundamental ES (Baer and Birgé 2018). The soil treaty and inclusion as one ES is a knowledge area where progress is needed to achieve genuine soil protection and sustainable long-term management. It is clear that the global-level treaties are the basis for national policy discussion and own implementation in the regional countries.

It is important to favor the interaction between researchers and politicians to study an integrative planning to ameliorate the economic and social conditions of the population. This planning must consider an adequate livestock management and other land uses in addition to the traditional cattle production. Currently, in northwest Patagonia, the livestock management is continuous or rotational in large paddocks with low or moderate stocking rate (0.1–0.3 sheep ha^{-1} $yr.^{-1}$) and depending on the forage supply of the natural grassland (Gaitán et al. 2009). Generally, research is focused on the improvement of the forage resource but not in other fundamental ecosystem services (e.g., soil fertility increase). For example, continuous moderate grazing management increased the soil carbon and nitrogen sequestration and forage biomass in rangelands of southern Patagonia (Oñatibia et al. 2015; Oñatibia and Aguiar 2016). More studies must be promoted in northwest Patagonian grasslands aimed at determining the livestock carrying capacity and its impact over vegetation and other resources, highlighting ES provision.

In terms of production and market, there is a lack of incentives for sustainable land use perspectives. The diversification of production through alternatives to traditional sheep production can be considered. For example, guanacos (*Lama guanicoe*) breeding can be a viable alternative. The fiber of this camelid can be worth 15 or 20 times more than sheep's wool; it lives an average of 15 to 20 years compared to 5 years for sheep and has the triple number of offspring (Villareal and Longo 2003). Breeding for captive or semi-captive production has been established in Santa Cruz, Chubut, and Rio Negro (Los Menucos and Pilcaniyeu) that can be taken as a model (Amaya and von Thüngen 2003).

Another alternative land use is the agrotourism, ecotourism, and rural tourism that currently had a little but promissory development in most western ranches of Río Negro Province (Coronato et al. 2015). For example, the community Rural Tourism Network "Cultura Rural Patagónica" comprises a group of farmer families of the west of Río Negro that develop tourism as a complement to their productive activities and promote their enterprises (Cultura Rural Patagónica 2020). Other alternative activities are pig farms, crops of flowers, fine fruit, or aromatic plants crops; however, these activities are not available everywhere and are situated in more productive lands (Coronato et al. 2015).

References

Amaya J, Von Thüngen J (2003) Cría de guanacos en semicautividad. Misceláneas 2:5–10

Andrade-Díaz MS, Sarquis JA, Loiselle BA, Giraudo A, Díaz-Gómez JM (2019) Expansion of the agricultural frontier in the largest South American Dry Forest: identifying priority conservation areas for snakes before everything is lost. PLoS One 14(9):e0221901

Baer S, Birgé H (2018) Soil ecosystem services: an overview. In: Reicosky D (ed) Managing soil health for sustainable agriculture, Fundamentals, vol 1. Burleigh Dodds Science Publishing, Cambridge, UK, pp 1–20

Baldi R (2012) Un animal que podría salvar a la Patagonia. https://www.bbc.com/mundo/noticias/2012/09/120920_argentina_guanacos_vs. Accessed 17 Sept 2020

Bava J, Loguercio G, Salvador G (2015) ¿Por qué plantar en Patagonia? Estado actual y el rol futuro de los bosques plantados. Ecol Austral 25:101–111

Becker G, Bran D, Cremona M, Acevedo C (2012) Impacto de las lluvias torrenciales en Comallo. Revista Presencia 59:47–49

Bengtsson J, Bullock JM, Egoh B, Everson C, Everson T, O'Connor T, Farrell PJ, Smith HG, Lindborg R (2019) Grasslands—more important for ecosystem services than you might think. Ecosphere 10:1–20

Bidinost F, Gaetano A, Alvárez R (2020) Informe de situación de los sistemas ganaderos y de la prospectiva de la productividad de los pastizales en la provincia de Río Negro. Instituto Nacional de Tecnología Agropecuaria, Buenos Aires, Argentina

Bodle R, Stockhaus H (2020) Improving international soil governance: analysis and recommendations. Final report,. Umweltbundesamt Publisher, Berlin, Germany

Bran D, Siffredi G, Ayesa J, López C, Umaña F, Gaitán J, López D, Franzese J (2006) Evaluación del estado de los recursos forrajeros de la Estancia San Ramón. Comunicación técnica # 106, Area recursos naturales, INTA Bariloche, Río Negro, Argentina

Bran D, Lopez C, Ayesa J, Gaitán J, Umaña F, Quiroga S (2015) Provincia de Río Negro. In: Casas R, Albarracín G (eds) El deterioro del suelo y del ambiente en Argentina. Fundación para la Educación, la Ciencia y la Cultura, Buenos Aires, Argentina, pp 213–223

Bran D, Velasco V (2017) Sistema de Monitoreo y Evaluación del Observatorio Nacional de Degradación de Tierras y Desertificación. Sitio Piloto Jacobacci, Río Negro, Argentina, ONDTyD

Carrera A, Vargas D, Campanella M, Bertiller MB, Sain CL, Mazzarino MJ (2005) Soil nitrogen in relation to quality and decomposability of plant litter in the Patagonian Monte, Argentina. Plant Ecol 181:139–151

Chartier M, Rostagno C (2006) Soil erosion thresholds and alternative states in northeastern Patagonian rangelands. Range Ecol Manage 59:616–624

Chartier M, Rostagno C, Pazos G (2011) Effects of soil degradation on infiltration rates in grazed semiarid rangelands of northeastern Patagonia, Argentina. J Arid Environ 75:656–661

Chartier M, Rostagno C, Videla L (2013) Selective erosion of clay, organic carbon and total nitrogen in grazed semiarid rangelands of northeastern Patagonia, Argentina. J Arid Environ 88:43–49

Colazo JC, Carfagno P, Gvozdenovich J, Buschiazzo D (2019) Soil Erosion. In: Rubio G, Lavado R, Pereyra F (eds) The soils of Argentina, World Soils Book Series. Springer, Cham, Switzerland, pp 239–250

Coppa R (2004) El deterioro de los pastizales patagónicos. Medioambiente 4:19–22

Cooper N, Brady E, Steen H (2016) Aesthetic and spiritual values of ecosystems: recognising the ontological and axiological plurality of cultural ecosystem 'services'. Ecosyst Serv 21:218–229

Coronato F, Fasioli E, Schweitzer A, Tourrand J (2015) Rethinking the role of sheep in the local development of Patagonia, Argentina. Rev Elev Med Vet Pays Trop 68:129–133

Coronato F (2015) Ovejas, territorio y políticas públicas en la Patagonia. Estudios del ISHiR 13:6–19

Cultura Rural Patagónica (2020). http://www.turismoruralbariloche.com/. Accessed 26 October 2020

Defossé G, Dentoni MC (1999) Estudio de Grandes Incendios: el caso de la Estancia San Ramón, en Bariloche - Río Negro, Argentina. PNMF, CIEFAP, Esquel, Chubut, Argentina

del Valle H, Elissalde N, Gagliardini D, Milovich J (1998) Status of desertification in the Patagonian region: assessment and mapping from satellite imagery. Arid Soil Res Rehabil 12:95–122
Dronova I (2019) Landscape beauty: a wicked problem in sustainable ecosystem management? Sci Total Environ 688:584–591
Easdale M, Madariaga M (2009) Zonas agroeconómicas homogéneas Patagonia Norte. Estudios socioeconómicos de la sustentabilidad de los sistemas de producción y recursos naturales Neuquén y Río Negro. Instituto Nacional de Tecnología Agropecuaria, Buenos Aires, Argentina
Easdale M, Aguiar MR (2012) Regional forage production assessment in arid and semi-arid rangelands: a step towards social-ecological analysis. J Arid Environ 83:35–44
Evans D, Orr P, Lazar S, Breton D, Gerard J, Ledbetter DH, Janosco K, Dots J, Batchelder H (2012) Human preferences for symmetry: subjective experience, cognitive conflict and cortical brain activity. PLoS One 7(6):e38966
FAO. Food and Agriculture Organization of the United Nations (2020) Ecosystem services. http://www.fao.org/ecosystem-services-biodiversity/en. Accessed 14 Sept 2020
Franzese J, Ghermandi L (2014) Early competition between the exotic herb *Rumex acetosella* and two native tussock grasses with different palatability and water stress tolerance. J Arid Environ 106:58–62
Franzese J, Urrutia J, García R, Taylor K, Pauchard A (2017) Pine invasion impacts on plant diversity in Patagonia: invader size and invaded habitat matter. Biol Inv 19:1015–1027
Fu B, Liu Y, Lu X, He C, Zeng Y, Wu B (2011) Assessing the soil erosion control service of ecosystems change in the Loess Plateau of China. Ecol Complexity 8:284–293
Gaitán J, Lopez C, Bran D (2009) Efectos del pastoreo sobre el suelo y la vegetación en la estepa patagónica. Ciencias del Suelo 27:261–270
Gaitán J, Oliva G, Bran D (2014) Vegetation structure is as important as climate for explaining ecosystem function across Patagonian rangelands. J Ecol 102:1419–1428
Gaitán J, Bran D, Oliva G (2019) Patagonian deserts. In: Goldstein M, DellaSala D (eds) Encyclopedia of the World's Biomes. Elsevier, Amsterdam, Netherlands, pp 163–180
Geist HJ, Lambin EF (2004) Dynamic causal patterns of desertification. Bioscience 54:817
Ghermandi L, Guthmann N, Bran D (2004) Early post-fire succession in northwestern Patagonia grasslands. J Veg Sci 15:67–76
Ghermandi L, Gonzalez S, Lescano MN, Oddi F (2013) Effects of fire severity on early recovery of Patagonian steppes. Int J Wildland Fire 22:1055–1062
Ghermandi L, Gonzalez S, Franzese J, Oddi F (2015) Effects of volcanic ash deposition on the early recovery of gap vegetation in Northwestern Patagonian steppes. J Arid Environ 122:154–160
Godagnone R, Bran D (eds) (2009) Inventario integrado de los recursos naturales de la Provincia de Río Negro. Geología, Hidrología, Geomorfología, Suelos, Clima, Vegetación y Fauna, Ediciones INTA, Buenos Aires, Argentina
Golluscio R, Deregibus A, Paruelo J (1998) Sustainability and range management in the Patagonian steppes. Ecol Austral 8:265–284
Guerra C, Maes J, Geijzendorfferc I, Metzger MJ (2016) An assessment of soil erosion prevention by vegetation in Mediterranean Europe: current trends of ecosystem service provision. Ecol Indic 60:213–222
Hansson H, Norberg C (2009) Cold matters, cultural perceptions of snow, ice and cold. Umeå University and the Royal Skyttean Society Publisher, Northern Studies Monographs, Umeå, Sweden
Herrero M, Wirsenius S, Henderson B, Rigolot C, Thornton P, Havlík P, de Boer I, Gerber P (2015) Livestock and the environment: what have we learned in the past decade? Ann Rev Environ Res 40:177–202
Lal R (2003) Soil erosion and the global carbon budget. Env Int 29:437–450
Lal R, Lorenz K, Huttl RF, Schneider BU, von Braun J (2013) Ecosystem services and carbon sequestration in the biosphere. Springer Science Business Media, Dordrecht, Netherlands
Lankoski J, Thiemb A (2020) Linkages between agricultural policies, productivity and environmental sustainability. Ecol Econ 178:e106809

León R, Aguiar M (1985) El deterioro por uso pastoril en estepas herbáceas patagónicas. Phytocoenologia 13:181–196

Li J, Okin G, Alvarez L, Epstein H (2007) Quantitative effects of vegetation cover on wind erosion and soil nutrient loss in a desert grassland of southern New Mexico, USA. Biogeochemistry 85:317–332

Li FR, Zhao WZ, Liu JL, Huang ZG (2009) Degraded vegetation and wind erosion influence soil carbon, nitrogen and phosphorus accumulation in sandy grasslands. Plant Soil 317:79–92

MARAS. Monitoreo Ambiental de Zonas Áridas y Semiáridas (2020). Instituto Nacional de Tecnología Agropecuaria. Buenos Aires, Argentina. http://sipas.inta.gob.ar/?q=maras-portal. Accessed 14 Sept 2020

Mazzoni E, Vázquez M (2010) Desertification in Patagonia. In: Latrubesse E (ed) Natural hazards and human-exacerbated disasters in Latin America, Elsevier, vol 13. New York, USA, pp 351–377

Nuñez P, Michel C, Leal Tejada P, Nuñez M (2020) Rural women's invisible work in census and state rural development plans: the Argentinean Patagonian case. Land 9:92

Oliva G, Bran D, Gaitán J, Ferrante D, Massar V, García Martínez G, Adema E, Enrique M, Dominguez E, Paredes P (2019) Monitoring drylands: the MARAS system. J Arid Environ 161:55–63

ONDTyD. Observatorio Nacional de Degradación de las Tierras y Desertificación (2020). Ministry of Environment and Sustainable Development. http://www.desertificacion.gob.ar. Accessed 14 Sept 2020

Oñatibia G, Aguiar M, Semmartin M (2015) Are there any trade-offs between forage provision and the ecosystem service of C and N storage in arid rangelands? Ecol Eng 77:26–32

Oñatibia G, Aguiar MR (2016) Continuous moderate grazing management promotes biomass production in Patagonian arid rangelands. J Arid Environ 125:73–79

Oddi F, Ghermandi L (2016) Fire regime from 1973 to 2011 in north-western Patagonian grasslands. Int J Wildland Fire 25:922–932

Parizek B, Rostagno CM, Sottini R (2002) Soil erosion as affected by shrub encroachment in northeastern Patagonia. J Range Manag 55:43–48

Paritsis J, Landesmann J, Kitzberger T, Tiribelli F, Sasal Y, Quintero C, Dimarco R, Barrios-García N, Iglesias A, Diez J, Sarasola M, Nuñez M (2018) Pine plantations and invasion alter fuel structure and potential fire behavior in a Patagonian forest-steppe ecotone. Forests 9:e117

Paruelo J, Aguiar M (2003) Impacto humano sobre los ecosistemas: El caso de la desertificación. Ciencia Hoy 77:48–59

Paruelo J (2015) ¿Conviene seguir fomentando las plantaciones forestales en el norte de la Patagonia Argentina? ¿Dónde? ¿Para qué? ¿A quién le conviene? Ecol Austral 25:112–118

Peri P (2011) Carbon storage in cold temperate ecosystems in southern Patagonia, Argentina. In: Atazadeh E (ed) Biomass and remote sensing of biomass. InTech Publisher, Zagreb, Croacia, pp 213–226

Peri PL, Lencinas MV, Bousson J, Lasagno R, Soler R, Bahamonde H, Martínez Pastur G (2016) Biodiversity and ecological long-term plots in Southern Patagonia to support sustainable land management: the case of PEBANPA network. J Nat Conserv 34:51–64

Peri PL, Lasagno RG, Martínez Pastur G, Atkinson R, Thomas E, Ladd B (2019) Soil carbon is a useful surrogate for conservation planning in developing nations. Sci Rep 9:1–6

Primmer E, Jokinen P, Blicharska M, Barton D, Bugter R, Potschin M (2015) Governance of ecosystem services: a framework for empirical analysis. Ecosyst Ser 16:158–166

Pulleman M, Creamer R, Hamer U, Helder J, Pelosi C, Peréz G, Rutgers M (2012) Soil biodiversity, biological indicators and soil ecosystem services-an overview of European approaches. Curr Opin Environ Sust 4:529–538

Reboratti C (2010) Un mar de soja: La nueva agricultura en Argentina y sus consecuencias. Revista de Geografía Norte Grande 45:63–76

Rostagno C, del Valle H (1988) Mounds associated with shrubs in aridic soils of northeastern Patagonia: characteristics and probable genesis. Catena 15:347–359

Simberloff D, Nuñez MA, Ledgard NJ, Pauchard A, Richardson D, Sarasola M, Van Wilgen B, Zalba S, Zenni RD, Bustamante R, Peña E, Ziller S (2010) Spread and impact of introduced conifers in South America: lessons from other southern hemisphere regions. Austral Ecol 35:489–504

Tribot AS, Deter J, Mouquet N (2018) Integrating the aesthetic value of landscapes and biological diversity. Proc R Soc B 285:e20180971

United Nations Convention to Combat Desertification - UNCCD (1994). https://www.unccd.int. Accessed 15 Sept 2020

United Nations – UN (1987–2015) Sustainable development goals. https://www.un.org/sustainabledevelopment/sustainable-development-goals. Accessed 14 Sept 2020

Valenta M, Golluscio R, Frey A (2020) Short-term responses to sheep grazing in a Patagonian steppe. Rangeland J 42:1–8

van Zanten B, Verburg P, Koetse M, van Beukering P (2014) Preferences for European agrarian landscapes: a meta-analysis of case studies. Land Urb Plan 132:89–101

Villareal F, Longo L (2003) Valoración económica del guanaco. Rev Facultad de Agronomía 23:59–69

Zhao Y, Liu Z, Wu J (2020) Grassland ecosystem services: a systematic review of research advances and future directions. Landsc Ecol. https://doi.org/10.1007/s10980-020-00980-3

Chapter 8
The Ecosystem Services Provided by Peatlands in Patagonia

Rodolfo J. Iturraspe and Adriana B. Urciuolo

Abstract Peatlands are distinctive wetlands of Patagonia that represent a valuable natural heritage due to its near-pristine conditions and the wide range of the ecosystem services (ES) they provide. The objective of this chapter is to review the state of knowledge about the ES of Patagonian peatlands and the related processes that explain them. Some of these services, such as carbon storage, carbon sink, flood control, and water supply, are outcomes of complex biotic and abiotic interactions that take place in these ecosystems. We identified and classified ES of Patagonian peatlands in nonmonetary terms, following the guidelines of the Common International Classification of Ecosystem Services (CICES) V5.1. Different peatlands may contrast in the quality or essence of respective ES; therefore, the analysis of the ES significance requires information from local studies based on scientific knowledge and field data. We supply evidence obtained from study cases and discuss the accuracy of current estimations of carbon storage in Patagonian peatlands, as well as the peat bogs efficiency on the flood and the erosion control.

Keywords Peatlands · Ecosystem services · Patagonia · Carbon store · Peatland hydrology

1 Introduction

Ecosystem services (ES) are defined as the contributions that ecosystems make to human well-being, according to the Common International Classification of Ecosystem Services – CICES (Haines-Young and Potschin 2017). As a response to the multiple worldwide evidence of degradation and unsustainable use of ecosystems, the concern with the valuation ofecosystem functions, goods, and services

R. J. Iturraspe (✉) · A. B. Urciuolo
Universidad Nacional de Tierra del Fuego, Antártida e Islas del Atlántico Sur, Ushuaia, Tierra del Fuego, Argentina

P. L. Peri et al. (eds.), *Ecosystem Services in Patagonia*, Natural and Social Sciences of Patagonia, https://doi.org/10.1007/978-3-030-69166-0_8

raised substantially in the last decades of the twentieth century, as well as publications with different approaches to this matter. Daily (1997) provided a comprehensive description of ES, and De Groot et al. (2002) proposed a conceptual framework and typology for describing, classifying, and valuing ecosystem functions, goods,and services.

The Millennium Ecosystem Assessment (MEA) was called for the United Nations, and more than 2000 authors and reviewers worldwide were committed. The MEA (2005) report emphasizes the idea that human well-being depends on ecosystems and proposed for ES a conceptual, methodological, and application-decision-making framework. Some years later, CICES resulted as a response to the need to adjust the MEA framework in the way to ES are described, and in the identification of the final services that link to the goods and benefits that are valued by people. The first version CICES V4.3 published in 2013 was upgraded afterward to the CICES V5.1 (Haines-Young and Potschin 2017).

Before the MEA(2005) report, Joosten and Clarke (2002) performed a first frame encompassing peatland values and functions.

"Peat is the incomplete decayed remains of the plants that grew on what was once the surface" (Clymo et al. 1998), and peatlands are lands that accumulate or have accumulated peat (Joosten and Clarke 2002). More extended definitions are available for peat and peatlands (e.g., Ivanov 1981; Rydin et al. 2006). A mire is a peatland where peat is being formed and accumulating. This word distinguishes a degraded peatland from that remains in an almost natural state. The majority of peatlands in Patagonia are mires; they extend 2000 km through the length of Patagonia, from the Cape Horn in the extreme south to the Region *X* in Chile and the Province of Neuquén in Argentina. Most of these peatlands have remote locations; however, Patagonia is undergoing dynamic transformations and progressive pressures over the environment that are threatening some of them.

Historically, in different regions of the world, people have ignored, minimized, or denied the benefits of peatlands and have regarded them as a difficulty for human settlement, terrestrial connectivity, and production. Consequently, in the past centuries, several countries of the Northern Hemisphere have drained, large extensions of bogs and fens, changing the land use to agriculture, forestry, and urbanization, among others.

There are no local studies about the public perception of peatland values, but it seems to be still low in Patagonia, both in Chile and in Argentina, where many people appreciate peatlands less than any other ecosystem, even some of them ignore what a peatland is.

Patagonian peatland-related knowledge has been improved through the last two decades, with the predominance of biological, ecological, hydrological, and paleoclimatic approaches. Comparatively, oriented studies to hold scientific evidence on the scope and the effectiveness of peatlands ES are scarce.

Several authors analyzed peatland ES in the framework of MEA and CICES for countries or regions of the Northern Hemisphere (e.g., Whitfield et al. 2011). We aim to replicate such analysis, focusing specifically on the ES of Patagonian mires, seeing their particular features and the social context that involves them.

The objective of this chapter is to review the state of knowledge about the ES of Patagonian peatlands, as well as discuss the related processes that explain particular features of them, with special attention in such correspond to the carbon cycle control and the hydrological regulation.

2 Peatlands Features and Distribution in Patagonia

Patagonia presents a wide variety of peatland types. Peat-forming plants, hydrology, climate, chemical proprieties of mineral soil, land geomorphology, and more factors determine particular characteristics that distinguish them and give place to many questions about its ecology, hydrology, ES, and management.

Some typical features of peatlands are:

- Peatlands are the result of a millenary process started after the ice retreat of the last glaciation.
- Older peatlands in Patagonia have basal ages over 18,000 years. The mean age of peatlands in the region is 12,400 years (Loisel 2015).
- Peatlands are wetlands recognized as such by the Ramsar Convention.
- Peatlands have been resilient to high variations of climatic cycles throughout the Holocene period; nevertheless, they are highly vulnerable to human activity.
- Peatlands are ecosystems that support living organisms of special biodiversity.
- Peatlands are highly vulnerable to changes in their hydrological regime.

A widespread distribution of peatlands, which is greater than in other temperate regions of the Southern Hemisphere, is a characteristic of Patagonia, where peatlands may be found on plains and mountains, from hyper-humid Pacific islands to the semiarid steppe. However, the best conditions for their development correspond to humid and oceanic environments, moderate slopes, and altitude below 600 m ASL. Some examples of places where vast peatland systems are concentrated are southeastern Tierra del Fuego (TDF) Argentina, Navarino withadjacent islands, western TDF between 53 40' and 54 20'S latitude, Brunswick Peni nsula and the Obstruction Sound zone.

Neither Argentina nor Chile has available complete inventories. It is necessary to improve or complete a regional evaluation of the peatland surface area (Loisel 2015, Vega-Valdés and Domínguez Díaz (2015). A survey by CONAF et al. (1999) later cited by Arroyo et al. (2005) determined a wetland extension of 32,000 km^2 in the Magallanes Region, 11,500 km^2 in Aysén, and 573 km^2 in Los Lagos, totalizing 44,000 km^2 in the Chilean Patagonia. The report explains that the majority of these wetlands are peatlands. This has motivated that the areas determined by CONAF as wetlands have been assumed as peatland areas, which is not strictly true. A remote sensing-based study by Ruiz and Doberty (2005) determined, in the XII Magallanes Region, a total peatland surface of 22,700 km^2 (in rounded numbers). Vega-Valdés and Domínguez Díaz (2015) adjusted it to 21,000 km^2, which is 11,000 km^2 less than the wetland extent measured by CONAF (1999, 2012) in the Magallanes Region. Supposing that the entire wetland area (12,000 km^2) assigned to the X and

XI regions corresponds to peatlands, the total in the Chilean Patagonia would be around 33,000 km^2, as a preliminary approximation. Not the entire referred wetland area corresponds to peatlands, and on the other hand, there might be unidentified peatlands below the forest canopy.

In Argentina, most peatlands are located in TDF, where they cover 2700 km2 (Iturraspe et al. 2012). Several authors (Carretero 2004, Perotti et al. 2005, Fuertes-Lasala et al. (2008), Chimner et al. 2011, Iturraspe 2016) indicate peatlands spread in low coverage rates along the Andean forest eco-region of the continental Argentinean Patagonia. The negative W-E precipitation gradient results in a bogs-fens-wet meadows succession. There is no documented information of this peatland extent, but it would reach between 250 and 300 km^2. Thus, the full extension of peatlands in the Argentinean Patagonia would be around 3000 km^2. Under the assumptions considered for the Chilean peatland area, the total peatland extension in Patagonia would be near 36,000 km^2, as a preliminary estimation based on the available documented data.

In addition to raised bogs and graminaceous fens, the special climate features of Patagonia determine unique wetland ecosystems that frequently accumulate peat.

The swamp forest (locally named *Tepual*) is a typical wetland from northern and central Chilean Patagonia, located in poorly drained lands dominated by *Tepualia stipularis*. These tree branches sprout profusely from the base and normally form a dense tangle 5–8 m high above the black surface of stagnant water, rich in organic matter (OM) (Veblen and Schlegel 1982). The horizontal growth develops complex structures accumulating large amounts of biomass, which leads to the formation of arboreal soils, and almost impenetrable nets of trunks (Bannister et al. 2017). These waterlogged OM deposits usually form a peat layer 0.30 to 1 m thick (Holdgate 1961).

Anthropogenic peatlands (*pomponales*) are *Sphagnum* wetlands, originated in the late nineteenth century, after burning or clear-cutting of wide forest areas of Chile, in places with poor drainage (Díaz et al. 2008). Moss fibers grow with very high rates in these ecosystems, but since its recent origin, peat accumulation has been weak or null.

Cushion bogs dominate hyperoceanic wind-exposed western areas in southern Chile, mixed with communities of sedges and graminoid fens (Pisano 1977).

In Argentina, *Sphagnum* bogs prevail in mountain valleys and transitional areas of TDF, while fens are frequent in the forest-steppe ecotone. Ninety percent of this peatland area is concentrated east of longitude 66°30'W. Cushion plants with dense roots have colonized deep *Sphagnum* bogs in coastal lands. Inland, *Sphagnum* bogs occupy wide mountain valleys, and mixed plant communities cover blanket bogs at rounded hills.

Wet meadows (*mallines, vegas*) are productive ecosystems on seasonally saturated soils by groundwater feeding. The water table level (WTL) is normally highly variable, but peat accumulation may occur if groundwater remains stable close to the surface. These wetlands offer water availability and habitat for biodiversity in the contrasting semiarid surrounding environment. Wet meadows cover 4–5% of the extra-Andean Patagonian steppe, but just some of them accumulate peat.

The Fig. 8.1 shows examples of different peatland environments.

Fig. 8.1 Contrasting peatlands of Patagonia. (1) Northern fen bounded by basalts, near Aluminé, Neuquen, Arg., 41° 05'S Lat. (2) Wet anthropogenic peatland in Chiloe. (3) Anthropogenic peatland near Tortel, Aysén, Ch. (4) Fen in the Baker river mouth, Tortel, Aysén, Ch. (5) Peat wet meadows near Chaltén, Santa Cruz, Arg. (6) *Sphagnum* bog with a beaver dam in the Karukinka Reserve, TDF Ch. (7) *Sph.* bog with Guaitecas Cypress (*Pilgerodendron uviferum*) near Obstruccion Sound, Magallanes, Ch. (8) *Sph.* bog in Carbajal Valley, near Ushuaia. TDF, Arg. (9) Wide bog with pools and gallery forest, López Valley, Peninsula Mitre, eastern TDF, Arg

3 Ecosystem Services of Peatlands in Patagonia

We summarize the ES of peatlands of Patagonia, following the Common International Classification of Ecosystem Services (CICES) version 5.1. CICES classification has a hierarchical structure, whose highest level is organized into three sections related to the type of contribution to human well-being that the ecosystems support. Such sections are (a) the provisioning of material and energy needs, (b) regulation and maintenance of the environment for humans, or (c) the non-material characteristics of ecosystems that affect physical and mental human states. Table 8.1 includes ES that specifically Patagonian peatlands supply, including goods and examples of benefits derived from them. After that, we analyze and discuss more significant ES provided by peatlands in the region.

Table 8.1 Peatland ecosystems services and goods following the CICES V.1 classification

Final service			Benefits provided by peatlands (examples)	
Section	Division	Group/class	Possibly compatible with peat accumulation	Peat consuming
Provisioning	Biomass	Cultivated terrestrial plants for nutrition, materials, or energy		Tree plantations in drained peatlands (i.e., Chiloe)
		Wild plants for nutrition materials or energy	Plants, flowers, fruits, or related products used in medicines or sold for medicinal purposes	
			Compost from plants materials Wood from wild trees and shrubs for fuel and materials (e.g., Chiloe) Peat moss biomass	
		Fibers and other material of wild plants for fertilizer/soil improvement		Peat as soil improver Peat as a substrate component Organic peat blankets Peat as fuel for personal use
		Fibers of wild plants for direct use or processing		*Sphagnum* fiber harvesting for export, family economy support
		Reared animals for nutrition, materials	Cattle, sheep livestock	
		Wild animals for nutrition and materials	Guanacos	
		Wind energy	Altitude of peatland areas in Chiloe makes them suitable for wind energy generation	Infrastructure can produce partial peatland damages
	Water	Water for nutrition, material or energy	Drinking water Water for irrigation	

(continued)

Table 8.1 (continued)

Final service			Benefits provided by peatlands (examples)	
Section	Division	Group/class	Possibly compatible with peat accumulation	Peat consuming
Regulating and maintaining services	Transformation of biochemical or physical inputs	Bioremediation	Detoxification in land/soils and freshwater	Waste water cleaning
		Filtration/ sequestration	Clean water supply though filtration of atmospheric pollutants	Waste water cleaning
	Regulation of physical, chemical, biological conditions	Control of erosion rates	Maintenance of soil and water quality for human/ productive activities	
		Buffering and attenuation of mass movement	Protection of people and goods Mitigation of hazards and extreme events damages Sediment retention Contrib. to the population safety	
		Flood control	Flood damage reduction and prevention	
		Fire protection	Wet and undrained mires may act as firebreaks	
		Maintaining habitats	Conservation of biodiversity	
		Regul. chemical condition of freshwater by living processes	.Drinking Water supply	
		Regul. chemical composition atmospheric C storage/sink	Contribution to mitigation of Global Climate change	

(continued)

Table 8.1 (continued)

Final service			Benefits provided by peatlands (examples)	
Section	Division	Group/class	Possibly compatible with peat accumulation	Peat consuming
Cultural services	Direct, in situ, and outdoor interactions with living systems that depend on presence in the environmental setting	Peatland features that enable activities promoting health or enjoyment through active or immersive interactions	Opportunity for outdoor recreation, walking/hiking, skiing. Tourist-related services	Trips by using ATVs, 4x4 cars, or intensive walking producing ecosystem damages
		Peatland features that enable activities promoting health or enjoyment through passive or observational interactions	Wildlife watching and environment and landscape that provide a sensory experience (i.e., flowers, plants, animals, etc.)	
		Peatland features that enable scientific investigation or the creation of traditional ecological knowledge	Research opportunities: stratigraphic archive function and pollen record Paleoclimate knowledge	
		Characteristics of peatlands that enable education and training	Opportunities to understand, communicate, and educate about mire values: Educational-guided tours, subject matter for wildlife programs, hydrological issues	
		Characteristics of peatlands that are resonant in terms of culture or heritage	Historic records, archeological artefacts preservation in peat	

(continued)

Table 8.1 (continued)

Final service			Benefits provided by peatlands (examples)	
Section	Division	Group/class	Possibly compatible with peat accumulation	Peat consuming
		Characteristics of peatlands that enable aesthetic experiences	Peatland landscape observation, appreciation of landscape for art and literature. Opportunities for recreational and tourism activities	
	Indirect, remote, often indoor interactions with living systems with no presence in the mire setting	Elements of peatlands used for entertainment or representation	Peatlands landscape that provides a sensory experience, which may lead to the benefit of inspiration for art or can be directly used in art (i.e., films, soundtracks, etc.)	
			Ex situ viewing experience of nature through different media	

3.1 Provisioning Services

3.1.1 Peat and Fiber Provision

The peat mining activity started in Patagonia in the 1970s in Carbajal Valley, near Ushuaia city, TDF (Iturraspe and Urciuolo 2004), which progressively increase in the following decades. Peat is used as a substrate component for industrial and home production of flowers, tobacco, and garden or fruit plants. Vertical gardens and organic blankets for oil spill treating are some additional uses. Peat extraction in Argentina is concentrated in 38 sites of TDF. In Chile, main places are TDF, the Province of Magallanes, and Los Lagos Region.

Peat extraction results in the drainage of peatlands that increases both the organic matter decomposition and CO_2 emissions due to the consequent peat aerobic conditions produced by the WTL lowering. The cutting of the upper layer eliminates the ecosystem's living plants; therefore, it annuls the bog function as a CO_2 sink.

Anthropogenic peatlands of northern Chilean Patagonia are subject to *Sphagnum* fiber harvesting. In contrast to the extracted peat, that is sold domestically, the moss harvested is totally exported. Chile commercialized in 2019 around 4000 tons of dry *Sphagnum* fiber for USD 15 million (Instituto Forestal 2020). Local families harvest fibers and sell them to collecting companies. Fiber harvesting has increased since the 1990s without environmental controls and with serious alterations on these ecosystems. The Ministerio de Agricultura, through Decreto 25/2017, enacted a specific regulation for this activity. Regeneration of *Sphagnum* fiber after harvesting is technically feasible under specific conditions and requirements, but the sustainability of this activity is uncertain because of the necessary expenses, care, and control to ensure moss regeneration.

3.1.2 Livestock Support

Fens and peat-wet meadows of Patagonia support diversity of plants and animals. Due to its significant biomass productivity, most of them have been subjected to the grazing of domestic livestock for over one century (Paruelo et al. 2004, Collantes et al. 2013). Cattle raising is the main rural productive activity in extra-Andean Patagonia, which was originally based on sheep and in the last decades combined with cows. Paruelo et al. (2004) report 4000 kg h^{-1} y^{-1} of aerial net primary production (ANPP) in meadows of the western portion of the extra-Andean Patagonian steppe (Argentina). In contrast, they specify 500 kg h^{-1} y^{-1} of ANNP in the surrounding semi-desert lands. Covering only 3.3% of the area, meadows contributed more than 12% of the total ANPP in this study area. Considering the traditional extensive farming practice, with rotational grazing in fenced paddocks, the stocking rate in each farm depends on the proportion of wet meadows that each one has.

Fens and meadows are natural habitats of guanaco (*Lama guanicoe*) which is the unique large native ungulate. Before sheep introduction, guanaco was the integral sustenance for indigenous communities that inhabited Patagonia, since they got from it food, coating, and materials to build hunting artefacts and basic housing.

3.1.3 Water Supply

Like other kinds of wetlands, mires present available surface water for wild animals and cattle. Peatland outflow contributes to maintaining runoff during no rain periods. Bog contributions can be effective for the short-term runoff but rarely in the long term (Evans et al. 1999, Holden and Burt 2003). Active bog water reservoirs have a limited capacity. Extended dry periods induce asymptotic WTL drops and near null outflows. Fens and further groundwater-fed peatlands can sustain water outputs in dry periods for the long term, depending on the water source, as it occurs in some fens and wet meadows of extra-Andean Patagonia. No dry season and no extended periods without rain are precipitation features of Sphagnum bog environments of Patagonia, such that it occurs in Ushuaia, Punta Arenas, and Coyhaique, where there are and more than 40% of rainy days in the year and moderate seasonality (Sarricolea and Martín-Vide 2012, Iturraspe and Schroder 1999). In this short-term context, bogs can be effective in attenuating runoff reduction.

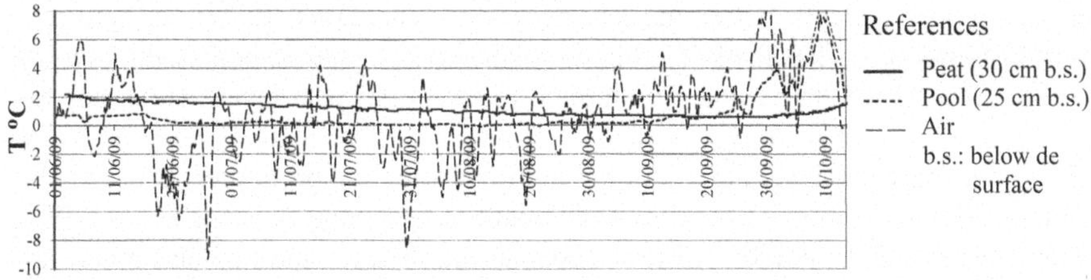

Fig. 8.2 Winter air, peat, and pool water temperature in Rancho Hambre bog, TDF (54°45'S, 67°50W, 125 m ASL). The pool froze in mid-June, at the same time that neighboring mineral soil (not plotted on the image). The lower acrotelm layer remained unfroze throughout the winter. Maximum depth of peat freezing was 17 cm, in late August. Water bog outflow persisted in winter

Water Provision in Wintertime

No permafrost exists in the region at the latitudinal range in which peatlands develop; instead, seasonal soil freezing occurs in southern Patagonia (e.g., TDF). Consequently, substantial runoff reduction causes difficulties to the water supply for urban and rural populations. Soil and peatland temperature monitoring in *Sphagnum* bogs of TDF shows the acrotelm horizontal flow remaining hydrologically active during most of the winter, while bog, pools, and external mineral soils are frozen (Fig. 8.2). The WTL persists near the surface, feed by meltwater; this way, peat bogs generate winter outflows that mitigate runoff reduction at the time that other water reservoirs are inactive.

Forest-peatland mosaics favor snow accumulation on peatlands by wind effect. Forest and peatlands have their own snowmelt timing, which extends the general melting period.

Peatlands water regulation favors provision for urban and rural populations, notably in Chiloé Island. The intensive *Sphagnum* fiber harvesting in Chiloé threatens the natural hydrological balance and the water availability for human consumption and other uses (Domínguez Díaz 2014). Punta Arenas and Ushuaia are examples of cities using water that peatlands partially provide.

3.2 *Regulating and Maintaining Services*

3.2.1 Flood Control

Wetlands play a central role in the water cycle, and peatlands are the dominant wetland type in much of Patagonia. Water content in peat is over 90% (Loisel and Yu 2013). However, although water is largely the main peat component, most of it remains immobilized in micro-pores or as a constituent of undecayed plant fibers. Water movement into peatland is restricted to the upper layer and depends on vegetation, decomposition degree, and compression; therefore, water conductivity decreases as the depth below the ground surface increases. Rainwater infiltrates very fast until it reaches the WTL, but no vertical flux occurs in bog's saturated layers. Horizontal water movement is limited to the upper peat layer, which has low decomposition and compression levels. Ivanov (1981) called diplotelmic to the mires that show two layers that differ in their hydrological properties: the acrotelm, which is the undecomposed and hydrologically active upper layer where the WTL fluctuates, and the catotelm, which is beneath it, always saturated and practically impervious. *Sphagnum*-raised bogs are typical diplotelmic mires in Patagonia. Fens and other kinds of peatland do not have a definite acrotelm layer. Moreover, a more complex vertical variation in peat properties could exist. In addition, natural pipes that allow water flux like underground streams can carry out water excess rapidly.

Peatlands, marshes, lakes, swamp forests, mineral soils, and other hydrological basin components are natural water reservoirs that contribute to the flood control. This reservoir's capacity varies in the time, as a function of their water storage, and their efficiency depends on the flood magnitude:

$$\text{Outflow} = \text{inflow} - \Delta S \quad \Delta S : \text{the storage change in a determined interval time}$$

No full reservoir has food control aptitude; thus, the efficacy in flood peaks mitigation varies significantly, according to peatland type, peatland cover rate, climatic and hydrological features, geomorphology, and other factors. That should be carefully interpreted, in order to avoid generalizations that have been a matter of the discrepancy between authors who argue that peatlands have high importance in flood mitigation and those who have obtained evidence to the contrary.

Peat bogs may store input water into the unsaturated upper layer and over its surface. Rain infiltration causes rapid WTL raising. The WTL elevation/precipitation rate in *Sphagnum* bogs of TDF is in average 3.3 ± 0.8, according to measures in several mires (Iturraspe 2010). Normal WTL in most *Sphagnum* bog lawns from TDF is about 17 cm below the surface (with lower values in hollows and higher ones in hummocks). So that signifies a 51 mm rain potential retention, that is right for TDF, where yearly precipitation on *Sphagnum* mires is 500–700 mm and daily rates, over 40 mm in 24 hours, are unusual.

Bog saturated areas produce water excess and surface flux that can keep in hollows and pools. The pool water level is always lower than the marginal WTL, which means a pool receives runoff from its micro water catchment and sub-surface flux from the acrotelm too. In wetter conditions, the output peatland response depends on the micro-hydrological mire systems connectivity. As the linkage is better, the surface water retention capacity decreases and the output flow increases. In the lowlands, the surface storage may be more significant that the inner storage.

For moderate to large storms preceded by relatively wet conditions, headwater bogs may have a little regulatory effect. Blanket peat catchments in the United Kingdom exhibit flashy regimes, and saturation excess overland flow dominates hillslope runoff (Evans et al. 1999, Holden and Burt 2003). The general blanket peat slope does not favor water retention on the surface, so peatland regulation is limited.

There are far fewer references from flood control by peatlands located in middle-low streams in gentle slopes. This is difficult to quantify due to the low rate of the peatland area with respect to the whole water basin area. Hydrological models are useful aids for understanding and quantifying the processes operating in a peatland. Furthermore, they can produce data on WTL and water discharge. However, numerical models are often oversimplified or misrepresent the complex structure of mire systems.

Discharge responses of patterned mires in Canadian subarctic regions are distinctive from those of other peatland types because the large capacity of the wetland pools delays the runoff (Quinton and Roulet 1998). Water movement in patterned peatlands is controlled by the nature and position of pools and ridges within the basin, and water volume retained in microrelief can be significant. These systems often have an endorheic behavior, except in special wet conditions. Peatlands in bottom valleys normally have a poor drainage capacity, so when it is exceeded by flow inputs, a part of the input volume keeps on flooded areas.

Iturraspe (2010) evaluated the water storage capability in Carbajal wetland, which occupies the middle-lower valley (150 m ASL) of Olivia River, near Ushuaia, TDF. It is a mire complex composed of patterned ombrotrophic raised bogs, mesotrophic fens, pools, lagoons, and rivers (Fig. 9.2). The main river runs along the wetland body and receives six tributaries from transversal mountain valleys that cross the mire.

This evaluation supposes a rain event in a standard wet antecedent condition, with a usual former WTL of 17 cm below the surface, 40 mm as the normal maximum daily precipitation in the year, and 24 m^3s^{-1} as the ordinary annual maximum flow of Olivia River. The calculated storage capability in the unsaturated peat layer, microrelief and pools, results able to hold almost all the local rain. Therefore, it is possible to suppose it produces negligible outputs. The main river and tributaries supply the major inputs to the peatland that inundate low wetland areas and temporally connected lagoons. This water mass was calculated using a digital terrain model. The total volume that holds in the wetland equals 35% of the runoff volume in 24 h. Applying the same procedure for an extraordinary rain-flood event, the related ratio is 19%. These results validate the reliable capacity of certain peatlands for flood mitigation, even appreciable in extreme events.

3.2.2 Erosion Control Rates and Sediment Transport

Peatland settings in the water basin determine different ways of sediment control. Typically, water erosion is an active process at the headwaters, given the high runoff energy. Natural peatlands have effective resistance to water erosion, and thus headwater areas covered by mires do not contribute with sediments to the fluvial systems. However, peatland degradation by drainage, intensive grazing, burning, or peat mining might cause peat erosion, with gullies reaching the mineral substrate. This way, water and wind mobilize and transport organic and mineral sediments.

Although these processes are more intense on the slopes, they can also occur on flat lands.

Floods cause high sediment transport, which results in the siltation of detention conveyance capacity. Sediment excess turns more expensive the water treatment for human consume. Lowland peatlands may contribute to sediment retention due to their limited capacity to drain incoming floods (Fig. 8.3).

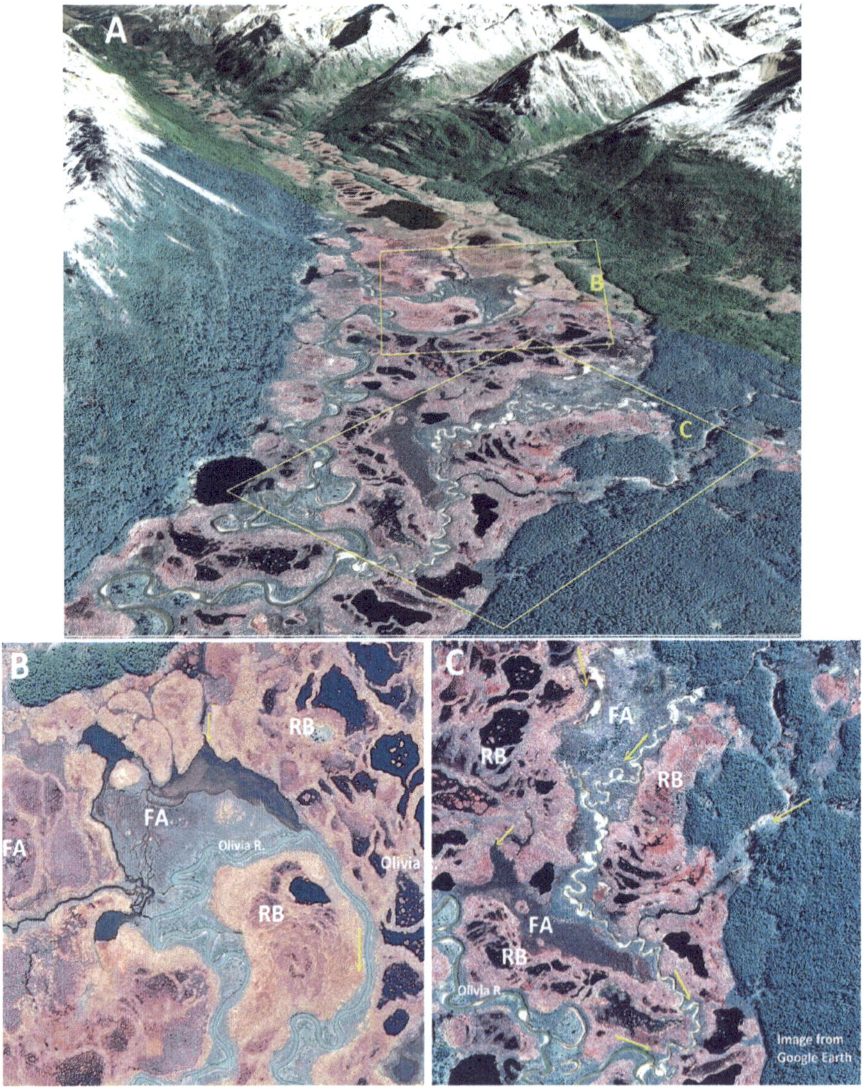

Fig. 8.3 The patterned wetland of the Carbajal Valley, TDF. (**a**) General view. (**b**) View of the surface water reservoirs in pools of raised bogs (RB) and floodable areas (FA). (**c**) A torrential tributary loses energy in the peatland and deposits sandy sediments (in white) in the riverbed and margins. Green areas adjacent to the stream are shrub and moss vegetation over sediments deposited during extraordinary floods

3.2.3 Regulation of Climate Change Through the Control of Carbon Cycle

The increasing concentration of greenhouse gases (GHG) in the atmosphere in the industrial era is the main factor in the radiative forcing of climate (the difference between incoming insolation and energy radiated back to space). The current concentration is the net result of the history of its past emissions and removals from the atmosphere (Solomon et al. 2007). For the 1995–2005 period, the growth rate of carbon dioxide (CO_2) in the atmosphere led to a 20% increase in its radiative forcing. That is indicative of the significance of CO_2 on the greenhouse effect and the meaning of the carbon (C) cycle over the climate change.

Soils are terrestrial C pools; therefore, soil C cycling is a key to develop efficient mitigation strategies of land use-induced C losses. In this context, peatlands are the largest terrestrial organic C store, because of the great quantity of C that they have accumulated since the end of the last glaciation. Mires are sinks for atmospheric C because decomposition is slowed down under anoxic under-logged conditions in prolonged absence of oxygen.

To evaluate the knowledge status of Patagonian peatlands as drivers of the C cycle, it is necessary to know the storage accumulated by them in the past and the rate at which they are accumulating it now (Clymo 1998).

The Peatland Carbon Store. How Much Carbon is Stored in Patagonian Peatlands?

A single peat coring allows measuring bulk density, OM, and C content at different layers to integrating these in the vertical profile. The regional C storage assessment requires these data for distinct kinds of peatlands, a proper estimation of peatlands extension, and the peat thickness-weighted average. This information is still insufficient for Patagonia, but first estimations have been improved through the last decade.

Yu et al. (2010) estimated 15 GtC (1 GtC = 1000 million tons carbon) considering a peatland extension of 45,000 km^2 in South Patagonia and C storage data from 16 peatlands. Loisel and Yu (2013) improved this result based on information from 24 sites to obtain a rate of 0.168 kg m^{-2} C and 7.6 GtC stored in Patagonia.

Finally, Loisel (2015) expanded the database to 52 peat coring in Patagonia and updated the C storage evaluation:

C density (CD): 28 kg/m^3
Thickness average (TA): 5.49 m
Mean C rate (CR = CD.TA): 154 kg/m^2
Total C = CR: 45,000 km^2 = 6.9 GtC

So that, C content in Patagonia would represent 1.26% from 547 GtC estimated by Yu et al. (2010) for the global C stored in 4.4 million km^2 peatland extension in the world. Loisel (2015) notices the uncertainty of the estimated peatland area as the main error factor.

In this sense, the peatland area south of 45°S (45,000 km^2) is overvalued and excludes peatlands of northern Patagonia. However, this error would decrease if this surface amount and the computed C storage were assigned to the entire Patagonian peatland area, that means considering 41°S as the northern limit of the study area.

On the other hand, the thickness average (5.49 m) has been also likely overestimated because it corresponds to a biased sample. Most of the peatlands in the database were selected for paleoclimate studies that required deep peat accumulation; so thin peatlands are not represented. The lowest basal depth data is 1.50 m, while seven sites in the database exceed 8 m thickness. Next, we mention general information related to peatland thickness in Patagonia that supports the idea of a lower peat thickness average.

- *Sphagnum* bogs have basal depths close the applied average value of 5.5 m, but bogs are not the dominant mire type. Vega-Valdés and Domínguez Díaz (2015) computed 2845 km^2 *Sphagnum* bog extension in the Region of Magallanes that is 13.5% of the peatland area in this region.
- Peat soundings in lowlands of Peninsula Mitre (Eastern TDF) where peatlands extend 2400 km^2 range from 2.6 to 5.7 m in depth, with around 4 m average (Grootjans et al. 2014). Peat depth at the headwaters of blanket peatlands is around 1.5 m.
- The inventory of a bog-fen-grassland-forest mosaic that extends 355 km^2 to the east of the Fagnano Lake (TDF) determined 2.6 m as mean thickness and 20% of peatland cover (Roig 2001).
- Rodríguez Martínez (2015) calculates 0.76 m as the basal deep average in Aysén, Chile, for ten representative peatlands in the Baker and Pascua basins. Maximal depth of each one ranges from 0.80 m to 4.17 m.
- Much peatlands corresponds to the Pacific Archipelago. In this area prevails hyperoceanic cushion bogs, which are 25–100 cm deep. Higher accumulations are limited to small level areas, scattered on these rugged and sloping terrains (Pisano 1977, Holdgate 1961, Roig & Faggy 1985, Domínguez Díaz et al. 2015).

These references suggest average peatland thickness in Patagonia would be likely lesser than 2.5 m, which would imply a C amount storage less than a half of what is estimated by Loisel (2015) and 1/5 of what is estimated by Yu et al. (2010).

A more accurate estimate of peatland coverage and peat deep average will improve the important advances reached in the evaluation of the regional C storage.

Peatlands as a Carbon Sink: The Current Carbon Balance

The current C balance in peatlands is the second point denoted by Clymo (1998) to find out the present accumulation C rates of peatlands in Patagonia. Peat accumulation rate is widely driven by the physiological state of the primary producers, and a limitation of photo-assimilation by external conditions can turn it into a source of

CO_2 (Naumov et al. 2020). Decomposition must be less than the net primary productivity (NPP) over time for peat to accumulate.

Peat coring dating at a level not far to the surface allows determining the recent C accumulation rate (RERCA) usually in the order of 100 years. RERCA can be considered as indicative of the current accumulation rate; however, the present productivity of peat bogs may differ due to the climatic instability proper of the last century.

Leon and Olivan (2014) published RERCA data for five sites in Chiloé: two oligotrophic *Sphagnum* bogs and three anthropogenic peatlands. Results range from 9.4 to 15.5 $gCm^{-2}\ y^{-1}$ for bogs and 32.8 to 58.2 $gCm^{-2}\ y^{-1}$ for anthropogenic peatlands.

Gas Fluxes Between Peatlands Surface and Atmosphere

Net C flux between ecosystems and atmosphere is a key to understanding the C balance. Photosynthesis is the primary C input way to ecosystems; soil CO_2 efflux is a large respiratory flux and a critical component of the global C cycle (Riveros et al. 2008). Primary productivity is the formation rate of the biomass created by plants through photosynthesis, resulting from the difference between the rates of photosynthesis and autotrophic respiration (Rydin et al. 2006). Respiration produced by roots and soil organisms is the primary pathway for CO_2 fixed by plants to return to the atmosphere (Peri et al. 2015).

Wetlands are likely the main natural source of the CH_4 in the atmosphere. Mires are sources of this GHG. As well as other wetlands, peatlands are good habitats for methanogenic archaea that form CH_4 during the OM decomposition. Methanogens require OM without oxygen, and permanent peatland's saturated layers comply with that.

Dissolved organic carbon (DOC) and particulate organic carbon (POC) are the results of the partial decay of OM in the peat matrix pore. Usually, DOC and POC are outputs of the peatland C balance to the groundwater, or to the surface flow.

Nitrous oxide (N_2O) emissions from soils are the result of the N cycling in the soil. N_2O is a GHG that has a low concentration in the atmosphere, but its influence on the radiative forcing is considerable (Moore 1994).

GHG Flux in Patagonia

A reduced number of studies have generated GHG flux field data from peatlands of Patagonia, and all of them correspond to the last decade. Table 8.2 shows a literature review of GHG emissions for Patagonia published until July 2020. We selected only specific papers that provided new data, acquired by in situ measuring in a closed chamber method (Hutchinson and Livingston 1993). We have homogenized the units of measurement.

Table 8.2 Literature review on GHG emissions in Patagonian peatlands. N_2O, mg N_2O $m^{-2}d^{-1}$; CO_2, g CO_2 m^{-2} d^{-1}; CH_4, mg CH_4 m^{-2} d^{-1}; GPP, TRE, NEE, g C $m^{-2}y^{-1}$. *AP* anthropogenic peatland. Negative values indicate ecosystem sink behavior

Region	Site	Lat S Long W	Alt. [masl]	Ecosystem	Period	Dominant veg	N_2O	CO_2	CH_4 Lawn	CH_4 Pool	CH_4 Mean	Reference
TDF-Arg	Moat	54°58' 66°44'	40	Cushion bog	Dec 2008 Nov 2009 Mar 2009	*Astelia pumila*, *Donatia fascicularis*			~0	~0	<1	Fritz et al. (2011)
						Clipped *Astelia pumila*, and *Donatia fascicularis*			2–19			
						Sphagnum patch			1–11			
	Valle Andorra	54° 45′ 68°20'	200	Raised bog		*Sphagnum magellanicum*			1–14		7.5	
Magallanes, Chile	SkyI	52°31' 72°08'	16	Cushion bog	Mar–Apr 2010	*Sphagnum*, *Empetrum*, *Astelia pumila*		2.4	< 3.2		< 3.2	Broder et al. (2012)
	SkyII	52°28' 71°56'		Raised Bog		*Sphagnum*, *Empetrum rubrum*		1.8	< 3.2		< 3.2	
	PBr2	53°38' 70°05'	15	Raised bog		*Sphagnum*, *Empetrum rubrum*		2.2	< 3.2		< 3.2	
TDF-Arg	Valle R. Pipo	54°50' 68°27'	40	Raised bog (N)	Jan 2015 (4 days)	*Sphagnum magellanicum*			49.0	5.4	13.1 to 21.8	Lehmann et al. (2016)
						Empetrum rubrum			3.4			
TDF-Arg	Harowen	54° 44′ 67°55'	150	Sph.Bog	Nov 2014 (6 days)	*Sphagnum magellanicum*	0.37	2.4	7.1		7.1	Veber et al. (2018)
	Laguna Esperanza	54° 21′ 67°45'	60	Managed Fen: moder. grazing		*Carex* spp., *Deschampsia antarctica*, *Hordeum* spp., *Alopecurus magellanicus*	−1.57	5.0	22.3		22.3	

Region	Site	Lat S Long W	Alt. [masl]	Ecosystem	Period	Dominant veg	N_2O	CO_2	CH_4			Reference
									Lawn	Pool	Mean	
TDF-Arg	Moat	54°58' 66°44'	40	Cushion Bog	Dec 2014–Mar 15; F-Mar 16	*Astelia pumila*			1.4			Münchberger et al. (2019)
						Donatia fascicularis			10.6	3.7		
						Sphagnum magellanicum			24.4			
Chiloé - Chile	Ancud	41° 50′ 73°48'	70	AP	Sep 2014–Aug 2015	*Sphagnum* spp.	−0.01 to 0.0	6.9			−0.3	Urrutia (2017)
				Temp. rain Forest		*Podocarpus nubigena Nothofagus nitida, Drimys winteri*	−0.04 to 0.1	14.1			−1.1	
TDF-Arg	Moat	54°58' 66°44'	40	Cushion bog	Feb 2016–Mar 2018	*Astelia pumila*		(g C $m^{-2}y^{-1}$) GPP:-777; TRE: 655;NEE:-122				Holl et al. (2019)
	Valle Río Pipo	54° 50′ 68°27'	40	Sph. bog	Mar 2016–Mar 2018	*Sphagnum magellanicum*		(g C $m^{-2}y^{-1}$) GPP: −393;TRE: 366; NEE: −27				
Chiloé - Chile	Ancud	41°52′ 73°40'	70	AP undisturbed	18 m.: Apr 2015–Oct 2016	*Sticherus cryptocarpus Baccharis patagonica*		(g C $m^{-2}y^{-1}$) GPP: −1053; TRE:1016 NEE:-135				Valdés Barrera et al.(2019)
				AP disturbed		*Sphagnum* spp. *Exotic* spp.		(g C $m^{-2}y^{-1}$) GPP:-981; TRE: 972; NEE:-32				

Fritz et al. (2011) published the first results on gas fluxes in peatlands of Patagonia, reporting close to null CH_4 emissions from *Astelia* cushion bogs (TDF) and comparing them with the flux in *a Sphagnum* bog. The distribution of cushion-forming *A. pumila* root density and associated O_2 supply strongly controlled the CH_4 production and consumption. However, Munchberger et al. (2019) noticed patches of *Donatia fascicularis* with a weak root density that accelerates CH_4 production and increases emissions to intermediate level by aerenchymatic roots.

Broder et al. (2014) reported low CH_4 and CO_2 flux rates in three *Sphagnum* bogs located around 50 km of Punta Arenas city. They suggest sea spray and a lack of essential trace metals, such as nickel, are likely factors that constraint CH_4 and CO_2 emission.

Lehman et al. (2016) analyzed CH_4 spatial variability in a pristine *Sphagnum* bog in TDF, with contrasting results: major emission was from *Sphagnum* lawns, 49.04 ± 25.7 mg m^{-2} day^{-1}, and the lower one from *Empetrum* fluxes, 3.97 ± 2.99 mg m^{-2} day^{-1}. The *Sphagnum* lawn emission is the higher CH_4 flux rate measured in Patagonia.

A transect from Quebec, Canada, to TDF for a comparative study of gas emissions from peatlands in America (Veber et al. 2018) included measures in TDF, Argentina, in a pristine bog and a managed fen. The CO_2 emission rate in the pristine bog of TDF (26.8 mg C h^{-1} m^{-2}) was the lowest one recorded in this transect.

Holl et al. (2019) determined in TDF the annual CO_2 net ecosystem exchange (NEE) in an *Astelia* cushion bog and in a *Sphagnum* bog. The NEE uptake was 4.5 times larger in the *Astelia* bog than in the moss-dominated bog. The balance of the last one was within the typical range of data from similar raised bogs of the Northern Hemisphere. These results would not be applicable to cushion bogs of the hyperoceanic Pacific shore due to differences in depth and peat structure.

In addition to the information corresponding to southern Patagonia, just mentioned, there are also GHG flux data from the north of the region. Urrutia (2017) carried out simultaneous measures in Chiloé, in temperate rain forest soils and anthropogenic peatland (AP), resulting in close to zero N_2O flux rates and weak negative CH_4 values for both, the forest and the peatland. Until now, all measures made in Patagonia demonstrated very low N_2O emission rates.

Valdés-Barrera (2019), also in Chiloé, determined CO_2 NEE, comparing managed with no managed areas of an AP. They found very low CO_2 NEE rates in the managed sector and a more significant CO_2 retention in the no managed one.

In the order of a better understanding of the general terrestrial C cycle, we compiled results of GHG flux measurements in soils of diverse Patagonian ecosystems (Table 8.3) to facilitate the comparison of peatland emissions with these.

Dissolved and Particulate Organic Matter

Water movement through the peat is an exit way of C to pools, surface streams, or groundwater. Holden et al. (2012) remark that pipes (natural tunnels on peat) are important pathways for fluvial C export. The quantification of the dynamic of water

Table 8.3 GHG soil emissions in no peatland ecosystems of Patagonia. Units: N_2O, mg $m^{-2}d^{-1}$; CO_2, g m^{-2} d^{-1}; CH_4, mg m^{-2} d^{-1}

Region	Site/purpose	Lat S Long W	Alt. [masl]	Ecosystem	Period	Dominant veg	N_2O	CO_2	CH_4	Reference
Coyhaique Chile	Comparing ecosystem management	45°25' 72°00'	730	Man.Nat pasture	Nov 2007–Nov 2009	Perennial grasses		5.1		Dube et al. (2013)
				Pinus Pond. plantation		*Pinus Pond.*		4.3		
				Silvopastoral		Perennial grasses		4.6		
Santa Cruz Arg	Comparing ecosystems	51°31' 70°04'	300	Grass Steppe (S)	2011–2013, seasonal	*Jarava chrysophylla*, *Carex* spp., *Poa* spp.		9.6		Peri et al. (2015)
		51°07' 70°59'	300	Shrub S		*J. tridens* grasses		8.2		
		51°13' 72°08'	350	Andean S		*Festuca pallescens*, *Carex* spp., *Poa* spp.		14.6		
	Comparing grazing intensity	51°33' 69°17'	150	Dry S moderated grazing		*Festuca gracillima* grasses		13.9		
		51°31' 69°31'	160	Hum. S High grazing		*Festuca gracillima, F. magellanica* grasses		9.8		
	Comparing land use	51°13' 72°16'	330	Primary forest		*Nothofagus antarctica*, herbs and shrubs		18		
		51°13' 72°15'	240	Silvopastoral		*Nothofagus antarctica*, *C. andina*, *Poa pratensis*		18		
		51°13' 72°08'	320	Open grassland		*Festuca pallescens*, *Carex* spp., *Poa* spp.		14.4		
N Patag. Arg.	Río Mayo	45°24' 70°17'	540	Vegetated patch	Jun 2015–May 2016	Shrubs and grasses		1.18		Carbonell Silletta et al. (2019)
				Bare soil patch				0.80		
				Average steppe				1.06		
TDF-Arg	Martial Valley, Ushuaia	54°47' 68°24'	600	Tree line	29 Nov–5 Dec 2011 and 1–6 Feb 2012	*Nothofagus antarctica*	−0.07	8.4	−1.70	Médice Firme Sá et al. (2019)
			660	Lower Andean tundra		Cushion veg: *Bolax gummifera*, *Empetrum rubrum*	0.01	16.7	−0.14	
			700	Upper Andean tundra		*Bolax gummifera Empetrum rubrum*	0.02	18.0	0.15	

AP Anthropogenic peatland, *S* steppe

discharge from bog complexes is significant for the C balance, because part of the organic C accumulated by primary production is decomposed, released, and dissolved in superficial bog water, to be subsequently exported to fluvial systems (Moore et al.1998).

There are not focused papers on DOC fluvial outputs from peatlands of Patagonia, neither in bogs lateral flux. However, some references are available: Garcia et al. (2017) determine 15.4 ± 6 mg l^{-1} as DOC average concentration in 26 pools of 2 *Sphagnum* bogs in Southern TDF, Argentina. Broder et al. (2012) report DOC concentration in the peat matrix from three *Sphagnum* bogs ranging from <30 mg l^{-1} to 160 mg l^{-1} increasing with depth. In eastern TDF, where mires dominate the landscape, dark brown-colored river waters with pH close to 5 indicate high rates of DOC fluvial transport from the peat soils to the sea (Iturraspe et al. 2012).

3.2.4 Maintaining Habitats

Patagonian peatlands maintain habitats of particular biodiversity that includes a range of rare and specialized plants and animals. These habitats present specific features that respond to the different peatland types. Peatlands are barriers to alien species invasions because their acidic and anoxic conditions constitute a strong restriction for not specilized plant development.

Domínguez Díaz et al. (2015) identified 126 species in 50 mires of Magallanes and none exotic, while a similar study found 8 invader aliens, between a total of 24 species, in a drained and abandoned peatland after peat extraction, 20 years ago (Domínguez Díaz et al. 2012). Graminoid fens and cushion bogs have a greater number of species than *Sphagnum* bogs, where mosses and hepatic flora prevail over vascular plants.

The richness of species increased in northern Patagonia. León et al. (2014) reported 129 species in *Sphagnum* bogs and swamp forests of *Tepualia stipularis* in Chiloé, which were distributed in 50 mosses, 52 liverworts, and 27 macrolichens.

Peat bogs are not a preferred habitat of large and medium mammals. However, fens and wet meadows of the extra-Andean region provide water and grazing to guanacos (*Lama guanicoe*) and are eventual hunting areas of pumas (*Puma concolor*), red foxes (*Lycalopex culpaeus*), and introduced grey foxes (*Lycalopex griseus*). These wetlands also support sheep and cow grazing. A particular anthropogenic case is an overpopulation of wild cows and horses in the isolated lands of Península Mitre, eastern TDF. These animals escaped from neighboring farms and bred there.

Bogs and fens could not stop beaver (*Castor canadensis*) spreading throughout TDF and adjacent islands. This introduced species can transform their peatland habitat, building dams that flood them.

Bird's presence in mires depends on habitat factors, such as food availability, wildness, and the structure of dominant vegetation (Riveros et al. 2015). Peatlands of Patagonia are wild environment, without significant human perturbations, with low productivity and very variable vegetation features. No bird species have peatland as unique habitat, but several of them develop the ability to adapt to it. No more than 25 species nest or remain in peat bogs of TDF (Schlater 2004). Riveros et al. (2015) identified 46 bird species in mires of the Magallanes Region. *Sphagnum* bogs have the lower bird richness, but bird diversity increases in *Sphagnum* bogs with shrubs. Focused studies in peatlands are scarce because bird cadasters are usually made with a regional scope.

Although the entomofauna in mires is less rich than in other ecosystems of the region, it verifies an important function in the transfer of energy, sustaining amphibian and bird populations (Jerez and Muñoz-Escobar 2015). The highest diversity corresponds to the order Coleoptera.

Peatlands include aquatic habitats, such as endorheic pools, lagoons, and streams, with specific biodiversity. Ortiz (2015) denotes four amphibious species in peatlands of southern Aysén and Magallanes regions; they do not show a situation of vulnerability, but peat extraction is not compatible with the preservation of them.

In Tierra del Fuego, García et al. (2017) recorded 29 taxa of acuatic microinvertebrates in two peat bog. The authors denote notably contrasts in the environmental characteristics of pools in the same peatland that explain differences in species richness and diversity among communities of microinvertebrates of these habitats. They highlight the importance of *Sphagnum* moss as a low diversity extreme environment that supports endemic species.

3.3 Cultural Services

Cultural ES emerge out of the relationships between ecosystems and humans (Fish et al. 2016); they have a strong connection with the roots of the population and the links between ecosystems and local traditions. These features assign to the cultural ES an evolutionary character.

3.3.1 Physical and Experiential Interactions with Natural Environments

The scenic value of landscapes, notably in cases of mire landscapes, depends on perception as a subjective social component. Pungas-Kohv et al. (2015) explain mire perception changes in Estonia over the twentieth century; initially, they have been considered useless and sometimes dangerous. Consequently, many peat bogs disappeared because of their land use change. Next, the industrial mentality is dominated that evaluated the natural environment in terms of direct resources exploitable for human purposes. The recognition in the last decades of ecological values, as well as the recreational potential of them, has changed the former perception to appreciate pristine, rare, and contrasting environments. On the other hand, current farmers and naturalists, for example, contrast in the way they view different kinds of landscapes; thus, the perception is cultural, as well. In comparison with Northern Europe, where mires were part of the history and the life of nations, the cultural factor in Patagonia, as a contribution to a positive peatland perception, is weak, in general terms. The experience seeing a bog from a viewpoint of someone, who has never explored anyone and does not know about its millenary age or its environmental services, is able only to feel aesthetical features of this landscape. Living and interactive experiences on the site signify a more substantial involvement of visitors than the simple contemplation (Pungas-Kohv et al. 2015). A long walk by a bog, with the attention on pools, animals, and vegetation tips (like small flowers, moss fibers, or carnivorous plants), is a way to impact on visitors.

Despite this context, and recognizing that peatlands differ on its aesthetic values, many peatland sites achieve, in Patagonia, objective scenic attributes, in terms of water, colors, singularity, wilderness, loneliness, etc.

Domínguez Díaz and Bahamonde (2012) analyzed methodologies to value the quality of peatland landscapes, through the case study of a bog near Punta Arenas, Chile. They recommend introducing techniques for landscape quality assessment to the planning of peatland management to avoid the scenic environmental degradation.

Peatlands are singular components of the Patagonian landscape, notably in the Andean Patagonian forest ecoregion. Glaciers, lakes, sea, native forest, mountains, and peatlands compose their unique and remarkable landscapes that contribute to the well-being of local inhabitants. Tourist agencies have not yet taken advantage of the wild mire's value as an attraction; meanwhile, mires work as anonymous actors in the landscape concert that nature play in the beautiful tourist places of Patagonia.

In mountain valleys of TDF (Argentina), *Sphagnum* bogs give support to tourist and recreational winter activities like cross-country skiing, snow racket walking, dog sledding, ice skating on natural pools, and, recently, speed riding.

Although the main interest of visitors focuses on the nearby mountain sky center, these winter sports on the bogs represent a great complement and improve the tourist offer. In winter 2017, TDF received 60,000 visitors that contributed to the local economy. That is a case example of the benefits that people may get from bogs, through sustainable use.

Several public or private reserves in Patagonia, with peatlands as a plain attraction, at the time that protect these nature heritages, promote the regional tourism. Some of them are the Tantauco Park (southern Chiloé), the Ramsar site in the Andorra Valley (TDF), and the Omora Ethnobotanical Park (Navarino Island, Chile). The last one offers a tourist circuit through a micro-forest of mosses, lichens, and hepatics, by using a magnifying glass.

3.3.2 Intellectual and Representative Interactions: Peatland as Data Archive for Scientific Investigation

Peatlands accumulate in situ most of the OM they produce and, with it, anything that falls on its surfaces, such as pollen or tephra. This way, they have preserved paleoenvironmental and paleoclimatic data over thousands of years.

This peatland characteristic enabled, one century ago, scientific researches in Patagonia, related to the history of the vegetation and climate evolution in the last 18,000 years.

In the 1920s, pollen preserved in peat bogs became a tool to study the Quaternary vegetation and climate change, leading to the expansion of this research line in Northern Europe. At that time, Carl Caldenius, who was studying glacial deposits in the Argentinean Patagonia, extracted two cores 150 cm long from a peatland located in TDF, at the eastern head of Fagnano Lake. He sent them to the Swedish scientist Von Post to be analyzed. The pollen diagram made by Von Post (Von Post 1929) from these peat samples represents the first vegetation history record from South America (Markgraf 2016).

Väinö Auer developed the first extensive peatland field works on paleoclimatic research, between 1928 and 1952 (Tuhkanen 1997). Auer's work demonstrated that pollen records obtained from bogs of Patagonia contained a sequence of paleoenvironmental changes for which chronological control was critical, in order to establish synchroneity of events during a period of major global paleoclimatic changes (Markgraf 1983).

Radiocarbon-dated basal peat determines a minimum age for previous geological processes, like ice retreating or sea-level drop. Peatlands preserve tephra layer sequences from volcanic eruptions in the Holocene.

Plant macrofossils analysis can be used to reconstruct vegetation development as a paleoclimate proxy. This method was used by the first time in Patagonia in Andorra bog-TDF (Mauquoy et al. 2004). Multiproxy analysis improves and complements paleoclimate data. Van Bellen et al. (2014) used testate amoeba (a globally dispersed unicellular protist, and abundant in *Sphagnum* bogs) as a proxy for reconstructing Holocene water table dynamics in bogs of TDF. New techniques, such as peat humification, biomarkers, stable isotopes, inorganic geochemistry, etc., offer complementary proxies, enhancing the scientific potential of peatlands to understand paleoenvironmental changes.

Peat coring is a tool for archaeology. Charcoal particles preserved in peat, out of reach of volcanic activity (e.g., in TDF), are indicative of anthropological fires, revealing ages of human occupation.

4 Conclusions

We compiled an arranged listing of peatland's ES in Patagonia, following the CICES V 5.1 structure, with the aim to assess the contribution of these ecosystems to human well-being, in order to offer useful information and references for the wise use of peatland in Patagonia.

Peatlands provide peat, moss fibers, and water, as well as the natural products that grow on its surface. Even though the current *Sphagnum* peat extraction has a relatively low rate, it is concentrated in areas near the population settlements, affecting the ES that they offer. Fiber harvesting is an extractive use that should be carefully planned and controlled. The fast spatial spreading of such interventions is threatening the future of *Sphagnum* ecosystems. Water provision, as well as several regulation services, is not easy to quantify.

The ES respond to complex processes that depend not only on peatland features but also the climate. We indicate effective case examples in Patagonia of water provision as well as flood and sediment control, but we emphasize that different peatlands may contrast in the kind of ES that they provide, as well as in their efficacy and importance. This remark has a special validity for Patagonia, due to the dissimilar peatland environments in this region.

The global meaning of peatlands in the regulation and mitigation of climate change stimulated last years the scientific interest in the C stock and GHG fluxes of these ecosystems in Patagonia. We reviewed the progress in the valuation of C storage, which has been estimated at 6.9 GtC in the last calculation by Loisel (2015). In our opinion, this result should be even improved through a more proper estimation of peatland thickness and extent.

The compiled information regarding GHG fluxes shows this research line is only 10 years old in Patagonia, with very scarce data still on this topic; GHG emissions have a great variability and depend on many variables. Notably, a significant part of the GHG data corresponds to particular peatland types of Patagonia, like the cushion *Astelia* bogs of TDF and the anthropogenic peatlands of Chiloe. Both denote near null CH4 emissions and a good efficiency as C sinks.

Patagonian peatlands have played a significant role in ratifying the global incidence and synchronicity of climatic instability episodes during the Late Glacial Interstadial and the Holocene, which were initially identified just for the Northern Hemisphere.

Mires also allow the study of contrasts and similarities on carbon accumulation – climate relationships in the past and the present – under unlikely environmental conditions than as that of the northern peatlands. Patagonia expands the context to improve the understanding of potential climate change effects on ES of the world's peatlands relating to the C cycle.

Most of the services and benefits described and analyzed in this chapter correspond exclusively to natural peatlands. Degraded peatlands are not C sinks but sources; its aesthetic value is usually negative; they don't assure protection from erosion, but rather induce ditch's formation, contributing to organic and inorganic sediments; drainage modifies its regulatory capacity and damages the natural biodiversity in favor of invasive species.

The majority of peatlands in Patagonia maintain its original state. That is their main attribute ensuring the efficacy of ES; however, such state has been disturbed in the environs of most populated areas.

Peatland management policies must turn on from debates in cross-sectorial decision-making, with an ecosystem approach, and apply ES concepts and examples of them. This way, it is possible to contrast the continuity of services and goods that these ecosystems provide to the entire society, versus occasional benefits reaching sometimes the community and other times a particular sector. To this end, scientists and professionals must continue to provide evidence-based references giving proven testimony of derived benefits from ES.

Most peatlands are owned by the state in both countries, usually, and especially in Chile, under a protected status. It should be noted that a majority of these mires are located far away from inhabited areas. Peatland services become effective when local persons are benefited from them. Several ES, like water provision for human use, flow regulation, erosion control, maintaining biodiversity, and aesthetic or cultural benefits, are examples of services that can only be given by peatlands located near inhabited areas, but these nearby wetlands are frequently out the protected areas.

The National System of Wild Protected Areas of Chile includes 71% of the existing peatland area in the Magallanes Region (Vega-Valdés and Domínguez Díaz 2015), 21% in Aysén, and 2% in Los Lagos. The last one, with lesser protection, is the most populated region of the Chilean Patagonia. As expected, as the population increases, the protected peatland area decreases. In Chiloe, many peatlands from private owners are subjected to a productive management with negative environmental consequences.

It is not enough to protect large expanses of peatlands in remote places; supplementary policies and regulations are necessary for a sustainable or at least a rational management of nearby peatlands that provide valuable local services.

In TDF, Argentina, most of the mires are in public lands, at the wild eastern side of the island. Several law projects were presented to the Provincial Parliament in the last 20 years, to protect a 1930 km^2 mire area, which represents 71% of the total in the Province. However, none of these law projects has had so far a positive treatment.

The National Mining Law was the unique regulation for peatland use in TDF (Argentina) until 2008. That year, the government formulated a participative strategy for the wise use of peatlands that among other actions included a moratorium on new concessions and delimited the zoning plan for peat mining, which remains currently in effect. No granting of new concessions outside this established area is permitted. In addition, applicable conditions in the peat-mining zone were established (Iturraspe and Urciuolo 2014). These measures ended a disordered proliferation of peat concession requests in inappropriate sites. At the time of the moratorium, these demands involved most of the peatlands accessible by road. Likewise, no peat extraction from fens is permitted in TDF, to maintain livestock use of them.

The rest of the Patagonian peatlands of Argentina has no specific regulations, in addition to the mining law, for peatland management. The Argentinean provinces have the competence in planning the use of its natural resources, among which, these ecosystems are not abundant, and even less, *Sphagnum* bogs, which are the most required for peat mining. Therefore, it would be significant to discuss into each provincial ambit about the convenience to avoid extractive activities, before first requests for concessions appear.

Chile is working on policies and measures that lead to better use of peatlands. Stricter requirements for peat mining and environmental impact studies have been established, and since 2014, new regulations for fiber harvesting have been enforced.

Despite these advances, it is still necessary to continue enhancing the peatland management, with proper policies, regulations, knowledge, and education. On this way, we present below some criteria and actions that can be considered.

Concessions for peat extraction should be subject to the opinion of other technical areas of the state, in addition to the mining authority (e.g., water resources, environment, etc.) to avoid possible loss of valuable ES from the competence of such areas.

Peat mining bad practices should be corrected. Drain closures should be progressively established, according to present planning and before the mining company leaves the bog, avoiding the abandonment of peat bogs with open drains.

Environmental impact studies for peat mining projects should consider the ES evaluation and its affectation.

Peatlands that allow a sustainable alternative use (i.e., tourism, education) should be preserved from peat extraction, as well as those with rarity attributes or special aesthetic value. It should be protected, notably peatlands located in water basins that provide water to populations and those which maintain flood or sediment control of rivers that run in the urban areas.

Peat draining should be avoided because it causes wetland degradation. In addition to it, dry peat is a highly combustible material, therefore drained bogs are potential sources of fire that spread to forest. Road's drainages in peatlands degrade extensive mire stretches. It should be avoided through suitable constructive ditch road specifications for peaty terrains.

Specific regulation for urban peatland use and conservation is essential.

Livestock in fens or wet meadows is normally owned by private farmers, who seek to optimize its production through sustainable management. However, they are often unaware of the fragility of these ecosystems. For this reason, training and assistance from the state agencies are important.

On the other hand and considering the concern of people as essential for ecosystem conservation, the government should promote education and dissemination policies on peatland values, particularly where they represent the dominant and distinctive wetland ecosystem.

References

Arroyo MTK, Pliscoff P, Mihoc M, Arroyo-Kalin M (2005) The Magellanic Moorland. In: Fraser L and Keddy P (eds) The World's Largest Wetlands. Ecology and Conservation, Cambridge Univ Press, Cambridge, p 424–445

Bannister J, Kremer K, Carrasco N, Galindo N (2017) Importance of structure for species richness and tree species regeneration niches in old-growth Patagonian swamp forests. Forest Ecol Manag 401:33–44. https://doi.org/10.1016/j.foreco.2017.06.052

Broder T, Blodau C, Biester H, Knorr KH (2014) Sea spray, trace elements, decomposition patterns as possible constraints on the evolution of CH_4 and CO_2 concentrations and isotopic signatures in oceanic ombrotrophic bogs. Biogeochemistry 122:327–342

Broder T, Blodau C, Biester H, Knorr K (2012) Peat decomposition records in three pristine ombrotrophic bogs in southern Patagonia, Biogeosciences, 9, 1479–1491. https://doi.org/10.5194/bg-9-1479-2012

Carbonell Silletta L, Cavallaro A, Kowal R et al (2019) Temporal and spatial variability in soil CO_2 efflux in the patagonian steppe. Plant Soil 444:165–176. https://doi.org/10.1007/s11104-019-04268-7

Carretero E (2004) Los turbales Patagónicos. En: Blanco D y de la Balse V (eds) Los turbales de la Patagonia: bases para su inventario y la conservación de su biodiversidad. Wetlands International, Buenos Aires, p 45–71

Chimner R, Bonvissuto GL, Cremona M et al (2011) Condiciones eco-hidrológicas de humedales a lo largo de un gradiente de precipitación en la Patagonia, Argentina. Ecol Austral 21(03):329–337

Clymo RS, Turunen J, Tolonen K (1998) Carbon accumulation in peatland. OIKOS 81:368–388. Copenhagen

Collantes M, Escartín C et al (2013) Grazing and grazing exclusion along a resource gradient in Magellanic meadows of Tierra del Fuego. Rangeland Ecol Manag 66(6):688–699. https://doi.org/10.2111/REM-D-11-00201.1

CONAF (2012) Catastro de los recursos vegetacionales nativos de Chile. Monitoreo de cambios y actualizaciones 1997-2011. Corporación Forestal Nacional, Santiago

CONAF, CONAMA, BIRF (1999) Catastro y Evaluación de Recursos Vegetacionales Nativos de Chile. Corporación Forestal Nacional, Santiago de Chile

Daily GC (ed) (1997) Nature's services: societal dependence on natural ecosystems. Island Press, Washington

De Groot R, Wilson M, Boumans R (2002) A typology for the classification, description and valuation of ecosystem functions, goods and services. Ecol Econ 41:393–408

Díaz MF, Larraín J, Zegers G, Tapia C (2008) Caracterización florística e hidrológica de turberas de la Isla Grande de Chiloé, Chile. Rev Chil Hist Nat 81:445–468

Domínguez Díaz E (2014) Manual de Buenas prácticas para el uso sostenido del musgo *Sphagnum magellanicum* en Magallanes, Chile. Boletín INIA 276. Inst. Investigaciones Agropecuarias-CRI Kampenaike. Punta Arenas

Domínguez Díaz E, Bahamonde N (2012) Manual de evaluación de turberas de *Sphagnum*. Caso de estudio efectos de la extracción de turba sobre el paisaje. Región de Magallanes. INIA 256, Punta Arenas

Domínguez Díaz E, Bahamonde N, Muñoz-Escobar C (2012) Efectos de la extracción de turba sobre la composición y estructura de una turbera de *Sphagnum* explotada y abandonada hace 20 años, Chile. Anales del Instituto de la Patagonia 40(2):3–45

Domínguez Díaz E, Vega-Valdés D, Dollenz O et al. (2015) Flora y vegetación de turberas de la Región de Magallanes En: Domínguez Díaz E, Vega-Valdés D (eds). Funciones y servicios ecosistémicos de las turberas en Magallanes. INIA 33. Inst. Invest. Agropecuarias. CRI Kampenaike. Punta Arenas, p 149–195

Dube F, Thevathasan N, Stolpe N et al (2013) Selected carbon fluxes in Pinus ponderosa-based silvopastoral systems, exotic plantations and natural pastures on volcanic soils in the Chilean Patagonia. Agrofor Syst 87:525–542

Evans MG, Burt TP, Holden J, Adamson JK (1999) Runoff generation and water table fluctuations in blanket peat: evidence from UK data spanning the dry summer of 1995. J Hydrol 221:141–160

Fish R, Church R, Winter M (2016) Conceptualising cultural ecosystem services: a novel framework for research and critical engagement. J Ecosyst Serv. https://doi.org/10.1016/j.ecoser.2016.09.002

Fritz C, Pancotto V, Elzenga J, Visser EJ, Grootjans AP, Pol A, Iturraspe RJ, Roelofs J, Smolders A (2011) Zero methane emission bogs: extreme rhizosphere oxygenation by cushion plants in Patagonia. New Phytol 190:398–408

Fuertes Lasala E, Prada C, Rolleri C (2008) Humedales Patagónicos de la Argentina. Flora criptogámica (briófitos y pteridófitos). Museo 3(22):85–89

Garcia P, García R, Marinone MC et al (2017) Aquatic microinvertebrate abundance and species diversity in peat bogs of Tierra del Fuego (Argentina). Limnology 18:85–96. https://doi.org/10.1007/s10201-016-0492-9

Grootjans A, Iturraspe R, Fritz C, Moen A, Joosten H (2014) Mires and mire types of Peninsula Mitre, Tierra del Fuego, Argentina. Mires Peat 14. Online: http://www.mires-and-peat.net/pages/volumes/map14/map1401.php

Haines-Young, R, Potschin MB (2017) Common International Classification of Ecosystem Services (CICES) V5.1 and guidance on the application of the revised structure. Available from www.cices.eu

Holden J, Burt T (2003) Runoff production in blanket peat covered catchments. Water Resour Res 39(7):1191. https://doi.org/10.1029/2002WR001956

Holden J, Smart RP, Baird AJ et al (2012) Natural pipes in blanket peatlands: major point sources for the release of C to the aquatic system. Glob Ch Biol 18(12):3568–3580

Holdgate M (1961) Vegetation and Soils in the South Chilean Islands. J Ecol 49(3):559–580

Holl D, Pancotto V, Heger A, Camargo S, Kutzbach L (2019) Cushion bogs are stronger carbon dioxide net sinks than moss-dominated bogs as revealed by eddy covariance measurements on Tierra del Fuego, Argentina. Biogeosciences 16:3397–3423

Hutchinson GL, Livingston GP (1993). Use of chamber systems to measure tracegas fluxes. Agric Ecosyst Eff Trace Gases Glob Clim Change 63–78

Instituto Forestal (2020) El Sector Forestal Chileno 2019. Ministerio de agricultura, Chile. https://wef.infor.cl/sector_forestal/sectorforestal.php

Iturraspe R (2010) Las turberas de Tierra del Fuego y el cambio climático global. Fund. Humedales – Wetlands International, Buenos Aires

Iturraspe R (2016) Patagonian peatlands (Argentina and Chile). In: Finlayson CM et al (eds) The Wetland Book. Springer, Dordrecht

Iturraspe R, Schroder C (1999) El Clima en el Canal Beagle. En: La vida material y social de los yamana. Orquera L, Piana E. (eds). EUDEBA, Buenos Aires, p 33–44

Iturraspe R, Urciuolo A (2004) Les tourbieres de la Terre de Feu en Argentine: un patrimoine naturel tres menace. Géocarrefour 79(4):143–152

Iturraspe R Urciuolo A (2014) Advances in Peatlands management in Tierra del Fuego, Argentina. In: Mitigations of climate change in agriculture series (9): Towards climate responsible peatlands management. FAO-ONU, p 64–66. http://www.fao.org/3/a-i4029e.pdf

Iturraspe R, Urciuolo A, Iturraspe RJ (2012) Spatial analysis and description of eastern peatlands of Tierra del Fuego, Argentina. In: Heikkilä R, Lindholm T (eds). Mires from pole to pole, The Finnish Environment 38, p 385–399

Ivanov (1981) Water movement in mirelands. Academic Press, London, U.K.

Jerez V y Muñoz-Escobar C (2015). Coleópteros y otros insectos asociados a turberas del páramo magallánico en la Región de Magallanes, Chile. En: Domínguez Díaz E, Vega-Valdés D (eds). Funciones y servicios ecosistémicos de las turberas en Magallanes. INIA 33:Inst Inv Agropecuarias. CRI Kampenaike. Punta Arenas, p 199–224

Joosten H, Clarke D (2002) Wise use of mires and peatlands. Background and principles including a framework for decision-making. International Mire Conservation Group & International Peat Society, Saarijarvi

Lehmann JR, Münchberger W, Knoth C et al (2016) High-resolution classification of South Patagonian peat bog microforms reveals potential gaps in Up-Scaled CH4 fluxes by use of unmanned aerial system (UAS) and CIR imagery. Remote Sens 8:173. https://doi.org/10.3390/rs8030173

León C and Oliván G (2014) Recent rates of carbon and nitrogen accumulation in peatlands of Isla Grande de Chiloé-Chile. Rev Chil Hist Nat, 87(1):1–7

León C, Oliván G, Larraín J et al (2014) Bryophytes and lichens in peatlands and Tepualia stipularis swamp forests of Isla Grande de Chiloé (Chile). Anales Jard Bot Madrid 71(1):e003. https://doi.org/10.3989/ajbm.2342

Loisel J (2015) Las turberas como sumideros de carbono. En: Domínguez Díaz E, Vega-Valdés D (eds). Funciones y servicios ecosistémicos de las turberas en Magallanes. INIA 33. Inst. Invest. Agropecuarias. CRI Kampenaike. P. Arenas p 297–315

Loisel J, Yu Z (2013) Holocene peatland carbon dynamics in Patagonia. Quat Sci Rev 69:125–141. https://doi.org/10.1016/j.quascirev.2013.02.023

Markgraf V (1983) Late and postglacial vegetational and paleoclimatic changes in subantarctic, temperate, and arid environments in Argentina. Palynology 7(1):43–70

Markgraf V (2016) First Pollen Record in South America Commentary: Die Zeichenschrift der Pollenstatistik In: Earth Science 4. https://doi.org/10.3389/feart.2016.00100

Mauquoy D, Blaauw M, van Geel B et al (2004) Late Holocene climatic changes in Tierra del Fuego based on multiproxy analyses of peat deposits. Quaternary R 61:148–158

MEA (2005) Millennium ecosystem assessment: ecosystems and human Well-being. Island Press, Washington DC

Médice Firme Sá M, Schaefer C, Loureiro C et al (2019) Fluxes of CO_2, CH_4, and N_2O in tundra-covered and Nothofagus forest soils in the Argentinian Patagonia. Sci Total Environ 659:401–409. https://doi.org/10.1016/j.scitotenv.2018.12.328

Moore TR (1994) Trace gas emissions from Canadian peatlands and the effect of climatic change. Wetlands 14(3):223–228

Moore TR, Roulet N, Waddington J (1998) Uncertainty in predicting the effect of climatic change on the carbon cycling of Canadian peatlands. Clim Chang 40(2):229–245. https://doi.org/10.1023/A:1005408719297

Münchberger W, Knorr K, Blodau C, Pancotto V et al (2019) Zero to moderate methane emissions in a densely rooted, pristine Patagonian bog - biogeochemical controls as revealed from isotopic evidence. Biogeosciences 16:541–559

Naumov AV, Kosykh N et al (2020) Network environment analysis of a model of carbon flows in a peat bog and fen. Mires Peat 26. https://doi.org/10.19189/MaP.2019.OMB.StA.1790

Ortiz JC (2015) Anfibios de las turberas del extremo austral de Chile. En: Domínguez E, Vega-Valdés D (eds). Funciones y servicios ecosistémicos de las turberas en Magallanes. INIA 33. Inst. Inv. Agropec. CRI Kampenaike. Punta Arenas, p 229–240

Paruelo JM, Golluscio R, Guerschman JP et al (2004) Regional scale relationships between ecosystem structure and functioning: the case of the Patagonian steppes. Glob Ecol Biogeogr 13:385–395

Peri PL, Bahamonde H, Christiansen R (2015) Soil respiration in Patagonian semiarid grasslands under contrasting environmental and use conditions. J Arid Environ 119(1):8. https://doi.org/10.1016/j.jaridenv.2015.03.008

Perotti M, Diéguez M, Jara F (2005) State of the knowledge of north Patagonian wetlands (Argentina): major aspects and importance for regional biodiversity conservation. Rev Chilena de Historia Natural 78:723–737

Pisano E (1977) Fitogeografía de Fuego-Patagonia Chilena.I. Comunidades vegetales entre las latitudes 52° y 56°S (Phytogeography of Fuego-Chilean Patagonia I- Plant communities between lat. 52° and 56° S). Anales Inst. de la Patagonia 8:121–250

Pungas-Kohv P, Keskpaik R et al (2015) Interpreting Estonian mires: common perceptions and changing practices. Fennia 193(2):242–259

Quinton W, Roulet N (1998) Spring and summer runoff hydrology of a subartic patterned wetland. Artic Alpine Res 30(3):285–294

Riveros D, McGlynn B, Epstein H, Welsch D (2008) Interpretation and evaluation of combined measurement techniques for soil CO_2 efflux: discrete surface chambers and continuous soil CO_2 concentration probe. J Geophys Res Atmos 113(G4):4027. https://doi.org/10.1029/2008JG000811

Riveros G, Kusch A, Cárcamo J, Domínguez E (2015) Avifauna en turberas Fuego- Patagónicas. En: Domínguez E, Vega-Valdés D (eds) Funciones y servicios ecosistémicos de las turberas en Magallanes. INIA 33. Inst Invest. Agropecuarias. CRI Kampenaike. Punta Arenas, p 245–275

Rodríguez Martínez AC (2015) Hydrogeomorphic classification of mire ecosystems within the Baker and Pascua Basins in the Region Aysén, Chilean Patagonia: a tool for their assessment and monitoring. Doctoral Thesis. Humboldt University, Berlin, 272p

Roig C (2001) Inventario de los turbales de la zona centro de la Provincia de Tierra del Fuego. Consejo Federal de Inversiones, Buenos Aires. http://biblioteca.cfi.org.ar/wp-content/uploads/sites/2/2001/01/42961.pdf.
Roig FA, Faggy A (1985) Transecta botánica de la Patagonia Austral. In: Boelke O, Moore DM, Roig FA (eds) CONICET. Instit. de la Patagonia-The Royal Society, Buenos Aires
Ruiz & Doberty (2005) Catastro y Caracterización de los Turbales de Magallanes. Sernageomin. BIP N 20196401-0, Punta Arenas
Rydin H, Jeglum J, Hooijer A (2006) The biology of peatlands. Oxford University Press, Oxford
Sarricolea P, Martín-Vide J (2012) Distribución espacial de las precipitaciones diarias en Chile mediante el índice de concentración a resolución de 1 mm, entre 1965-2005. En: Cambio climático, extremos e impactos. Cuadrat JM et al. (eds). Asoc Española de Climatología, Salamanca, p 631–639
Schlater R (2004) Fauna de turberas de la XII Región y Tierra del Fuego. En: Blanco D, de la Balse V (eds) Los Turbales de la Patagonia. Bases para su inventario y la conservación de su diversidad. Wetlands International, Buenos Aires, p 107–112
Solomon S, Qin D, Manning M et al. (2007) Technical summary. In: Climate change 2007: the physical science basis. Contribution of Working Group I to the Fourth Assessment Report of the IPCC. Solomon S, et al. (eds). Cambridge University Press
Tuhkanen S (1997) Tierra del Fuego as a target for research of Finnish researches of nature. Anales Inst. Patagonia, Serie Cs. Naturales 25:105–116
Urrutia PD (2017) Estimation of greenhouse gas fluxes from soils of a forest and an anthropogenic peatland in Ancud, Chiloé. Degree Thesis, Univ. Chile, Sant. de Chile
Valdés-Barrera A, Kutzbach L, Celis-Diez J, Armesto J (2019) Effects of disturbance on the carbon dioxide balance of an anthropogenic peatland in northern Patagonia. Wetlands Ecol Manag. https://doi.org/10.1007/s11273-019-09682-3
Van Bellen S, Mauquoy D, Payne R et al (2014) Testate amoebae as a proxy for reconstructing Holocene water table dynamics in southern Patagonian peat bogs. J Quat Sc 29(5):463–474. https://doi.org/10.1002/jqs.2719
Veber G, Kull A, Villa J, Maddison M, Paal J, Oja T, Iturraspe R et al (2018) Greenhouse gas emissions in natural and managed peatlands of America: case studies along a latitudinal gradient. Ecol Eng 114:34–45. https://doi.org/10.1016/j.ecoleng.2017.06.068
Veblen T, Schlegel F (1982) Reseña ecológica de los bosques del sur de Chile. Bosque 4(2):73–115
Vega-Valdés D, Domínguez E (2015) Análisis espacial de la distribución geográfica de las Turberas de *Sphagnum* en la Región de Magallanes y la Antártica Chilena. En: Domínguez E y Vega-Valdés D (eds) Funciones y servicios ecosistémicos de las turberas en Magallanes. INIA, 33. Punta Arenas, p 43–77
Von Post L (1929) Die Zeichenschrift der Pollenstatistik. Geol Fören För 51:543–565. https://doi.org/10.1080/11035892909449566
Whitfield S, Reed M, Thomson K et al (2011) Managing peatland ecosystem services: current UK policy and future challenges in a changing world. Scottish Geogr J 127(3):209–230
Yu ZC, Loisel J, Brossea D et al (2010) Global peatland dynamics since the last glacial maximum. Geoph Res Letters 37:L13402. https://doi.org/10.1029/2010GL043584

Chapter 9
Restoration for Provision of Ecosystem Services in Patagonia-Aysén, Chile

Carlos Zamorano-Elgueta and Paulo C. Moreno

Abstract The Chilean Patagonian forest is one of the most important productivity biomes in the world and is part of the largest area of intact temperate forests in the southern hemisphere. In Western Patagonia, in the Aysén Region, the natural regeneration of the 4 million ha of forest burned has been poor or null, with soils exposed to intensive hydric and eolic erosion processes, in addition to the invasion of exotic species and overgrazing. These factors of deforestation and degradation of natural vegetation represent a serious and permanent threat with unknown consequences on the landscape capacity to provide ecosystem services. To mitigate or reverse these processes, ecosystem restoration is now recognized as a global priority. Ecosystem services (ES) are increasingly used as a tool to approach these challenges by integrating both ecological and social values in ecological restoration at large scales. Given economic resources to undertake ecological restoration at large scales are often scarce, a tool to effectively prioritize sites for restoration and enhance multiple ES supply and human well-being is critical. In this chapter we address a conceptual model of prioritization of restoration areas for the provision of ES and an approach to their quantification in Patagonia-Aysén. We address a conceptual prioritization model of restoration areas and an approach for quantification of ES provision. We propose to integrate socio-ecological variables, considering the social feasibility of restoration activities, improving human health and well-being, as well as biodiversity and nature conservation values. This method is based on the following sequential steps: (i) defining target areas for restoration and socio-ecological criteria; (ii) assessing the ecological suitability and the social feasibility of restoration; and (iii) combining suitability and feasibility maps to identify priority areas for forest restoration. After defining priority areas for forest restoration and implement restoration actions, the next challenge corresponds to determine their effects on ES. Through

C. Zamorano-Elgueta (✉)
Department of Natural Science and Technology, Universidad de Aysen, Coyhaique, Chile

Universidad de Chile, Centro de Ciencia del Clima y la Resiliencia – CR2, Santiago, Chile
e-mail: carlos.zamorano@uaysen.cl

P. C. Moreno
Centro de Investigación de Ecosistemas de la Patagonia – CIEP, Coyhaique, Chile
e-mail: paulo.moreno@ciep.cl

P. L. Peri et al. (eds.), *Ecosystem Services in Patagonia*, Natural and Social Sciences of Patagonia, https://doi.org/10.1007/978-3-030-69166-0_9

measuring and monitoring, ES will be possible to lead to better planning decisions to support both biodiversity conservation and ecosystem service delivery, among others. This method can support policy guidance and promote initiatives to reverse or mitigate the degradation processes that affect forest ecosystems and the associated environments, in general, and Patagonia, in particular.

Keywords Forest fires · Degradation · Productivity loss · Socio-ecological criteria · Ecosystem recovery

1 Introduction

The Chilean Patagonian forest is one of the most important productivity biomes in the world and is part of the largest area of intact temperate forests in the southern hemisphere (Astorga et al. 2018). In Western Patagonia, in the Aysén Region, the predominant land cover corresponds to evergreen and deciduous native forest dominated by *Nothofagus pumilio* (lenga) and *N. antarctica* (ñirre) (Lara et al. 2005), totaling over 41% of the native forest surface area of Chile and 60% of the Chilean forests under a national protection system, such as national parks, reserves, or natural monuments (Gysling et al. 2018). However, these forests have been historically exposed to intense and chronic disturbances (e.g., grazing, selective logging, forest conversion to agriculture and other intensive uses, as well as invasion of exotic species and fires), which pose critical threats to their long-term conservation (Zamorano-Elgueta et al. 2014; Promis and Allen 2017). The natural regeneration of the 4 million ha of forest burned in Western Patagonia has been poor or null, with soils exposed to intensive hydric and eolic erosion processes, in addition to the invasion of exotic species (e.g., *Pinus contorta*, *Rosa rubiginosa*, *Cirsium vulgare*) and overgrazing (Quintanilla 2008). Natural regeneration of these forests has resulted in dense second-growth forests around the margins of the burned areas, while a majority of fire areas have transitioned into grassland. In other areas deforested by fires, local authorities initiated afforestation projects using fast-growing exotic coniferous species in the 1970s and 1980s, including *P. contorta*, *P. sylvestris*, and *P. ponderosa* (McIntire and Fajardo 2011). These plantations cover an area that in total does not exceed 40,000 ha. Consequently, the land cover changes have transformed part of the Patagonian landscape from continuous old-growth *Nothofagus* forests to a mosaic of several land cover types, including native forests at different successional stages, grasslands, coniferous plantations, and barren areas (McIntire and Fajardo 2011). Moreover, in Western Patagonia the most recent fires, called "megafires," destroyed 15,000 ha of forests and steppes in few days in the locality of Cochrane (González et al. 2020). These new factors of deforestation and degradation of natural vegetation represent a serious and permanent threat with unknown consequences on the landscape capacity to provide ecosystem services.

To mitigate or reverse these processes, ecosystem restoration is now recognized as a global priority (Aronson and Alexander 2013). Globally, more than two billion hectares of forestlands that have been either cleared or degraded require restoration (Laestadius et al. 2011). The focus of restoration ecology has greatly evolved from aiming at reconstructing pristine and reference sites (McDonald et al. 2016) to an objective-oriented strategy (Dufour and Piégay 2009) aiming at maximizing ecological functions and services. This change in focus is due to a virtual lack of undisturbed reference sites (Dufour and Piégay 2009) and because ecological restoration projects need to integrate society values in the restoration process (Choi 2004). Ecosystem services (ES) are increasingly used as a tool to approach these challenges by integrating both ecological and social values in ecological restoration at large scales (Trabucchi et al. 2012). Given economic resources to undertake ecological restoration at large scales are often scarce (Neeson et al. 2016), a tool to effectively prioritize sites for ecological restoration and enhance multiple ES supply and human well-being is critical (Comín et al. 2017). Limited funding, however, requires that any forest restoration initiative be based on careful identification of priority areas to maximize the socio-ecological benefits of available resources (Perring et al. 2015). These priorities can be based on the area's biodiversity values, vulnerability to threats, contribution to landscape-scale processes, or some combination of all of these factors (Zhu et al. 2015).

All large ecological restoration and management efforts have the difficult task of determining if project goals and objectives are being met (Doren et al. 2009; Kershner et al. 2011). Thousands of species, habitats, ecological conditions, and processes will be directly or indirectly affected by restoration actions and could potentially be monitored to track overall restoration progress. Although monitoring a wide set of species and habitats as indicators across the ecosystem may be considered ideal in some cases (Carignan and Villard 2002), the limits of available funding and political will for monitoring expenditures create the need for a practical approach that takes into account the size of the ecosystem, funding constraints, and political complexities. Ecological indicators act as measures of the overall health of the ecosystem and provide insight on the condition of the different natural environments (Doren et al. 2009) and can be monitored to track ecosystem management goals and objectives (Doren et al. 2009; Kershner et al. 2011).

Most authors agree that the term ES was incorporated into society in the early 1980s and that the magnitude of studies and publications increased in the early 1990s (Burkhard and Maers 2017). Due to previous initiatives, between 2001 and 2005, the United Nations requested a Millennium Ecosystem Assessment (MEA 2005). We identified this initiative as the first in-depth analysis of ES, with more than 1300 experts around the planet, whose focus was to find the connections between ecosystems and human well-being using the ES as a proxy. The ES associated with MEA were the provision of clean water, food, and forest products, the regulation of flows, and natural resources. The main results of this initiative indicate that in the last 50 years, the speed of change in ecosystems was higher than before.

There are various classifications of ES (MEA 2005 and TEEB 2010, among others), and in 2009, and revised in 2013, the Common International Classification of

ES (CICES) was proposed (Burkhard and Maers 2017). Most of these classifications divide the ES in "supporting" (nutrient cycle, soil formation, primary production, etc.), "provisioning" (food, freshwater, wood and fiber, fuel, etc.), "regulating" (climate, floods, diseases, water purification, etc.), and "cultural" (aesthetic, spiritual, educational, recreational, etc.) (MEA 2005). A crucial aspect of ES is that they are associated with human well-being as elements of security, materials necessary for life, health, good social relations, and freedom of choice and action. This association implies that humanity is strongly dependent on the proper functioning of ecosystems and their natural capital, which translates into a constant flow of SE to the society (Burkhard and Maers 2017).

One aspect that we highlight is the concept of biodiversity, wherein many studies it is excluded as ES. Nevertheless, it is incorporated as a structural aspect that defines the natural capital of some ecosystems, mainly due to the information presented by the natural systems, which in turn plays a vital role in supporting those ES. Examples of species as support are primary production and pollination for food production, decomposition processes in the soil nutrient cycle, and regulation of natural disasters such as hurricanes, increasing the resilience of ecosystems. Furthermore, biodiversity has an ethical valuation or a value per se due to cultural or instrumental aspects (Burkhard and Maers 2017).

In this chapter we address a conceptual model of prioritization of restoration areas for the provision of ES and an approach to their quantification in Patagonia-Aysén. This method can support policy guidance and promote initiatives to reverse or mitigate the degradation processes that affect the natural forest ecosystems and the associated environments, in general, and Patagonia, in particular.

2 Restoring the Provision of ES: Where to Start?

Identifying priority areas would increase the efficiency and impact of available resources to design, plan, and establish forest restoration programs, where interventions will produce the greatest benefits. The prioritization problem has been addressed in a large variety of different proposals (Mittermeier et al. 1998). To identify priority areas for forest restoration, conservation scientists and planners face the challenge of developing ecologically sound methods that can be applied at different spatial scales so that the results are relevant to land use planning (Moilanen et al. 2011). Multi-criteria analysis (MCA) methods provide a useful tool for addressing such problems. These methods involve the holistic evaluation and comparison of possible alternatives against a set of different objectives and criteria (Veronesi et al. 2017). By coupling MCA methods with Geographic Information System (GIS), spatial MCA tools are able to translate the outcomes of analyses into readily understandable maps of restoration priorities at the landscape scale (Borda-Niño et al. 2017).

Here we propose a conceptual approach to define priority areas for restoration of ES integrating socio-ecological variables, considering the social feasibility of resto-

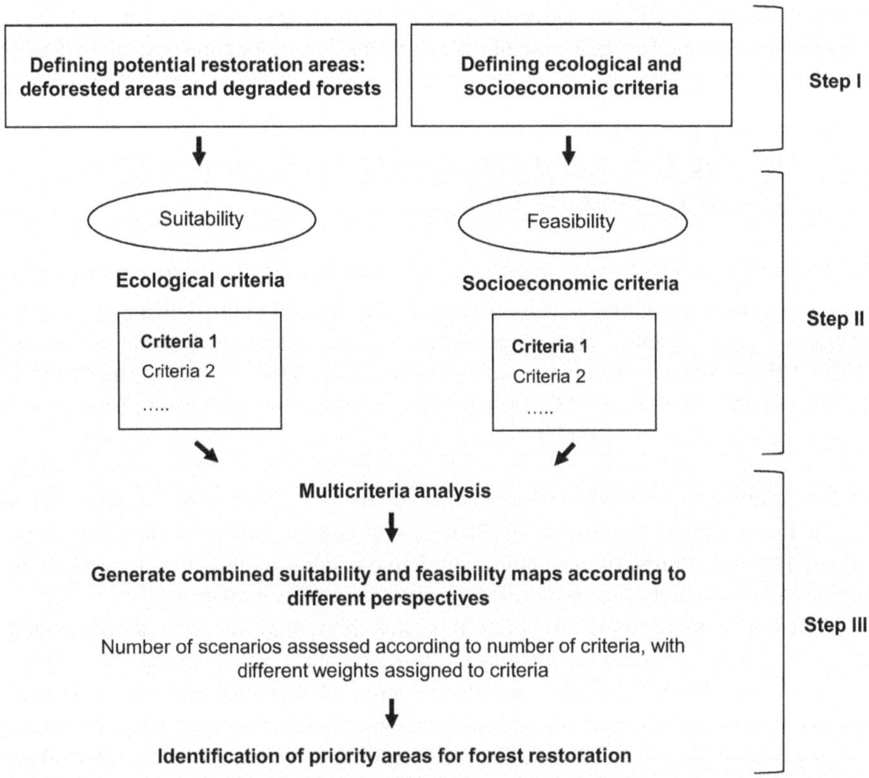

Fig. 9.1 Flow chart of the methodological approach proposed, including the following steps: (i) defining potential areas for forest restoration and ecological and socioeconomic criteria; (ii) assessing the ecological suitability and the social feasibility of restoration; and (iii) combining suitability and feasibility maps to identify priority areas for forest restoration

ration activities, improving human health and well-being along with biodiversity and nature conservation (Posner et al. 2016). This method is based on the following sequential steps: (i) defining target areas for restoration and socio-ecological criteria; (ii) assessing the ecological suitability and the social feasibility of restoration; and (iii) combining suitability and feasibility maps to identify priority areas for forest restoration (Fig. 9.1).

In order to make the results of an ecological evaluation operational, they must be conveyed to decision-makers in the most efficient and transparent way (Geneletti 2004). This means that the framework adopted during the evaluation (i.e., the criteria and indicators that have been selected) must be made explicit, so as to allow tracking of the influence of each factor on the evaluation results. Moreover, all possible scenarios resulting from the evaluation must be considered, to redirect further discussion toward the conflicting aspects only. This is optimally achieved by resorting to a decision support system (DSS), which can be defined as an interactive computer-based system that can help to utilize data and models to solve a decision problem. Coupling GIS and DSS is becoming a common strategy to deal with decision problems related to environmental planning and land allocation (Geneletti 2004).

2.1 *Defining Target Restoration Areas and Criteria: A Case of Study in Southern Chile*

In 2015, Chile included the land use, land use change, and forestry (LULUCF) sectors as a central component of its nationally determined contributions (NDCs) to mitigate climate change. These commitments were updated and consider, for the 2020–2030 period, (i) sustainable management and recovery of 200,000 ha of native forest; (ii) afforestation of 200,000 ha; and (iii) emission reduction in the forestry sector associated with degradation and deforestation of the native forest by 25% with respect to average emissions in the period 2001–2013 (Gobierno de Chile 2020). Additionally, the NDC update considers the development of a National Plan for the Restoration of Landscapes (NPRL), comprising 1 million ha, including those affected by forest fires. In this context, to define restoration priorities represents an urgent task, especially in support of environmental decision-making.

To define restoration priority areas, typically, those areas targeted are deforested areas, due to the difficulty of identifying degraded areas at a landscape scale (Orsi and Geneletti 2010). The accurate identification of degraded areas is important because restoring such areas can be faster and cheaper than restoring deforested areas (Vásquez-Grandón et al. 2018), allowing disturbed ecosystems to recover to their previous high-carbon state (Lewis et al. 2019). Recovery times can be accelerated by planting native species, and the area under natural regeneration expanded using legislation and incentives, such as those pioneered in Costa Rica (Lewis et al. 2019). Because of this, here we propose to consider both deforested land and degraded forests as target areas for forest restoration. In general, methods for monitoring the current state and changes of forest ecosystems are based on satellite-borne or airborne remote sensing imagery (GOFC-GOLD 2015). One of the main advantages of using remote sensing data and procedures is that these have the potential to be decidedly instrumental in the assessment of forest degradation and deforestation processes at a much lower costs than any other methods (Mascaro et al. 2011; Olander et al. 2012). However, these methods focused mainly on deforestation because it is easier to measure, as compared to forest degradation (Plugge and Köhl 2012).

Forest degradation has been defined as a change process that negatively impacts the characteristics of a forest such that the value and production of its goods and services decline and maintain only limited biological diversity (CBD 2005). The drivers and intensity of degradation varied by region, but the impact of forest loss and degradation can be felt at all scales, from global climate change to declining economic value of forest resources and biodiversity and threatened local livelihoods (Mitchell et al. 2017).

Remote sensing methods for the identification of degraded forests are expensive, as they rely on very high-resolution satellite imagery or LIDAR datasets to detect changes in canopy cover or aboveground biomass. These methods are often ineffective at detecting below-canopy degradation, such as the removal of understory vegetation or lack of forest species recruitment (Mitchell et al. 2017). In this chapter, we therefore proposed an alternative approach, building on previous research from a study area, which demonstrated that socioeconomic factors can be used to predict forest degradation. Degradation is the result of anthropogenic disturbances, which can include invasive species, firewood collection, and livestock grazing (causing soil compaction as well as trampling and removal of understory), among others (FAO 2009). For example, within the Cordillera de la Costa in southern Chile, it had been demonstrated that socioeconomic factors strongly predict both the likelihood and intensity of cattle ranching and selective logging, two of the major regional drivers of forest degradation in rural areas (Zamorano-Elgueta et al. 2014). High frequency and intensity of these alterations are typically associated with rural small properties (i.e., less than 200 ha as defined by the Chilean laws) owned by "campesinos" (the Spanish name for rural people living in small-sized properties with a subsistence economy) due to the need to achieve levels of production to ensure family subsistence. On these small properties, cattle ranching and selective logging have been shown to severely impact the regeneration capacity of many plant species (Zamorano-Elgueta et al. 2014). If forests are permanently disturbed by low-intensity disturbances such as cattle grazing and selective logging, their composition will be profoundly altered by loss of biodiversity and changes in the dominance of different species. Effects of altered habitat conditions on forest regeneration could lead to less phenotypic diversity in characteristics such as fruit type, seed mass by area unit, and flowering period. These changes could generate unknown impacts on functional ecosystem properties and on the ecosystem's response to disturbance (Fisher and Christopher 2007). In this context, there is an urgent need to quantify and predict the effects of disturbance on biodiversity patterns to guide conservation efforts and the management of ecological resources (Mouillot et al. 2012). This evidence could be integrating/complementing using ecosystem functions, e.g., linked to total nutrient pools or biological productivity in a study area. Furthermore, linked this evidence to spatial features such as land tenure and forest successional stages, it is possible to represent in a territory conservation states of forests ecosystems (Fig. 9.2). Once defined the restoration target, the next step corresponds to assess the suitability and feasibility of the territory for forest restoration in terms of the objectives defined.

2.2 *Assessing the Ecological Suitability and the Social Feasibility of Restoration*

In order to assess the relevance for forest restoration of a study area, a set of evaluation criteria, and relevant indicators to measure them, must be selected. Quite a number of previous works in evaluation schemes for prioritization of restoration

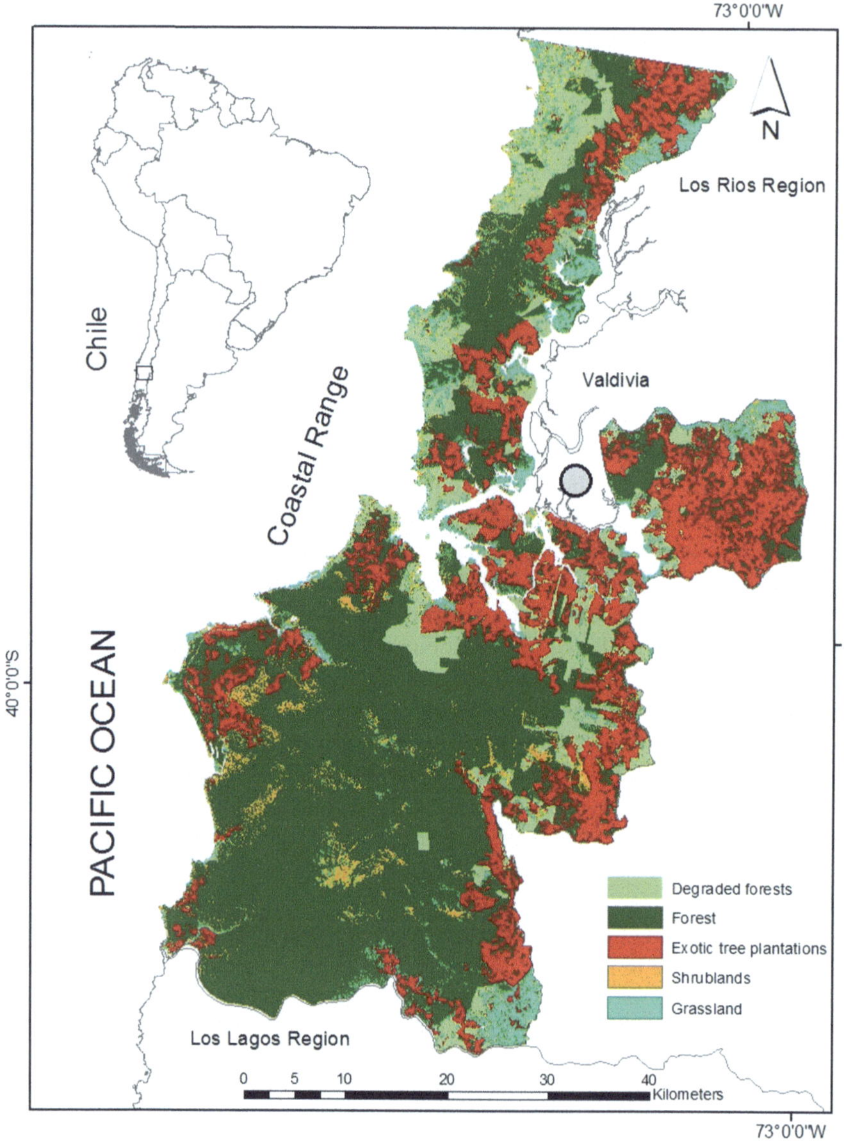

Fig. 9.2 Spatial distribution of estimated degraded forest and major land cover types in the Coastal Range of Región de los Ríos, southern Chile

area to local and landscape are available in the literature (Zamorano-Elgueta 2014; Oakleaf et al. 2017; Maxwell et al. 2019). Commonly, employed criteria to measure ecological aspects are related to landscape metrics, e.g., patch density, mean shape index, patch dimension, isolation, exposure to external disturbances, and urban pressure. On the other hand, economic constraints determine what is realistic. Financial constraints define the degree of realism of a proposal in a particular socio-economic context and depend on public acceptance of a project (Guida-Johnson and Zuleta 2017). At the same time, the degree to which the public embraces a project may be influenced by the ratio between costs and perceived benefits. In this context, social aspects are associated to several criteria as distance of roads or population density. However, in most cases, social criteria constituted restrictive factors (Guida-Johnson and Zuleta 2017). To assess the ecological suitability and the social feasibility of restoration, ES knowledge represents a decisive role. It is used to support the formulation and structuring of the decision problem to identify criteria for screening, ranking, and spatial targeting of the alternatives. It also assume a design role, when it is used to set the basis for implementation tools, including the definition of standards and policy targets and the design of regulations, certifications, pricing, and incentives. It was also employed to establish damage compensations (Barton et al. 2018). In general, the ecological criteria should represent not only ecosystem relevance but also social needs. The criteria must be based on biotic and social aspects with clearer interpretation and reliable and available data. Each selected criterion is spatially represented through a map, e.g., using typical functionalities of raster-based GIS, such as distance operators and spatial filters. The GIS package ILWIS 4.0 provides powerful tools for analysis and transformation of raster data. A detailed step-by-step description can be found in Geneletti (2002) and Zamorano-Elgueta (2014).

In the Coastal Range of the Región de los Ríos, southern Chile, Zamorano-Elgueta (2014) defined several socio-ecological criteria to identify priority areas for restoration of ES based on natural forests. Ecological criteria were selected among those related to biodiversity (potential flora richness, distance from well-conserved forests, distance from protected areas), soil protection (erosion regulation), and water (water provision) variables. On the other side, social criteria were defined among land tenure, accessibility (soil slope, land cover, distance from roads), and exposure to external disturbance variables.

2.3 Generating Suitability and Feasibility Maps

The starting point is the evaluation matrix, which can contain the steps, possible alternatives, and the criteria under which they have to be evaluated. Each cell of the matrix can fill with a score representing the performance of the corresponding alternative with respect to the relevant criterion. In the case of landscape restoration-related decision problems, the matrix can be termed prioritization matrix and contains the different project solutions as alternatives and the different type of variables as criteria. The criterion scores consist in raw measurements expressed by different scales or units (monetary units, physical units, etc.). In order to be relatable to the degree of "desirability" of the alternatives under analysis, such scores need to be transformed from their original units into a value scale (Geneletti 2002). This is the role of the value assessment, through which the criterion scores lose their dimension and become an expression of the achievement of the evaluation objectives. As a result, all the scores are transformed into a given value range (e.g., between 0 and 1). This operation is performed by generating a value function, e.g., a curve that expresses the relationship between the criterion scores and the correspondent value scores (Geneletti 2002, 2004; Zamorano-Elgueta 2014). Value functions are the mathematical representation of human judgment on criterion scores and transform the original scores of a given criterion into dimensionless values between 0 and 1, where 0 corresponds to minimum desirability and 1 to maximum desirability (Geneletti 2005). The value functions can be selected and belong to different main categories, depending on whether the relationship between desirability and the original score was linear (e.g., erosion regulation, water provision) or quadratic (e.g., distance from roads) and whether maximum desirability was attained at high or low original scores.

The different evaluation criteria are usually characterized by different importance levels, which need to be included into the evaluation. This is obtained by assigning a weight to each criterion (prioritization). A weight can be defined as a value assigned to a criterion that indicates its importance relatively to the other criteria under consideration (Geneletti 2004). For example, Zamorano-Elgueta (2014) considers four scenarios, both for suitability and feasibility, which simulated different perspectives (Table 9.1). Three scenarios assigned a greater weight to one specific criterion, whereas the fourth one assigned the same weight to all criteria. A survey of the methods developed to support the weight assignment can be found in Van Herwijnen (1999). Once the weights are assigned to each criterion, the aggregation can be performed.

This is done by using a decision rule that dictates how best to order the alternatives, on the basis of the data on the alternatives (criterion scores) and on the preferences of the decision-makers (criterion assessment and weights). The most widely used decision rule is the weighted linear combination. An overall score is calculated for each alternative by first multiplying the valued criterion scores by their appropriate weight and then summing the weighted scores for all criteria. Another method is the "concordance analysis," which assesses the ranking by pairwise comparison of

Table 9.1 Assignment of weights to the different criteria under different perspectives of suitability and feasibility (Zamorano-Elgueta 2014)

		Perspective			
		Criteria 1	Criteria 2	Criteria 3	Balanced
Suitability feasibility	Criteria 1	0.50	0.25	0.25	0.30
	Criteria 2	0.25	0.50	0.25	0.30
	Criteria 3	0.25	0.25	0.50	0.30

the alternatives. The last step in the procedure is represented by the sensitivity analysis. It aims at determining the robustness of the ranking with respect to the uncertainties in the assigned weights, value functions, and scores, as well as to changes in the aggregation method. The information available to the decision-makers is often uncertain and imprecise, owing to measurement and conceptual errors. Finally, sensitivity analysis considers how such errors can affect the result of the evaluation.

2.4 Identifying Priority Areas for Forest Restoration

Individual criterion maps are combined by means of multi-criteria analysis (MCA). First, binary maps are created for each perspective by assigning a value of 1 to the most suitable/feasible pixels and a value of 0 for all the other pixels. Second, the binary maps for suitability and feasibility are added, generating maps with pixels that were never identified as a priority, and those pixels were identified as priorities under all the considered perspectives. Third, priority areas for restoration are combined to those that had been selected as priorities under all perspectives, both suitability and feasibility (Fig. 9.3).

After defining priority areas for forest restoration and implement restoration actions, the next challenge corresponds to determine their effects over provision of ES. Measuring and monitoring ES are a key step in any restoration plan because they (i) lead to better planning decisions to support both biodiversity conservation and ecosystem service delivery; (ii) identify and inform management strategies to enhance economic sustainability and human well-being; (iii) provide information on additional benefits from traditional approaches to biodiversity conservation; (iv) identify those affected by land use management decisions and also help spread the costs and benefits more fairly among stakeholders; and (v) provide information to raise awareness and build public and government support for evidence-based policy and management decisions (Peh et al. 2013).

3 How to Monitor Restoration Activities? Quantifying Ecosystem Services

Costanza et al. (1997) incorporate within its analysis for each type of ecosystem 17 ES associated with specific ecosystem functions, e.g., CO_2/O_2 balance is a gas regulation service associated with the chemical regulation function of the atmosphere. MEA (2005) incorporates 24 ES, grouping them by provisioning, regulating, and cultural. A Chilean study includes four ES, such as timber production, nature-based recreation, soil fertility maintenance, and freshwater supply (Nahuelhual et al. 2007). De Groot et al. (2012), in a global overview, include the value of ES of 10 biomes, with over 320 publications screened. They presented for tropical and

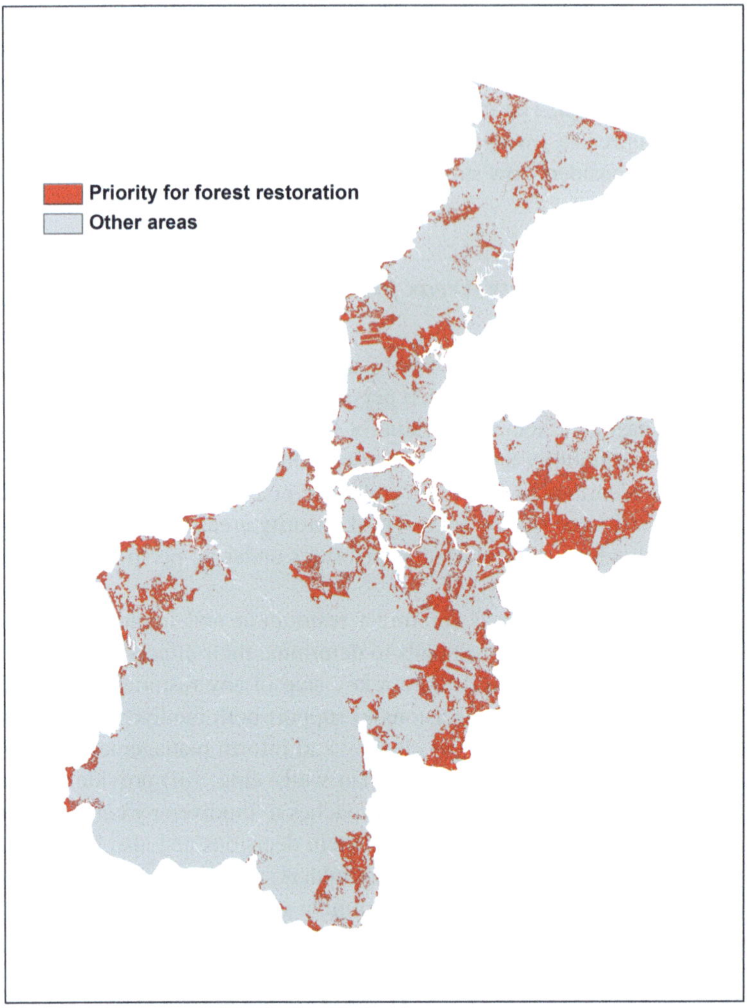

Fig. 9.3 Priority areas for restoration of deforested areas and degraded forests combining the highest value pixels from both suitability and feasibility maps. Selected areas correspond to pixels that presented suitability or feasibility values under the four scenarios assessed

temperate biomes, the most common ES, such as timber and freshwater, carbon sequestration, cultural recreation and tourism, and biodiversity as natural capital. In the context of evaluating results from forest ecosystem restoration activities, we note that at least the most common ES found by De Groot et al. (2012) should be included, plus the inclusion of soil formation or erosion control. This group of ES that we could call main, due to the advance in the incorporation of value in society, involves provision, regulation, support, and cultural services, which integrates the entire range of ES in the analysis.

Finally, the last consideration to incorporate in the discussion about ES quantification is the impossibility of individual evaluation of each one of them, this due to the concepts of competition and synergies that exist, creating unique dynamics for each ecosystem types (Terrado et al. 2016). To incorporate synergies and trade-offs in the valuation of ES implies estimating the marginal value of each of them compared to the others, with different econometric methods for estimating them (Susaeta et al. 2019). We identified the understanding of these dynamics between ES as a basic need, since it is crucial for decision-makers, both at the national and at the producer level (Cademus et al. 2014; Susaeta et al. 2019). Several studies have analyzed the dynamics of competition and synergies of different ES, where the production or regulation of water, maintenance of biodiversity, timber production, and carbon storage are the primary services investigated (Cademus et al. 2014; Delphin et al. 2013; Susaeta et al. 2019) (Table 9.2).

Table 9.2 Summary of provisioning, regulating, and cultural ES defined by MEA (2005)

Service	Subcategory
Provisioning services	
Food	Crops
	Livestock
	Capture fisheries
	Aquaculture
	Wild foods
Fiber	Timber
	Cotton, hemp, silk
	Wood fuel
Genetic resources	
Biochemicals, natural medicines, pharmaceuticals	
Freshwater	
Regulating services	
Air quality regulation	
Climate regulation	Global
	Regional and local
Water regulation	
Erosion regulation	
Water purification and waste treatment	
Disease regulation	
Pest regulation	
Pollination	
Natural hazard regulation	
Cultural services	
Spiritual and religious values	
Aesthetic values	
Recreation and ecotourism	

Changes in SE provision and interactions are the result of anthropogenic influences, such as restoration, and biophysical stressors, such as forest degradation and natural disturbances. Due to this, we consider that knowing the optimal production of several services of an ecosystem, the dynamics of trade-offs and synergies, and what are the forcing agents that are involved in the different levels of the provision should be a priority for decision-makers. Indeed, we identified as an excellent approach to identify those that are the most important in maximizing ES efficiency. The aforementioned involves a spatial-temporal conception in the quantification of the provision of ES, spatial such as watersheds that allow incorporating decisions to maintain these service flows in the long term. We believe that this spatial-temporal point is crucial due to the global changes and their effects on ES provision.

3.1 Methodologies Available: How to Quantify?

We offer here a light review of the most used methods for each ES selected as crucial.

3.1.1 Timber

The quantification of volume timber in a forest is, without doubts, the most conventional and familiar method compared to other ES because it is associated with market values for a long time in history. There are several mechanisms to calculate the timber volume in a forest ecosystem, such as taper functions, individual and stand volume models, merchantable logs breakdown equations, yield tables, growth and yield simulators, etc. The basis for the majority of the methods is through field information as forest inventories, where it is possible to obtain tree and stand parameters, to calculate the timber volume. Nowadays, remote sensing technics are an alternative to quantify timber of a forest ecosystem generating options for estimation in isolated or extensive areas. However, due to forest variability, sampling makes it possible to obtain more reliable information about forest resources and their physical environments at a specified point in time and at a reasonable cost. Susaeta et al. (2019) for accounting timber used the Forest Inventory and Analysis (FIA) of the US Department of Agriculture, occupying the timber volume proportioned by the FIA plots.

3.1.2 Carbon Sequestration

Carbon sequestration is an ES classified most of the time as global climate regulation (MEA 2005). The value of this service is in the reduction of CO_2 concentration in the atmosphere to avoid the Earth's temperature which continues to rise, thus causing incommensurable changes on the climate of the planet. A crucial concept related to carbon sequestration is temporality, so we need to account for the carbon stocks increment in the time. Here, we found two approaches to estimate carbon changes, the gain loss and the stock change method (IPCC 2019), where the selection will depend on the information to use and the mechanism or carbon market identified; both directly correlate with the long-term monitoring expectation. In the case of restoration activities, it is necessary to determine if the baseline scenario was a forest land or not. That is to say if the monitoring is focused on "forest land remaining as forest land" or a "land converted to forest land." Both are possible to be considered as reducing emissions from deforestation and forest degradation (REDD+) activities. Here, it is an excellent option to use the national forest definition for describing a forest land. From a public initiative, the nationally determined contributions (NDCs) to mitigate climate change should be other mechanism to accounting the carbon sequestration from restoration activities (see Sect. 2.1).

The carbon in a forest should be measured related to the pool where it is accumulated. Here, we have five carbon pools as aboveground biomass, belowground biomass, litter, dead wood, and soil carbon (Penman et al. 2003; IPCC 2019). A Key Category Analysis should be conducted to address the significant pools to monitor as REDD+ activities. The in situ information of carbon stock is difficult to obtain and high cost, but we have three data levels to use in our quantification as options. First, it is possible to use the Intergovernmental Panel on Climate Change (IPCC) default approach known as Tier 1, where there are simplified some assumptions about calculations and selection of carbon pools. Second, national data and IPCC methodologies are used (Tier 2) being similar to Tier 1 but apply more specific factors of emission and removal and other parameters associated with the country. Finally, Tier 3 uses particular data, factors, and settings of the country being more flexible than the other two Tiers systems. Therefore it is possible to use in a wide range of restoration activities.

3.1.3 Freshwater

Water as a provisioning service is used for drinking, industry, irrigation, hydro energy, etc. (MEA 2005), but in others, it is related to other uses, e.g., for recreational angling as a social service (Boyd and Banzhaf 2007). However, most of the cases, water yield is the ecosystem service accounted (Nahuelhual et al. 2007; de Groot et al. 2012; Susaeta et al. 2019). Measurements about the effects through activities restoration into water yield must incorporate a baseline and the monitoring on the water production.

Two methods are utilized to determine water yield, direct and indirect approaches. The first consists in several measurements of caudal in the time to estimate the water production (Oyarzún et al. 1998; Little et al. 2008; Donoso et al. 2014; Cuevas et al. 2018). The second uses showed indirect relationships among several parameters, such as precipitation, evapotranspiration, leaf area index of a stand, etc. (Cademus et al. 2014; Susaeta et al. 2019). The option to use one method or another is defined by the data and functions available for the study area and monitoring budget, among other aspects.

The stream-flow monitoring must be focused follow each study goal, where the total amount of water production is not a unique parameter to obtain. There are essential differences in discharge curves after a precipitation event (Frêne et al. 2019), between seasons (Neira 2005), even exiting a diary cycle (Cuevas et al. 2018). Therefore, the caudal interval measurement must be small as possible to understand the watershed behavior. Sometimes water yield is not the unique parameter for addressing freshwater ES. For drinking water, for example, it is necessary to evaluate the nutrients or contaminants on the streamflow to detect danger concentrations for human consumption (Donoso et al. 2014). On the other hand, we are interested in nutrients as exports to other systems as marine ecosystems because they are a crucial component in primary productivity (Torres et al. 2020) or measuring eutrophication in lagoons (Rodríguez-Gallego et al. 2017). Studies on biogeochemical fluxes are associated to the monitoring of dissolved organic matter (DOM) and organic and inorganic forms of nitrogen, phosphorous, dissolved silica (DSi), etc. (Oyarzún et al. 1998; Little et al. 2008).

3.1.4 Soils

Several ES can be addressed related to soils, such as nutrient cycling, erosion control, and soil formation (Costanza et al. 1997). At least one of them should be incorporated in the monitoring, depending on the goals of restoration. For example, if the study goal is the internal cycling, storage, processing, and inputs of nutrients should be measured, nitrogen fixation, phosphorous and other elements. Otherwise, if the objective is the retention of soil within an ecosystem, it can be estimated with evaluations of loss of substrate by wind, runoff, or other removal processes. Finally, if the activities of restoration are for the increasing of soil formation, the measurements should be focused on the accumulation of organic material and weathering of rocks (Costanza et al. 1997).

3.1.5 Recreational and Tourism

Nature-based recreation is a cultural ecosystem service (MEA 2005) related to physical and emotional interactions on the natural environment (Vallecillo et al. 2019). Areas with restoration activities in a future should increase their aptitude as recreational use on forests, either for eco-tourism, sport fishing, or other outdoor activities. Several studies exist incorporating valuation methods for nature-based recreation, grouped into three categories: stated preference, revealed behavior, and a mix of them. Stated preference is focused on identifying the offer to the visitors or potential opportunities for recreation. Some techniques are choice experiments, participatory mapping, and contingent valuation. Revealed behavior methods identify the demand or what the people do in natural areas, where travel cost is the typical way of determining the site's recreational value. A third group incorporate the demand, the potential, and actual flows to assess the active recreational ecosystem service (Nahuelhual et al. 2007; Vallecillo et al. 2019). The spatially explicit method incorporates the three components of ES, such as the possible service or opportunities for recreation provided by nature, the demand of the population considering long-distance travel or daily entertainment, and the estimate of the actual flow from biophysical assessment to monetary terms through valuation techniques (Vallecillo et al. 2019).

3.2 Experiences of Provision Quantification on Ecosystem in Aysén Region

Few examples of provision quantification exist in the Aysén Region, even less if we incorporate the valuation of these services. Some of them are focused on one or two ES. Still, we have presented a compendium of examples, models, and data, among other information useful for those interested in developing a quantification of ES through restoration activities.

Data concerning to quantify timber volume is on several sources from forest inventories until volume models. Two official forest monitoring sample plots are established in the region, the "inventory and monitoring of native forest ecosystems" of the Chilean Institute of Forest Research (INFOR) and the "wood-based biomass energy inventory" of the National Forest Service (CONAF). The first of them has volume information spatially represented for the Aysén Region on maps and the inventory data (INFOR 2020).

There are some functions for specific species and productive sites. *Nothofagus pumilio* (lenga) was the most studied species with several volume models or other biometric functions (Alvarez and Grosse 1978; Alfaro 1982; Donoso 1993; INFOR 2003a, b; Cruz Johnson et al. 2005; Martin et al. 2018; Salinas et al. 2019). Beside this, it is possible to find models for *N. dombeyi*, *N. betuloides*, *N. antarctica*, and *Laureliopsis phillipiana* (Drake et al. 2003; Martin 2007; Torres 2017). However, some models are site-specific, and many species are not considered on these studies, with an existing lack of crucial information to obtain the total volume by species, even worse for timber volume by-products. Information about carbon sequestration as a global regulating ecosystem service in the Aysén Region is scarce and partial, covering just some tree species, but with deficient baseline, and limited models, for shrubs or shade-tolerant species from evergreen forests. The IPCC Tier 1 (Penman et al. 2003) presents default values for cold temperate wet and warm temperate wet climate associated with deciduous and evergreen forests, respectively. The data correspond to the average annual increment in aboveground biomass, root-to-shoot ratios, litter carbon stocks, and natural mortality rates, among others.

INFOR (2009) developed a pilot study under the CDM Kyoto Protocol through *Pinus ponderosa* reforestation on degraded areas of Coyhaique Province. That project contains several local information and conversion factors (Bahamondez et al. 2009). Beside this, local biomass functions are available for *N. pumilio*, *N. antarctica*, *Pseudotsuga menziesii*, *Pinus ponderosa*, and *P. contorta*. Specifically with biomass models of aboveground, branches, foliage, live root, and bark (Gayoso et al. 2002; Gayoso 2013; CONAF 2013; Torres 2017). National greenhouse gas inventory presents emission and reduction factors, but the information is at country level (MMA 2014). A carbon sequestration potential study of different land uses was developed in Mano Negra area (Dube et al. 2011).

Related to freshwater provision, the Dirección General de Aguas has monitoring of water yield with +25 stations in Aysén. Unfortunately, changes in water yield from restoration activities must be monitored immediately contiguous to the study area. Thus, quantification of water yield downstream is not relevant to associate an effect, positive or negative. Monitoring of headwater micro-watersheds is being carried on Aysén identifying relations between anthropogenic disturbances, water yield, and nutrients export (B. Reid, unpublished) and wetlands (Aguilar 2018). An eco-social evaluation study was carried on the Aysén basin with a reduction of living cost of 148 USD per month incorporating the freshwater and timber services to the accounting of rural households (Delgado et al. 2013). Ponce et al. (2011) used the contingent valuation method to quantify the impacts of hydropower project "Hidroaysen" incorporating willingness to pay for people living in five Chilean cities.

A detailed soil characterization of Aysén was developed by Hepp and Stolpe (2014) with baseline information about properties and cartography. Yarrow and Leitao (2007) developed a SWAT model for the Aysén basin to understand the nitrogen cycle from tree growth and litter production. Other studies incorporate baseline information about soils in Aysén, such as carbon dynamics (Dube et al. 2009; Stolpe et al. 2010), modeling (Bachmann-Vargas 2013), anthropogenic disturbances (Fajardo and Gundale 2015), and volcanic ash soils (Vandekerkhove et al. 2016).

Nature-based recreation is associated with national protected areas (SNASPE) in Aysén, incorporating studies related to valorization, wildlife conservation, noise perception, planning, etc. (Gale and Ednie 2019; Gale et al. 2019, 2020). However, more detailed studies of stated preference and revealed behavior methods should be carried out in this region.

Revising the state of the art about quantification of ES in the Aysén Region, scientists from Patagonian Ecosystems Investigation Research Center (CIEP) are working on a project for a better understanding of provision, valuation, and modeling of essential services for human well-being. The project is funded by Fondo de Investigación del Bosque Nativo (FIBN), and the goal is to describe and quantify the provision of at least four ES, such as the provision of wood and clean water, the carbon storage as a climate-regulated service, the soil from a support point of view, and also species richness as natural capital. For this, 12 watersheds are being evaluated in 4 areas with different rainfall and temperatures, associated with 2 types of forest, deciduous and evergreen, and with a human intervention gradient (Fig. 9.4). Based on optimization and econometric methods, the provision of services will be valued for use in the National Native Forest Law as an instrument of promotion through a valuation table. Finally, this project will consolidate the previous results into a Bayesian network model for the provision of ES in Aysén forests incorporating trade-off, synergies, and the most determining stressors. The results could be applied

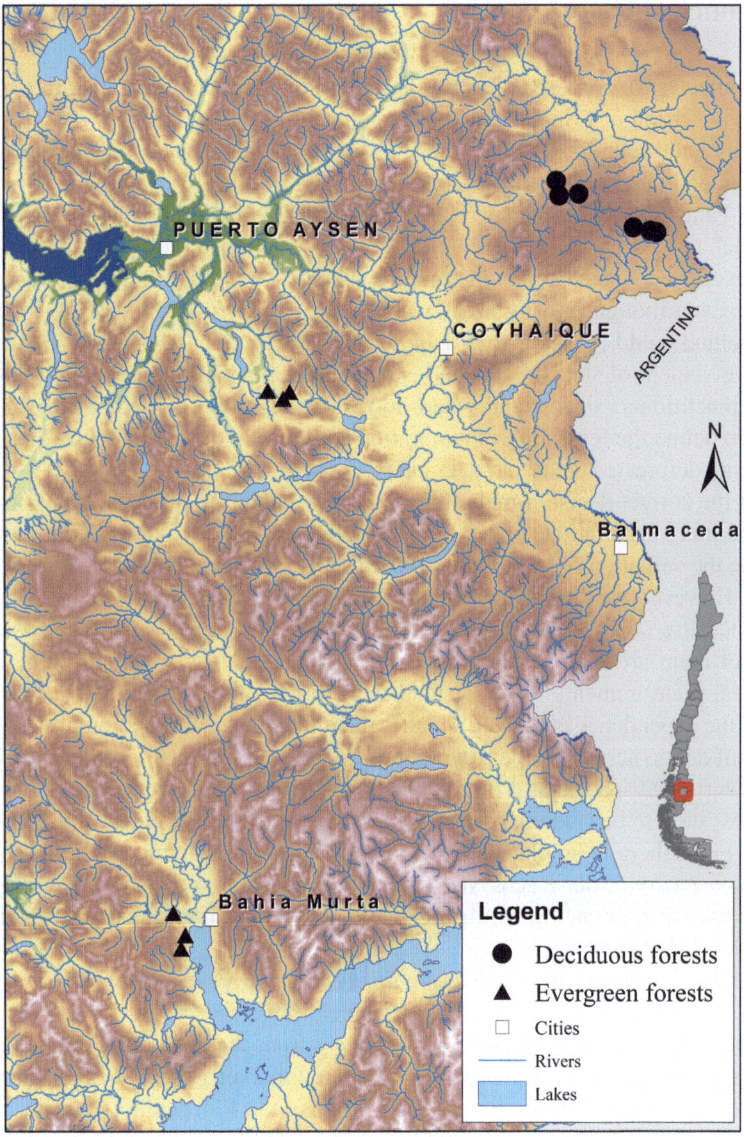

Fig. 9.4 Location of 12 watersheds monitored by CIEP on the Aysén Region

to the majority forest watersheds south of the 40th parallel, the most extensive area with forests in the country. Besides, this territory has limited conditions for money market products, such as wood, and where an adequate valuation of the ES is vital to increase the current perceived value of the forest by the owners.

4 Conclusions

In this chapter we proposed a conceptual model of prioritization of restoration areas for the provision of ES and an approach to quantify ES in Patagonia-Aysén. These methods could support policy guidance and promote initiatives to reverse or mitigate the degradation processes that affect forest ecosystems in any ecosystem, in general, and in Patagonia, in particular. These methods provide an integral and innovative approach to identify priority areas for forest restoration at the landscape scale and to assess the ecological suitability and the socioeconomic feasibility of forest restoration. Even though priority areas were defined following different perspectives, this method could be adapted to achieve particular goals and/or modified through the involvement of stakeholders and expert judgment. This approach will not only allow practitioners understand where to restore according to ecological variables but also define the feasibility of restoration activities in the medium and long term, including deforested areas and degraded forests.

For the correct decision-making in the management of our ecosystems, such as incorporating restoration or avoiding degradation, it is of utmost importance to include the valuation of the services that their natural capital offers us for our well-being. However, we know that those practices are not standard in decision-makers. First, because of the nonexistent quantifications of these ES, or if they exist, only some of them are incorporated tangentially in the analysis. Second, there are no temporal measurements to understand these flows of materials and information. Third, the dependence of these ES makes the individual quantification biased due to trade-off and synergies in their dynamics. Fourth, the ES must be quantified based on a determined space that is directly associated with a logical conceptual model of the ecosystem to be studied, which is often incompatible with the needs of the producer. Fifth, there is no clarity of the stressors and the maximum marginal provisioning capacity of these ecosystems. Finally, there are no tools in the market that incorporate or promote the willingness to pay for those ES, which are essential to human well-being.

The shortcomings expressed before involve that native forest conservation problems are often due to the low perceived value of this resource for landowners. The perception is even lower in areas further away from the main commercial centers and with a higher cost of production of money market goods, such as timber or plywood. Meanwhile, these remote zones, south of the 40th parallel in Chile, are the areas with the most considerable amount of native forests that need to be conserved, incorporating the integral value of the ES that they deliver to society. Otherwise, the only option for the owner will be the degradation of the forest until it ceases to be, or simply change the use of the land through local burning or, even worse, mega-fires. Besides, this territory concentrates the majority of Chile's water resources: precipitation (68%), lakes (64% per area), lagoons (84%), rivers (75% of runoff), and glaciers (98% of the water equivalents, 21,993 km^2) (Reid et al. 2019). Also, lakes are highlighted worldwide for their area, depth, and transparency/water quality, and some of the most important rivers in the Pacific Ocean, which export freshwater and nutrients on an extensive area of fjord ecosystems (Reid et al. 2019). Hence, any forest degradation process will give rise to problems on the provision of freshwater, involving other ecosystems and economic activities, such as artisanal fisheries.

Acknowledgments This work was supported by "U-Semilla 2017 project. Universidad de Aysén" (C.Z.E.), Center for Climate and Resilience Research (CR2 – ANID/FONDAP/15110009) (C.Z.E.), Programa Regional CONICYT R17A10002 Laboratorio Ecoclimático (P.M.), and Chilean Native Forest Research Fund (FIBN-CONAF) (project 033/2019) (P.M.).

References

Aguilar M (2018) Caracterización de humedales dentro de microcuencas con distintos niveles de intervención en la Región de Aysén. Undergraduate thesis. http://cybertesis.uach.cl/tesis/uach/2018/fifa283c/doc/fifa283c.pdf

Alfaro O (1982) Estudio de crecimiento en Lenga en las masas forestales más importantes de la XI Región. Undergraduate thesis. https://bibliotecadigital.infor.cl/handle/20.500.12220/739

Alvarez S, Grosse H (1978) Antecedentes generales y análisis para el manejo de Lenga en Alto Mañihuales, Aysén. Undergraduate thesis. https://bibliotecadigital.infor.cl/handle/20.500.12220/2320

Aronson J, Alexander S (2013) Ecosystem restoration is now a global priority: time to roll up our sleeves. Restor Ecol 21:293–296

Astorga A, Moreno PC, Reid B (2018) Watersheds and trees fall together: an analysis of intact forested watersheds in Southern Patagonia (41-56° S). Forests 9:385

Bachmann-Vargas P (2013) Ecosystem services modeling as a tool for ecosystem assessment and support for decision making process in Aysén region, Chile. Master thesis. https://www.researchgate.net/publication/273457371

Bahamondez V, Barros S, Stuven M, Caroline M, Moreno PC, Villalobos EL (2009) Estudio de línea base proyecto forestación bajo mecanismo de desarrollo limpio (MDL) del protocolo de Kyoto sobre cambio climático región de Aysén, Chile. https://bibliotecadigital.infor.cl/handle/20.500.12220/17391. Accessed 15 Oct 2020

Barton DN, Kelemen E, Dick J et al (2018) (Dis) integrated valuation – Assessing the information gaps in ecosystem service appraisals for governance support. Ecosyst Serv 29:529–541

Borda-Niño M, Hernández-Muciño D, Ceccon E (2017) Planning restoration in human modified landscapes: new insights linking different scales. Appl Geogr 83:118–129

Boyd J, Banzhaf S (2007) What are ecosystem services? The need for standardized environmental accounting units. Ecol Econ 63:616–626

Burkhard B, Maers J (eds) (2017) Mapping ecosystem services. Pensoft Publishers, Sofia

Cademus R, Escobedo FJ, McLaughlin D, Abd-Elrahman A (2014) Analyzing trade-offs, synergies, and drivers among timber production, carbon sequestration, and water yield in *Pinus elliottii* forests in southeastern USA. Forests 5:1409–1431

Carignan V, Villard MA (2002) Selecting indicator species to monitor ecological integrity: a review. Environ Monit Assess 78:45–61

CBD (2005) Report of the inter-sessional (Second) meeting of the AHTEG on the review of implementation of the programme of work on forest biological diversity. https://www.cbd.int/meetings/TEGFORIS-01. Accessed 15 Oct 2020

Choi Y (2004) Theories for ecological restoration in changing environment: toward futuristic restoration. Ecol Res 19:75–81

Comín FA, Miranda B, Sorando R, Felipe-Lucia M, Jiménez J, Navarro E (2017) Data from: prioritizing sites for ecological restoration based on ecosystem services. J Appl Ecol. https://doi.org/10.1111/1365-2664.13061

CONAF (2013) Compendio de funciones alométricas para la estimación de la biomasa de especies forestales presentes in Chile. http://bosques.ciren.cl/handle/123456789/26280. Accessed 12 Oct 2020

Costanza R, d'Arge R, De Groot R, Farber S, Grasso M, Hannon B, Limburg K, Naeem S, O'neill RV, Paruelo J (1997) The value of the world's ecosystem services and natural capital. Nature 387:253

Cruz Johnson P, Honeyman Lucchini P, Caballero Tapia C (2005) Propuesta metodológica de ordenación forestal, aplicación a bosques de lenga en la XI Región. Bosque 26(2):57–70

Cuevas JG, Arumí JL, Zúñiga-Feest A, Little C (2018) An unusual kind of diurnal streamflow variation. J Hydrol Hydromech 66:32–42

De Groot R et al (2012) Global estimates of the value of ecosystems and their services in monetary units. Ecosyst Serv 1:50–61

Delgado LE, Sepúlveda MB, Marín VH (2013) Provision of ecosystem services by the Aysén watershed, Chilean Patagonia, to rural households. Ecosyst Serv 5:102–109

Delphin S, Escobedo F, Abd-Elrahman A, Cropper W Jr (2013) Mapping potential carbon and timber losses from hurricanes using a decision tree and ecosystem services driver model. J Environ Manag 129:599–607

Donoso C (1993) Bosques templados de Chile y Argentina. Variación, Estructura y Dinámica. Ecología Forestal. Editorial Universitaria. Santiago, Chile

Donoso PJ, Frêne C, Flores M, Moorman MC, Oyarzún CE, Zavaleta JC (2014) Balancing water supply and old-growth forest conservation in the lowlands of south-central Chile through adaptive co-management. Landsc Ecol 29:245–260

Doren RF, Trexler JC, Gottlieb AD, Harwell MC (2009) Ecological indicators for system wide assessment of the greater everglades ecosystem restoration program. Ecol Indic 9:2–16

Drake F, Emanuelli P, Acuña Carmona EA (2003). Compendio de funciones dendrométricas de bosque nativo. http://bibliotecadigital.ciren.cl/handle/123456789/10558. Accessed 15 Oct 2020

Dube F, Zagal E, Stolpe N, Espinosa M (2009) The influence of land-use change on the organic carbon distribution and microbial respiration in a volcanic soil of the Chilean Patagonia. For Ecol Manag 257:1695–1704

Dube F, Thevathasan NV, Zagal E, Gordon AM, Stolpe NB, Espinosa M (2011) Carbon sequestration potential of silvopastoral and other land use systems in the Chilean Patagonia. In: Kumar B, Nair P (eds) Carbon sequestration potential of agroforestry systems, Advances in agroforestry, vol 8. Springer, Dordrecht, pp 101–127

Dufour S, Piégay H (2009) From the myth of a lost paradise to targeted river restoration: forget natural references and focus on human benefits. River Res Appl 25:568–581

Fajardo A, Gundale MJ (2015) Combined effects of anthropogenic fires and land-use change on soil properties and processes in Patagonia, Chile. For Ecol Manag 357:60–67

FAO (2009) Towards defining degradation. FRA Working Paper 154. Rome

Fisher B, Christopher T (2007) Poverty and biodiversity: measuring the overlap of human poverty and the biodiversity hotspots. Ecol Econ 62:93–101

Frêne C, Dörner J, Zúñiga F, Cuevas JG, Alfaro FD, Armesto JJ (2019) Eco-hydrological functions in forested catchments of southern Chile. Ecosystems 23:307–323

Gale T, Ednie A (2019) Can intrinsic, instrumental, and relational value assignments inform more integrative methods of protected area conflict resolution? Exploratory findings from Aysén, Chile. J Tour Cult Chang. https://doi.org/10.1080/14766825.2019.1633336

Gale T, Ednie A, Beeftink K (2019) Worldviews, levels of consciousness, and the evolution of planning paradigms in protected areas. J Sustain Tour 27(11):1609–1633

Gale T, Ednie A, Beeftink K, Adiego A (2020) Beyond noise management: exploring visitors' perceptions of positive emotional soundscape dimensions. J Leis Res. https://doi.org/10.1080/00222216.2020.1749912

Gayoso J (2013) Funciones alométricas para la determinación de existencias de carbono forestal para la especie *Nothofagus obliqua* (Mirb.) Oerst. Corporación Nacional Forestal
Gayoso J, Guerra J, Alarcón D (2002) Medición de la capacidad de captura de carbono en bosques de Chile y promoción en el mercado mundial. Universidad Austral de Chile, Valdivia
Geneletti D (2002) Ecological evaluation for environmental impact assessment, NGS Series 301. Netherlands Geographical Studies, Utrecht. https://webapps.itc.utwente.nl/librarywww/Papers/phd_2002/geneletti.pdf. Accessed 15 Oct 2020
Geneletti D (2004) A GIS-based decision support system to identify nature conservation priorities in an alpine valley. Land Use Policy 21:149–160
Geneletti D (2005) Formalizing expert's opinion through multi-attribute value functions. An application in landscape ecology. J Environ Manage 76:255–262
Gobierno de Chile (2020). Chile's nationally determined contribution. Update 2020. https://www4.unfccc.int/sites/ndcstaging/PublishedDocuments/Chile%20First/Chile%27s_NDC_2020_english.pdf. Accessed 10 Oct 2020
GOFC-GOLD (2015) A sourcebook of methods and procedures for monitoring and reporting anthropogenic greenhouse gas emissions and removals associated with deforestation, gains and losses of carbon stocks in forests remaining forests, and forestation. Wageningen University. Accessed 10 Sept 2020
González ME, Sapiains R, Gómez-González S, Garreaud R, Miranda A, Galleguillos M, Jacques M, Pauchard A, Hoyos J, Cordero L, Vásquez F, Lara A, Aldunce P, Delgado V, Arriagada, Ugarte AM, Sepúlveda A, Farías L, García R, Rondanelli RJ, Ponce R, Vargas F, Rojas M, Boisier JP, Carrasco C, Little C, Osses M, Zamorano-Elgueta C, Díaz Hormazábal I, Ceballos A, Guerra E, Moncada M, Castillo I (2020) Incendios forestales en Chile: causas, impactos y resiliencia. http://www.cr2.cl/incendios/. Accessed 10 Oct
Guida-Johnson B, Zuleta GA (2017) Riparian rehabilitation planning in an urban–rural gradient: integrating social needs and ecological conditions. Ambio. https://doi.org/10.1007/s13280-016-0857-7
Gysling A, Álvarez V, Soto D, Pardo E, Poblete P (2018) Anuario Forestal 2018. (Bol. Tec. N°163), Instituto Forestal, Ministerio de Agricultura, Santiago, Chile. INFOR
Hepp K, Stolpe NB (2014) Caracterización y propiedades de los suelos de la Patagonia Occidental (Aysén). Instituto de Investigaciones Agropecuarias, Centro de Investigación INIA Tamel Aike, Aysén, Chile
INFOR (2003a) Informe de ajuste de Funciones de Volumen y Altura para lenga (*Nothofagus pumilio*) y coihue de magallanes (*Nothofagus betuloides*). Proyecto "Modelos de Gestión Sostenible para Incorporar Bosques de Lenga de Aysén a la Producción Nacional" INFOR, FDI-CORFO
INFOR (2003b) Informe procedimientos y resultados modelos de ahusamiento. Proyecto "Modelos de Gestión Sostenible para Incorporar Bosques de Lenga de Aysén a la Producción Nacional" INFOR, FDI-CORFO
INFOR (2009) Manual Elaboración proyectos forestación / reforestación bajo mecanismo de desarrollo limpio en Chile. A/R MDL protocolo de Kyoto. Informe técnico N°176. Ministerio de Agricultura, Santiago
INFOR (2020) National forest inventory. https://ifn.infor.cl/index.php. Accessed 7 Oct 2020
IPCC (2019) In: Calvo Buendia E, Tanabe K, Kranjc A, Baasansuren J, Fukuda M, Ngarize S, Osako A, Pyrozhenko Y, Shermanau P, Federici S (eds) Refinement to the 2006 IPCC guidelines for national greenhouse gas inventories. IPCC, Switzerland
Kershner J, Samhouri JF, James CA, Levin PS (2011) Selecting indicator portfolios for marine species and food webs: a Puget Sound case study. PLoS One. https://doi.org/10.1371/journal.pone.0025248
Laestadius L, Saint-Laurent C, Minnemeyer S, Potapov P (2011) A world of opportunity: the world's forests from a restoration perspective. The global partnership on forest landscape restoration. World Resources Institute, South Dakota State University and the International Union for the Conservation of Nature

Lara A, Villalba R, Wolodarsky-Franke A, Aravena J, Luckman B, Cuq E (2005) Spatial temporal variation in *Nothofagus pumilio* growth at tree line along its latitudinal range (35°40'-55°S). J Biogeogr 32:879–893

Lewis SL, Wheeler CE, Mitchard ETA, Koch A (2019) Regenerate natural forests to store carbon. Nature 568:25–28

Little C, Soto D, Lara A, Cuevas JG (2008) Nitrogen exports at multiple-scales in a southern Chilean watershed (Patagonian Lakes district). Biogeochemistry 87(3):297–309

Martin K (2007) Funciones de volumen y ahusamiento para *Nothofagus dombeyi* (Coigüe común), y *Laureliopsis phillipiana* (Tepa), en el Valle Mirta, localidad de La Junta, XI región de Aysén. Undergraduate thesis. http://cybertesis.uach.cl/tesis/uach/2007/fifm379f/doc/fifm379f.pdf

Martin M, Pilquinao B, Müller-Using S, Bahamóndez C, Guiñez R, Acuña B, Bava J, Loguercio G (2018) Diagramas de manejo de densidad para bosques de lenga de Aysén y Magallanes con fines de producción de madera. Instituto Forestal

Mascaro J, Detto M, Asner GP, Muller-Landau HC (2011) Evaluating uncertainty in mapping forest carbon with airborne LiDAR. Remote Sens Environ 115:3770–3774

Maxwell SL, Reside A, Trezise J, McAlpine CA, Watson JEM (2019) Retention and restoration priorities for climate adaptation in a multi-use landscape. Glob Ecol Conserv. https://doi.org/10.1016/j.gecco.2019.e00649

McDonald T, Gann GD, Jonson K, Dixon KW (2016) International standards for the practice of ecological restoration – including principles and key concepts. Society for Ecological Restoration, Washington, DC

McIntire EJB, Fajardo A (2011) Facilitation within species: a possible origin of group selected superorganisms. Am Nat 178:88–97

MEA (2005) Millennium ecosystem assessment. https://www.millenniumassessment.org/en/Reports.html. Accessed 10 Oct 2020

Mitchell AL, Rosenqvist A, Mora B (2017) Current remote sensing approaches to monitoring forest degradation in support of countries measurement, reporting and verification (MRV) systems for REDD+. Carbon Balance Manag. https://doi.org/10.1186/s13021-017-0078-9

Mittermeier RA, Myers N, Thomsen JB, da Fonseca GAB, Olivieri S (1998) Biodiversity hotspots and major tropical wilderness areas: approaches to setting conservation priorities. Conserv Biol 12:516–520

MMA (2014) Primer informe bienal de actualización de Chile ante la convención marco de las Naciones Unidas sobre el Cambio Climático. https://mma.gob.cl/cambioclimatico/primer-informe-bienal-de-actualizacion-de-chile/. Accessed 10 Oct 2020

Moilanen A, Aderson BJ, Eigenbrod F, Heinemeyer A, Roy DB, Gillings S, Armsworth PR, Gaston KJ, Thomas CD (2011) Balancing alternative land uses in conservation prioritization. Ecol Appl 21:1419–1426

Mouillot D, Graham NAJ, Villéger S, Mason NWH, Bellwood DR (2012) A functional approach reveals community responses to disturbances. Trends Ecol Evol 28:167–177

Nahuelhual L, Donoso P, Lara A, Nuñez D, Oyarzun C, Neira E (2007) Valuing ecosystem services of Chilean temperate rainforests. Environ Dev Sustain 9(4):481–499

Neeson TM, Smith SDP, Allan JD, McIntyre PB (2016) Prioritizing ecological restoration among sites in multi-stressor landscapes. Ecol Appl 26:1785–1796

Neira E (2005) Producción de agua en dos microcuencas con diferente cobertura de renovales de *Nothofagus* en la cordillera de los Andes (San Pablo de Tregua). Master thesis. http://www.forestal.uach.cl/postgrado/magister-recursos-forestales/tesis.php

Oakleaf JR, Matsumoto M, Kennedy CM, Baumgarten L (2017) LegalGEO: Conservation tool to guide the siting of legal reserves under the Brazilian Forest Code. Appl Geogr 86:53–65

Olander LP, Galik CS, Kissinger GA (2012) Operationalizing REDD+: scope of reduced emissions from deforestation and forest degradation. Curr Opin Environ Sustain 4:661–669

Orsi F, Geneletti D (2010) Identifying priority areas for Forest Landscape Restoration in Chiapas (Mexico): an operational approach combining ecological and socioeconomic criteria. Landsc Urban Plan 94:20–30

Oyarzún CE, Godoy R, Sepúlveda A (1998) Water and nutrient fluxes in a cool temperate rainforest at the Cordillera de la Costa in southern Chile. Hydrol Process 12(7):1067–1077

Peh KSH, Balmford AP, Bradbury RB, Butchart SHM, Brown C, Hughes FMR, Stattersfield AJ, Thomas D, Walpole M, Birch JC (2013) Measuring and monitoring ecosystem services at the site scale: a practical toolkit. Ecosyst Serv 5:51–57

Penman J, Gytarsky M, Hiraishi T, Krug T, Kruger D, Pipatti R, Buendia L, Miwa K, Ngara T, Tanabe K, Wagner F (eds) (2003) IPCC good practice guidance for land use, land-use change and forestry. IGES, Japan

Perring MP, Standish RJ, Price JN, Craig MD, Erickson TE, Ruthrof KX, Whiteley AS, Valentine LE, Hobbs RJ (2015) Advances in restoration ecology: rising to the challenges of the coming decades. Ecosphere. https://doi.org/10.1890/ES15-00121.1

Plugge D, Köhl M (2012) Estimating carbon emissions from forest degradation: implications of uncertainties and area sizes for a REDD+ MRV system. Can J For Res 42:1996–2010

Ponce RD, Vásquez F, Stehr A, Debels P, Orihuela C (2011) Estimating the economic value of landscape losses due to flooding by hydropower plants in the Chilean Patagonia. Water Resour Manag. https://doi.org/10.1007/s11269-011-9820-3

Posner S, McKenzie E, Ricketts TH (2016) Policy impacts of ecosystem services knowledge. Proc Natl Acad Sci 113:1760–1765

Promis A, Allen RB (2017) Tree seedlings respond to both light and soil nutrients in a Patagonian evergreen-deciduous forest. PLoS One. https://doi.org/10.1371/journal.pone.0188686

Quintanilla PV (2008) Perturbaciones a la vegetación nativa por grandes fuegos de 50 años atrás, en bosques Nordpatagónicos. Caso de estudio en Chile Meridional. Anales de Geografía 28:85–104

Reid B, Astorga A, Madriz I (2019) Estado de conocimiento y conservación de la biodiversidad de los ecosistemas dulceacuícolas de la Patagonia. White paper para Pew Charitable Trust

Rodríguez-Gallego L, Achkar M, Defeo O, Vidal L, Meerhoff E, Conde D (2017) Effects of land use changes on eutrophication indicators in five coastal lagoons of the Southwestern Atlantic Ocean. Estuar Coast Shelf Sci 188:116–126

Salinas S, Riquelme F, Acuña B, Uribe M (2019) Crecimiento de renovales de Lenga (*Nothofagus pumilio* (Poepp. & Endl.) Krasser) en el sur austral de Chile. https://bibliotecadigital.infor.cl/handle/20.500.12220/29204. Accessed 14 Oct 2020

Stolpe NB, Dube F, Zagal E (2010) Calibration of CO2FIX to native forest, pine plantation, and pasture on a volcanic soil of the Chilean Patagonia. Acta Agric Scand Sect B Soil Plant Sci 60(3):235–244

Susaeta A, Sancewich B, Adams D, Moreno PC (2019) Ecosystem services production efficiency of longleaf pine under changing weather conditions. Ecol Econ 156:24–34

Terrado M, Momblanch A, Bardina M, Boithias L, Munné A, Sabater S, Solera A, Acuña V (2016) Integrating ecosystem services in river basin management plans. J Appl Ecol 53:865–875

Torres C (2017) Funciones de biomasa y volumen para *Nothofagus antarctica* (Forster) Oerst y *Pinus contorta* (Douglas) ex Loud en la comuna de Coyhaique, Región Aysén del General Carlos Ibáñez del Campo. Undergraduate thesis. http://cybertesis.uach.cl/tesis/uach/2017/fift693f/doc/fift693f.pdf

Torres R, Reid B, Frangópulos M, Alarcón E, Márquez M, Häusermann V, Fösterra G, Pizarro G, Iriarte JL, González HE (2020) Freshwater runoff effects on the production of biogenic silicate and chlorophyll-a in western Patagonia archipelago (50–51° S). Estuar Coast Shelf Sci. https://doi.org/10.1016/j.ecss.2020.106597

Trabucchi M, Ntshotsho P, OFarrell P, Comín FA (2012) Ecosystem service trends in basin-scale restoration initiatives: a review. J Environ Manag 111:18–23

Vallecillo S, La Notte A, Zulian G, Ferrini S, Maes J (2019) Ecosystem services accounts: valuing the actual flow of nature-based recreation from ecosystems to people. Ecol Model 392:196–211

Van Herwijnen M (1999) Spatial decision support for environmental management. Doctoral thesis. https://research.vu.nl/en/publications/spatial-decision-support-for-environmentalmanagement

Vandekerkhove E, Bertrand S, Reid B, Bartels A, Charlier B (2016) Sources of dissolved silica to the fjords of northern Patagonia (44-48 S): the importance of volcanic ash soil distribution and weathering. Earth Surf Process Landf 41(4):499–512

Vásquez-Grandón A, Donoso PL, Gerding V (2018) Forest degradation: when is a forest degraded? Forests 9:726

Veronesi F, Schito J, Grassi S, Raubal M (2017) Automatic selection of weights for GIS based multicriteria decision analysis: site selection of transmission towers as a case study. Appl Geogr 83:78–85

Yarrow M, Leitao PC (2007) Simulating *Nothofagus* forests in the Chilean Patagonia: a test and analysis of tree growth and nutrient cycling in SWAT. In: Abstracts of the 4th International SWAT Conference, Delft, The Netherlands, 2–6 July 2007

Zamorano-Elgueta C (2014) Definición de áreas prioritarias para la restauración forestal en la Cordillera de la Costa de la Región de Los Ríos, Chile. Doctoral dissertation. https://ebuah.uah.es/dspace/handle/10017/22516

Zamorano-Elgueta C, Cayuela L, Rey Benayas JM, Donoso PJ, Geneletti D, Hobbs RJ (2014) The differential influences of human-induced disturbances on tree regeneration community: a landscape approach. Ecosphere. https://doi.org/10.1890/ES14-00003.1

Zhu M, Hoctor T, Volk M, Frank KI, Zwick PD, Carr MH, Linhoss AC (2015) Spatial conservation prioritization to conserve biodiversity in response to sea level rise and land use change in the Matanzas River Basin, Northeast Florida. Landsc Urban Plan 144:103–118

Chapter 10
The North American Beaver Invasion and the Impact Over the Ecosystem Services in the Tierra del Fuego Archipelago

Alejandro Huertas Herrera, Mónica D. R. Toro Manríquez, María Vanessa Lencinas, and Guillermo Martínez Pastur

Abstract North American beaver is the most iconic and studied invasive species of Tierra del Fuego. The scientific literature has focused on highlighting the beaver negative impacts on the ecosystems, which have resulted in reduced actions such as eradication. However, the beaver problem in relation to ecosystem services (ES) has been little described or analyzed. This chapter provides a beaver habitat suitability map and discusses the potential positive and/or negative relationships between beaver invasion and ES (provisioning, cultural, supporting and regulating). We found that ES approaches can provide evidence of the beaver role for the people at different archipelago's vegetation zones. We conclude by outlining how the comprehensive assessment of ES can be a tool that allows to understand and to guide decision-making for the beaver issue.

Keywords Exotic species · Species loss · Passive and active restoration · Monitoring · Ecosystem assessment

A. Huertas Herrera (✉) · M. V. Lencinas
Centro Austral de Investigaciones Científicas (CADIC), Consejo Nacional de Investigaciones Científicas y Técnicas (CONICET), Ushuaia, Tierra del Fuego, Argentina
e-mail: ahuertasherrera@cadic-conicet.gob.ar

M. D. R. Toro Manríquez
Departamento de Ciencias Agropecuarias y Acuícolas, Universidad de Magallanes (UMAG), Punta Arenas, Magallanes, Chile

G. Martínez Pastur
Consejo Nacional de Investigaciones Científicas y Técnicas, Centro Austral de Investigaciones Científicas y Técnicas, Ushuaia, Tierra del Fuego, Argentina

P. L. Peri et al. (eds.), *Ecosystem Services in Patagonia*, Natural and Social Sciences of Patagonia, https://doi.org/10.1007/978-3-030-69166-0_10

1 The Beaver Issue in Tierra del Fuego Archipelago and the Associated Impacts

The Tierra del Fuego archipelago (73,000 km^2) is a group of islands in the southern part of Patagonia (52°–56° SL to 72°–63° WL) shared between Argentina and Chile, where the largest terrestrial area is the binational territory of the main island of Tierra del Fuego. The vegetation zones range from grasslands, shrublands, and peatlands to temperate deciduous and evergreen forests (Moore 1983). Here, the North American beavers (*Castor canadensis*) were introduced to Argentina from Canada in 1946, to develop a fur industry. However, the fur industry was unsuccessful, and the beavers were left to their fate. In a short time, they generated an unprecedented ecological and economic disaster (Choi 2008; Anderson et al. 2009).

Beavers are known as extraordinary ecosystem engineers (Wright et al. 2002). They use woody material inputs (mainly branches and trunks of trees) to build their lodges and dams, changing the environment to fulfill their life cycle (Lizarralde 2008), which means important ecological alterations, e.g., modification of the hydrology of watercourses by dams (Naiman et al. 1988; Martínez Pastur et al. 2006; Lizarralde et al. 2008). Studies about the impact of beavers on ecosystems have been documented, e.g., in riparian forests, they erode riverbanks and increase the surface of slow aquatic environments (Lizarralde et al. 1996; Anderson et al. 2009; Garcia and Rodríguez 2018). This impact produces the retention of sediments and organic matter in the affected areas that change the nutrient cycle and the decomposition dynamics and influence the physical-chemical characteristics of the water (Lizarralde et al. 2004; Garcia and Rodríguez 2018; Rodríguez et al. 2020), e.g., dissolved oxygen is higher in streams of primary unmanaged forests (Lizarralde et al. 1996; Simanonok et al. 2011). Under these alterations, the structure and function of the benthic macroinvertebrate communities in the streams decrease, regarding the areas that have not been affected (e.g., primary unmanaged forest) (Simanonok et al. 2011). It has also been found that disturbances caused by beaver modify the structure of the aquatic biota favoring algae such as *Diatom*, *Cyanophyta*, and *Chlorophyta* (Lizarralde et al. 2008; Rodríguez et al. 2020) and introduced salmons (Lizarralde et al. 2008; Arismendi et al. 2020). In addition, new nesting sites for migratory birds have emerged (Lizarralde et al. 2008; Coronato et al. 2015). For some species of forest birds, such as black woodpeckers (*Campephilus magellanicus*), it has been revealed that favorable habitat conditions may exist near small beaver dams and surrounded by ancient forests, probably due to an increase in the availability of wood-boring larvae (Soto et al. 2012). Besides this, impacts on the topography have been documented, where the streams that flow down the mountainsides are dammed, modifying the flow of water and sediments towards the lower sectors of the water basins, producing soil erosion (Coronato et al. 2015), and causing significant geomorphological alterations uplands stream valleys in mountainous areas and wetlands (Coronato et al. 2003; Lizarralde et al. 2004). Other studies suggest that beavers negatively affect archaeological sites when building their dams near them (Parmigiani et al. 2015), as well as contemporary human constructions, such as routes, sewers, bridges, and tourist walkways, among others (Olave 2008;

Pietrek et al. 2015). However, the social dimension integrated to the beaver issue also attributed utilitarian, aesthetic, and humanistic values, where exists a remarkable awareness of the topic and motivation for finding solutions among those who live in the closest vicinity of invasive species (Schüttler et al. 2010).

Lacking natural predators, the beaver population grew exponentially (from the 25 pairs of beavers in 1946 to at least 100,000 in the 2000s) and spread throughout the archipelago (Anderson et al. 2009; Davis et al. 2016). The beavers' status captured scientific, technical, and political attention quickly (Anderson et al. 2015), in part because (i) the archipelago's vegetation zones are one of the last best-preserved in the planet (Watson et al. 2018), and (ii) by geography, these are thinly separated to the continental mainland, which is threatened by beaver spread (Jaksic et al. 2002; Schüttler et al. 2019). The first records of the beaver's presence in the Chilean territory dated from the 1960s (Anderson et al. 2009), on the main island of Tierra del Fuego, where it advanced from Argentina ±100 km in line to reach the west side of Fagnano Lake (e.g., ~6.7 km $year^{-1}$), as in the Navarino Island, where it advanced ±50 km in a straight line towards the south crossing the Beagle Channel (e.g., ~3.3 km $year^{-1}$). Currently, beavers are established in all the vegetation zones, occupying ecosystems from deciduous and evergreen *Nothofagus* forests (>70% of the total affected area) to grasslands and shrublands in the steppe. In an exhaustive inventory based on satellite images and ground truth, a total of 206,203 beaver dams were estimated (100,951 in Argentina and 105,252 in Chile) (Huertas Herrera et al. 2020).

Unlike the beavers' natural habitat in North America, where forest species can naturally regenerate after a beaver cut, the archipelago's native *Nothofagus* forests lack adaptive mechanisms and reproductive strategies (e.g., vegetative regeneration) to cope with the impact of the beaver, except for the ñire (*Nothofagus antarctica*) (Peri et al. 2013). In the riparian forests, mostly of lenga (*N. pumilio*) and/or mixed with guindo or coihue de Magallanes (*N. betuloides*), the flooded areas near to a dam are converted into long-term stable meadows (Martínez Pastur et al. 2006; Henn et al. 2014), where the trees cannot survive (e.g., due to excessive soil moisture) (Toro Manríquez et al. 2018). In many parts of the archipelago, the ecological impact of the beaver is striking due to the trees that have died standing because of drowning when the beaver dam is built. Obviously, the greatest economic and market opportunity losses (e.g., timber, carbon sequestration) lie with the forests that the beaver destroyed. Probably, the major loss and/or change of ES focused on the *Nothofagus* forests. For example, in the Argentinean part of the archipelago, the humans have exploited nearly ~40,000 ha of native forest (e.g., timber), an area equivalent to that destroyed by beavers since the beginning of the invasion (~35,000 ha) (Henn et al. 2016). Direct eradication plans have been attempted in the archipelago (Lizarralde et al. 2008). For instance, with a bounty for dead beaver, where anyone carrying a hunted beaver was expected to receive a payment (Malmierca et al. 2011). However, this kind of incentive can lead to the so-called cobra effect. In fact, when local authorities stopped payments for dead beavers, hunters stopped hunting and the beaver population started to grow again (Lizarralde et al. 2008). In other words, this wonderful mammal, which is capable of building dams visible from the space and is a flagship species in the Northern Hemisphere (e.g., a national symbol of Canada), has become a problematic invasive species in Argentina and Chile.

The objective herein was to analyze the relationship between the North American beaver invasion and the ecosystem services (provisioning, cultural and supporting and regulating), with special emphasis on the comprehensive understanding of the positive and/or negative potential effects of beavers on ecosystem services, to propose and promote use and conservation strategies that improve the solution of the current beaver problem in the Tierra del Fuego archipelago.

2 Beaver Habitat Suitability and the Ecosystem Services

Quantifying the presence of invasive species is usually the starting point to help with the location of eradication or control efforts (Pietrek et al. 2017; Schüttler et al. 2019). So, it is essential to understand how beaver dams are organized within the natural environment (Huertas Herrera et al. 2020). Figure 10.1 shows a habitat

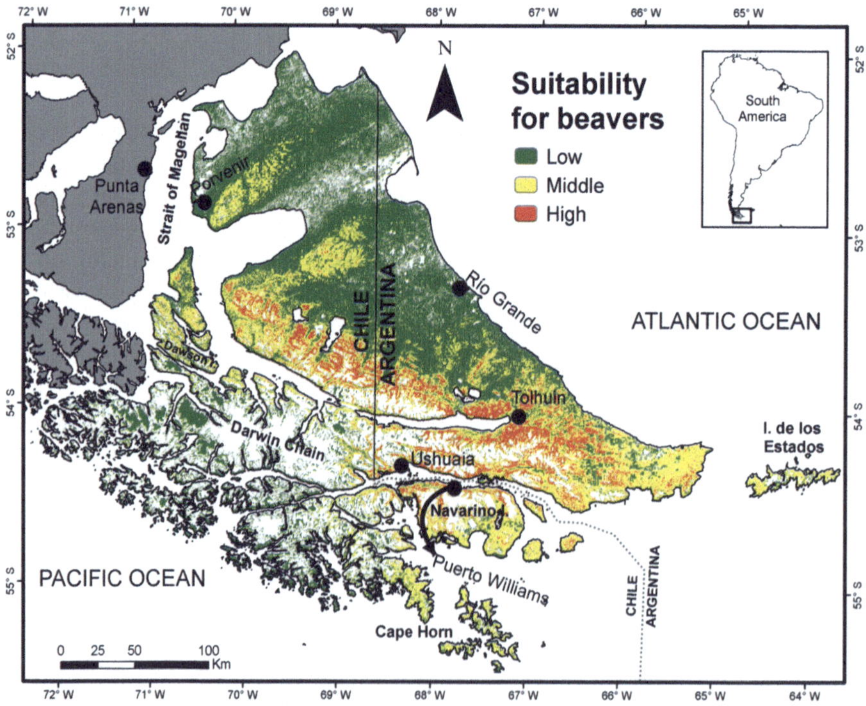

Fig. 10.1 Habitat suitability for beavers. The suitability criteria that favor the distribution of beaver were obtained from Huertas Herrera et al. (2020). The weighting scale ranged from 1 to 3 (0–100%), where 3 (>75%) and 1 (<35%) are the highest and lowest suitability rating, respectively (white = unsuitable, green = low, yellow = middle, red = high). The zone that was not analyzed is in gray

suitability model, whose ecological data typically serves as a basis for fit models that allow to anticipate how the invasion will advance, according to the preferences that the beavers have when colonizing new territories, and to design strategies to stop the advance throughout the archipelago, e.g., the yellow and red areas could help to predict the potential areas of invasion. Hence, this habitat suitability information can be used as a tool to deploy effort in a control/eradication operation (e.g., which sites are best for the beaver spread) and/or create alternative scenarios with different ecosystem services (ES) assumptions (e.g., where to conserve vegetation zones that are more threatened by the beaver spread); in this case, ES such as supporting and regulating (e.g., maintaining local biodiversity), provisioning (e.g., tree regeneration in stands) and cultural (e.g., ecotourism) have relevance and notoriety.

The preferable vegetation zones for the beaver are the wooded ones, mainly lenga forests, while the transition ecosystems, such as the steppe-forest ecotone, are less preferred. Other extreme ecosystems such as tundra and steppes also showed signs of preference. However, the beavers showed plasticity to different environments, being able to build beaver dams in desolate steppes or to the limit of vegetation in the high mountains. The ES approaches can be crucial to design strategies linking with the ecological knowledge of the species, whose habitat preferences can be seen as management opportunities (e.g., where is beaver located and how can its population spread and grow), and not as an illusory truth effect. Research related to beaver invasion and the ES they support has been poorly explored or disclosed, contrary to years of alert its impacts. There are topics in which further understanding is sorely needed, e.g., (i) broadening the spectrum of studies related to the beaver and integrating the opportunities or profits it may generate (not just focus on the losses or impacts but also benefit); (ii) the generation of precise and punctual strategies where the beavers can generate greater problems and risks of invasion to non-yet-invaded areas (it can be supported by habitat models like the one in Fig. 10.1); and (iii) the dedicated efforts to include creative and useful strategies for the society that seek to achieve balanced solutions among the specific ES involvement.

3 Ecosystem Services Approach as a New Tool to Understand the Beaver Issue in Tierra del Fuego

Since the European colonization (eighteenth century), the archipelago of Tierra del Fuego has undergone important socio-ecological changes that translate into an ES dynamics in the last centuries. Here, the indigenous peoples, conformed by groups of hunter-gatherer, had a different perception of nature and the benefits it offered, which have been described as linked to local human nature ties (Gusinde 1951), while the Europeans embodied their vision of the territory (e.g., cattle ranching) (Gea Izquierdo et al. 2004), motivated by the maximization of immediate economic performance (Anderson et al. 2015; Peri et al. 2016). In this context, the ES that can be characterized and managed in the archipelago derives from the western worldview of the territory (Anderson et al. 2015), which is different from the eighteenth to the twenty-first centuries, e.g., during the twenty-first century, this worldview appears to be more conservationist and tied to sustainable development, which includes the ethical treatment of biodiversity (Haider and Jax 2007; Ballari et al. 2020).

At the beginning of the twentieth century, the beaver introduction was completely accepted and even celebrated (Anderson et al. 2015). The socioeconomic and geopolitical marginalization of the archipelago facilitated this end (the introduction of beavers) since the area was a remote and sparsely populated territory. It was only recently (at the end of the twentieth century) that scientists and technicians showed the beaver environmental impacts, focusing on the provision (e.g., riparian forest change and/or loss) and supporting and regulating (e.g., macroinvertebrate fauna reduction) ES affectation (Anderson et al. 2009, 2017). Of course, the presence of the beaver in a human territory implies that people can perceive them in different ways (Schüttler et al. 2010), e.g., conservationist and productivity visions are in conflict. For that matter, although the beaver has been an environmental disaster and several technical and scientific efforts have been carried out to study it and/or eradicate it, where significant international cooperation has been done, today the presence of the species is tangible in the society and vegetation zones of the archipelago.

Based on the Common International Classification of Ecosystem Services (CICES) (Haines-Young and Potschin 2018), the beaver's issue in the archipelago seems to be integrated positively and negatively with many ES (Fig. 10.2). Contrary to the supporting and regulating ES (negatively affected), several cultural ES could be derived from the beavers (both negatively and positively affected), for instance, the most important ski center in southern Patagonia: Cerro Castor; the publicity about the sighting the beaver dams; the souvenirs made in reference to this charismatic rodent; the giant beaver costume to let tourists take pictures on the street; and the Tierra del Fuego National Park beaver dam's viewpoint, among others. In other words, the beaver is a charismatic species despite being exotic, which generates an attraction on people that results in the identification or appropriation of its name and figure, which is adopted many times as an emblem and even to name places and/or undertakings. This shows to what extent the common population does not consider it as a negative influence on the nature that surrounds them, but as an integral part of it.

The studies and understanding of the ES that beavers provide and affect can help to interpret the role of this invasive species in the archipelago ecosystems and plan actions that allow the species to be managed in such a way that the impacts it generates can be minimized, and maybe even living with them since control and/or eradication plans may fail if just biological and/or environmental factors (e.g., ecological niches, trophic networks) are taken into account. Therefore, it is imperative to generate empirical evidence in ES assessment and supporting their integration into the archipelago's economic and management policies.

As was explained earlier, in the archipelago the beaver has invaded the major vegetation zones that offer several ES and so generates mixed feelings in society (Anderson and Valenzuela 2014). The tourism sector (e.g., a cultural ES) has been benefited by the beaver (Fig. 10.3a, c, e, f), being a species that is used within the

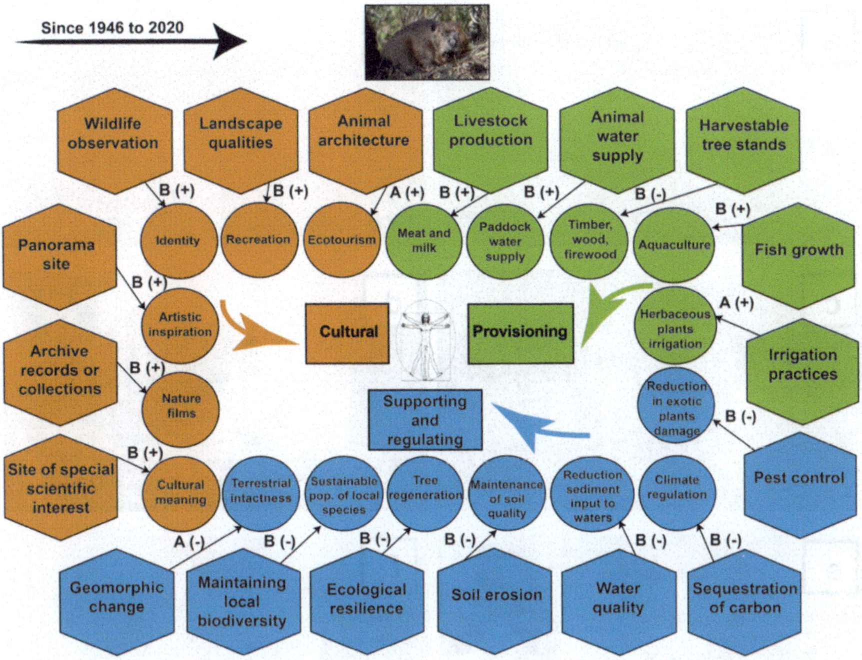

Fig. 10.2 Identification and trend (+ or −) of the ecosystem services (ES) in the Tierra del Fuego archipelago based on Common International Classification of Ecosystem Services (CICES) Version 5.1. Orange color represents cultural ES; blue color represents supporting and regulating ES; and green color represents provisioning ES. The capital letters A and B mean abiotic or biotic ES, respectively. Categories for CICES: (i) specific ES involvement (squares), (ii) example service (circles), (iii) example goods and benefits (hexagons). The black arrows mean connections between categories. In the upper photograph, a beaver is illustrated. Vitruvian Man (source: www.leonardo-davinci.net) represents the ideal ratio of ES for the people

products that are offered (e.g., sighting of the beaver dams, souvenirs) (Fig. 10.3d, e). In addition, several ranchers in the ecotone between forests and grasslands (Fig. 10.3b) appreciate the beaver dams that generate bodies of water for cattle or sheep, because their dams act to decrease the flow of a stream creating ponds that recharge the floodplain (e.g., provides yearlong water to ranches). Besides, with the beaver dams, some arid steppes may become greener for livestock forage, considering beaver dams as provision ES that favors livestock production (e.g., water supply, forage). However, other ranchers may consider beavers as troubles because they reduce the amount of water available for livestock management. The forest industry has been affected since, on the one hand, part of the forest destined for production is devastated and (Henn et al. 2016), on the other, the beaver may affect the road networks (e.g., bridges), generating losses in infrastructure (Olave 2008). Here, the *Nothofagus* forests (state or private lands) offer important ES to the local people (e.g., wood, fodder for livestock, recreation) (Martínez Pastur et al. 2016, 2017), besides offering supporting and regulating services such as sediment retention and

Fig. 10.3 Example of cultural ecosystem services obtained from beavers and/or beaver dams: (**a**) beaver dams close to a camping area; (**b**) beaver dams for livestock in the ecotone; (**c**) sighting of the beaver dams; (**d**) carved wood beaver; (**e**) tour agency poster to advertise beaver watching; (**f**) and street art in Ushuaia City (colorful beaver mural)

water purification (Anderson et al. 2009; Simanonok et al. 2011; Martínez Pastur et al. 2016) and cultural services such as aesthetic values, ecotourism, and spiritual values (Martínez Pastur et al. 2016). However, the beavers could alter the composition of several plant assemblages (Martínez Pastur et al. 2006; Henn et al. 2016), which play a crucial role in the supporting and regulating ES, e.g., the herbaceous plants are angular organisms in forests and grasslands since plants contribute to biodiversity in general (Lencinas et al. 2008, 2011), interact with the initial phases of the installation and growth of tree species (Beckage et al. 2000) and thereby determine much of the energy flow and cycling of nutrients (O'Brien et al. 2007), and respond in complex ways to natural and anthropogenic disturbances (Gilliam 2007; Huertas Herrera et al. 2018).

Positive trends of the distinct ES (and categories) can be found in the different vegetation zones (Table 10.1). The *Nothofagus* forests are one of the most important in the provision ES (e.g., wood), taking into account the sociocultural context in which are immersed, e.g., these forests that are part of protected, fiscal, and private

Table 10.1 Identification and trend (+ or –) of the ecosystem services (ES) derived from beavers in the vegetation zones of the Tierra del Fuego archipelago. The ES criteria were based on Common International Classification of Ecosystem Services (CICES) Version 5.1

Specific ES involvement-example service	Example goods and benefits	Vegetation zone					Literature examples for ES
		Deciduous	Mixed-evergreen	Ecotone	Steppe	Tundra	
Cultural (abiotic)							
Animal architecture	Ecotourism	+	+	+	+	+	
Cultural (biotic)							
Landscape qualities	Recreation	+	+	+	+	+	
Wildlife observation	Identity	+	+	+	+	+	Soto et al. (2012)
Panorama site	Artistic inspiration	+	+	+	+	+	
Archive records or collections	Nature films	+	+	+	+	+	
Site of special scientific interest	Cultural meaning	+	+	+	+	+	Anderson et al. (2009, 2017)
Supporting and regulating (abiotic)							
Geomorphic change	Terrestrial intactness	–	–	–	–	–	Coronato et al. (2003), Henn et al. (2016), Latham et al. (2017)
Supporting and regulating (biotic)							
Maintaining local biodiversity	Sustainable populations of useful or iconic species	–	–	–	–	–	Martínez Pastur et al. (2006), Crego et al. (2016), Schüttler et al. (2019)
Ecological resilience	Tree regeneration in stands	–	–	–	NA	NA	Martínez Pastur et al. (2006), Henn et al. (2014), Toro Manríquez et al. (2018)

(continued)

Table 10.1 (continued)

Specific ES involvement-example service	Example goods and benefits	Vegetation zone					Literature examples for ES
		Deciduous	Mixed-evergreen	Ecotone	Steppe	Tundra	
Soil erosion	Maintenance of soil quality	–	–	–	–	–	Coronato et al. (2003), Grootjans et al. (2010), Henn et al. (2014)
Water quality	Reduction of sediment input to waters	–	–	–	+	–	Simanonok et al. (2011), Henn et al. (2016)
Sequestration of carbon	Climate regulation	–	–	–	–	–	Martínez Pastur et al. (2006), Papier et al. (2019), Huertas Herrera et al. (2020)
Pest control	Reduction in exotic plant damage	–	–	–	–	–	Martínez Pastur et al. (2006), Anderson et al. (2009), Crego et al. (2016), Toro Manríquez et al. (2018)
Provisioning (abiotic)							
Irrigation practices	Herbaceous plant irrigation	–	–	+	+	–	
Provisioning (biotic)							
Fish growth	Aquaculture	+	+	+	+	+	Lizarralde et al. (2008), Arismendi et al. (2020)
Harvestable tree stands	Timber, wood, firewood	–	–	–	NA	NA	Martínez Pastur et al. (2006), Simanonok et al. (2011)
Animal water supply	Paddock permanent water supply	+	+	+	+	+	
Livestock production	Meat and/or milk sold on farm or in shops	+	+	+	+	+	

areas that fulfill the fundamental role of conserving local biodiversity and timber and non-timber forest resources (Table 10.1). However, in terms of economic value for the people, this provision ES may be of less interest to different social actors compared to the direct benefits resulting from tourism and recreation. A key factor here could be the number of people and the short-term profits and benefits from tourism against, for instance, the long-term forest management.

4 Is it Possible to Decrease the Trade-Offs of the Beaver Impact? Recommendations for Policy Makers

The ES approach provides guidance to decision-makers to explore the benefits and costs of some invasive species to generate their own resources for the species control, as the case of beavers, e.g., the people created a work niche around the beaver independent of national and international resources and/or funds. Different options for dealing with the beaver (eradicate, control, tolerate, promote) have been proposed from the viewpoint of environmental ethics and the biological conservation, where the ethical arguments do not decrease the need for sound scientific data but may even increase this demand (Haider and Jax 2007). Specifically, the eradication is promoted because mostly negative impacts generated by the beaver are considered. However, using the ES approach, it emerges that the beaver also generates services and benefits for society. Perhaps it is convenient to use all the knowledge derived from scientific research and technical trials to generate guidelines for the appropriate use of ES linked to the beaver. In other words, efforts to the beaver issue should not be reduced to the control or eradication alternatives, but rather set on the table market alternatives that generate sustenance for the people, such as tourism. In this context, it is expected that a greater demand for ES can arise from enterprising people (e.g., entrepreneurs), as well as a greater development and/or implementation of cultural and provision ES. The decision-makers (e.g., technical advisors at regional and international level, NGOs, community leaders, and legislative and political advisers) can use both ecological information (habitat quality model) not only to understand the beaver biological component but also to have a regional overview of the opportunities and risks of ES immersed in the different vegetation zones (e.g., to enacted regional laws). Notably, the people in the archipelago have increased exponentially since the introduction of the beaver, e.g., only in the Argentine sector of the main island, the population increase from 5045 people in the 1940s (when the beaver was introduced) to over 173,000 in the 2020s (INDEC 2015). Beavers nowadays can be a marketable product not only in the archipelago but throughout Patagonia, as there are hundreds of thousands of visitors who come to this part of the world.

In addition, the ES approach could be useful to understand the social and environmental dimensions related to the beaver issue, beyond the biological and technical considerations that have been developed (e.g., control and/or eradication of the beaver). Consequently, it would be essential to incorporate in the control and/or eradication plans the fundamental factors that explain the relationships in the ES and the beavers. The issue with the beaver does not lie in a species and nature relationship where biological trophic networks are describable and calculable alone, but the temporal social perspective linked to the beavers must be included. Based on the construction of models that illustrate the biological and spatially explicit preferences of the species, the data of the variables of the ES should be linked (e.g., important provisioning ecosystem services). Nevertheless, the beavers amplify the impact of crucial ES (e.g., provision of wood, forest harvesting) (Henn et al. 2016), leading provincial and national governments to view beaver as a potential environmental and economic risk. In the legislation of Argentina and Chile, the riparian forests (those that have been mainly affected by beavers) are protected, as they are considered of very high conservation value. It should be noted that also many of these environments invaded by the beaver are embedded in natural parks, supported by conservation laws. Assuming that a binational policy between both countries was based on the development of biodiversity conservation strategies, this approach can be useful to explore the risk of the biophysics and socioeconomic dimensions of ES considering environmental and conservation policies.

References

Anderson CB, Valenzuela AEJ (2014) Do what I say, not what I do. Are we linking research and decision-making about invasive species in Patagonia? Ecol Austral 24:193–202

Anderson CB, Martínez Pastur G, Lencinas MV, Wallem PK, Moorman MC, Rosemond AD (2009) Do introduced North American beavers Castor canadensis engineer differently in southern South America? An overview with implications for restoration. Mammal Rev 39:33–52

Anderson CB, Martínez Pastur G, Lencinas MV, Soler R, Guillozet K, Santo A, Henn J, Zagarola JP, Kreps G (2015) Más que un ingeniero de ecosistemas: Impactos ambientales, restauración ecológica y la dimensión social del casto. La Lupa 7:4–9

Anderson CB, Roulier C, Pizarro C (2017) Perspectivas de actores clave respecto del acuerdo binacional entre Argentina y Chile sobre la erradicación del castor norteamericano y la restauración de los ecosistemas afectados. Bosque 38:555–562

Arismendi I, Penaluna BE, Jara CG (2020) Introduced beaver improve growth of non-native trout in Tierra del Fuego, South America. Ecol Evol 10:9454–9465

Ballari SA, Roulier C, Nielsen EA, Pizarro CJ, Anderson CB (2020) A review of ecological restoration research in the global south and north to promote knowledge dialogue. Conserv Soc 18:298–310

Beckage B, Clark JS, Clinton BD, Haines BL (2000) A long-term study of tree seedling recruitment in southern Appalachian forests: the effects of canopy gaps and shrub understories. Can J For Res 30:1617–1631

Choi C (2008) Tierra del Fuego: the beavers must die. Nature 453:968

Coronato A, Escobar J, Mallea C, Roig C, Lizarralde M (2003) Características geomorfológicas de ríos de montaña colonizados por Castor canadensis en Tierra del Fuego, Argentina. Ecol Austral 13:15–26

Coronato A, Escobar J, Deferrari G (2015) Un poco de agua es suficiente: Ríos, arroyos y turbales modificados por el castor. La Lupa 7:35–39

Crego RD, Jiménez JE, Rozzi R (2016) A synergistic trio of invasive mammals? Facilitative interactions among beavers, muskrats, and mink at the southern end of the Americas. Biol Invasions 18:1923–1938

Davis EF, Valenzuela AE, Murcia S, Anderson CB (2016) Habitat use by invasive North American beaver during intermediate and long-term colonization periods in southern Patagonia. Mastozoología Neotropical 23:51–61

García VJ, Rodríguez P (2018) Efecto del castor en el metabolismo del perifiton y en variables limnológicas de ríos y arroyos fueguinos. Ecol Austral 28:593–605

Gea-Izquierdo G, Martínez Pastur G, Cellini JM, Lencinas MV (2004) Forty years of silvicultural management in southern Nothofagus pumilio primary forests. For Ecol Manage 201:335–347

Gilliam FS (2007) The ecological significance of the herbaceous layer in temperate forest ecosystems. Bioscience 57:845–858

Grootjans A, Iturraspe R, Lanting A, Fritz C, Joosten H (2010) Ecohydrological features of some contrasting mires in Tierra del Fuego, Argentina. Mires Peat 6:1–15

Gusinde M (1951) Hombres primitivos en la tierra del fuego: (de investigador a compañero de tribu). Escuela de Estudios Hispano-Americanos de Sevilla, Sevilla, p 398

Haider S, Jax K (2007) The application of environmental ethics in biological conservation: a case study from the southernmost tip of the Americas. Biodivers Conserv 16:2559–2573

Haines-Young R, Potschin MB (2018) Common international classification of ecosystem services (CICES) V5.1 and guidance on the application of the revised structure. Nottingham, UK, p 53

Henn JJ, Anderson CB, Kreps G, Lencinas MV, Soler R, Martínez Pastur G (2014) Determining abiotic and biotic factors that limit transplanted Nothofagus pumilio seedling success in abandoned beaver meadows in Tierra del Fuego. Ecol Restor 32:369–378

Henn JJ, Anderson CB, Martínez Pastur G (2016) Landscape-level impact and habitat factors associated with invasive beaver distribution in Tierra del Fuego. Biol Invasions 18:1679–1688

Huertas Herrera A, Cellini JM, Barrera M, Lencinas MV, Martínez Pastur G (2018) Environment and anthropogenic impacts as main drivers of plant assemblages in forest mountain landscapes of Southern Patagonia. For Ecol Manage 430:380–393

Huertas Herrera A, Lencinas MV, Toro Manríquez M, Miller J, Martínez Pastur G (2020) Mapping the status of the North American beaver invasion in the Tierra del Fuego archipelago. PLoS One 15:e0232057

INDEC (2015) Instituto Nacional de Estadística y Censos de la República Argentina. Estimaciones de población por sexo, departamento y año calendario. Años 2010–2025. Informe técnico y cuadros. N° 38 Serie análisis demográfico, p 110

Jaksic FM, Iriarte JA, Jiménez JE, Martínez DR (2002) Invaders without frontiers: cross-border invasions of exotic mammals. Biol Invasions 4:157–173

Latham ADM, Warburton B, Byrom AE, Pech RP (2017) The ecology and management of mammal invasions in forests. Biol Invasions 19:3121–3139

Lencinas MV, Martínez Pastur G, Solán R, Gallo E, Cellini JM (2008) Forest management with variable retention impact over bryophyte communities of *Nothofagus pumilio* understory. Forstarchiv 79:77–82

Lencinas MV, Martínez Pastur G, Gallo E, Cellini M (2011) Alternative silvicultural practices with variable retention to improve understory plant diversity conservation in southern Patagonian forests. For Ecol Manage 262:1236–1250

Lizarralde M (2008) Presencia y ecología de castores en la Tierra del Fuego, Argentina. In: Saavedra B, Silva C (eds) Actas del Taller Internacional de Control de Castores en el Archipiélago Fueguino. Wildlife Conservation Society, Santiago, pp 13–23

Lizarralde M, Deferrari A, Alvarez SE, Escobar J (1996) Effects of beaver (*Castor canadensis*) on the nutrient dynamics of the Southern Beech forest of Tierra del Fuego (Argentina). Ecol Austral 6:101–105

Lizarralde M, Escobar J, Deferrari G (2004) Invader species in Argentina: a review about the beaver (*Castor canadensis*) population situation on Tierra del Fuego Ecosystem. Interciencia 29:352–356

Lizarralde M, Escobar J, Deferrari G, Fasanella M (2008) El castor austral. Investigación y ciencia 379:58–64

Malmierca L, Menvielle M, Ramadori D, Saavedra B, Saunders A, Volkart NS, Schiavini AC (2011) Eradication of beaver (*Castor canadensis*), an ecosystem engineer and threat to southern Patagonia. In: Veitch CR, Clout MN, Towns DR (eds) Island invasives: eradication and management. IUCN, Gland, pp 87–90

Martínez Pastur G, Lencinas MV, Escobar J, Quiroga P, Malmierca L, Lizarralde M (2006) Understorey succession in *Nothofagus* forests in Tierra del Fuego (Argentina) affected by *Castor canadensis*. Appl Veg Sci 9:143–154
Martínez Pastur G, Peri PL, Lencinas MV, García-Llorente M, Martín-López B (2016) Spatial patterns of cultural eco-system services provision in Southern Patagonia. Landsc Ecol 31:383–399
Martínez Pastur G, Peri PL, Huertas Herrera A, Schindler S, Díaz-Delgado R, Lencinas MV, Soler R (2017) Linking potential biodiversity and three ecosystem services in silvopastoral managed forests landscapes of Tierra del Fuego, Argentina. Int J Biodivers Sci Ecosyst Serv Manag 13:1–11
Moore D (1983) Flora of Tierra del Fuego. Anthony Nelson, London
Naiman RJ, Johnston CA, Kelley JC (1988) Alteration of North American streams by beaver: the structure and dynamics of streams are changing as beaver recolonize their historic habitat. Bioscience 38:753–762
O'Brien MJ, O'Hara KL, Erbilgin N, Wood DL (2007) Overstory and shrub effects on natural regeneration processes in native *Pinus radiata* stands. For Ecol Manage 240:178–185
Olave RJ (2008) Consideraciones económicas de la invasión del castor en el archipiélago de Tierra del Fuego. In: Silva CA, Saavedra B (eds) Actas del Taller Internacional para el Control de Castores en la Patagonia. Wildlife Conservation Society, Chile, pp 63–71
Papier CM, Poulos HM, Kusch A (2019) Invasive species and carbon flux: the case of invasive beavers (*Castor canadensis*) in riparian *Nothofagus* forests of Tierra del Fuego, Chile. Clim Change 153:219–234
Parmigiani V, Alvarez Soncini MC, Clemente Conte I (2015) Los castores y la arqueología. La Lupa 7:14–19
Peri PL, Lencinas MV, Martínez Pastur G, Wardell-Johnson GW, Lasagno R (2013) Diversity patterns in the steppe of Argentinean southern Patagonia: environmental drivers and impact of grazing. In: Morales Prieto MB, Traba Diaz J (eds) Steppe ecosystems: biological diversity, management and restoration. Nova Science Publishers Inc, New York, pp 73–95
Peri PL, Lencinas MV, Bousson J, Lasagne R, Soler R, Bahamonde H, Martínez Pastur G (2016) Biodiversity and ecological long-term plots in Southern Patagonia to support sustainable land management: the case of PEBANPA network. J Nat Conserv 34:51–64
Pietrek AG, Escobar J, Feldman M (2015) Biología del invasor más famoso de Tierra del Fuego. La Lupa 7:26–31
Pietrek AG, Escobar J, Fasola L, Roesler I, Schiavini A (2017) Why invasive Patagonian beavers thrive in unlikely habitats: a demographic perspective. J Mammal 98:283–292
Rodríguez P, González Garraza G, García V, Granitto M, Escobar E (2020) Beaver dam effect on phytoplankton and periphyton composition and hydrology in streams from Tierra del Fuego (Argentina). Hydrobiologia 847:1461–1477
Schüttler E, Rozzi R, Jax K (2010) Towards a societal discourse on invasive species management: a case study of public perceptions of mink and beavers in Cape Horn. J Nat Conserv 19:175–184
Schüttler E, Crego RD, Saavedra-Aracena L, Silva-Rodríguez EA, Rozzi R, Soto N et al (2019) New records of invasive mammals from the sub-Antarctic Cape Horn Archipelago. Polar Biol 42:1093
Simanonok MP, Anderson CB, Martínez Pastur G, Lencinas MV, Kennedy JH (2011) A comparison of impacts from silviculture practices and North American beaver invasion on stream benthic macroinvertebrate community structure and function in *Nothofagus* forests of Tierra del Fuego. For Ecol Manage 262:263–269
Soto GE, Vergara PM, Lizama ME, Celis C, Rozzi R, Duron Q, Hahn IJ, Jiménez JE (2012) Do beavers improve the habitat quality for Magellanic Woodpeckers? Bosque 33:271–274
Toro Manríquez M, Promis A, Huertas Herrera A, Martínez Pastur G (2018) Influencia del micrositio y la exposición en la regeneración de bosques de *Nothofagus pumilio* afectados por *Castor canadensis* en Tierra del Fuego: un análisis exploratorio. Bosque 39:431–440
Watson JEM, Evans T, Venter O, Williams B, Tulloch A, Stewart C et al (2018) The exceptional value of intact forest ecosystems. Nat Ecol Evol 2:599–610
Wright JP, Jones CG, Flecker AS (2002) An ecosystem engineer, the beaver, increases species richness at the landscape scale. Oecologia 132:96–101

Chapter 11
Social Links for a Nexus Approach from an Ecosystem Services Perspective in Central-East Patagonia

Virginia Alonso Roldán

Abstract The integral approach of several development goals related to water, energy, food and health has been gaining scientist's attention in the last decade, and it is known as nexus. The ecosystem service (ES) approach and network analysis can contribute to fill gaps linking the individual components of nexus by assessing two indissoluble components of social-ecological systems: (i) mechanisms underpinning ES and their interrelations and (ii) the integrated governance schemes matching these interrelations. These components need to be considered for management action success and to foresee policy results. This chapter explores the relation between existing ES and social actors of the Comarca VIRCH-Valdés by using two-mode networks in order to examine opportunities and threats of the governance system to get simultaneously multiple benefits. An SNA approach is applied to characterize the interrelations among different ES associated with the nexus approach components and to assess the match between natural and social relations. Implication of governance schemes and possible pathways to improve partnership for nexus approach are discussed.

Keywords Social network analysis · Ecosystem services · Governance · Two-mode networks · Central Patagonia (Argentina)

V. Alonso Roldán (✉)
Grupo de Investigación en Gestión, Desarrollo Territorial y Ambiente (GesDTA), Facultad Regional Chubut, Universidad Tecnológica Nacional (UTN), Puerto Madryn, Chubut, Argentina

Instituto para el Estudio de Ecosistemas Continentales Patagónicos (IPEEC), Consejo Nacional de Investigaciones Científicas y Técnicas (CONICET), Puerto Madryn, Chubut, Argentina

P. L. Peri et al. (eds.), *Ecosystem Services in Patagonia*, Natural and Social Sciences of Patagonia, https://doi.org/10.1007/978-3-030-69166-0_11

1 Systemic Approaches: Nexus and Ecosystem Services

The planet's ability to continue supplying ecosystem services (ES) on which humanity depends is limited and declining due to demographic and socio-economic development pressures at different scales (MEA 2005; IPBES 2019). There is an increasingly urgent need to align conservation and development, as seen in global policies such as the Sustainable Development Goals (SDGs) established by the United Nations in 2015 (United Nations 2015). The integral approach of several development goals, mainly those related to water, energy and food but also including health, has been gaining scientist's attention in the last decade, and it is known as the nexus approach.

The nexus approach highlights the interdependence of these components and the natural resources that underpin that security (Hoff 2011). Since the Bonn 2011 Nexus Conference that promoted the nexus approach as a global research agenda, scientific work has been mainly focused on (1) interpreting nexus approach; (2) developing models in order to detect and interpret synergies and trade-offs, guiding the development of intersectoral policy; and (3) empirical research aimed to identify drivers and document collaboration among stakeholders (Zhang et al. 2019). Despite progress modelling biophysical subsystem, there are still gaps linking the individual components of nexus with valuation and decision-making models (Kling et al. 2017).

The ecosystem services (ES) approach is a promising platform to achieve this alignment given that it is a whole system-aware approach that highlights the interlinkage between social and ecological realms (Costanza et al. 2017; Díaz et al. 2018; Tallis et al. 2009) as well as synergies and trade-offs among different components of the nexus (Pahl-Wostl 2019). Moreover, the inclusion of other ES and co-benefits in the assessment and management of multipurpose landscapes can improve sustainability (Hanes et al. 2018). However, and besides the fact that Hoff's (2011) seminal paper at the Bonn conference included the ES concept, further research has not been consistent in aligning nexus and the ES approach.

Due to the complex relationships between social and ecological factors that depend on the social, cultural, economic and environmental configuration, the valuations and the cause-and-effect relationships around ES vary between social-ecological systems and between focus groups of actors whose contribution is then extremely valuable for achieving sustainable development (Tauro et al. 2018). Therefore, although configuration is system- specific, two indissoluble components of social-ecological systems that need to be considered for management actions to succeed and to foresee policy results are (i) the mechanisms underpinning ES and their interrelations and (ii) integrated governance schemes matching these interrelations (Dee et al. 2017; Leck et al. 2015; Alonso Roldán et al. 2015; Alonso Roldán et al. 2019). Given that governance and management systems are organized within sectoral boundaries, major structural transformations are required (Pahl-Wostl 2019). Different barriers to decision-making have been highlighted in communication and collaboration (Howarth and Monasterolo 2016; Leck et al. 2015), and pathways to overcome them and achieve the integration vision are needed (Al-Saidi and

Elagib 2017). Network analysis applied to an ES approach can contribute to fill these gaps in nexus research and operationalization (Pahl-Wostl 2019).

This chapter will examine ecological and social relations among nexus components within a Patagonian case study by means of a network analysis from an ES perspective. In the next sections, I will describe the methodological approach and the social-ecological system. Then I will examine ecological and social relations while testing if social links reflect ecological relations between ES associated with the different nexus components. Finally, possible pathways to improve partnership for nexus approach are discussed.

2 Methods

2.1 *The Methodological Approach: Social Network Analysis*

Social network analysis (SNA) has proven to be a useful tool in studying and explaining social phenomena to provide an innovative framework to analyse the social dimension of social-ecological systems (e.g. Bodin and Crona 2009; Crona and Bodin 2010; Ramirez-Sanchez and Pinkerton 2009). Key findings emerging from this work are the important interplay between social capital and leadership for effective resource governance. Examples from documented transitions in natural resource governance show that networks of contacts between user groups and scientists are important for increasing exchange of information, leading to changed mindsets and deeper understanding of critical issues facing management (Meijerink and Huitema 2010). In addition, network analysis could play an important role in understanding interrelations among multiple ES, detecting bundles as emergent properties and supporting decision-making based on ecological processes while integrating social and economic dimensions (The QUINTESSENCE Consortium 2016). However, almost all studies have taken into account only a few number of ES (Bodin and Crona 2008), and there are very few examples of multilayer network analysis (The QUINTESSENCE Consortium 2016; Dee et al. 2017). In this context, the research by Rathwell and Peterson (2012) is noteworthy because of addressing direct and indirect interactions of institution sharing interests in different ES at the watershed scale using two-mode networks.

In SNA, while one-mode data records ties between nodes of one class, two-mode data records ties between two sets of nodes of different classes, and the corresponding networks are called two-mode networks (Borgatti 2009). In our case study, two-mode networks allow us to record the relations between ecological and social factors, being ES and social actors the two modes or classes of nodes considered. Two-mode networks reflect association of social actors through ES that could be compared to associations established directly between social actors which are recorded in one-mode networks.

This chapter explores the relation between existing ES of the study area and social actors by using two-mode networks in order to examine opportunities and threats of the governance system to get simultaneously multiple benefits. An SNA approach is applied to characterize the interrelations among different ES associated with the nexus approach components and to assess the match between natural and social relations.

The characterization of natural nexus was based on a theoretical network of relations among ES. The list of ES was built based on CICES classification considering socio-economic activities and social-ecological characteristics of the area. The identification was confirmed through interviews with social actors who testified the benefits captured. Based on Anderson et al. (2019) and CICES v5.1 (Haines-Young and Potschin 2018), these ES were related to Nature's Contributions to People defined within the conceptual framework of IPBES, the SDGs "zero hunger" (SDG 2), "good health and well-being" (SDG 3), "clean water and sanitation" (SDG 6) and "affordable and clean energy" (SDG 7) and from there to the components of nexus approach, namely, food, health, water and energy. I built a theoretical-hypothetical network linking ES by talking with local experts and compiling the literature (Alonso Roldán et al. 2015). I considered possible synergies or trade-offs because of the influence of the same process over different ES, or consequences of the actions using an ES over other ES, by changing conditions in the system that conduct to a higher demand or diminish the supply in other ES.

The social link to ES was characterized through two-mode network use and governance with actor-ES relations (Fig. 11.1). The actors considered were organizations or institutions representing a group of individuals which are ES users or involved in ES management within the study area. I built the lists of actors based on available documents of creation and activity regulation of organizations, provincial and municipal laws and scientific and grey literature. I researched and registered relationship between actors and ES (actor-ES) and complemented the actor's list using structured interviews with representatives from each actor (see details in Alonso Roldán et al. 2015). To characterize two-mode networks, I calculated density, actor degree and ES degree as basic properties.

The match between natural and social relations can be evaluated comparing the characteristics, structure and fragmentation of networks. Are the main ES which affect others the main ES in terms of governance and use by social actors? Are the ES theoretically related also related through social actors using them or participating in their governance? Do intersectoral relations of social actors reflect relations among ES representing the components of the nexus approach? In order to answer these questions, (1) two-mode networks were compared between each other and with the ES-ES theoretical-hypothetical network; (2) one-mode ES-ES networks were derived from two-mode networks and their structure was compared with the ES-ES theoretical-hypothetical network by means of a Quadratic Assignment Procedure (QAP) correlation (Prell 2012), testing the association between the two networks; and (3) the E-I index (external-internal index; Krackhardt and Stern 1988) was calculated for ES-ES, actor-actor use and governance networks blocked by nexus elements and tested if were smaller than expected by chance with

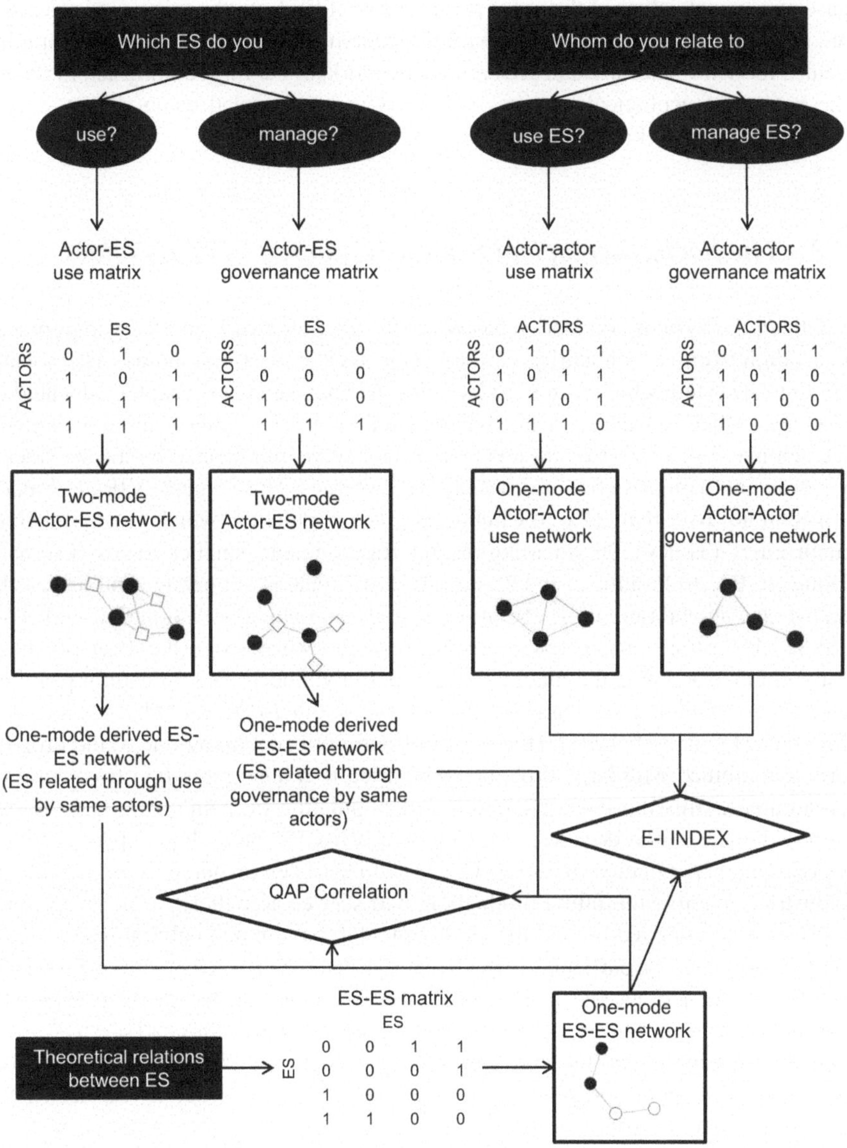

Fig. 11.1 Diagram showing the sources of information to build social networks and the analysis performed

permutation tests of 1000 iterations (Fig. 11.1). The E-I index measures the extent to which macrostructures, like the blocking by environment, "cluster" the interaction patterns of nodes which fall within them comparing the numbers of ties within groups and between groups (Hanneman and Riddle 2005). The index ranges from −1 (all ties are internal to the group) to +1 (all ties are external to the group). In this case, we expect values of the index from −1 to 0 if ES from the same nexus element are more related among them than with ES from the other nexus element and similar values for actor-actor use and governance networks if the links among actors reflect the pattern of ecological relations. All SNA was performed using statnet suit of packages (Handcock et al. 2008) in R (R Core Team 2016).

2.2 *The Social-Ecological System: Comarca VIRCH-Valdés*

I defined the system boundaries based on the ES framework and the geographical and administrative boundaries of the region tacked as a case study. The social-ecological system which forms the basis for the analysis in this chapter is located in Comarca VIRCH-Valdés, in the NE of Chubut province and Central Patagonia (Argentina, Fig. 11.2). *Comarca* is an administrative unit created by the provincial government to improve regional productive strategies. The Comarca VIRCH-Valdés concentrates over than 24% of Chubut province population, which mostly inhabits main cities (Trelew, Puerto Madryn and Rawson) and smaller towns (Gaiman, Dolavon, Puerto Pirámides and 28 de Julio).[1] Climate is temperate semi-arid, with an average annual rainfall of 250 mm and high interannual variation (Paruelo et al. 1998). Most rural area consists of private properties where extensive sheep ranching for wool production is the main economic activity. Sheep feed on natural pastures and shrubs which are characteristics of the southern Monte Phytogeographic Province (León et al. 1998). However, this activity is declining due to the drop of prices combined with large droughts, which led to a huge migration from rural to urban areas, stressing the concentration of population in main cities. In addition, during the last four decades, the Comarca VIRCH-Valdés has experienced an increase in its population of 380%. This growth has been favoured by an increase in industrial, mainly aluminium production, and service activities. At the same time, tourism was promoted around the environmental, natural and cultural resources of the region, being of particular importance the Península Valdés Natural Protected Areas, a world heritage site, and Punta Tombo. As a consequence, population growth and industrial activity were accompanied by productive activities and services associated with tourism and urbanization: hotel and restaurant services, transportation and communications and production of goods related to the supply of electricity, gas, water and construction. Accelerated urban growth has shown deficiencies in planning with consequences in the supply of ES such as flood prevention (due to clearing and modification of hydrology) and water quality (due to contamination by effluent spills due to lack of sewage networks) and strong pressure on the demand for energy and water provision services for human consumption (Bilmes et al. 2016; González and Benseny 2013; Kaminker and Velásquez 2015). The uncontrolled tourism development in areas of sustainable use has also generated conflicts with

[1] Last National Census data (2010) available at http://www.estadstica.chubut.gov.ar/home

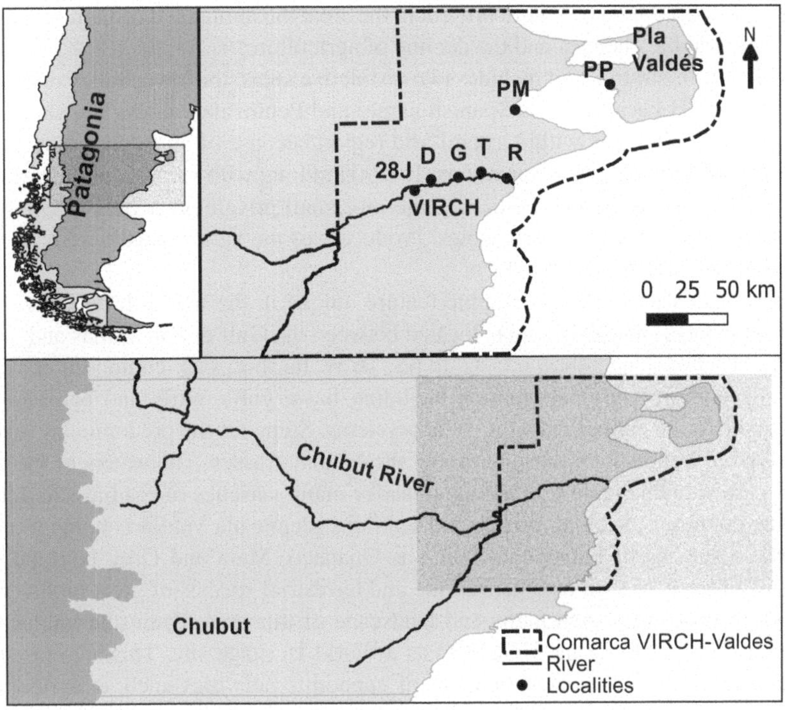

Fig. 11.2 Map showing the location of study area (Comarca VIRCH-Valdés) and Chubut province which comprise most of the Chubut River valley. With black-filled circles are indicated the localities within the study area identified with their names' initials 28J = 28 de Julio, D = Dolavon, G = Gaiman, T = Trelew, R = Rawson, PM = Puerto Madryn and PP = Puerto Pirámides

activities related to the provision of food from wild populations, such as the coastal harvesting of shellfish on different beaches in Península Valdés (Uriburu 2014). On the other hand, the region presents a marked development of primary activities related to agricultural, livestock and fishing activities that are in the process of change and adaptation seeking productive diversification. Livestock activity in the VIRCH has been reinforced with the take-off of feedlot since the valley presents favourable environmental and sanitary characteristics compared to other meat-producing regions. Aquaculture production and the exploitation of marine resources such as algae were added to traditional fishing activities.

Despite the high level of industrial activity, currently in this region, there is also the highest unemployment rate in Patagonia, which is why new alternatives that generate employment continue to be promoted. In this context, changes in production practices and the development of activities in various economic sectors without territorial attachment, as well as the changes in land use that they imply, may present compromises between different human activities due to their effects on ES, generating undesired outcomes affecting people well-being. Actually, as has been mentioned, some of the causes already identified by the MEA (2005) which usually cause the degradation of ES are present in the area: the unplanned economic growth, the demographic changes and the decline of agriculture.

This administrative unit includes two distinctive areas: the lower valley of Chubut River (VIRCH because of its Spanish name) and Península Valdés. The valley is a highly productive area within a semi-arid region because of an extensive irrigation system based on Chubut River waters. It is 90 km long with variable width between 7 and 10 km. This area consists in numerous small private properties whose main productions are fruits and vegetables. Producers of the valley usually associate for process and trade of their products.

Peninsula Valdes is a geographic feature unique in the world. It has an area of approximately 350,000 ha and is located between the Gulf of San Matias and Golfo Nuevo at 42 ° 00 ′42° 48′S, 63° 32′ to 65° 16′W. Its shores are composed of varied coastal geomorphological features including bays, gulfs, cliffs and beaches that contribute to the aesthetical value of ecosystems. Steppe is the predominant vegetation type of Peninsula Valdes, with low shrubs and grasses. The waters of the peninsula are valuable natural breeding areas for many varieties of sea birds and large marine mammals, such as whales and sea lions. Peninsula Valdes is home to characteristic species of Patagonian fauna as Guanaco, Mara and Grey Fox and 181 species of birds as well as other marine and terrestrial species of great biodiversity value. To protect the rich fauna and landscape of this area, Peninsula Valdes was recognized in 1999 by the UNESCO as a World Heritage site. This area provides ideal conditions for the development of economic activities such as agriculture, tourism, whale watching, commercial and recreational fishing and surfing, all of which generate significant economic and social benefits for the country. For example, in 2019 over 227,000 tourists visited Península Valdés.[2]

3 Results and Discussion

3.1 *Natural Nexus*

The VIRCH-Valdés region, according to the aforementioned (Sect. 2.2), is a social-ecological system that, due to its variety of socio-economic activities and land uses, can provide valuable information to fill gaps in the knowledge of the interactions between ES and the capacity to provide services for different types of land use and coverage. Based on socio-economic activities and social-ecological characteristics of the area, I identified 25 ES relevant for the Comarca VIRCH-Valdés. These ES contribute to the four SDGs considered (SDG 2, 3 6 and 7) and therefore to the components of nexus approach food, health, water and energy (Table 11.1).

Desertification is one of the main processes interlinking multiple ES and the four nexus components in Comarca VIRCH-Valdés (Fig. 11.3). Reduction in vegetation

[2] https://peninsulavaldes.org.ar/ingresos-a-peninsula-valdes/ [Accessed 02/03/2021]

Table 11.1 Ecosystem services identified as relevant and designated according to CICES 4.3

ID	Section	Name	Nexus	SDG	NCP	CICES4.3	CICESV5.1
S01	Provisioning	Crops and animals reared for nutritional purposes	F	2 & 3	12	1.1.1.1 & 1.1.1.2	1.1.1.1 & 1.1.3.1
S02	Provisioning	Food products from wildlife	F	2 & 3	12, 13, 14	1.1.1.3 & 1.1.1.4	1.1.5.1 & 1.1.6.1
S03	Provisioning	Food products from fishing and algae	F	2 & 3	12, 13, 14	1.1.1.3 & 1.1.1.4	1.1.5.1 & 1.1.5.2 & 1.1.6.1
S04	Regulation	Control of erosion rates	F	2	8	2.2.1.1	2.2.1.1
S05	Regulation	Lifecycle maintenance, habitat and gene pool protection	F	2	1, 2	2.3.1	2.2.2
S06	Regulation	Soil formation and composition	F	2	8	2.3.3	2.2.4
S07	Provisioning	Fibres and other materials from flora and fauna	H	3	13–14	1.2.1.1	1.1.1.2 & 1.1.3.2 1.1.5.2 & 1.1.6.2
S08	Provisioning	Mineral substances used for material purposes	H	3	–	4.2	
S09	Regulation	Mediation of wastes or toxic substances of anthropogenic origin by living processes	H	3 & 2	8, 10, 3	2.1.1 & 2.1.2.1	2.1.1
S10	Regulation	Dilution by atmosphere, freshwater and marine ecosystems of waste, toxics and other nuisances	H	3	–	2.1.2.2	5.1.1.1
S11	Regulation	Mediation of smell/noise/visual impacts	H	3	9	2.1.2.3	2.1.2
S12	Regulation	Ventilation and transpiration	H	3		2.2.3.2	2.2.6.2
S13	Regulation	Pest and disease control	H	3	10	2.3.2	2.2.3
S14	Cultural	Physical and experiential interactions with environmental settings	H	3	6, 16	3.1.1	3.1.1
S15	Cultural	Intellectual and representative interactions with environmental settings	H	3	6, 13, 15, 16, 17	3.1.2	3.1.2 & 3.2.1.3
S16	Cultural	Spiritual, symbolic and other interactions with environmental settings	H	3	6, 17	3.2.1	3.2.1.1 & 3.2.1.2
S17	Provisioning	Drinking water	W	6	–	1.1.2	4.2.1.1 & 4.2.2.1

(continued)

Table 11.1 (continued)

ID	Section	Name	Nexus	SDG	NCP	CICES4.3	CICESV5.1
S18	Provisioning	Water for non-drinking purposes	W	6	–	1.2.2	4.2.1.2 & 4.2.2.2
S19	Regulation	Buffering and attenuation of mass movement	W	6	9	2.2.1.2	2.2.1.2
S20	Regulation	Hydrological cycle and water flow maintenance	W	6 & 2	6	2.2.2.1	2.2.1.3
S21	Regulation	Flood and storm protection	W	6	6 & 9	2.2.2.2 & 2.2.3.1	2.2.1.3 & 2.2.1.4
S22	Regulation	Maintenance of water conditions	W	6	7	2.3.4	2.2.5
S23	Provisioning	Biomass-based energy sources	E	7	–	1.3.1	1.1.1.3 & 1.1.5.3
S24	Provisioning	Renewable abiotic energy sources	E	7	–	4.3	4.2.1.3 & 4.3.2.3 & 4.3.2.4
S25	Regulation	Maintenance of atmospheric composition and climate regulation	E	7 & 6	3, 4	2.3.5	2.2.6

For each ES correspondent nexus component (F = food, H = health, W = water and E = energy), Sustainable Development Goals (SDG), Nature Contribution to People (NCP) and codes for CICES4.3 and 5.1 categories are identified

cover due to sheep raising for meat and wool production (Campanella et al. 2016a; Segesso et al. 2019) or wind farm establishment for wind power production reduces the supply of Control of erosion rates ES. It in turn affects water quality and quantity for human consumption and use within VIRCH watershed (Liberoff et al. 2019) and water availability for plants and its flow in the rest of the comarca (Campanella et al. 2018). Changes in soil cover and eolic erosion of soil can have multiple impacts on human health and security (Goudie 2014). Security impacts relate land use changes and desertification to flood protection (Bilmes et al. 2016). Beyond the immediate effects on population's health, environmental degradation due to desertification could risk future disease control due to effects over species that showed potential medicinal use (Cenzano and Arslan 2020; Marani et al. 2017) or already contribute to healing uses by native cultures (Castillo and Ladio 2017). Finally, the desertification loop returns over food production given that eroded soils produce fewer pastures (Campanella et al. 2016b) reducing meat production. However, it could be a positive relation between erosion rates and marine productivity due to dust transportation by wind from Patagonia to the Southern Atlantic Ocean (Crespi-Abril et al. 2018). All these effects of desertification are expected to increase due to climate change, which is also synergistic with environmental degradation through grazing that diminishes C stock (Larreguy et al. 2017).

On the opposite way, the regeneration pathway presents multiple co-benefits. The restoration of vegetation cover should produce synergistic effects in soil

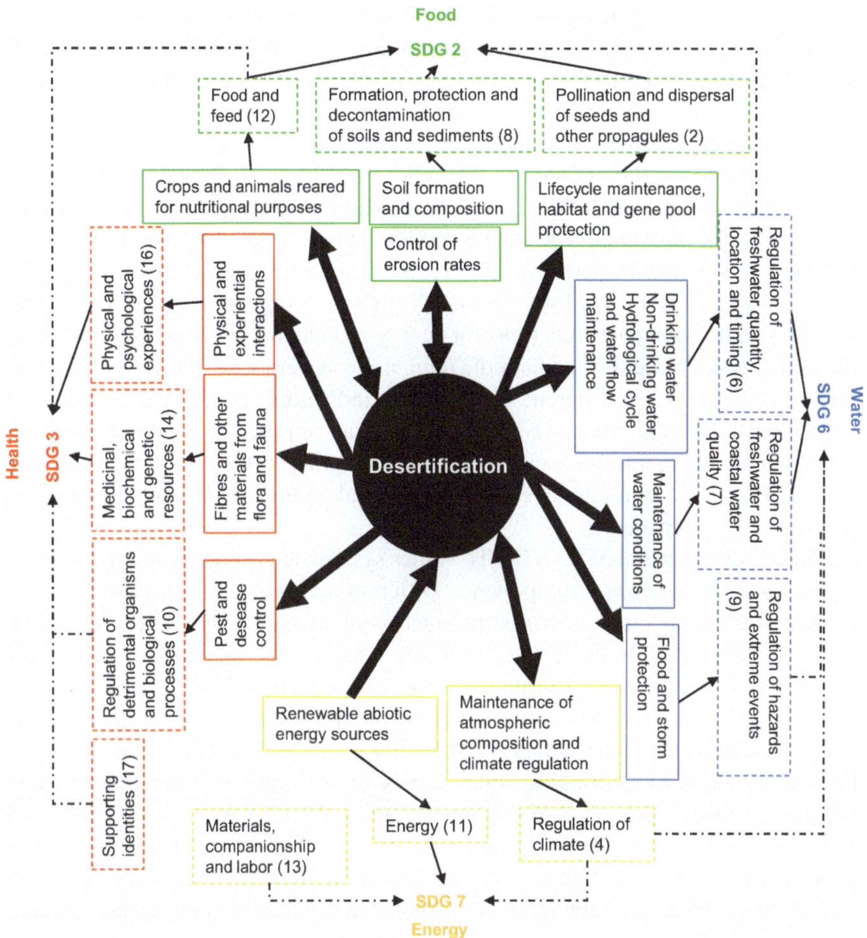

Fig. 11.3 Diagram showing how nexus components (exterior layer) were associated with ES (inner layer in solid line) through SDG and Nature Contributions to People (middle layer in dashed line). Only some ES are showed that are related through desertification as the main process linking the four nexus components

conservation, life cycle maintenance and habitat protection, water quality maintenance, climate regulation, flood preventions and protection of health in several aspects including the benefits of cultural services related to physical interactions with natural environments. The repopulation with native ungulates also contributes to habitat restoration and maintenance given that their populations present self-regulatory mechanism derived from resource defence that may prevent overgrazing (Marino et al. 2015).

The aforementioned interactions operate at different scales, and also spatial configuration needs to be taken into account to understand system functioning and consequences for management. Energy-related ES has omnidirectional effect (climate regulation), or its benefits are diffused across the system and beyond because of the configuration of the national power distribution network. Besides, at local scale energy-related ES interact through the mentioned desertification processes with ES providing in situ benefits (Fisher et al. 2009) as soil formation and water flow regulation. The water consumed in all the cities of the comarca and used in the productive valley is produced in the source of Chubut River basin about 800 km to the west (Fig. 11.2, Pessacg et al. 2020). Therefore, drinking water and water for other uses are examples of directional ecosystem services (Fisher et al. 2009) with the provision area outside the system, requiring regional governance (i.e. watershed governance). Water flow to the lower basin (VIRCH) is regulated by de Florentino Ameghino dam, which has functions for energy production and flood control. Many interactions take place downstream the dam at the system scale. However, Liberoff et al. (2019) found that different water quality indicators are affected by land use/land cover at different scales. Overall, the interplay of processes acting at different scales and involving multiple nexus components highlights the need for coordinated actions of actors at different levels (local, regional, national and international) and intersectoral governance.

Spatial configuration of the VIRCH-Valdés system determines particular interactions among ES and nexus components. Different activities and associated ecosystem services on the cities and food production areas within the valley depend on water and interact due to the location of water intake points, cities over the riverside, farm areas and kaolin clay exploitations along the river in VIRCH. Therefore, sediments from kaolin clay exploitation, barren soil and food production areas, in the upper extreme on the valley, affect water quality in intake points of the cities that are downstream, as well as drinking water supply by affecting the operation of water treatment plants (Kaless et al. 2008; Liberoff et al. 2019). In turn, cities positioned upstream of other food production and recreation areas affect water quality because of storm water runoff through impervious surfaces and industrial effluents (Liberoff et al. 2019). If land and water uses were organized differently in the space, relationships among ES and nexus components could be different.

The aforementioned are only the main relationships involving the four components of nexus, but there are multiple pairwise relationships between the different ES involved in each sector and between them (density 0.345). Density may be lower if some relationships are not functionally significant or higher if new activities in the region establish new relationships between benefits and natural processes.

The network shows multiple relations among ES from food, health, water and energy groups (Fig. 11.4). The main ES affecting other ES was Hydrological cycle and water flow regulation, associated with the water group, followed by several ES from the food group (Table 11.2). The theoretically most vulnerable ES being affected by numerous ES was lifecycle maintenance, habitat and gene pool protection, from the food group, but also in the top five of vulnerable ES were represented the water and health group (Table 11.2). Habitat and gene pool protection services have been reported to have synergistic relationship with soil formation services (Lee and Lautenbach 2016) which is in line with the desertification process aforementioned and, therefore, its key role in VIRCH-Valdés system.

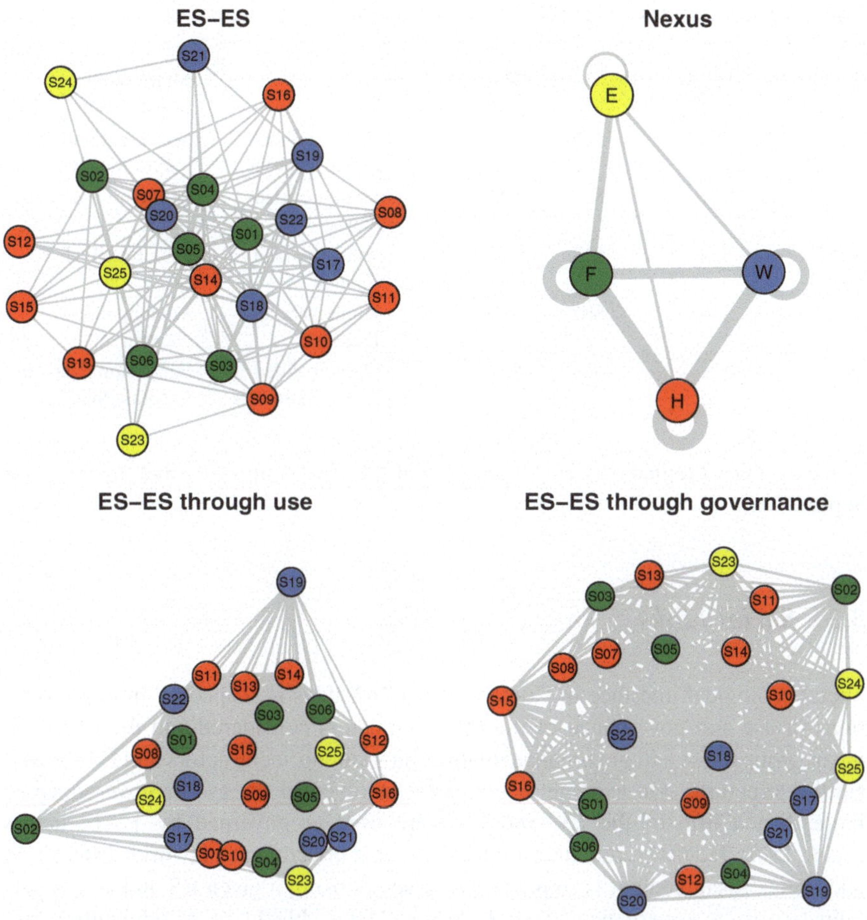

Fig. 11.4 Theoretical network between ecosystem services (ES-ES), nexus network adding up the links by group of ES, one-mode ES-SE networks derived from two-mode actor-ES networks of use and governance (ES-ES through use and ES-ES through governance) showing nodes associated with components of nexus approach: food (green), health (red), water (blue) and energy (yellow)

At group level the theoretical network showed that food and health groups of ES affect more ES than other groups. It is important to highlight that ES associated with health are mainly regulating and cultural services (Table 11.1). These services, as well as the element health of the nexus approach, are less studied and assessed than others. However, regulating services tend to be synergistic with other regulating services and cultural services (Lee and Lautenbach 2016). Therefore, actions aimed to secure supply of ecosystem services related to health can produce multiple co-benefits on provision of water and food.

Relations of direct effects were as frequent between ES of different nexus components as within these groups (E-I index = 0.218, $p = 0.89$). Therefore, management

Table 11.2 Ecosystem services (ES) and actors with the highest degree use and governance networks

ES affecting more ES	ES affected by more ES	Main ES use	Main ES governance	Main actors (use)	Main actors (governance)
S20	S05	S15	S18	Municipalidad Puerto Madryn	Municipalidad Puerto Madryn
S05	S01	S09	S05	Municipalidad Rawson	Gobierno provincial
S01	S18	S18	S09	Municipalidad Gaiman	CCT CENPAT (Research)
S04	S14	S01	S10	Municipalidad Trelew	Municipalidad Gaiman
S07	S22	S24	S22	Municipalidad Puerto Piramides	Fundacion Patagonia Natural (NGO)

of the social-ecological system to secure multiple SDGs must involve intersectoral approaches.

3.2 *Social Links*

Social-ecological interface in Comarca VIRCH-Valdés is shaped by multiple relations of use and governance of the 60 social actors considered and the ES of all nexus groups. Use and governance relationships among social actors and ES showed similar structures. On average, actors were involved in the use or governance of more than a fifth of the ES, and ES were used or governed by 12–14 actors (Table 11.3). However, the main actors in use and governance networks differed. In addition, a disagreement is found between actors using a given ES and actors participating in its governance (rho = 0.06 p = 0.63; Table 11.2). Municipalities use many ES because they represent the community in each town, but as governmental institution, they are the authority over few ES. In general, they do not implement policies to regulate the use or impacts over ES under management within their range (i.e. to prevent diminishing of water retention by soil or pollution conducted by storm drains improving urban planning). Instead, within actors, provincial government, an NGO and a research institute are the most active actors participating in ES governance. Moreover, these three institutions were central actors in actor-actor networks (Alonso Roldán et al. 2015), indicating that they may lead and coordinate interactions of other groups of actors related to a single or a smaller group of ES. These groups could include associations of producers and chambers of commerce whose linkage with ES is restricted by their activities and stakes. Additionally, some of the most used ES were not the ones with more actors involved in their governance (Table 11.2). This situation could cause a mismatch between users' stakes and management decisions. The incorporation of users in discussion of rules is critical to induce compliance of rules and encourage adaptation and change processes (Dietz et al. 2003).

Table 11.3 Density and mean degree scores of actors-ES use and governance networks. Standard deviation of mean degree indicated between parentheses

	A-SE use	A-SE governance
Density	0.24	0.22
Actor degree	6.03 (7.04)	5.32 (5.95)
ES degree	14.48 (6.94)	12.76 (4.49)

The social-ecological links showed a mismatch with theoretical ecological links in the three aspects analysed. First, the main ES affecting others were not the main ES in terms of more social actors involved in their governance (rho = 0.27 $p = 0.18$; Table 11.2). Instead, the ES that could be considered more vulnerable (affected by more ES) were in general the ES with more social actors involved in their use (rho = 0.45 $p = 0.02$) and governance (rho = 0.561 $p = 0.003$; Table 11.2). As an example, the ES Fibres and other materials from flora and fauna (S07) affect many ES together with Control of erosion rates (S04), which is congruent with the desertification process derived by overgrazing described before; however, not many were involved in its governance. This result could indicate a trend to remediate the effects (like water quality, S22, and supply, S18) but not the origin of degradation of ES which frequently starts with overexploitation of land in arid ecosystems (MEA 2005). A similar situation was reported in Canadian watershed management, where agricultural municipalities were engaged in fewer activities concerning water quality even when agricultural activities lead to most problems related to water (Rathwell and Peterson 2012).

Second, direct links among ES were weakly correlated with links derived from common actors using them or participating in their governance (QAP = 0.19 $p = 0.027$ and QAP = 0.28 $p = 0.001$, respectively). It shows that related ES are not used or managed by the same actors, highlighting the importance of interactions of social actors associated with related ES and intersectoral approaches. The communication among actors related to different elements of the system is an important component of social capital (Pretty 2003) that could contribute to prevent or detect effects of activities in one environment over the other and to coordinate actions (Bodin et al. 2006; Bodin and Crona 2008).

Third, intersectoral relations of social actors do not reflect relations among ES representing the components of the nexus approach. Relations among social actors classified in the four groups of nexus components reflected the same pattern regarding use (E-I index = 0.089, $p < 0.001$) and governance of ES (E-I index = 0.098, $p < 0.001$), showing fewer ties among groups than expected by chance, and less than ecological relations among groups. These results added to the second mismatch with ecological links suggest that feedback information linking different and related ES could be disregarded, missing opportunities for adaptive management. In this context the misfit between social and ecological relations hampers the ability to take into account complexities in ES interactions, diminishing the success probability of policies and management practices to avoid undesired impacts (Liu et al. 2007).

In addition, a previous study (Alonso Roldán et al. 2015) found that actor-actor relations are weakly correlated with those derived from actor-ES relations, meaning that actors with common interests about ES are not necessarily working together.

This result highlights a mismatch between incentives defining relations among actors and the ecosystem-based management, in concordance to the attention addressed to consequences rather than causes mentioned earlier. The absence of connections and structures grouping actors with stakes in the same ES could hinder the development of tacit knowledge, reducing the collective capacity to adapt to change (Bodin et al. 2006; Bodin and Crona 2008) and agreement achievement on resource regulation (Bodin et al. 2014). In contrast, the connections among actors related to different ES could enhance the information flow, a positive aspect of heterogeneity in governance structures (Duit et al. 2010; Dietz et al. 2003; Sandström and Rova 2010). However, it is important to highlight that to detect possible synergies and trade-offs, actors linked to related ES should be connected.

Despite the social structural drawbacks, the system shows a great potential as natural and social capital. The social capital, beyond network links, is presented as collaborating work in each node because actors considered here (60) are groups of individuals working with common interests. In addition, cultural services related to intellectual interaction were among the most used, because there are several research institutes and universities as actors but also because several actors stated they have organized training and awareness activities. This is a desirable scenario to develop adaptive management actions and to shift the system to power-balanced structures of co-management, both supported by information sharing and social learning as component of social capital (Pretty 2003; Pahl-Wostl 2007). Furthermore, the central position of knowledge production institutions shown in the governance structure is important given the relation between ES provision and knowledge leadership (Kenward et al. 2011).

3.3 *Insights to Improve Partnership for Nexus Approach*

Previous characterization of actor-actor use and governance links in Comarca VIRCH-Valdés showed that networks were cohesive and had low density and high centralization in structure and power distribution, given that actors with most links also were linking other actors (Alonso Roldán et al. 2015). Despite low density, the cohesion of both networks could indicate that trust and willingness to cooperate exist, basic components of social capital (Pretty 2003). Thus present connections should be the basis to strengthen relationships and institutions, building new governance schemes. The low density is probably because actors of the same type or those sharing interests are not working together, both issues that a shift in governance towards a systemic approach should change. A review over several case studies within Latin America found that the establishment of long-lasting institutional instruments and the involvement of intermediaries connecting sectors are complementary pathways to improve integrated governance (Alonso Roldán et al. 2019). Results in this chapter and some experiences in the study area are in line with these complementary pathways.

A shift of the system towards more balanced and participative structures could be led by central actors, mainly the provincial government which is in that position because it is the legal authority and possesses domain of natural resources according to the law. The provincial government has established participation instances which may stimulate connections among users and co-management, like the fishing and water committees and the desertification network. The Chubut River Watershed Committee is a new forum, not yet consolidated, whose operation has been irregular (Pascual et al. 2020). The technical institutions which are part of the watershed committee, taking action independently, detected that there are very few permanent forums and governance schemes that bring together the different social actors and sectors, especially those that have divergent interests regarding the use and management of water (Olivier et al. 2018). During a meeting of territorial actors of the lower basin of the Chubut River, it was proposed to strengthen the structure and function of the watershed committee and to actively generate new forums for the treatment of water problems (Pascual et al. 2019). In turns, the desertification network involved different provincial government agencies, including those related to food production, water management and environment, knowledge production institutions and associations of agricultural producers. However, several foundational members leave due to restrictions in applications of recommendations of the network to policy related to the lack of a statute and the designation by law of an authority for conservation of soils. During a workshop organized by the desertification network involving multiple stakeholders (including a company that owns a wind power field in the region for the first time), deficiencies of the current law and little political will of the government were detected as the main obstacles to operationalize actions for the prevention of desertification and counteract processes of soil degradation (Red de desertificación 2019). Although these experiences are sectoral or focused on a single problem or resource, involve the participation of stakeholders of other sectors, showing that the establishment of long-lasting institutional instruments (i.e. protocols and decision workflows determined by law) is needed to create an institutional framework for sector interaction long-lasting and meaningful in terms of governance.

Besides the role of provincial government, the relevance of an NGO and a research institution in governance, two-mode networks are remarkable. These actors had also central positions in actor-actor governance network (Alonso Roldán et al. 2015) indicating a heterogeneous governance structure of the system in terms of institutions and functions. This feature could be counterbalancing some disadvantages of centralization given that heterogeneity is part of new governance models to lead with complex problems (Dietz et al. 2003; Duit et al. 2010). Moreover, these institutions can play an important role as intermediaries fostering systemic approaches. This key role prompting the implementation of institutional instruments, mentioned earlier, and facilitating communication and joint action among stakeholders from different sectors has been reported in several cases (Cowell and Lennon 2014; Alonso Roldán et al. 2019).

A third component moving the system to intersectoral structures and approaches is conceptual. Using ES as a boundary concept could improve integration between sectors as it provides a common language, promotes collaboration on a common task (sustainability) and highlights a common interest (Abson et al. 2014; Mollinga 2010; Steger et al. 2018; Pahl-Wostl 2019). In addition, an ES trade-off analysis can help to introduce ecosystem relations to stakeholders related to their own interests by modelling change scenarios (Alonso Roldán et al. 2019). This approach is useful to making the relationships among ecosystem structure, function and services explicit (De Groot et al. 2002), which is important for achieving integrated management, even when some ecological relations are already intuitive for stakeholders (Arkema et al. 2015).

4 Concluding Remarks

Merging ES and nexus approaches could be beneficial. ES approach enriches nexus analysis considering co-benefits and processes relating multiple components and benefits. In addition, ES concept can be used as a bounding concept to facilitate intersectoral relations among social actors. Conversely, nexus approach could guide ES analysis, highlighting main interrelations, synthesis of results and communication to policy.

Sustainable Development Goals and nexus approach are global policies that need to be rooted in local and regional processes to be meaningful given that main decision-making takes place in territorial scale and social-ecological systems characteristics and configuration condition interplay of nexus components. In the examined social-ecological system, desertification was the main process linking ES associated with the four components of nexus.

Social network analysis was a useful tool to detect inconsistencies between ecological relations and governance structures, and particularly two-mode networks helped to identify central ES from the biophysical perspective, for stakeholders, and gaps in governance of highly used ES. From these results and local intersectoral participative experiences, insights to improve partnership for nexus approach could be recognized. Structural characteristics related to resilience and management success like heterogeneity in governance networks offer opportunities for information flow, social capital building and the construction of management strategies that benefit multiple stakes. Centralization of power in few actors from different sectors can also represent an opportunity for structural improvement through institutional strengthening and network reorganization if key actors stimulate intersectoral interaction developing bridging functions. Reorganization processes would also address a better fit of relations among actors to ecological patterns, strengthening relations among actors with stakes in the same ES or related ES in order to prevent the main threats of the actual governance system: inability to agree management actions and to take into account complexities in ES interactions, poor coordination and loss of valuable information to enhance synergies and prevent trade-offs.

Acknowledgements This chapter was written within the framework of the PUE-IPEEC-2016 22920160100004.

References

Abson DJ, von Wehrden H, Baumgärtner S, Fischer J, Hanspach J, Härdtle W, Heinrichs H, Klein A, Lang D, Martens P (2014) Ecosystem services as a boundary object for sustainability. Ecol Econ 103:29–37

Alonso Roldán V, Villasante S, Outeiro L (2015) Linking marine and terrestrial ecosystem services through governance social networks analysis in Central Patagonia (Argentina). Ecosyst Serv 16:390–402

Alonso Roldán V, Galván DE, Lopes PFM, López J, Sanderson Bellamy A, Gallego F, Cinti A, Rius P, Schröter B, Aguado M, Muñoz Barriga A, Pittman J, Ávila-Flores G, López-Gómez CP, Góngora ME (2019) Are we seeing the whole picture in land-sea systems? Opportunities and challenges for operationalizing the ES concept. Ecosyst Serv 38:100966. https://doi.org/10.1016/j.ecoser.2019.100966

Al-Saidi M, Elagib NA (2017) Towards understanding the integrative approach of the water, energy and food nexus. Sci Total Environ 574:1131–1139. https://doi.org/10.1016/j.scitotenv.2016.09.046

Anderson CB, Seixas CS, Barbosa O, Fennessy MS, Díaz-José J, Herrera-F B (2019) Determining nature's contributions to achieve the sustainable development goals. Sustain Sci 14(2):543–547

Arkema KK, Verutes GM, Wood SA, Clarke-Samuels C, Rosado S, Canto M, Rosenthal A, Ruckelshaus M, Guannel G, Toft J (2015) Embedding ecosystem services in coastal planning leads to better outcomes for people and nature. Proc Natl Acad Sci 112(24):7390–7395

Bilmes A, Pessacg N, Alvarez M, Brandizi L, Cuitiño JI, Kaminker S, Bouza PJ, Rostagno CM, Núñez de la Rosa D, Canizzaro A (2016) Inundaciones en Puerto Madryn: relevamiento y diagnóstico del evento del 21 de Enero de 2016. Informe Técnico CCT CONICET-CENPAT

Bodin Ö, Crona BI (2008) Management of natural resources at the community level: exploring the role of social capital and leadership in a rural fishing community. World Dev 36(12):2763–2779

Bodin Ö, Crona BI (2009) The role of social networks in natural resource governance: what relational patterns make a difference? Glob Environ Chang 19(3):366–374

Bodin Ö, Crona B, Ernstson H (2006) Social networks in natural resource management: what is there to learn from a structural perspective? Ecol Soc 11(2). http://www.ecologyandsociety.org/vol11/iss2/resp2/

Bodin Ö, Crona B, Thyresson M, Golz AL, Tengö M (2014) Conservation success as a function of good alignment of social and ecological structures and processes. Conserv Biol 28(5):1371–1379

Borgatti SP (2009) 2-Mode concepts in social network analysis. In: Encyclopedia of complexity and system science, vol 6, pp 8279–8291

Campanella MV, Bisigato AJ, Rostagno CM (2016a) Plant production along a grazing gradient in a semiarid Patagonian rangeland, Argentina. Plant Ecol 217(12):1553–1562. https://doi.org/10.1007/s11258-016-0668-8

Campanella MV, Rostagno CM, Videla LS, Bisigato AJ (2016b) Interacting effects of soil degradation and precipitation on plant productivity in NE Patagonia, Argentina. Arid Land Res Manag 30(1):79–88. https://doi.org/10.1080/15324982.2015.1053629

Campanella MV, Rostagno CM, Videla LS, Bisigato AJ (2018) Land degradation affects shrub growth responses to precipitation in a semiarid rangeland of north-eastern Patagonia (Argentina). Austral Ecol 43(3):280–287. https://doi.org/10.1111/aec.12563

Castillo L, Ladio A (2017) Traditional veterinary solutions for herders living in limited and changing conditions: a case study of "crianceros" of Central Northern Patagonia, Argentina. J Arid Environ 145:90–101. https://doi.org/10.1016/j.jaridenv.2017.06.001

Cenzano AM, Arslan I (2020) Comprehensive and quantitative profiling of lipid molecular species by LC-ESI-MS/MS of four native species from semiarid Patagonian Monte. Plant Physiol Biochem 146:447–456. https://doi.org/10.1016/j.plaphy.2019.11.028

Cowell R, Lennon M (2014) The utilisation of environmental knowledge in land-use planning: drawing lessons for an ecosystem services approach. Eviron Plann C Gov Policy 32(2):263–282

Crespi-Abril AC, Barbieri ES, Villalobos LG, Soria G, Paparazzo FE, Paczkowska JM, Gonçalves RJ (2018) Perspective: continental inputs of matter into planktonic ecosystems of the Argentinean continental shelf—the case of atmospheric dust. In: Plankton ecology of the Southwestern Atlantic. Springer, Cham, pp 87–99

Crona B, Bodin Ö (2010) Power asymmetries in small-scale fisheries: a barrier to governance transformability? Ecol Soc 15(4). http://www.ecologyandsociety.org/vol15/iss4/art32/

De Groot RS, Wilson MA, Boumans RM (2002) A typology for the classification, description and valuation of ecosystem functions, goods and services. Ecol Econ 41(3):393–408

Dee LE, Allesina S, Bonn A, Eklöf A, Gaines SD, Hines J, Jacob U, McDonald-Madden E, Possingham H, Schröter M (2017) Operationalizing network theory for ecosystem service assessments. Trends Ecol Evol 32(2):118–130

Dietz T, Ostrom E, Stern PC (2003) The struggle to govern the commons. Science 302(5652):1907–1912

Duit A, Galaz V, Eckerberg K, Ebbesson J (2010) Governance, complexity, and resilience. Global Environmental Change 20(3):363–368. https://doi.org/10.1016/j.gloenvcha.2010.04.006

Fisher B, Turnera RK, Morling P (2009) Defining and classifying ecosystem services for decision making. Ecol Econ 68:643–653

González G, Benseny G (2013) Consecuencias ambientales del crecimiento demográfico y turístico en Puerto Madryn, Argentina

Goudie AS (2014) Desert dust and human health disorders. Environ Int 63:101–113. https://doi.org/10.1016/j.envint.2013.10.011

Haines-Young R, Potschin MB (2018) Common International Classification of Ecosystem Services (CICES) V5. 1 and guidance on the application of the revised structure. European Environment Agency (EEA). Available online: https://ciceseu/. Accessed on 7 June 2018

Handcock MS, Hunter DR, Butts CT, Goodreau SM, Morris M (2008) statnet: Software tools for the representation, visualization, analysis and simulation of network data. J Stat Softw 24(1):1548

Hanneman RA, Riddle M (2005) Introduction to social network methods. University of California, Riverside

Hoff H (2011) Understanding the nexus: Background paper for the Bonn2011 Nexus Conference: the water, energy and food security nexus. In: Nexus Conference: the water, energy and food security nexus. Stockholm Environment Institute, Bonn

Howarth C, Monasterolo I (2016) Understanding barriers to decision making in the UK energy-food-water nexus: the added value of interdisciplinary approaches. Environ Sci Pol 61:53–60

IPBES (2019) Summary for policymakers of the global assessment report on biodiversity and ecosystem services of the Intergovernmental Science-Policy Platform on Biodiversity and Ecosystem Services. IPBES secretariat, Bonn

Kaless G, Matamala F, Monteros B, Greco W (2008) Cambios hidrológicos y morfológicos en el Río Chubut aguas abajo de la presa Florentino Ameghino

Kaminker SA, Velásquez RY (2015) Programa de mejoramiento de barrios en la patagonia central: regularización de la informalidad urbana en Puerto Madryn, Chubut. Cuaderno urbano 18(18):89–110

Kenward R, Whittingham M, Arampatzis S, Manos B, Hahn T, Terry A, Simoncini R, Alcorn J, Bastian O, Donlan M (2011) Identifying governance strategies that effectively support ecosystem services, resource sustainability, and biodiversity. Proc Natl Acad Sci 108(13):5308–5312

Kling CL, Arritt RW, Calhoun G, Keiser DA (2017) Integrated assessment models of the food, energy, and water nexus: a review and an outline of research needs. Ann Rev Resour Econ 9(1):143–163. https://doi.org/10.1146/annurev-resource-100516-033533

Krackhardt D, Stern RN (1988) Informal networks and organizational crises: an experimental simulation. Soc Psychol Q 51:123–140

Larreguy C, Carrera AL, Bertiller MB (2017) Reductions of plant cover induced by sheep grazing change the above-belowground partition and chemistry of organic C stocks in arid rangelands

of Patagonian Monte, Argentina. J Environ Manag 199:139–147. https://doi.org/10.1016/j.jenvman.2017.04.086
Leck H, Conway D, Bradshaw M, Rees J (2015) Tracing the water–energy–food nexus: description, theory and practice. Geogr Compass 9(8):445–460
Lee H, Lautenbach S (2016) A quantitative review of relationships between ecosystem services. Ecol Indic 66:340–351. https://doi.org/10.1016/j.ecolind.2016.02.004
León RJC, Bran D, Collantes M, Paruelo JM, Soriano A (1998) Grandes unidades de vegetación de la Patagonia extra andina. Pp. 125-144. En: Ecosistemas patagónicos (Oesterheld, M.; M.R. Aguiar y J.M. Paruelo, eds.). Ecol Austral 8(2):75–308
Liberoff AL, Flaherty S, Hualde P, García Asorey MI, Fogel ML, Pascual MA (2019) Assessing land use and land cover influence on surface water quality using a parametric weighted distance function. Limnologica 74:28–37. https://doi.org/10.1016/j.limno.2018.10.003
Liu J, Dietz T, Carpenter SR, Alberti M, Folke C, Moran E, Pell AN, Deadman P, Kratz T, Lubchenco J (2007) Complexity of coupled human and natural systems. Science 317(5844):1513–1516
Marani MM, Perez LO, de Araujo AR, Plácido A, Sousa CF, Quelemes PV, Oliveira M, Gomes-Alves AG, Pueta M, Gameiro P, Tomás AM, Delerue-Matos C, Eaton P, Camperi SA, Basso NG, de Souza de Almeida Leite JR (2017) Thaulin-1: the first antimicrobial peptide isolated from the skin of a Patagonian frog Pleurodema thaul (Anura: Leptodactylidae: Leiuperinae) with activity against Escherichia coli. Gene 605:70–80. https://doi.org/10.1016/j.gene.2016.12.020
Marino A, Rodríguez V, Pazos G (2015) Resource-defense polygyny and self-limitation of population density in free-ranging guanacos. Behav Ecol 27(3):757–765. https://doi.org/10.1093/beheco/arv207
Meijerink S, Huitema D (2010) Policy entrepreneurs and change strategies: lessons from sixteen case studies of water transitions around the globe. Ecol Soc 15(2). http://www.ecologyandsociety.org/vol15/iss2/art21/
Millennium Ecosystem Assessment (2005) Ecosystems and human well-being: biodiversity synthesis. World Resources Institute, Washington, DC
Mollinga PP (2010) Boundary work and the complexity of natural resources management. Crop Sci 50(Supplement_1):S-1–S-9
Olivier T, Aigo J, Pascual M (2018) Gobernanza hídrica en el Valle Inferior del Río Chubut. Reporte Ejecutivo
Pahl-Wostl C (2007) Transitions towards adaptive management of water facing climate and global change. Water Resour Manag 21(1):49–62
Pahl-Wostl C (2019) Governance of the water-energy-food security nexus: a multi-level coordination challenge. Environ Sci Pol 92:356–367. https://doi.org/10.1016/j.envsci.2017.07.017
Paruelo J, Beltrán A, Jobbágy E, Sala OE, Golluscio RA (1998) The climate of Patagonia: general patterns and controls on biotic processes. Ecol Austral 8:85–101
Pascual M, Ambrosio M, Rimoldi P, Pessacg N, Liberoff AL, Aigo J, Olivier T, Sainz Trápaga J, Malnero H, Kaless G (2019) Memoria del Primer Encuentro de Actores Territoriales de la Cuenca inferior del Río Chubut, Trelew, 24 de abril de 2019
Pascual M, Olivier T, Brandizi L, Rimoldi P, Malnero HA, Kaless G (2020) Análisis de Factibilidad de Fondo Agua Cuenca del Río Chubut, Argentina
Pessacg N, Flaherty S, Solman S, Pascual M (2020) Climate change in northern Patagonia: critical decrease in water resources. Theor Appl Climatol 140(3):807–822. https://doi.org/10.1007/s00704-020-03104-8
Prell C (2012) Social network analysis: history, theory and methodology. Sage, Los Angeles
Pretty J (2003) Social capital and the collective management of resources. Science 302(5652):1912–1914
R Core Team (2016) R: A language and environment for statistical computing. R Foundation for Statistical Computing, Vienna. https://www.R-project.org/
Ramirez-Sanchez S, Pinkerton E (2009) The impact of resource scarcity on bonding and bridging social capital: the case of fishers' information-sharing networks in Loreto, BCS, Mexico. Ecol Soc 14(1). http://www.ecologyandsociety.org/vol14/iss1/art22/

Rathwell KJ, Peterson GD (2012) Connecting social networks with ecosystem services for watershed governance: a social-ecological network perspective highlights the critical role of bridging organizations. Ecol Soc 17(2). http://dx.doi.org/10.5751/ES-04810-170224

Red de desertificación (2019) Taller regional sobre conservación de suelos. Trelew, 8 de Agosto de 2019

Sandström A, Rova C (2010) Adaptive co-management networks: a comparative analysis of two fishery conservation areas in Sweden. Ecol Soc 15(3). http://www.ecologyandsociety.org/vol15/iss3/art14/

Segesso L, Carrera AL, Bertiller MB, Saraví Cisneros H (2019) Soluble phenolics extracted from Larrea divaricata leaves modulate soil microbial activity and perennial grass establishment in arid ecosystems of the Patagonian Monte, Argentina. Plant Ecol 220(4):441–456. https://doi.org/10.1007/s11258-019-00926-z

Steger C, Hirsch S, Evers C, Branoff B, Petrova M, Nielsen-Pincus M, Wardropper C, van Riper CJ (2018) Ecosystem services as boundary objects for transdisciplinary collaboration. Ecol Econ 143:153–160

Tauro A, Gómez-Baggethun E, García-Frapolli E, Lazos Chavero E, Balvanera P (2018) Unraveling heterogeneity in the importance of ecosystem services: individual views of smallholders. Ecol Soc 23(4). https://doi.org/10.5751/ES-10457-230411

The QUINTESSENCE Consortium (2016) Networking our way to better ecosystem service provision. Trends Ecol Evol 31(2):105–115. https://doi.org/10.1016/j.tree.2015.12.003

United Nations (2015) Transforming our world: the 2030 agenda for sustainable development. UN, New York

Uriburu BA (2014) Conflictos socioambientales. Una experiencia de gestión comunitaria de recursos naturales. Cuadernos de Antropología (11):57–68

Zhang P, Zhang L, Chang Y, Xu M, Hao Y, Liang S, Liu G, Yang Z, Wang C (2019) Food-energy-water (FEW) nexus for urban sustainability: a comprehensive review. Resour Conserv Recycl 142:215–224. https://doi.org/10.1016/j.resconrec.2018.11.018

Chapter 12
Salmon Farming: Is It Possible to Relate Its Impact to the Waste Remediation Ecosystem Service?

Sandra L. Marín, Ángel Borja, Doris Soto, and Daniela R. Farias

Abstract Waste remediation (waste decomposition) is one of the ecosystem services (ES) salmon farming relies upon to reduce organic waste from the uneaten food and fish faeces through the metabolism of fauna and aerobic and anaerobic microbial communities. Our objective was to infer the changes that may occur in ecosystem capacity for reducing organic matter by firstly verifying that an impact has occurred using the AZTI's Marine Biotic Index (AMBI) and then relating that impact with changes in the ES of waste remediation reported in the literature. We relied on the combination of (i) the Driver, Activity, Pressure, change of State, Impact (Wellbeing), Response (Measure) (DAPSI(W)R(M)) approach as a causal model to establish relationships; (ii) the AMBI, as an impact indicator; and (iii) waste remediation capacity as the ES potentially affected by salmon farming. Data came from sites with different salmon farming influence located in Los Lagos and Aysén regions, Chile. Regression analyses were used to establish causal relationships between (i) total organic matter and oxygen availability, (ii) oxygen availability and AMBI and (iii) dominance of ecological groups and oxygen availability gradient. Ecological indices were estimated for each site. DAPSI(W)R(M) allowed to determine that food provision as the driver of salmon farming activity produced an environmental pressure (organic matter), which caused a significant reduction in

S. L. Marín (✉)
Aquaculture Institute, Universidad Austral de Chile, Puerto Montt, Chile

Research Center Dynamics of High Latitude Marine Ecosystems, IDEAL, Faculty of Sciences, Universidad Austral de Chile, Valdivia, Chile
e-mail: smarin@uach.cl

Á. Borja
AZTI, Marine Research, Basque Research and Technology Alliance (BRTA), Pasaia, Spain

D. Soto
Interdisciplinary Center for Aquaculture Research, INCAR, Universidad de Concepción, Concepción, Chile

D. R. Farias
Aquaculture Institute, Universidad Austral de Chile, Puerto Montt, Chile

P. L. Peri et al. (eds.), *Ecosystem Services in Patagonia*, Natural and Social Sciences of Patagonia, https://doi.org/10.1007/978-3-030-69166-0_12

oxygen availability, indicator of state change. The main impact was due to a decrease in the oxygen availability. High representation of species of the ecological groups IV (e.g. *Nassarius coppingeri*) and V (e.g. *Capitella capitata*) in the most impacted sites suggests increased organic matter processing. The ecosystem is coping with the increase of total organic matter by changing macrofauna composition at expenses of a decrease in biological diversity and an increase in species dominance. Overall, the results indicated that salmon farming impacts could be related to the ES of waste remediation.

Keywords Benthic systems · AMBI · Macrofauna · DAPSI(W)R(M) · Chilean Patagonia

1 Introduction

Salmon farming is a type of aquaculture in which salmon species are reared in a cycle that includes the production and development of eggs in freshwater until they are physiologically apt to be transferred to the fattening stage at sea (Aquaculture Stewardship Council 2012). The duration of this latter stage depends on the growth rate of the fish and the desired harvesting size, after which the fish are processed and shipped to their destination markets. The rapid expansion of salmon farming around the world has posed well-known environmental challenges (Bannister et al. 2014; Gebauer et al. 2017; Gjelland et al. 2014; Miranda et al. 2018; Niklitschek et al. 2013; Osmundsen et al. 2020; Vilata et al. 2010; Wang et al. 2012). The fattening phase of the salmon production cycle occurs mostly in floating cages located in protected coastal areas (Ticina et al. 2020). Since these sheltered locations usually have limited water exchange rates, their carrying capacities are often limited, and consequently the local environment may be affected by the release and accumulation of farm waste products (Pillay 2004).

From an ecosystem services (ES) approach, aquaculture plays a double role. On the one hand, it is recognised as a provider of food and as an ES valued for its contribution to food security (Beveridge et al. 2013). Also, aquaculture of filter feeders (e.g. mussels, oysters) may provide ES such as water cleaning and nutrient cycling (Wang et al. 2016). On the other hand, aquaculture causes adverse effects on other provisioning, regulating and cultural ES, whose intensity depends on the farmed species, the spatial scale of the operation, the total production, the duration of the activity and the hydrodynamic conditions and water exchange rates of the specific sites (Borja et al. 2009). Outeiro et al. (2015) indicated that the provision of food as an ES from salmon farming was associated with large economic benefits in Chiloé, Chile, but it was also causing a reduction in regulating and cultural services.

Given the expanding scenario of salmon farming in Chile, it is urgent to use the knowledge and data available to understand the relationships between impact and ecosystem functioning through the ES approach. The interaction of salmon farming with the biotic and abiotic environment is diverse and complex (Buschmann et al. 2009). Particularly, the addition of dissolved (mainly nitrogen and phosphorus) and particulate waste (mainly organic matter) from fish excretion and uneaten feed and fish faeces, with respect to the marine environment, may modify the abundance of the organisms that use this type of waste as food (Soto and Norambuena 2004). The term waste is often defined as "substances present in the marine environment which would not otherwise be there in the absence of anthropogenic activity and/or is present at a higher level than typical levels" (Hinga et al. 2005). Although, dissolved and particulate wastes play an important role in sustaining the biodiversity and trophic webs of the systems, when they are added at a higher rate than that which organisms can use, or other abiotic processes can reduce, they may become contaminants (Pearson and Rosenberg 1978) and potentially affect ecosystem functions. The capacity of the organisms to use wastes and transform them into non-harmful substances is an ES whose simplest descriptor is "decomposing waste" (Haines-Young and Potschin-Young 2018). This particular pathway through which waste is decomposed is part of a wider set of pathways by which different types of waste products are removed named waste remediation (Watson et al. 2016). Thus, salmon farming relies on this ES to reduce the inorganic and organic waste generated during the fattening stage of salmon production.

The effect of organic waste from salmon farming on soft-bottom benthic systems is evaluated in different countries through the chemical and physical attributes of the sediment and through ecological indices such as diversity, richness and indicator species, as well as biotic indices (Borja et al. 2009; Keeley et al. 2013). Usually, the evaluation is made close to the farm and at a control site. One of the principles of the ecosystem approach to aquaculture states that "aquaculture should be developed in the context of ecosystem functions and services with no degradation of these beyond their resilience capacity" (FAO 2011). In this context, using the ES approach would allow to describe changes in the functionality of the ecosystems as relevant processes that may be affected by waste deposition from salmon farming.

Salmon farming in Chile is focused mainly on the production of Atlantic salmon (*Salmo salar*). In 2018, Chile ranked eighth in global aquaculture production and second in salmon production, generating 20% of world salmon production (FAO 2020). That year, a total 953,300 ton of salmon were produced in the regions of Los Lagos (38.5%), Aysén (49.5%) and Magallanes (12.0%) (SUBPESCA 2020). Chilean environmental regulation for aquaculture states the need for maintaining dissolved oxygen in sediment and the water column at adequate levels to ensure aerobic metabolism (degradation of organic waste by heterotrophic organisms using oxygen) (Fishing and Aquaculture General Act 1990). Environmental performance of marine sites is based on the compliance of acceptability limits for organic matter, and redox potential and pH simultaneously, absence/presence of microorganism cover and/or dissolved oxygen in the water column. This evaluation focuses on chemical variables associated with sediment but does not include any indicator associated with the ecosystem functioning. For soft-bottom subtidal systems, the macrofauna has been a common biota compartment used to evaluate the impact of anthropogenic activities since they integrate the changes, inform about ecosystem functioning and enable to predict what the system state would be if pressure continues (Keeley et al. 2020).

In this context, the objective of this chapter is to infer the changes that may occur in the processes involved in the remediation of deposition of organic matter from salmon farming, as an ES of soft-bottom benthic ecosystems by first verifying salmon farming impact on the ecosystem using the AZTI's Marine Biotic Index (AMBI) (Borja et al. 2000). Then, we want to relate this information to available bibliography regarding macrofauna capacity to reduce the excess of organic matter. To do this we will integrate different approaches: (i) Driver, Activity, Pressure, change of State, Impact on environment and human Welfare (effects on ES), Response and management Measures (DAPSI(W)R(M)) as a causal model of relationships between components of the system of interest to link anthropogenic pressures to societal benefits (Elliott et al. 2017); (ii) AMBI to both determine salmon farming impacts on the benthic ecosystem and to infer the potential changes associated with the process of waste remediation as it represents the changes of ecological groups of species along a gradient of an environmental pressure, in this case soft-bottom organic enrichment; and (iii) waste remediation as the ES and the processes responsible for this ES that may be affected by salmon farming.

2 Conceptual Approaches to Relate Salmon Farm Impact and the ES of Waste Remediation

One of the most documented effects of salmon farming is the organic enrichment of the sediment under and in the vicinity of cages (Chang et al. 2013; Hargrave et al. 1997; Keeley et al. 2013, 2019, 2020). However, when this effect represents an impact, it needs to be properly defined to implement effective regulations associated with managing the environmental performance of salmon farming. In this context, the DAPSI(W)R(M) framework provides a useful approach.

DAPSI(W)R(M) Under this framework, food provision is the **D**river of the salmon farming **A**ctivity, which produces changes in the ecosystem structure and functioning, and we have defined the deposition of organic matter to the sea bottom as the environmental **P**ressure (Fig. 12.1). This organic matter comes from fish faeces and from feed not consumed and may cause a change in the **S**tate of the system of interest, which in this study is the soft-bottom benthic system. Our hypothesis states that the effects of the organic matter environmental pressure will be observed in sediment oxygen availability (measured as the redox potential), as an indicator of system **S**tate change, which should decrease as organic matter increases, because organisms are using it for degrading the extra-organic matter. Thus, bottom sediment will change from an oxic situation to an anoxic one (Fig. 12.1). Furthermore, the magnitude of benthic ecosystem changes needs to be evaluated to determine whether the actual changes constitute an impact on the ecosystem. To do this we propose AMBI to measure **I**mpact. Those impacts on the benthic system and the

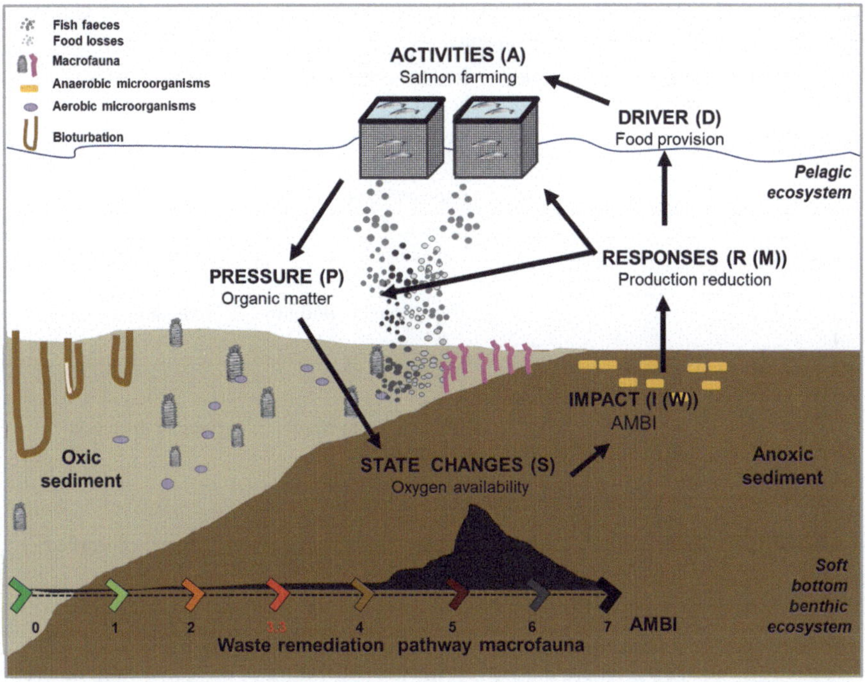

Fig. 12.1 Integration of conceptual frameworks used to relate salmon farming impact on soft-bottom ecosystems to the ecosystem service (ES) of waste remediation. Driver, Activity, Pressure, change of State, Impact on environment and human Welfare (effects on ES), Response and management Measures (DAPSI(W)R(M)), to establish causal relationships between system components, AZTI Marine Biotic Index (AMBI) as impact indicator related to ecosystem service of waste remediation

water column can result in effects to human **W**elfare (e.g. through salmon and other fish production reduction, socio-economic impacts to local population, changes in the ES provided by benthic fauna, such as organic matter remediation and cycling, etc.). This will need **R**esponses to revert the situation, which can include changes in the production after the monitoring alert on the impacts (e.g. moving cages to other locations, reducing the amount of production, temporal closures, etc.). Usually, these responses are implemented through management **M**easures, which can include modifications in the legislation (e.g. monitoring and assessment protocols, targets to be achieved) or others (e.g. socio-economic measures for producers).

AMBI It characterises the impact of an environmental pressure as a function of the change in the representativity of five ecological groups (EG, from I to V). These are groups of macrofauna species with different degrees of tolerance to the impact according to their adaptive strategies, being EGI-sensitive species, EGII indifferent, EGIII tolerant, EGIV second-order opportunistic and EGV first-order opportunistic. The impact is measured through a scale from 0 (no impact) to 7 (extreme impact)

Table 12.1 AZTI's Marine Biotic Index (AMBI) and its relationship with pollution classification of the site and health of the benthic community

AMBI	Dominating ecological group	Benthic community health	Site disturbance classification
0.0 < AMBI ≤0.2	I	Normal	Undisturbed
0.2 < AMBI ≤1.2		Impoverished	
1.2 < AMBI ≤3.3	III	Unbalanced	Slightly disturbed
3.3 < AMBI ≤4.3		Transitional to pollution	Moderately disturbed
4.3 < AMBI ≤5.0	IV–V	Polluted	
5.0 < AMBI ≤5.5		Transitional to heavy pollution	Heavily disturbed
5.5 < AMBI ≤6.0	V	Heavily polluted	
6.0 < AMBI ≤7.0	Azoic	Azoic	Extremely disturbed

Modified from Borja et al. (2000)

(Table 12.1, Fig. 12.1), where 3.3 represents the beginning of the ecosystem deterioration. According to the AMBI scale, the dominance of EG IV or V indicates sites moderately to heavily disturbed and communities from polluted, transitional to polluted and to heavily polluted. A more detailed description of EG is given below (Borja et al. 2000):

(i) Group I. Species very sensitive to organic enrichment and present under unpolluted conditions (initial state). They include specialist carnivores and some deposit-feeding tubicolous polychaetes.
(ii) Group II. Species indifferent to enrichment, always present in low densities with non-significant variations over time (from initial state, to slightly unbalanced). These include suspension feeders, less selective carnivores and scavengers.
(iii) Group III. Species tolerant of excess organic matter enrichment. These species may occur under normal conditions, but their populations are stimulated by organic enrichment (slightly unbalanced situations). They are surface deposit-feeding species, such as tubicolous spionids.
(iv) Group IV. Second-order opportunistic species, occurring in areas with high organic matter deposition (slightly to pronounced unbalanced situations). Mainly small-sized polychaetes: subsurface deposit-feeders, such as cirratulids.
(v) Group V. First-order opportunistic species (pronounced unbalanced situations). These are deposit-feeders, which proliferate in areas with high organic matter and reduced sediments.

Waste remediation It is a regulating service that has received different names and slightly different definitions depending on the ES classification framework (see Watson et al. 2016). We use the definition indicated by Watson et al. (2016): "The removal of waste products from a given environment by ecosystem processes that act to reduce concentrations of wastes by the mechanisms of cycling/detoxification, sequestration/storage and export". In this study, we focused on organic matter as waste, which is produced by salmon farming and deposited on soft-bottom ecosystems. The process reducing the organic matter that will be analysed is the cycling, and particularly the cycling made by macrofauna. Cycling is made not only by macrofauna but also by microorganisms and other species inhabiting these areas (Fig. 12.1). Less than 1% of organic matter reaching the marine environment remains in the marine sediment due to the efficient remineralisation in the water column and on the sea bottom by fauna and microorganisms (Burdige 2007).

Waste remediation in salmon farming ecosystems Waste remediation by macrofauna species (aerobic pathways) occurs all along the enrichment gradient, until organic matter input exceeds the capacity of macrofauna to process this waste and produce energy for metabolism. In this sense, as they will use most or all the oxygen available, this will cause localised anoxia, resulting in benthic malfunctioning with highly degraded sediments devoid of macrofauna (Brooks and Mahnken 2003; Gowen and Bradbury 1987; Valdemarsen et al. 2012). Under this situation, the ES of waste remediation is altered to the point in which the main process of organic matter cycling has been changed from an aerobic pathway to an anaerobic one, reducing the capacity of the ecosystem to provide the ES and producing significant changes in the trophic web and energy flow. In fact, one of the criteria to define carrying capacity for salmon farming sites is the assimilative capacity of the soft-bottom communities, which corresponds to the capacity for waste remediation until oxygen is not available. Such capacity has been measured as grams of carbon per square metre per day (gC m^2 d^{-1}) (Keeley et al. 2020) or grams of nitrogen per square metre per day (gN m^2 d^{-1}) depending on the element of the waste being studied.

Over the course of a production cycle in a salmon farm, soft-bottom benthic system is exposed to an increasing load of particulate organic matter because of the increase in fish biomass and feed. Consequently, an increasing capacity to process this extra-organic matter is required, and this can be achieved more efficiently through the species succession occurring as waste increases. For example, *Capitella capitata* (EGV) is an opportunistic species found at high abundances in sites with high depositional rate of organic matter. *C. capitata* in proximity to salmon cages can process carbon (as a measure of organic matter decomposition) through the gut passage at a rate of 44 gC m^{-2} d^{-1} (Keeley et al. 2020). These authors concluded that the metabolic activity of macrofauna accounted for most of the measured waste assimilation, and the ability to do so was ultimately attributed to physical properties of the site.

This information allows inferring that aerobic waste remediation capacity would exhibit changes, with higher capacities under the presence of opportunistic species (EG V), that is, when the AMBI value is ≥5, and will disappear under azoic condition (Fig. 12.1). It is important to note that the increase of EG V species is the manner in which that ecosystem is responding to an anthropogenic pressure that has already caused an impact revealed by AMBI, as AMBI ≥3.3 indicates the beginning of the benthic ecosystem degradation. The cost of this response could be associated with changes in the trophic web, energy flow or other ES, and therefore a response from managers and/or society may be required (Fig. 12.1).

3 Methodology

Study area and data collection The study was carried out in the Chilean regions of Los Lagos and Aysén. In each region, two influenced and two less influenced zones by salmon farming were selected (Fig. 12.2). Influenced zones were selected for

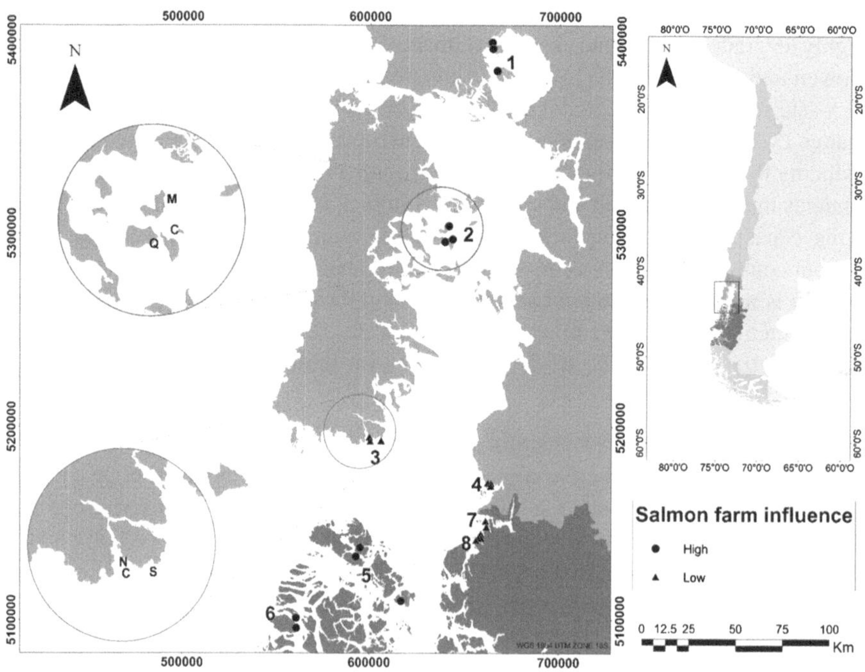

Fig. 12.2 Study area. Los Lagos region in light grey and Aysén region in dark grey. Three sampling sectors were selected per each of the four zones in the two regions. Zone 1, Reloncaví Estuary; Zone 2, Chiloé; Zone 3, Corcovado Gulf; Zone 4, Tic Toc Bay; Zone 5, Melinka North; Zone 6, Central Melinka; Zone 7, Raúl Marín Balmaceda North; Zone 8, Raúl Marín Balmaceda South. Zones 2 and 3 are zoomed to show the specific sampling sites (M, Meulín; Q, Quenac; T, Teuquelín; N, Corcovado North; C, Corcovado; S, Corcovado South)

being in proximity to salmon farms, and less influenced zones were areas geographically distanced from a farm by 3.2–11.8 km. Three sectors were selected in each zone where three sampling sites were located. A total of 72 sites were sampled during November 2016 and May 2017. At each site, redox potential in triplicate was obtained, as well as the percentage of total organic matter (TOM%; only one sample per station). The benthic macrofauna was sampled in triplicate at each station using a 0.1 m^2 Van Veen-type grab. In laboratory, the species were identified at the lowest possible taxonomical level.

DAPSI(W)R(M) framework: Pressure, State and Impact relationships Data from the 72 sampling sites were used to determine the relationships between the indicator of environmental pressure (TOM%) and the indicator of state (redox potential) using non-linear and linear regression and compared by Akaike Information Criteria (AIC). Lineal regression was used to relate redox and AMBI, the latter as indicator of system impact. Mann-Whitney test was used to compare TOM% and redox potential between influenced and less influenced. Further analyses were performed to determine the association of the EG to reduction of oxygen availability by mean of regression analysis between percentage of individuals belonging to each EG, as the dependent variable, and redox potential, as the independent variable. Reduction in oxygen availability and the presence of contaminants by anaerobic degradation of organic matter result in environmental conditions that can be tolerated by opportunistic species, which are included in the EG IV and V.

Salmon farm impact (AMBI) and waste remediation To infer the association between impact, measured by AMBI, and waste remediation, represented by the dominance of EG, two of the eight sampled zones were chosen: Chiloé (influenced by salmon farming) and Corcovado Gulf (less influenced), both in Los Lagos region (Fig. 12.2). The sectors in each zone were characterised regarding the reference farm when possible (distance, production the year before sampling and total organic particulate waste for that production, current velocity), as suggested by Borja et al. (2009). To estimate particulate waste, first the total feed given to fish was estimated using the economic conversion factor estimated to Chilean salmon industry (1.3, in Nahuelhual et al. (2019)) and the production of the reference farm. A food loss of 3% was assumed (Wang et al. 2012), and to estimate organic matter from faeces, the coefficient of apparent digestibility reported by Cheshuk et al. (2003) was used (0.8). Thus, TOM was calculated as the sum of that in food loss and faeces. Number of individuals, number of taxa, Shannon-Wiener diversity index and Simpson dominance index are presented for each sampling site, as well as AMBI and EG.

4 Results

DAPSI(W)R(M) framework: Pressure, State and Impact relationships Total organic matter was significantly larger in those influenced sites than less influenced ($F_{(1,112)} = 4.38$, $p = 0.040$). Redox potential was significantly lower in influenced sites than less influenced ($F_{(1,149)} = 304.7$, $p < 0.001$). The best fit of the relationship between redox potential and the changes in TOM% was a negative exponential function (Fig. 12.3a). AMBI values significantly decrease as redox increases (Fig. 12.3b). AMBI values ≥3.3 indicate a deterioration of the habitat and the benthic community, which is more frequently observed in values close to 0 mV and negative, thus suggesting an impact on the benthic system state. However, a wide range of redox values were associated with AMBI values higher than 3.3, which is reflected by the low value of the determination coefficient, suggesting that much of the variability associated with AMBI is not explained by redox. The 67% of the

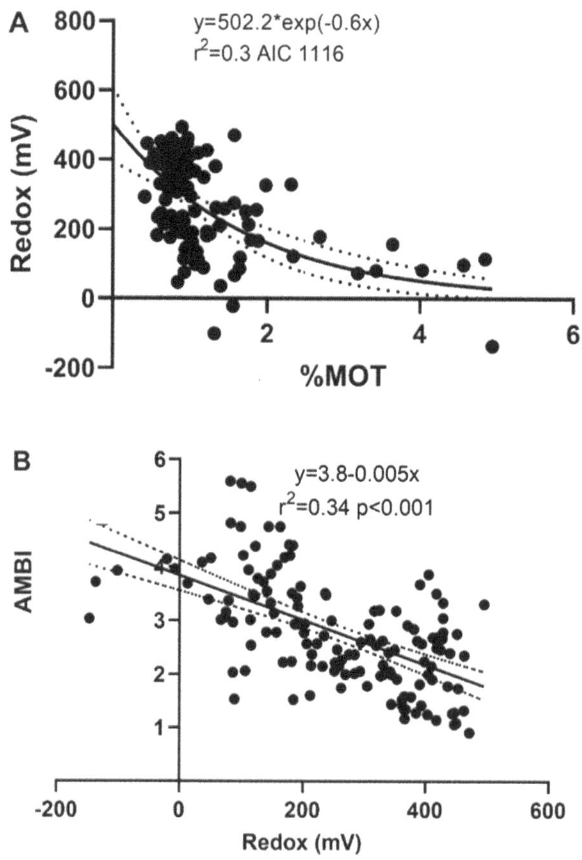

Fig. 12.3 Relationships between redox potential and percentage of total organic matter (TOM) (**a**) and relationship between AMBI and redox potential (**b**)

sampled sites with AMBI values ≥3.3 was located on influenced zones, predominantly in Los Lagos region, and only 6% of sites presented an AMBI value ≥5.

The variation in the percentage of individuals assigned to different EG throughout the redox potential gradient in sediment reflects the expected change in the representation of these EGs in accordance with the reduction in oxygen availability in sediment (Fig. 12.4a–f) due to excess organic matter (Fig. 12.3a). EG I and IV, in particular, relate significantly to the changes in redox potential in sediment, consistent with their life strategies, species sensitive to organic enrichment (EG I) and second-order opportunists (EG IV) (Fig. 12.4a, d). EG II species were common and

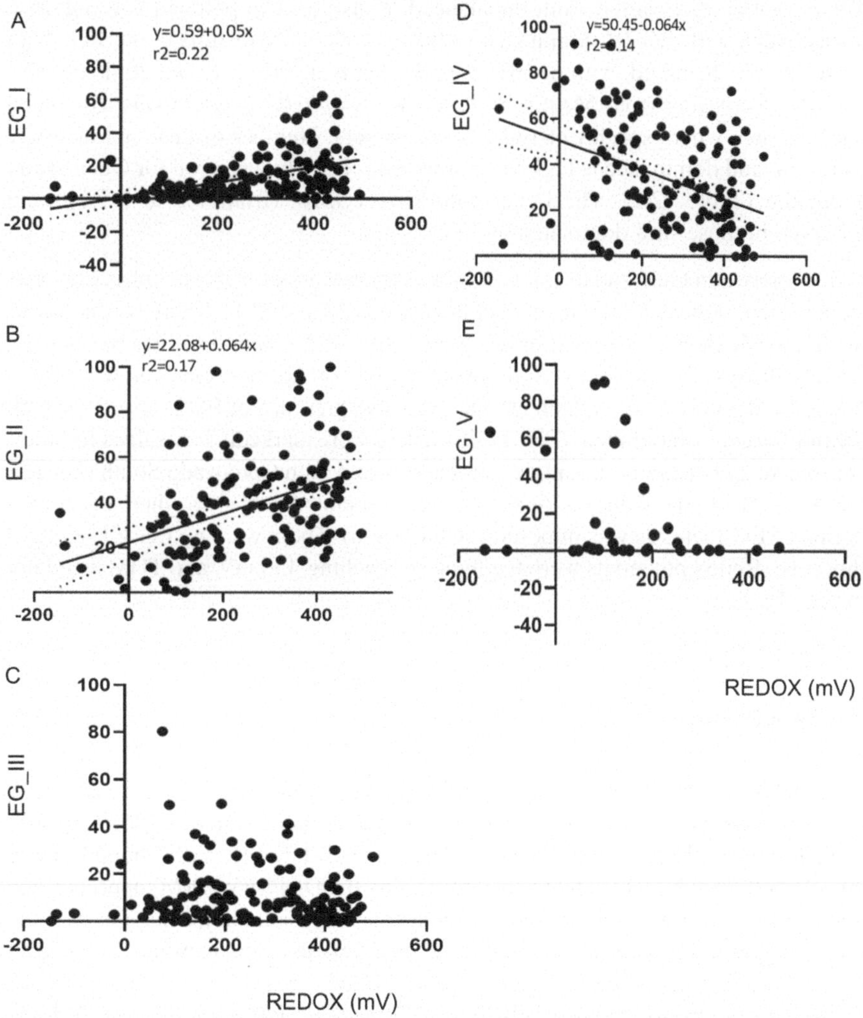

Fig. 12.4 Relationships between the percentages of individuals belonging to each of the five ecological groups (EG) along the redox potential gradient (**a**–**e**)

dominant in a wide range of values of redox potential (Fig. 12.4b). EG III species were present throughout the entire gradient in different percentages consistent with their tolerance for organic enrichment (Fig. 12.4c). EG V species (first-order opportunists) were found at few stations (22.7%); for this reason the analysis was performed only at stations where they were present. Species of EG V were observed all along the redox potential, but the highest percentages were observed between −200 and 200 redox potential.

Salmon farm impact (AMBI) and waste remediation Sampled sites presented an impact gradient. Less impacted site was Meulín, followed by Quenac and then Teuquelín, indicating slightly, moderately and heavily disturbed sites, respectively. Their communities varied from unbalanced, transitional to polluted and polluted, respectively, with EG IV dominance in those moderately disturbed and EG V in those heavily disturbed, with AMBI values higher than 5 (Fig. 12.5a). Redox potential was decreasing from Meulín to Teuquelín. *Capitella capitata* was associated with all sites where species of EG V were present. Number of taxa and diversity were low, and dominance is high in sampled sites heavily disturbed for Quenac and Teuquelín (Table 12.2). The larger values of TOM% (from 1% to 3.5%) were observed in those sites dominated by EG V.

In Corcovado Gulf, the distance from reference farms for the sampled sites varied between 3880 and 5360 m and depth between 28 and 51 m. There was no information available about harvest in this zone. Most of the sampling sites presented a slightly disturbed condition with dominance of EG II species. One site was classified as moderately disturbed, due to the exclusive presence of EG II species, which are indifferent to enrichment (Fig. 12.4b). There were markedly low values for number of taxa and diversity in samples sites associated with Corcovado South (E1–E3) (Table 12.2). Higher values of diversity were associated with sites where the contribution of EG I species was important (Corcovado). There were no EG V species in this zone. Redox potentials were the highest, reaching 450 mV, and TOM% did not exceed 1%.

5 Discussion

Salmon farming has been associated with several environmental impacts, and one of the main sources of impact is organic matter deposition (Aslam et al. 2020; Brooks et al. 2003; Buschamnn et al. 2006). We have analysed the effect of salmon farming on soft-bottom subtidal benthic ecosystem using the DAPSI(W)R(M) approach and AMBI to establish whether changes of state caused by salmon farming on the benthic ecosystem translate into an impact. Causal relationships between (i) environmental pressure and system state indicators (TOM%-redox) and (ii) system state and impact indicators (redox-AMBI) were statistically significant (Fig. 12.3). Most of the impacted sites were associated with sites influenced by salmon farming (67%), as found previously in the area (Quiroga et al. 2013). Ecosystem response to organic enrichment is the result of complex interactions, even at small spatial scales (Keeley et al. 2020; Tomasetti et al. 2016), which makes it difficult to generalise about ecosystem response. AMBI results observed in Chiloé suggest that the largest impact observed in Teuquelín was due to the activity of the reference farm, in contrast to the other two sectors where reference farms were under the fallowing period (Fig. 12.5). Interactions among site depth, distance to the reference farm, current

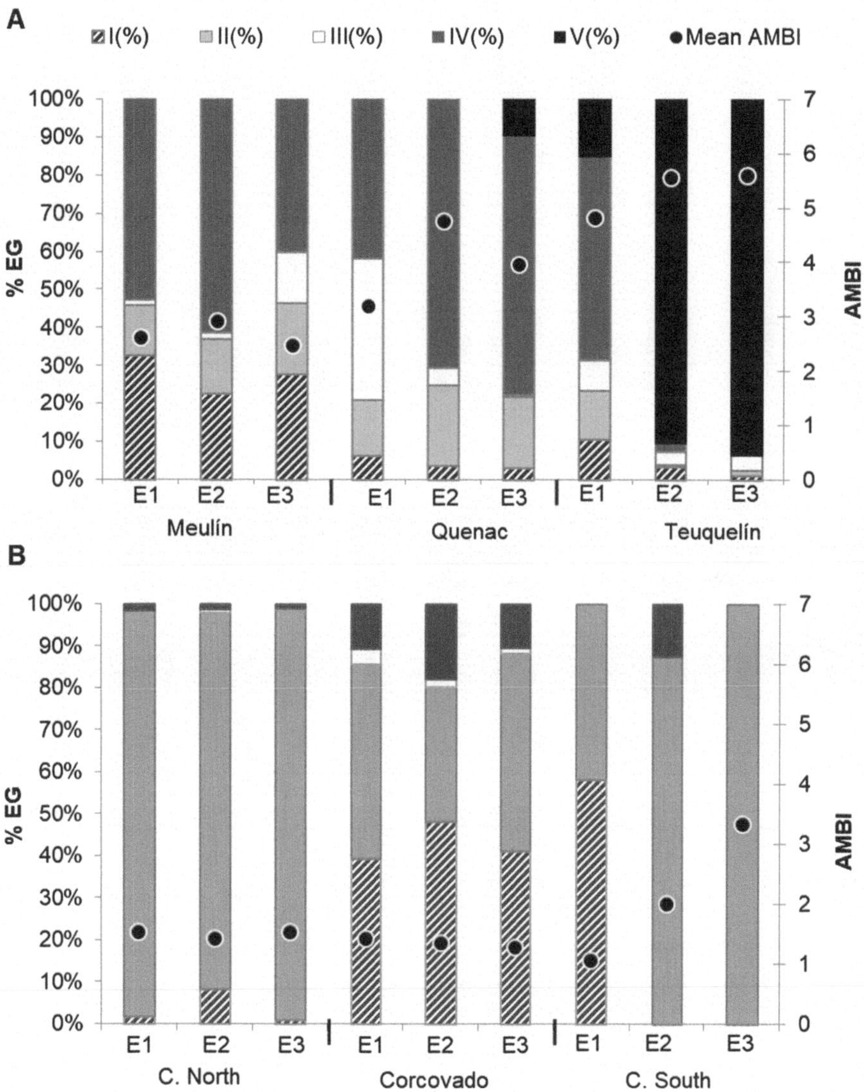

Fig. 12.5 Percentage of individuals per ecological group (EG) and mean value of AMBI (black circle) for sampled sites in (**a**) Chiloé, influenced by salmon farm, and (**b**) Corcovado Gulf, less influenced. Sampled sites in each sector are indicated by E1 to E3

Table 12.2 Information for sampled sites in Chiloé (sectors Meulín, Quenac and Teuquelín) and Corcovado Gulf (Corcovado North, Corcovado and Corcovado South). Sampled sites in each sector are indicated by E1 to E3

Sample site	Harvest[a] (ton)	Depth (m)	Distance to farm (m)	Redox potential (mV) (±sd)	Ecological group dominant	Abundance (nr individuals)	Richness (nr taxa)	Shannon-Wiener diversity (bits ind^{-1})	Simpson dominance
MeulínE1	4343	36.5	315	294.7 (50.0)	IV	2240	22	3.02	0.2
MeulínE2	4343	36	113	370.2 (58.6)	IV	1690	20	3.06	0.19
MeulínE3	4343	31.2	279	424.7 (35.4)	IV	2170	32	3.74	0.13
QuenacE1	2902	21.8	217	324.8 (108.7)		1640	21	3.45	0.13
QuenacE2	2902	23.1	134	162.5 (38.3)	IV	1970	26	3.27	0.21
QuenacE3	2902	29	103	111.69 (71.2)	IV	1380	14	2.59	0.27
TeuquelínE1	3281	39.5	160	82.5 (75.5)	IV	870	16	3.11	0.18
TeuquelínE2	3281	61.4	140	99.9 (36.1)	V	2150	11	0.77	0.82
TeuquelínE3	3281	56.8	145	81.5 (48.6)	V	1350	11	0.94	0.77
Corcovado SouthE1		30	3880	361.5 (17.3)	II	9650	14	0.74	0.82
Corcovado SouthE2		35.2	3880	361.4 (99.9)	II	3810	17	2.22	0.37
Corcovado SouthE3		34.5	3880	184.5 (103.7)	II	9050	15	0.65	0.85
CorcovadoE1		28	4585	393.1 (36.9)	II	340	11	3.15	0.13
CorcovadoE2		35	4585	364.5 (26.2)	I	960	16	3.19	0.13
CorcovadoE3		51	4585	443.8 (10.9)	II	1230	13	2.99	0.15
Corcovado NorthE1		42.6	5360	446.9 (16.0)	I	250	6	2.29	0.24
Corcovado NorthE2		43.7	5360	404.3 (30.1)	II	150	4	1.56	0.39
Corcovado NorthE3		36	5360	428.6 (21.5)	II	210	3	0.55	0.82

[a]Informed by the National Service of Fisheries and Aquaculture for the reference salmon farm

velocity and the amount of waste may have influenced the observed impact on sites located in Meulín and Quenac. Sites in Meulín were exposed to the highest particulate organic waste but were also the most distant to the reference farms. Current velocity allows us to infer a more dispersive site, in comparison to the other two. In Quenac, sites were exposed to smaller amount of waste but were closer to the reference farm, and depth for those sites was lower. Borja et al. (2009) indicated that depth, current velocity and distance to the farm were significant to predict the impact of a given production (ton).

The less influenced zone presented mostly slightly perturbed sites in which EG III should have dominated instead of EG II. This may be due to the poor species identification in sites not exposed to aquaculture. Unfortunately, a considerable larger effort is assigned to sampling and identification of macrofauna in impacted areas (Quiroga et al. 2013). Overall, this zone was under the AMBI limit; therefore, we should not expect benthic ecosystem degradation, which agrees with the a priori classification of influenced and less influenced sites.

The emphasis placed in the macrofauna component of soft-bottom benthic system in this analysis responds to the importance of this group in the provision of the ES of waste remediation. Keeley et al. (2019) demonstrated high level of responsiveness and therefore resilience in the biochemical and biological properties of the near-farm sediments, emphasising the role of the extremely prolific populations of opportunistic worms, such as those belonging to the EG IV and V. The presence of these species allows the benthic system to cycle excess of organic matter, but the specific rates may vary according to specific situation. The estimated rate of carbon degradation for *Arenicola marina* in a Norwegian salmon farm was 6.1 gC m^2 d^{-1} and for *C. capitata* was 5.4 gC m^2 d^{-1} (Keeley et al. 2020), and the same species in a different farm processed 16.4 gC m^2 d^{-1} and 44.2 gC m^2 d^{-1}, respectively (Keeley et al. 2020). In general, deposit-feeding polychaetes have demonstrated their potential for reducing efficiently 20–85% of organic matter, 31–91% of nitrogen and 65–91% of organic carbon (Fang et al. 2016; Gómez et al. 2019; Honda and Kikuchi 2002; Marques et al. 2017; Nederlof et al. 2020; Pajand et al. 2017; Palmer 2010). In this regard, the opportunistic species of the genus *Capitella* are efficient extractive species that reduce organic matter from underneath salmon cages sediments (Nederlof et al. 2020).

The increase in waste remediation by opportunistic species occurs in a benthic ecosystem that, according to AMBI classification, is heavily perturbed (Borja et al. 2000). Thus, the response of the ecosystem to cope with the excess of organic matter is a change in the community structure that can be observed by the changes in the EG dominance (Fig. 12.1) but also in a reduction of community diversity and high dominance (Table 12.2), which in turn may affect supporting ES. The evidence and success of the ES of waste remediation may be observed directly from physicochemical properties (environmental degradation, mineralisation rates, organic matter load across time and space) (Aslam et al. 2020). However, in the absence of data, specific responses of living organisms can act as indicators in response to waste (Watson et al. 2016). In this context, AMBI provides information about salmon farming impact that allows to make inferences regarding the demand for this ES and the mechanisms by which is provided: colonisation of soft-bottom ecosystems by EG IV and V species and their increase in abundance are the mechanisms through which the ES of waste remediation can still be provided under an increasing demand for processing the excess of organic matter input by salmon production. Adaptative strategies of these species allow them to persist and reproduce in the presence of toxic effect of compounds derived from the anaerobic assimilation of wastes and hypoxia. In fact, species belonging to EG V resist more severe hypoxic conditions (Brooks and Mahnken 2003) and are indicative of a reduced environment (Borja et al. 2000), where anaerobic microorganisms dominate the degradation of organic matter (Jessen et al. 2017).

Carbon processing capacity of opportunistic species suggests that waste remediation, measured as gC m^2 d^{-1}, increases along the enrichment gradient to peak, probably at the opportunistic species peak. This peak would be followed by a reduction of their abundance and further on in the gradient they would disappear because they would not be able to cope with anoxic and contaminated sediments (AMBI ≥ 6) (Fig. 12.1). However, since this ES is carried out by aerobic and anaerobic organisms, waste remediation will continue, but since microorganisms do not perturb the sediment, bacterial breakdown of complex molecules is relatively slow and inefficient in the absence of benthic fauna (Van Colen et al. 2012). Thus, the ES is not lost under low availability/absence of oxygen, but its capacity to reduce the incoming organic matter is severely affected because of the change from an aerobic to an anaerobic degradation (Jessen et al. 2017).

Waste remediation provided by marine ecosystems is a public good in terms of its use for disposal and remediation of waste (Watson et al. 2016). Several anthropogenic activities use these ecosystems, and because of that, it becomes difficult to distinguish the source of a given impact. This raises the concern about waste management of companies and other businesses that derive benefits from this ES at the local level. Salmon industry could include their own monitoring along the fattening stage of the production cycle, besides that established by the country regulation. Such a monitoring should include biotic indices such as AMBI, which would indicate when the demand for processing organic matter is changing the benthic community to stages where the anaerobic waste remediation may dominate the pathway for proving waste remediation. This would allow them to implement management measures opportunely.

Salmon industry should also engage in innovative approaches to reduce the impact associated with organic matter deposition in the sea as it is made in the production of salmon smolt in Recirculating Aquaculture Systems (RAS). In this stage of the salmon production cycle, sludge management may resemble the waste remediation provided by the ecosystem. Sludge is reduced using technology with an estimated cost of USD 3.8 for producing 100 smolt of 150 g (R. Pavez, personal communication). At a global level, industry has made efforts to reduce the impacts by incorporating technology in their systems to mitigate environmental pressure, thus improving feed digestibility, food composition and feeding technology, as Measures in the DAPSI(W)R(M) framework. As a result, a reduction in nutrient excess, food waste and deposition in sediments has been observed (Wang et al. 2012).

Another measure, not well explored yet, is the Integrated Multi-Trophic Aquaculture (IMTA). One type combines three trophic levels such as shellfish, finfish and seaweed (Chopin and Bastarache 2004; Neori et al. 2004), but the scale of mariculture needed to mitigate pollution is sometimes unrealistic (Gentry et al. 2020). Although mussel farming reduces suspended solids, it also increments organic matter in sediment (through faeces and pseudo-faeces), which modifies both the benthic assemblages and sediment chemistry (reducing oxygen level) (Bergström et al. 2020; Chamberlain et al. 2001). A different approach has been raised from the knowledge about the functionality of different macrofauna communities. For example, polychaetes from EG V have been suggested as extractive species for bioremediation, where *Capitella* sp. was proposed as a model species for IMTA systems for the first time by Tsutsumi et al. (2005) and Kinoshita et al. (2008). Recently, Nederlof et al. (2019, 2020) demonstrated that polychaetes species such as *Capitella* sp. and *Ophryotrocha craigsmithi* are high-quality marine resources with great potential for bioremediation. Another study evaluated a native polychaete species from Norway, *Ophryotrocha* sp., as a model for benthic cultivation in open water systems, which showed that this species can consume salmon farm wastes and guarantee bio-mitigation (Jansen et al. 2019). Using macrofauna species as bioremediation has not been evaluated on salmon farms, neither in Chile or elsewhere, and although reported results are interesting, this type of approach may perform differently depending on the physicochemical conditions of the sediments and the hydrodynamic of the area (Keeley et al. 2020).

Marine aquaculture is an integral part of the growing coastal economy and requires balancing the rights and responsibilities in using and preserving the marine ecosystem. National regulations play an important role in achieving the balance between these rights and responsibilities through a clear definition of their environmental sustainability goals, along with the implementation of the strategy to reach these goals. Environmental sustainability goals should be focused on the functionality of the ecosystems for them to continue providing ES. In this context DAPSI(W)R(M) constitutes a useful approach to design both a monitoring programme for evaluating environmental performance of salmon farming and the responses to a non-compliance regarding the accepted impact on soft-bottom benthic ecosystem.

Environmental standards stablished in the Chilean regulation for salmon farming are based on TOM% and redox potential, which according to the analysis made in this study using the DAPSI(W)R(M) approach are environmental pressure and change of state indicators, respectively. Although redox potential varied between influenced and less influenced sites, these changes by themselves do not confirm impact, because they are not showing changes in ecosystem functionality. In fact, taking into considerations the acceptability limit for redox potential in Chilean regulation (Resolución Exenta 3612 2009), only 8% of the sites would have been defined as impacted, and none if using TOM% acceptability limit, in contrast to the 67% estimated using AMBI. Chilean regulatory institutions should test the conceptual relationships proposed in this study through the DAPSI(W)R(M) approach using information from the different geographic areas where salmon farming operates and define the impact index based on the ES approach. This study provided information to consider the ES of waste remediation for soft-bottom benthic ecosystems as an impact index. In a context of cost-efficient monitoring, AMBI provides enough information to infer about the possible changes that may be occurring regarding the ES of waste remediation. In addition, the 3.3 limit for AMBI would prevent reaching a level of impact that makes the ecosystem functionality change from an aerobic pathway to an anaerobic pathway for waste remediation. Special attention should be placed on the trade-off between the ES of waste remediation provided by macrofauna and other attributes of the ecosystem that may affect supporting ES over which other society interest relies, for example, biodiversity.

Complementary to these strategies, national regulations (responses and measures in the DAPSI(W)R(M) approach) still are the principal tool to maintain the ecosystem integrity in those coastal areas where anthropogenic activities are developed. In this context, managers and regulators should implement larger spatial scales monitoring programmes (measures) than those used for measuring impact of aquaculture farms. These monitoring programmes, as responses and measures, should be focused on waste remediation capacity as it may be impacted by different activities synergically.

Acknowledgements The data used in this analysis were obtained through a project sponsored by the Fisheries and Aquaculture Research Fund of the Chilean Fisheries and Aquaculture Undersecretary's Office entitled "Soft-bottom subtidal benthic communities and the characterization of their environmental conditions using biological indices (AMBI) and physical and chemical variables in southern Chile (stage one)". Thanks to IDEAL Center (FONDAP 15150003), FONDECYT 1190207 and CoastCarb for funding the formulation of this chapter as well as to Jonathan Arcos for designing the study area map.

References

Aquaculture Stewardship Council (2012) Aquaculture Stewardship Council (ASC) Salmon standard: version 1.0 June 2012. Aquaculture Stewardship Council, Utrecht. 102 pp

Aslam SN, Venzi MS, Venkatraman V, Mikkelsen Ø (2020) Chemical assessment of marine sediments in vicinity of Norwegian fish farms – a pilot study. Sci Total Environ 732:139130

Bannister RJ, Valdemarsen T, Hansen PK, Holmer M, Ervik A (2014) Changes in benthic sediment conditions under an Atlantic salmon farm at a deep, well-flushed coastal site. Aquac Environ Interact 5(1):29–47

Bergström P, Durland Y, Lindegarth M (2020) Deposition of shells modify nutrient fluxes in marine sediments: effects of nutrient enrichment and mitigation by bioturbation below mussel farms. Aquac Environ Interact 12:315–325

Beveridge MM, Thilsted SH, Phillips MJ, Metian M, Troell M, Hall SJ (2013) Meeting the food and nutrition needs of the poor: the role of fish and the opportunities and challenges emerging from the rise of aquaculture. J Fish Biol 83:1067–1084

Borja A, Franco J, Perez V (2000) A marine biotic index to establish the ecological quality of soft-bottom benthos within european estuarine and coastal environments. Mar Pollut Bull 40(12):1100–1114

Borja A, Rodríguez JG, Black K, Bodoy A, Emblow C, Fernandes TF, Forte J, Karakassis I, Muxika I, Nickell TD, Papageorgiou N, Pranovi F, Sevastou K, Tomassetti P, Angel D (2009)

Assessing the suitability of a range of benthic indices in the evaluation of environmental impact of fin and shellfish aquaculture located in sites across Europe. Aquaculture 293:231–240
Brooks KM, Mahnken CVW (2003) Interactions of Atlantic salmon in the Pacific northwest environment: II. Organic wastes. Fish Res 62:255–293
Brooks KM, Stierns AR, Mahnken CVW (2003) Chemical and biological remediation of the benthos near Atlantic salmon farms. Aquaculture 219:355–377
Burdige DJ (2007) Preservation of organic matter in marine sediments: controls, mechanisms, and an imbalance in sediment organic carbon budgets? Chem Rev 107(2):467–485
Buschamnn A, Riquelme VA, Hernández-González MC, Varela D, Jiménez JE, Henríquez LA, Vergara PA, Guiñez R, Filún L (2006) A review of the impacts of salmonid farming on marine coastal ecosystems in the southeast Pacific. ICES J Mar Sci 63:1338–1345
Buschmann A, Cabello F, Young K, Carvajal J, Varela D, Henríquez LA (2009) Salmon aquaculture and coastal ecosystem health in Chile: analysis of regulations, environmental impacts and bioremediation systems. Ocean Coast Manag 52:243–249
Chamberlain J, Fernandes TF, Read PA, Nickell TD, Davies IM (2001) Impacts of biodeposits from suspended mussel (Mytilus edulis L.) culture on the surrounding surficial sediments. ICES J Mar Sci 58:411–416
Chang BD, Page F, Losier RJ (2013) Variables affecting sediment sulfide concentrations in regulatory monitoring at salmon farms in the Bay of Fundy, Canada. Aquac Environ Interact 4:67–80
Cheshuk BW, Purser GJ, Quintana R (2003) Integrated open-water mussel (Mytilus planulatus) and Atlantic salmon (Salmo salar) culture in Tasmania, Australia. Aquaculture 218:357–378
Chopin T, Bastarache S (2004) Mariculture in Canada: finfish, shellfish and seaweed. World Aquacult 6
Elliott M, Burdon D, Atkins JP, Borja A, Cormier R, de Jonge VN, Turner RK (2017) "And DPSIR begat DAPSI(W)R(M)!" - a unifying framework for marine environmental management. Mar Pollut Bull 118(1–2):27–40
Fang J, Zhang J, Jiang Z, Du M, Liu Y, Mao Y, Gao Y, Fang J (2016) Environmental remediation potential of Perinereis aibuhitensis (Polychaeta) based on the effects of temperature and feed types on its carbon and nitrogen budgets. Mar Biol Res 12(6):583–597
FAO (2011) Enfoque ecosistémico a la acuicultura. FAO Orientaciónes Técnicas para la pesca responsable. In Desarrollo de la acuicultura, Roma. 60 pp
FAO (2020) The state of world fisheries and aquaculture. Sustainability in actions. Rome
Fishing and Aquaculture General Act. 1990 (1990) Ley General de Pesca y Acuicultura, Subsecretaría de Pesca y Acuicultura, Valparaíso. 201 pp
Gebauer P, Paschke K, Vera C, Toro JE, Pardo M, Urbina M (2017) Lethal and sub-lethal effects of commonly used anti-sea lice formulations on non-target crab Metacarcinus edwardsii larvae. Chemosphere 185:1019–1029
Gentry RR, Alleway HK, Bishop MJ, Gillies CL, Waters T, Jones R (2020) Exploring the potential for marine aquaculture to contribute to ecosystem services. Rev Aquac 12:499–512
Gjelland KØ, Serra-Llinares RM, Hedger RD, Arechavala-Lopez P, Nilsen R, Finstad B, Uglem I, Skilbrei OT, Bjørn PA (2014) Effects of salmon lice infection on the behaviour of sea trout in the marine phase. Aquac Environ Interact 5:221–233
Gómez S, Hurtado CF, Orellana J (2019) Bioremediation of organic sludge from a marine recirculating aquaculture system using the polychaete Abarenicola pusilla (Quatrefages, 1866). Aquaculture 507:377–384
Gowen RJ, Bradbury NB (1987) The ecological impact of salmon farming in coastal waters: a review. Oceanogr Mar Biol Annu Rev 25:563–575
Haines-Young R, Potschin-Young MB (2018) Revision of the common international classification for ecosystem services (CICES V5.1): a policy brief. One Ecosystem e27108:3
Hargrave BT, Phillips GA, Doucette LI (1997) Assessing benthic impacts of organic enrichment from marine aquaculture. Water Air Soil Pollut 99:641–650

Hinga KR, Ahmed MT, Lewis N (2005) Waste processing and detoxification. Ecosystems and human well-being: current state and trends: findings of the condition and trends working group. 1: 417pp

Honda H, Kikuchi K (2002) Nitrogen budget of polychaete Perinereis nuntia vallata fed on the feces of Japanese flounder. Fish Sci 68:1304–1308

Jansen HM, Hansen PK, Brennan N, Dahlgren T, Fang J, Nederlof MA, Strohmeier T, Sveier H, Strand Ø (2019) Enhancing opportunistic polychaete communities under fish farms: an alternative concept for integrated aquaculture. Aquac Environ Interact 11:331–336

Jessen GL, Lichtschlag A, Ramette A, Pantoja S, Rossel PE, Schubert CJ, Struck U, Boetius A (2017) Hypoxia causes preservation of labile organic matter and changes seafloor microbial community composition (Black Sea). Mar Ecol 3:e160189

Keeley N, Forrest BM, Macleod CK (2013) Novel observations of benthic enrichment in contrasting flow regimes with implications for marine farm monitoring and management. Mar Pollut Bull 66:105–116

Keeley N, Valdemarsen T, Woodcock S, Holmer M, Husa V, Bannister RJ (2019) Resilience of dynamic coastal benthic ecosystems in response to large-scale finfish farming. Aquac Environ Interact 11:161–179

Keeley N, Valdemarsen T, Strohmeier T, Pochon X, Dahlgren T, Bannister RJ (2020) Mixed-habitat assimilation of organic waste in coastal environments – It's all about synergy! Sci Total Environ 699:134281

Kinoshita K, Tamaki S, Yoshioka M, Srithongouthai S, Kunihiro T, Hama D, Ohwada K, Tsutsumi H (2008) Bioremediation of organically enriched sediment deposited below fish farms with artificially mass-cultured colonies of a deposit-feeding polychaete Capitella sp. I. Fish Sci 74:77–87

Marques B, Calado R, Lillebø AI (2017) New species for the biomitigation of a super-intensive marine fish farm effluent: combined use of polychaete-assisted sand filters and halophyte aquaponics. Sci Total Environ 599–600:1922–1928

Miranda C, Godoy FA, Lee MR (2018) Current status o fthe use of antibiotics and the antimicrobial resistance in the Chilean Salmon farms. Front Microbiol 9:1284

Nahuelhual L, Vergara X, Jullian C, Carmona A, Campos G (2019) Servicios ecosistémicos marino-costeros en la Región de Magallanes y la Antártica Chilena. Centro de Investigación: Dinámica de Ecosistemas Marinos de Altas Latitudes, IDEAL. 40 pp

Nederlof MAJ, Jansen HM, Dahlgren TG, Fang J, Meier S, Strand Ø, Sveier H, Verdegem MCJ, Smaal AC (2019) Application of polychaetes in (de)coupled integrated aquaculture: production of a high-quality marine resource. Aquac Environ Interact 11:221–237

Nederlof MAJ, Fang J, Dahlgren TG, Rastrick SPS, Smaal AC, Strand Ø, Sveier H, Verdegem MCJ, Jansen HM (2020) Application of polychaetes in (de)coupled integrated aquaculture: an approach for fish waste bioremediation. Aquac Environ Interact 12:385–399

Neori A, Chopin T, Troell M, Buschamnn A, Kraemer GP, Halling C, Schpigela M, Yarish C (2004) Integrated aquaculture: rationale, evolution and state of the art emphasizing seaweed biofiltration in modern mariculture. Aquaculture 231:361–391

Niklitschek E, Soto D, Lafon S, Molinet C, Toledo P (2013) Southward expansion of the Chilean salmon industry in the Patagonian Fjords: main environmental challenges. Rev Aquac 4:1–24

Osmundsen TC, Amundsen VS, Alexander KS, Asche F, Bailey J, Finstad B, Olseng MS, Hernández K, Salgado H (2020) The operationalisation of sustainability: sustainable aquaculture productionas defined by certification schemes. Glob Environ Chang 60:1–8

Outeiro L, Häussermann V, Viddi F, Hucke-Gaete R, Försterra G, Oyarzo H, Kosiel K, Villasante S (2015) Using ecosystem services mapping for marine spatial planning in southern Chile under scenario assessment. Ecosyst Serv 16:341–353

Pajand ZO, Soltani M, Bahmani M, Kamali A (2017) The role of polychaete Nereis diversicolor in bioremediation of wastewater and its growth performance and fatty acid composition in an integrated culture system with Huso huso (Linnaeus, 1758). Aqua Res:1–9

Palmer P (2010) Polychaete-assisted sand filters. Aquaculture 306:369–377

Pearson T, Rosenberg R (1978) Macrobenthic succession in relation to organic enrichment and pollution of the marine environment. Oceanogr Mar Biol Annu Rev 16:229–241
Pillay TVR (2004) Aquaculture and the environment, Second edn. Blackwell Scientific Publications, Oxford
Quiroga E, Ortiz P, Reid B, Gerdes D (2013) Classification of the ecological quality of the Aysén and Baker Fjords (Patagonia, Chile) using biotic indices. Mar Pollut Bull 68:117–126
Resolución Exenta 3612 (2009) Resolución Exenta 3612, Aprueba resolución que fija las metodologías para elaborar la caracterización preliminar de sitio (CPS) y la información ambiental (INFA). F.y.T. Ministerio de Economía, editor. Subsecretaría de Pesca y Acuicultura
Soto D, Norambuena R (2004) Evaluation of salmon farming effects on marine systems in the inner seas of southern Chile: a large-scale mensurative experiment. J Appl Ichthyol 20:493–501
SUBPESCA (2020) Vol. 2020. Subsecretaria de Pesca y Acuicultura, Chile
Ticina V, Katavic I, Grubišic L (2020) Marine aquaculture impacts on marine biota in oligotrophic environments of the Mediterranean Sea – a review. Front Mar Sci 7:217
Tomasetti P, Gennaro P, Lattanzi L, Mercatali I, Persia E, Vani D, Porrello S (2016) Benthic community response to sediment organic enrichment by Mediterranean fish farms: case studies. Aquaculture 450:262–272
Tsutsumi H, Kinoshita K, Srithongouthai S, Sato A, Nagata S, Inoue A, Yoshioka M, Ohwada K, Hama D (2005) Treatment of the organically enriched sediment below the fish farm with the biological activities of artificially mass-cultured colonies of a small deposit feeding Polychaete, Capitella sp. I. Benthos Res 60(1):25–38
Valdemarsen T, Wendelboe K, Egelund JT, Kristensen E, Flindt MR (2012) Burial of seeds and seedlings by the lugworm Arenicola marina hampers eelgrass (Zostera marina) recovery. J Exp Mar Biol Ecol 410:45–52
Van Colen C, Rossi F, Montserrat F, Anderson MGI, Gribsholt B, Herman PMJ, Degraer S, Vincx M, Ysebaert T, Middleburg JJ (2012) Organism-sediment interactions govern post-hypoxia recovery of ecosystem functioning. PLoS One:e49795 7(11)
Vilata J, Oliva D, Sepúlveda M (2010) The predation of farmed salmon by South American sea lions (Otaria flavescens) in southern Chile. ICES J Mar Sci 67:475–483
Wang X, Mork Olsen L, Reitan L, Olsen Y (2012) Discharge of nutrient wastes from salmon farms: environmental effects, and potential for integrated multi-trophic aquaculture. Aquac Environ Interact 2:267–283
Wang L, Fan Y, Yan C, Gao C, Xu Z, Liu X (2016) Assessing benthic ecological impacts of bottom aquaculture using macrofaunal assemblages. Mar Pollut Bull 114(1):258–268
Watson SCL, Paterson DM, Queirós AM, Rees AP, Stephens N, Widdicombe S, Beaumont NJ (2016) A conceptual framework for assessing the ecosystem service of waste remediation: in the marine environment. Ecosyst Serv 20:69–81

Chapter 13
Using the Ecosystem Services Approach to Understand the Distributional Effects of Marine Protected Areas in the Chilean Patagonia

María José Brain and Laura Nahuelhual

Abstract Gains in biodiversity from marine conservation might not correlate with a fair distribution of benefits across different social actors. In this chapter, we analyze the case of the marine protected area of multiple uses (MPA-MU) Seno Almirantazgo in the Magallanes region, created by the Chilean Government in 2018. We apply the ecosystem services (ES) lens to analyze perceived distributional effects of the implementation of the MPA-MU. Social actors' perceptions revealed three main scenarios: (i) the MPA-MU will generate benefits derived from the enhancement of specific ES; (ii) those benefits will not be distributed equally across social actors, where perception identified fishers as the potential "losers" and tour operators as the main "winners"; and (iii) changes in ocean access rights are perceived as the main barrier preventing an equitable distribution of monetary and non-monetary benefits. These perceptions are linked to three different dimensions of environmental justice (distribution, procedure, recognition), which are largely omitted in the conservation planning and particularly in Chile. The ES lens can be a useful tool to implement actions that include these dimensions in early stages of marine protected areas planning. However, such inclusion requires a large transfor-

---

M. J. Brain
Programa de Magíster en Ecología Aplicada, Facultad de Ciencias, Universidad Austral de Chile, Valdivia, Chile

Centro de Investigación en Dinámica de Ecosistemas Marinos de Altas Latitudes (IDEAL), Universidad Austral de Chile, Valdivia, Chile

L. Nahuelhual (✉)
Instituto de Economía, Universidad Austral de Chile Centro de Investigación: Dinámica de Ecosistemas Marinos de Altas Latitudes Instituto Milenio en Socio-Ecología Costera, Valdívia, Chile

P. L. Peri et al. (eds.), *Ecosystem Services in Patagonia*, Natural and Social Sciences of Patagonia, https://doi.org/10.1007/978-3-030-69166-0_13

mation of the institutional settings that nowadays influence marine conservation rules in the Chilean Patagonia.

Keywords Marine conservation · Marine governance · Environmental justice · Marine ecosystem services · Magallanes region

1 Introduction

Marine protected areas (MPA) are legally designated areas for the long-term conservation of biodiversity and associated ecosystem services (ES) (Day et al. 2012). Whereas primarily created for biodiversity, international policy and many national regulations require accounting for human populations' well-being and a fair sharing of costs and benefits from protected area designation (Rodríguez-Rodríguez et al. 2019).

Marine protected areas are likely the most important way to conserve marine biodiversity across the planet (Davies et al. 2017). As conservation institutions, they constitute a socially constructed set of rules for the allocation of access to and use of natural resources among different social actors (Mascia and Claus 2009). Changes in access may be positive or negative for local communities and may be felt differently by different social actors' groups (Schreckenberg et al. 2016). In developing countries, this can lead to local livelihood costs, which may not be distributed equally, while the benefits are shared globally or at least at supra-livelihood scales (Oldekop et al. 2016).

In Chile there are four types of MPA: (i) marine parks and (ii) marine reserves, which comprise essentially ocean waters, (iii) nature sanctuaries, and (iv) marine and coastal protected areas of multiple uses (MPA-MU), which may also contain patches of land. All types of MPA are created and regulated by the Ministry of the Environment of the Chilean Government. The first two types are regulated by the Subsecretaría de Pesca y Acuicultura (SUBPESCA), who is in charge of providing the background information for their destination, leaving the custody in the hands of the Servicio Nacional de Pesca y Acuicultura (SERNAPESCA). Additionally, there are inland waters associated with the Sistema Nacional de Areas Protegidas del Estado (SNASPE), which are under the supervision of the Corporación Nacional Forestal (CONAF). Marine protected areas are selected for their conservation values, and the uses must be conducted under a sustainable management of the marine biodiversity. Administrative and regulatory measures are clearly established to regulate the access to fishing and other activities such as tourism in order to prevent negative ecosystem impacts.

Chile's marine protected areas comprise 1,680,855 km^2 (approximately 40% of Chile's Exclusive Economic Zone, EEZ[1]). Of this area, 1,669,581 km^2 are outside the territorial sea of Chile but within its EEZ. The remaining 11,274 km^2 are under protection within the Chilean territorial sea. In turn, Patagonia comprises 151,552 km^2 of marine protected areas, where 140,383 km^2 are outside the territorial Chilean sea (Diego Ramírez Islands Marine Park and Drake Pass) and 11,169 km^2 are within it. Thus, of the total area within the territorial sea, 90% are in Patagonia (nine different MPA).

At present, we know little about how local social actors perceive that the creation of one MPA can affect their well-being, despite the important role that they can generate in conserving particular species and habitats. The ES approach can be very useful to this endeavor, helping policy-makers and managers to understand the realized benefits and costs of the conservation efforts. Specifically, the ES lens may help foresee the potential trade-offs that can emerge among social actors (McShane et al. 2011; Rees et al. 2013; Daw et al. 2015) which involve "gains in one ES or group of people, resulting in losses for others" (Daw et al. 2015, p. 6950). For instance, MPA may enhance benefits for recreational fishers or tourism operators, but might displace small-scale fishers who have traditionally relied upon those areas for their livelihoods (Davies et al. 2018).

In this chapter, we analyze the case of the MPA-MU Seno Almirantazgo, in the Magallanes region, Chile. The Chilean Government designated the MPA-MU in 2018 with the main purpose of protecting emblematic and endangered habitats and species. We apply the ES lens to analyze perceived distributional effects of the implementation of the MPA-MU. We assert that an early understanding of social actors' perceptions regarding ES types, benefits, and access barriers can help inform the design of future management plans while preventing injustices among the social actors (Bennett and Dearden 2014; Cinner et al. 2014; Pascual et al. 2016).

2 The MPA-MU Seno Almirantazgo

The MPA-MU (Ministry of Environment D.S. 11/18) is located in the Timaukel municipality, Province of Tierra del Fuego, in the Magallanes region of Chile, comprising 76,400 ha (Fig. 13.1). The conservation objects ("objetos de conservación"[2]) that guided its creation are flagship species such as the southern elephant seal (*Mirounga leonina*), the leopard seal (*Hydrurga leptonyx*), and the black-browed

[1] The Exclusive Economic Zone (EEZ) is a sea zone prescribed by the 1982 United Nations Convention on the Law of the Sea (UNCLOS) over which a sovereign state has special rights regarding the exploration and use of marine resources, including energy production from water and wind.

[2] The conservation objects or elements are natural systems, communities, rivers, species of flora and fauna, and others, whose importance lies in the fact that they are used as criteria for making decisions about the conservation, management, and protection of natural resources.

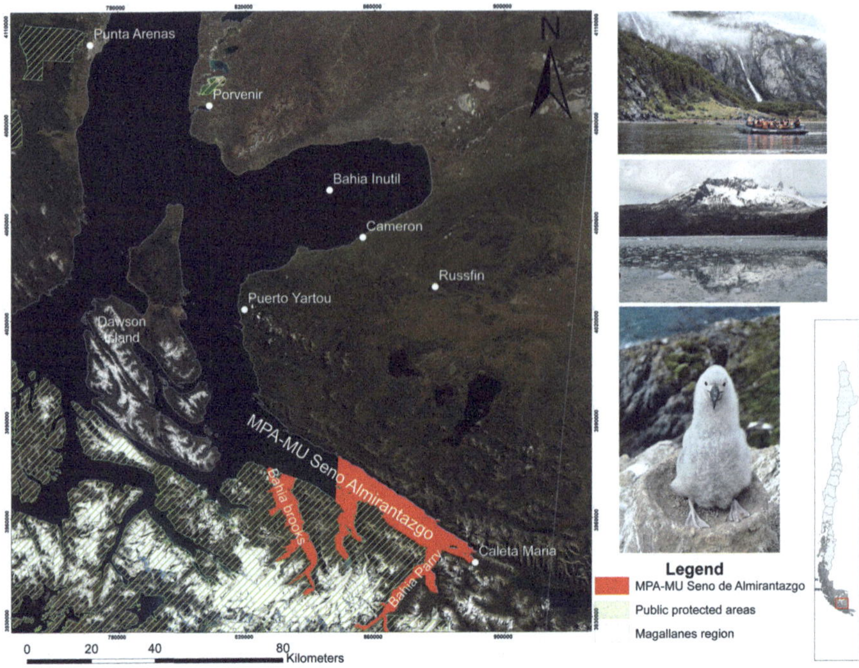

Fig. 13.1 Location of the MPA-MU Seno Almirantazgo (in red). Public protected areas neighboring the MPA-MU are shaded in green. (Photo credits: Daniela Droguett)

albatross (*Thalassarche melanophris*). At present, its management plan is being elaborated by the Ministry of Environment of Chile and the Wildlife Conservation Society (https://chile.wcs.org) using Open Standards protocols and methodologies, which include the collection of opinions and interests from the different social actors that converge in the area (CMP 2013).

Activities that take place in the area include the artisanal extraction of benthic resources, such as the Patagonian scallop (*Zygochlamys patagonica*) and the southern scallop (*Austrochlamys natans*), as well as many tourism activities (Vila et al. 2017) managed by large tour operators that carry over 2000 passengers every season. Scallop artisanal fishers operating in Seno Almirantazgo reside in the cities of Porvenir and Punta Arenas, 31 and 50 nautical miles away from the area, respectively (SHOA 1997), and stay in the area only during the fishing season. Nature-based tourism[3] is oriented to high-income visitors and is operated by few companies, e.g., Australis Cruises Company is the largest one in terms of tourists' landings in the past 20 years (Kirk et al. 2018).

[3] Nature tourism – responsible travel to natural areas, which conserves the environment and improves the welfare of local people. It is tourism based on the natural attractions of an area. Examples include birdwatching, photography, stargazing, camping, hiking, hunting, fishing, and visiting parks.

3 Interview Design and Application to Assess Social Actors' Perceptions

This chapter builds on part of the information collected by Brain et al. (2020), and therefore, methodological details can be found there. Here, we briefly reproduce the qualitative part of the methodology used in that study. We used an initial list of potential social actors and the "snowball" technique, through which we reached to 86 interviewees between March and December of 2018, including 34 artisanal fishers, 19 researchers from NGOs and universities, 14 state representatives, and 19 tour operators.

The interviews (semi open-ended) collected data on (i) ES and perceived benefits, (ii) level of dependence on ES and benefits, and (iii) potential access barriers for the ES. We handed the interviewees a pre-elaborated list of ES based on the Oceans and Coasts TEEB typology (TEEB 2010) and benefits based on MEA (2005) and the Intergovernmental Science-Policy Platform on Biodiversity and Ecosystem Services (IPBES) (nature's contributions to people) (Díaz et al. 2018) (Table 13.1). We asked the respondents to identify and prioritize ES and benefits from this list, as well as their level of dependence on them. We also asked respondents to identify access barriers, which were previously typified based on the Theory of Access by Ribot and Peluso (2003). These authors developed a concept of access and examined a broad set of factors that differentiate access from property. They defined access as the ability to derive benefits from things, broadening from property's classical definition as the right to benefit from things. Access, following this definition, is more similar to a bundle of powers than to property's notion of a bundle of rights (Ribot and Peluso 2003). Social actors' priorities and dependency regarding ES and benefits as well as barriers were analyzed using analysis of frequency.

4 Identification and Prioritization of Ecosystem Services and Benefits by Social Actors

Social actors recognized that the MPA-MU could potentially provide all the ES presented in Table 13.1. From that list, artisanal fishers prioritized food provision, followed by maintenance of genetic diversity, and information for cognitive development. The prioritization of food provision by fishers is in line with previous findings that reveal the higher importance that social actors place on direct use ES (Kari and Korhonen-Kurki 2013; Iniesta-Arandia et al. 2014). Provisioning ES are easier to relate to local economic activities and economic well-being, as stated by fishers: "It is our surviving wage," "It is our main economic activity" (artisanal fisher, Bahía Mansa, September 2018). Consequently, fishers most often highlighted basic materials for a good life (Fig. 13.2) as the main benefits associated with the prioritized ES.

Table 13.1 List of potential ES and benefits presented to the interviews

Service category	Ecosystem services (TEEB 2010)	Benefits (MEA + IPBES)
Provision	Food provision	Basic materials for a good life
	Medicinal resources	
	Ornamental resources	
Regulation	Moderation of extreme events	Security
		Health
		Non-direct contribution
		Basic materials for a good life
		Security
	Waste assimilation	Health
	Pest control	–
Supporting (Habitat in TEEB 2010)	Maintenance of life cycles of migratory species	Non- direct contribution
		Basic materials for a good life
		Leisure
		Existence value
		Non-direct contribution
	Maintenance of genetic diversity	Non-direct contribution
		Existence value
		Basic materials for a good life
		Leisure
Cultural	Opportunities for recreation and tourism	Leisure
		Basic materials for a good life
		Leisure
	Aesthetic information	Symbolic meaning
		Spiritual value
		Basic materials for a good life
		Spiritual value
		Existence value
	Sense of place	Identity
		Freedom of choice and action
		Education
		Identity
		Basic materials for a good life
	Spiritual experience	Spiritual value

(continued)

Table 13.1 (continued)

Service category	Ecosystem services (TEEB 2010)	Benefits (MEA + IPBES)
		Identity
		Existence value
	Information for cognitive development	Education
		Basic materials for a good life
		Education
		Non-direct contribution
		Education

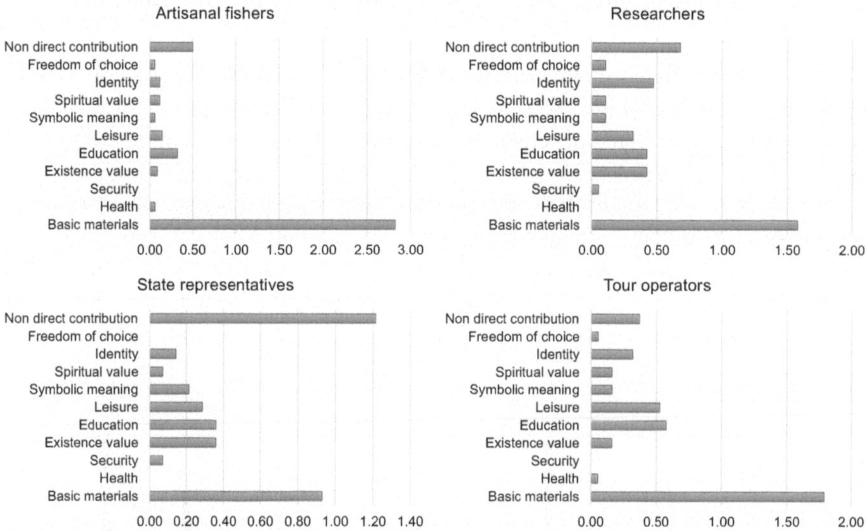

Fig. 13.2 Benefits identified more frequently by each social actor, in relation to the prioritized ecosystem services

State representatives prioritized the food provision and the maintenance of genetic diversity, followed by maintenance of the life cycle for migratory species. Among the benefits, the state representatives prioritized non-direct contributions to well-being, such as the intrinsic value of the species.

Researchers gave equal priority to food provision and maintenance of genetic diversity, followed by information for cognitive development. Regarding benefits, the researchers prioritized basic materials for a good life and existence values.

Tour operators prioritized the maintenance of genetic diversity, followed by information for cognitive development and food provision; and among the benefits, they prioritized basic materials for a good life.

The prioritization of maintenance of genetic diversity in all respondent groups can be attributed to the exposure of regional social actors to conservation speeches by NGOs and environmental authorities. As in other countries, NGOs and governmental conservation agendas are still primarily focused on biological conservation, habitat protection, emblematic species, and heritage preservation (Martín-López et al. 2009), which continue to be the prevalent ecological indicators to judge the public conservation success (Araos and Ther 2017). This is also reflected in the way that the press media announced the creation of the MPA-MU in 2018, highlighting the value of the Seno Almirantazgo as "A key location for species such as the southern elephant seal, the leopard seal and the black-browed albatross" (Ministry of Environment, Chile, 2018).

5 Level of Dependence on Ecosystem Services and Benefits

The level of dependence on prioritized ES varied from the absence of uses to extremely dependent (Fig. 13.3). As expected, the highest frequency of responses for very high and extreme dependency corresponded to fishers with regard to food provision.

Fishers perceived themselves and were perceived by the other social actors as the most dependent group of ES and their benefits in the Seno Almirantazgo, since their fishing sites are located within the MPA-MU. An artisanal fisher said "The scallop is the only resource that can be extracted in summer. We do not know what will

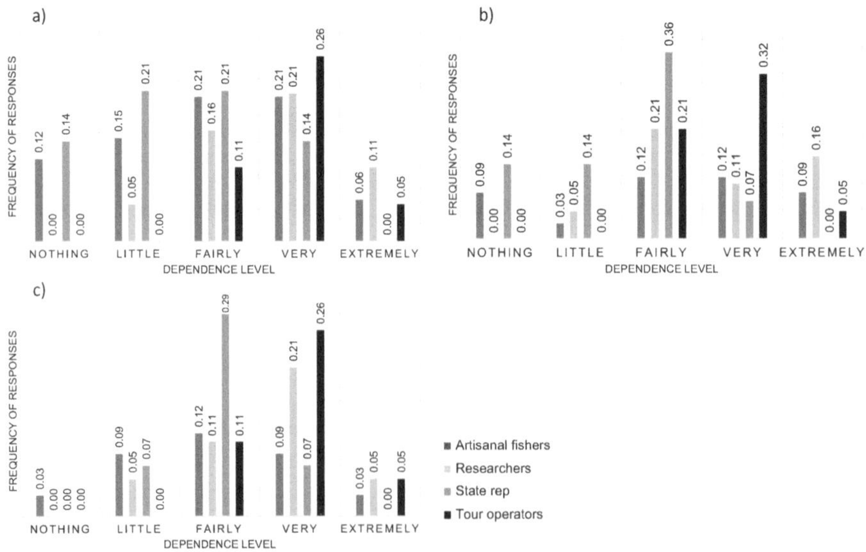

Fig. 13.3 Dependence level on prioritized ES for each group of social actors (dark gray, artisanal fishers; light gray, researchers; gray, state representatives; black, tour operators) as indicated by the relative frequency of response of food provision service, maintenance of genetic diversity service, and information for cognitive development

happen, but if there is strict protection, the current identity of the place will disappear" (Artisanal fisher, Bahía Santa María, October 2018). The level of dependence may partially explain the more pessimistic view of fishers regarding the future of fishing in the area, which contrasts with the optimistic expectations of the state representatives regarding all ES provision. The degree of involvement of state representatives in the design and management of the MPA-MU can explain their standpoint. Likewise, the large-scale tour operators know from experience that the implementation of the MPA-MU can specially benefit them, given the tourist attractions that will become available (e.g., sighting of particular fauna like the southern elephant seal), allowing the sale of attractive tourist packages for international wealthy tourists. Instead, the development of local tourism initiatives based on the MPA-MU is difficult or even unfeasible given the size and remoteness of the MPA-MU, as stated by some fishers: "To be able to develop tourism I would have to make a very large investment, to increase the size of my boat" (Artisanal fisher, Bahía Chilota, October 2018). Instead, in MPA with easier access, fishers can aspire to diversify their income through tourism activities such as tours (Lopes et al. 2015), without incurring major investments.

6 Perceived Access Barriers Generated by the MPA-MU

Generally, the respondents agreed that the implementation of the MPA-MU could create access barriers (Fig. 13.4). Fishers and tour operators most frequently perceived the "entitlements" as a potential barrier (62% and 79%, respectively), understood as limited rights to extract resources or as restrictions to enter the area. When the ability to benefit from something derives from rights attributed by law, custom, or convention, contemporary theorists have usually called it "property" (MacPherson 1978). The following testimony illustrates fishers' perceptions regarding access barriers: "My benefits in relation to the provision service [food from fisheries] and the sense of identity will decrease because there will be strict control to enter the area," and "it will not reduce conflicts between users, it will create them" (Artisanal fisher, Bahía Mansa, September 2018). Law-based property rights include access via permits and licenses, but in the case of the MPA-MU, it is not yet clear which kind of restrictions via property rights could be imposed. Clam fishers already hold permits to extract the species, but the MPA-MU could create another kind of restrictions, which are unknown to fishers.

Constraints established by the specific political-economic and cultural contexts mediate the ability to benefit from resources. This brings into play a number of what Ribot and Peluso (2003) call "structural and relational access mechanisms," among them the access to knowledge and access to authority or decision-makers (Ribot and Peluso 2003). The researchers interviewed perceived "knowledge" as the major barrier (84%), in the sense that a minimum of knowledge/education is needed to access the benefits derived from the MPA-MU. In turn, state representatives perceived "closeness to decision-makers" as the major barrier (79%) for them and other social

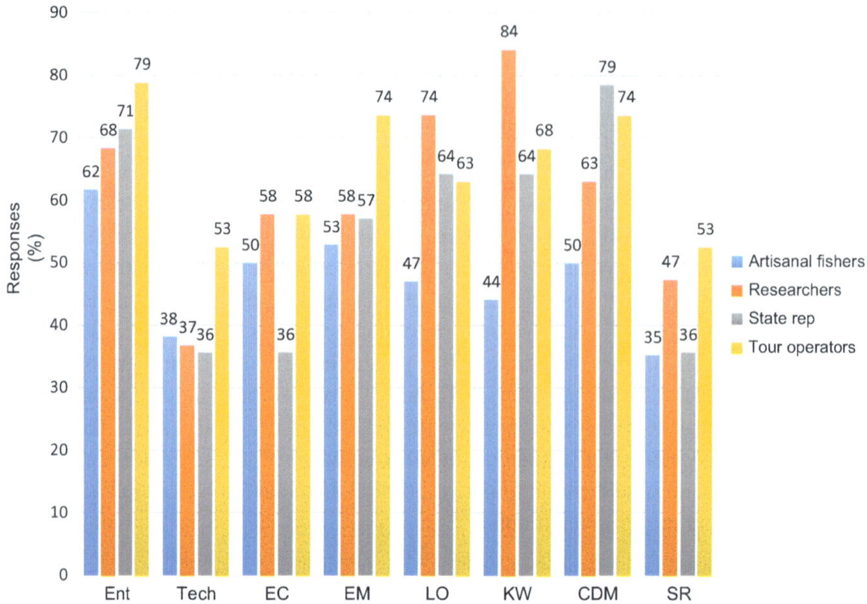

Fig. 13.4 *X*-axis represents the access barriers (Ent entitlements, Tech technology, EC economic capital, EM exclusive market, LO labor opportunity, KW knowledge, CDM closeness to decision-makers, SR social relations), identified by each social actor (orange, artisanal fishers; yellow, researchers; green, state representative; red, tour operators). *Y*-axis represents the frequency of response for each barrier

actors to be able to access the benefits. Access to authority shapes an individual's ability to benefit from resources. Privileged access to the individuals or institutions with the authority to make and implement laws can strongly influence who benefits from the resource in question (Ribot and Peluso 2003).

Because of the interaction between ES and benefits prioritized, the level of dependence, and the access barriers, the MPA-MU can eventually produce different well-being outcomes. Half of the fishers responded that tourism operators would be the primary beneficiaries of the creation of the MPA-MU, although 21% saw themselves as beneficiaries as well (Table 13.2). Researchers and state representatives recognized a wider range of potential beneficiaries, including tourist entrepreneurs, universities, and future generations. Finally, 47% of tour operators responded that they would be the primary beneficiaries of the creation of the MPA-MU, followed by future generations.

Table 13.2 Main beneficiaries identified by each type of respondent

Main potential beneficiary identified by respondents	Percentage of responses from each social actor group regarding MPA-MU primary beneficiary			
	Artisanal fishers	Researchers	State representatives	Tour operators
No beneficiaries	3	0	0	0
Artisanal fishers	21	16	21	5
Industrial fishers	3	0	0	0
Indigenous communities	6	5	0	0
Tourist entrepreneurs	50	26	14	47
Universities and research centers	6	21	21	16
Local economy	0	5	7	0
Future generations	0	16	21	21
Public services	0	0	7	5
Municipalities	6	0	0	0
NGOs	0	11	7	5
Big investors	6	0	0	0

7 Approaching Perceptions and Removing Barriers for a Just Governance

The increase in marine conservation worldwide is widely recognized as a way to achieve ambitious goals of biodiversity and sustainable development (Pendleton et al. 2018). Marine conservation is generally perceived by society as a policy that favors social welfare, while it aims at protecting high-value and at-risk ecosystems. However, this idealization of conservation often generates assumptions that are born from a global perspective and decontextualized from the local reality, ignoring possible adverse effects on local communities and/or direct users of the area to intervene (Grimmel et al. 2019). Marine protected areas are often met with reluctance by affected social actors and in some instances outright objection (e.g., Lopes et al. 2015). Some argue that this is due to insufficient understanding of the functions of MPA. Others suggest that it could be because of a perception that they are losing more than they are gaining. It is also possible that social actors are generally supportive of the idea but think that the MPA should be located elsewhere (Jentoff et al. 2012). Whereas dissimilar perceptions are an essential part of democratic systems, the governance of MPA faces the challenge of bringing them closer, clarifying them, or changing them for the better (Brain et al. 2020).

The perceptions from social actors are closely related to the three dimensions of environmental justice, which are central to acknowledging the achievement of the Sustainable Development Goals (SDGs) (Menton et al. 2020). It is widely recognized that SDG14 (conserve and sustainably use the oceans) is not simply about expanding MPA, but about how this contributes to human well-being, particularly to the people who most depend on marine resources for their livelihoods. MPA governance in the context of marine justice (Martin et al. 2019) urges states and conservation managers to address and answer the following questions:

(a) What are the benefits from marine conservation and how are they to be distributed? Who bears the costs and how are those costs to be prevented or diminished through management actions?

These questions have to do with the distributional dimension of environmental justice, that is, the fairness of distribution of benefits and burdens between different groups, including current and future generations, of the outcomes of conservation actions (Bennet et al. 2020). Our results show that social actors perceive various benefits derived from the ES that the MPA-MU can contribute to increase, but they also perceive a division between winners and losers. The current planning of MPA in general and the MPA-MU in particular does not offer clear answers to these distributional questions. On the one hand, this is because inequality and exclusion continue to be two structural conditions of the development pathways of neoliberal capitalism in Chile and they define the possibilities of a real transformation of marine economies and livelihoods (Araos and Ther 2017). On the other hand, the influence of the human dimension on marine conservation is often underestimated. This particularity results in a limited integration of the social sciences to understand and make effective conservation decision-making during planning, implementation, and management (Bennett et al. 2017; Christie et al. 2017). Besides this, there is still an expectation among both scientists and conservation managers that species can be conserved behind fences away from people, rather than coexisting alongside humans. While policy-makers have started to try to include people in conservation decisions, one criticism of these approaches is that they put ecosystem variables first and then add a limited number of social variables (Mehta and Kellert 1998; Grimmel et al. 2019). This means that significant social factors that contribute to conservation decision-making and responses are overlooked (Stoffle et al. 2010; Christie et al. 2017). This is reflected, for example, in the values of the MPA-MU that are highlighted by government officials, managers, and the press: "A key location for species such as the southern elephant seal, the leopard seal and the black-browed albatross, among others" (Ministry of Environment, Chile, 2018).

To achieve success in a context of distributive justice, conservation planning needs to integrate social research and indicators regarding perceived costs and benefits, values, and attitudes of a wide range of potential social actors (Mascia et al. 2003; Blicharska et al. 2016; Araos and Ther 2017). In this sense, the ES approach can contribute with key elements to address this challenge, if it is recognized as an operational framework in the initial stages of planning.

At present, there is no evidence from the MPA-MU or other MPA in the country of the inclusion of explicit distributive considerations in management plans. Moving towards distributive considerations requires a change in the conception of conservation at the institutional level. This change requires a recognition of the role of shifting property rights, power asymmetries, individual capabilities, and resource dependency in shaping synergies, trade-offs, and equity in conservation outcomes and utilizing multi-consequential frameworks to reinforce the well-being of vulnerable groups (Gill et al. 2019).

(b) Who ought to participate and in what? What level of participation guarantees legitimacy?

These questions have to do with the procedural dimension of environmental justice. Just participation concerns the inclusive and effective participation of all relevant actors and groups in rule and decision-making for conservation policies and programs (Bennett et al. 2020). Social actor participation is a way to legitimate some trade-offs over others, though participation has various meanings and therefore is likely to produce different outcomes (Gustavsson et al. 2014). Participation, along with funding, is one of the most important gaps for conservation in Chile (Gelcich et al. 2015). Participation in conservation planning processes in Chile continues to be passive (imposed) or by consultation, as other authors have described in other latitudes of the country (e.g., Oyanedel et al. 2016) and around the world (Gustavsson et al. 2014). The creation of MPA in Chile, like most conservation policies, obeys the "command and control" approach that relies on top-down laws, regulations, sanctions, and designations. This approach is usually synonymous with low participation at the MPA designation and planning stages.

In order to reconcile perceptions on benefits, costs, and barriers, it is crucial to create spaces for exchange and to ensure an equal representation of all actors in the governing body. Yet, participation has proven particularly challenging in the case of the MPA-MU, where fishers are the great absentees (Vila et al. 2017). The low participation may be due to a lack of conviction of its importance on the part of key social actors such as fishers, which relates to the importance that they and other interviewees give to knowledge access barriers and closeness to decision-makers (Fig. 13.4).

(c) Are all interests accounted for and how can they be handled in the long run?

This question also has to do with recognition justice, which refers to the acknowledgment and representation of the rights, cultures and identities, values and visions, knowledge systems, and livelihoods of local groups in conservation planning and management (Bennett et al. 2020). Biodiversity conservation is linked with recognition injustices for three main reasons. Firstly, protected areas are spatially associated with cultural diversity and with people whose knowledge and environmental governance institutions are vulnerable to being marginalized. Secondly, mainstream conservation management strategies, such as the ones that predominate in Chile, are influenced by culturally specific ideas about what works and about what counts as evidence of what works. MPA often assume a lack of social relations in the sea and overlay them within a scientific matrix of expertise; this expertise determines which spaces should be set aside according to experts in the field of marine conservation, often from a biological science perspective. MPA can then be read as spaces which scientists, as designers of conservation, are withdrawing from public use, regardless of historical and present use values (Chmara-Huff 2014). Thirdly, the dominant designs about how conservation should be done become a basis for the misrepresentation and misrecognition of indigenous and local people (Martin et al. 2016).

Chile has become the leader in marine conservation in South America by achieving Aichi Target 11 in 2017. Despite this achievement, there is an absence of important aspects regarding justice and how to ensure it. In this context, current socio-environmental research on marine conservation in Chile indicates that it is a crucial moment to address the gaps identified above to give legitimacy to conservation efforts, aiming at the effective management of marine spaces. The ES approach can serve this purpose by helping to foresee distributive aspects of conservation and providing sound indicators for equity monitoring.

Acknowledgments This research was funded by FONDAP Grant n° 15150003 (Chile). We thank Ximena Vergara and Gonzalo Campos for their assistance in interview application and figures preparation.

References

Araos F, Ther F (2017) How to adopt an inclusive development perspective for marine conservation: preliminary insights from Chile. Curr Opin Environ Sustain 24:68–72

Bennett NJ, Dearden P (2014) Why local people do not support conservation: community perceptions of marine protected area livelihood impacts, governance and management in Thailand. Mar Policy 44:107–116

Bennett NJ, Teh L, Ota Y, Christie P, Ayers A, Day JC, Franks P, Gill D, Gruby RL, Kittinger JN, Koehn JZ, Lewis N, Parks J, Vierros M, Whitty TS, Wilhelm A, Wright K, Aburto JA, Finkbeiner EM, Gaymer CF, Govan H, Gray N, Jarvis RM, Kaplan-Hallam M, Satterfield T (2017) An appeal for a code of conduct for marine conservation. Mar Policy 81:411–418. https://doi.org/10.1016/j.marpol.2017.03.035

Bennett NJ, Calò A, Di Franco A, Niccolini F, Marzo D, Domina I et al (2020) Social equity and marine protected areas: perceptions of small-scale fishermen in the Mediterranean Sea. Biol Conserv 244:108531

Blicharska M, Orlikowska EH, Roberge JM, Grodzinska-Jurczak M (2016) Contribution of social science to large scale biodiversity conservation: a review of research about the Natura 2000 network. Biol Conserv 199:110–122

Brain MJ, Nahuelhual L, Gelcich S, Bozzeda F (2020) Marine conservation may not deliver ecosystem services and benefits to all: insights from Chilean Patagonia. Ecosyst Serv 45:101170

Chmara-Huff F (2014) Marine protected areas: territorializing objects and subjectivities. EchoGéo 29

Christie P, Bennett NJ, Gray NJ, Wilhelm TA, Lewis NA, Parks J et al (2017) Why people matter in ocean governance: incorporating human dimensions into large-scale marine protected areas. Mar Policy 84:273–284

Cinner JE, Daw T, Huchery C, Thoya P, Wamukota A, Cedras M, Abunge C (2014) Winners and losers in marine conservation: fishers' displacement and livelihood benefits from marine reserves. Soc Nat Resour 27(9):994–1005

Conservation Measures Partnership (CMP) (2013) Open Standards for the Practice of Conservation Version 3.0/April 2013. 252019Available online: http://cmp-openstandards.org/wp-content/uploads/2014/03/CMP-OS-V3-0- Final.pdf. Accessed: May

Davies TE, Maxwell SM, Kaschner K, Garilao C, Ban NC (2017) Large marine protected areas represent biodiversity now and under climate change. Sci Rep 7(1):9569

Davies TE, Epstein G, Aguilera SE, Brooks CM, Cox M, Evans LS et al (2018) Assessing trade-offs in large marine protected areas. PLoS One 13(4):e0195760

Daw TM, Coulthard S, Cheung WW, Brown K, Abunge C, Galafassi D, Peterson GD, McClanahan TR, Omukoto JO, Munyi L (2015) Evaluating taboo trade-offs in ecosystems services and human well-being. Proc Natl Acad Sci 112(22):6949–6954

Day J, Dudley N, Hockings M, Holmes G, Laffoley DDA, Stolton S, Wells SM (2012) Guidelines for applying the IUCN protected area management categories to marine protected areas. IUCN

Díaz S, Pascual U, Stenseke M, Martín-López B, Watson RT, Molnár Z et al (2018) Assessing nature's contributions to people. Science 359(6373):270–272

Gelcich S, Peralta L, Donlan CJ, Godoy N, Ortiz V, Tapia-Lewin S et al (2015) Alternative strategies for scaling up marine coastal biodiversity conservation in Chile. Marit Stud 14(1)

Gill DA, Cheng SH, Glew L, Aigner E, Bennett NJ, Mascia MB (2019) Social synergies, tradeoffs, and equity in marine conservation impacts. Annu Rev Environ Resour 44:347–372

Grimmel H, Calado H, Fonseca C, de Vivero JLS (2019) Integration of the social dimension into marine spatial planning–theoretical aspects and recommendations. Ocean Coast Manag 173:139–147

Gustavsson M, Lindström L, Jiddawi NS, de la Torre-Castro M (2014) Procedural and distributive justice in a community-based managed marine protected area in Zanzibar, Tanzania. Mar Policy 46:91–100

Iniesta-Arandia I, García-Llorente M, Aguilera PA, Montes C, Martín-López B (2014) Socio-cultural valuation of ecosystem services: uncovering the links between values, drivers of change, and human well-being. Ecol Econ 108:36–48

Jentoft S, Pascual-Fernandez JJ, De la Cruz Modino R, Gonzalez-Ramallal M, Chuenpagdee R (2012) What social actors think about marine protected areas: case studies from Spain. Hum Ecol 40(2):185–197

Kari S, Korhonen-Kurki K (2013) Framing local outcomes of biodiversity conservation through ecosystem services: a case study from Ranomafana, Madagascar. Ecosyst Serv 3:e32–e39

Kirk C, Rozzi R, Gelcich S (2018) El turismo como una herramienta para la conservación del elefante marino del sur (mirounga leonina) y sus habitats en tierra del fuego, reserva de la biosfera cabo de hornos, Chile. Magallania (Punta Arenas) 46(1):65–78

Lopes PF, Pacheco S, Clauzet M, Silvano RA, Begossi A (2015) Fisheries, tourism, and marine protected areas: conflicting or synergistic interactions? Ecosyst Serv 16:333–340

Macpherson CB (ed) (1978) Property, mainstream and critical positions, vol 214. University of Toronto Press, Toronto

Martin A, Coolsaet B, Corbera E, Dawson NM, Fraser JA, Lehmann I, Rodriguez I (2016) Justice and conservation: the need to incorporate recognition. Biol Conserv 197:254–261

Martin JA, Gray S, Aceves-Bueno E, Alagona P, Elwell TL, Garcia A et al (2019) What is marine justice? J Environ Stud Sci 9(2):234–243

Martín-López B, Montes C, Ramírez L, Benayas J (2009) What drives policy decision-making related to species conservation? Biol Conserv 142(7):1370–1380

Mascia MB, Claus CA (2009) A property rights approach to understanding human displacement from protected areas: the case of marine protected areas. Conserv Biol 23:16–23

Mascia MB, Brosius JP, Dobson TA, Forbes BC, Horowitz L, McKean MA, Turner NJ (2003) Conservation and the social sciences. Conserv Biol 17(3):649–650

McShane TO, Hirsch PD, Trung TC, Songorwa AN, Kinzig A, Monteferri B et al (2011) Hard choices: making trade-offs between biodiversity conservation and human well-being. Biol Conserv 144(3):966–972

Mehta JN, Kellert SR (1998) Local attitudes toward community-based conservation policy and programmes in Nepal: a case study in the Makalu-Barun Conservation Area. Environ Conserv 25:320–333

Menton M, Larrea C, Latorre S, Martinez-Alier J, Peck M, Temper L, Walter M (2020) Environmental justice and the SDGs: from synergies to gaps and contradictions. Sustain Sci:1–16

Millennium Ecosystem Assessment (2005) Ecosystems and human well-being: synthesis. Island Press, Washington, DC
Oldekop JA, Holmes G, Harris WE, Evans KL (2016) A global assessment of the social and conservation outcomes of protected areas. Conserv Biol 30(1):133–141
Oyanedel R, Marín A, Castilla JC, Gelcich S (2016) Establishing marine protected areas through bottom-up processes: insights from two contrasting initiatives in Chile. Aquat Conserv Mar Freshwat Ecosyst 26(1):184–195
Pascual M, Rossetto M, Ojea E, Milchakova N, Giakoumi S, Kark S et al (2016) Socioeconomic impacts of marine protected areas in the Mediterranean and Black Seas. Ocean Coast Manag 133:1–10
Pendleton LH, Ahmadia GN, Browman HI, Thurstan RH, Kaplan DM, Bartolino V (2018) Debating the effectiveness of marine protected areas. ICES J Mar Sci 75(3):1156–1159
Rees SE, Attrill MJ, Austen MC, Mangi SC, Rodwell LD (2013) A thematic cost-benefit analysis of a marine protected area. J Environ Manag 114:476–485
Ribot JC, Peluso NL (2003) A theory of access. Rural Sociol 68(2):153–181
Rodríguez A, Arcos JM, Bretagnolle V, Dias MP, Holmes ND, Louzao M et al (2019) Future directions in conservation research on petrels and shearwaters. Front Mar Sci 6:94
Schreckenberg K, Franks P, Martin A, Lang B (2016) Unpacking equity for protected area conservation. Parks 22(2):11–26
Servicio Hidrográfico y Oceanográfico de la Armada de Chile (SHOA) (1997). Tablas de distancia. 4ta edición, actualizada al 2008
Stoffle RW, Minnis J, Murphy A, Van Vlack K, O'Meara N, Smith T, McDonald T (2010) Two-MPA model for siting a marine protected area: Bahamian case. Coast Manag 38(5):501–517
TEEB (2010) The Economics of Ecosystems and Biodiversity (TEEB): ecological and economic foundations. Earthscan, London
Vila A, Püschel N, Rodríguez M, Guijón R, Kusch A (2017) Propuesta Área Marina Costera Protegida de Múltiples Usos Seno Almirantazgo. Tierra del Fuego, Región de Magallanes y de la Antártica Chilena. Informe técnico. Wildlife Conservation Society –Chile. Santiago. 124 p. + anexos

Chapter 14
Sociocultural Valuation of Ecosystem Services in Southern Patagonia, Argentina

Pablo L. Peri, Santiago Toledo, Yamina M. Rosas, Leonardo Huertas, Evangelina Vettese, and Guillermo Martínez Pastur

Abstract Understanding sociocultural values towards ecosystem services (ES) facilitates a decision-making process across multiple management objectives. The aim of this chapter was to analyse the stakeholders' perceptions of ES, wellbeing and connectedness to nature at regional level in Southern Patagonia (Santa Cruz Province, Argentina). For this, we designed a questionnaire and conducted 451 face-to-face semi-structured interviews, in which 168 corresponded to local residents and 283 to foreign visitors. Ecosystem services were classified depending on the degree of perceived importance and vulnerability for wellbeing. From this, 12 ES (5 provisioning, 6 cultural and 1 regulating) were perceived as important for wellbeing. Analysing the perceptions of vulnerable ES by each local stakeholder, we found that both groups of locals and decision-makers perceived provisioning services (mainly livestock, fresh water, timber, fishing and shellfish) and regulating (erosion control, habitat for species and climate regulation) as important ones. Survey respondents generally indicated a high level of connectivity with nature

P. L. Peri (✉)
National Agricultural Technology Institute, Río Gallegos, Argentina
e-mail: peri.pablo@inta.gob.ar

S. Toledo
Universidad Nacional de la Patagonia Austral (UNPA), Consejo Nacional de Investigaciones Científicas y Técnicas (CONICET), Río Gallegos, Santa Cruz, Argentina

Instituto Nacional de Tecnología Agropecuaria (INTA), Río Gallegos, Santa Cruz, Argentina
e-mail: peri.pablo@inta.gob.ar

Y. M. Rosas
Laboratorio de Recursos Agroforestales, Centro Austral de Investigaciones Científicas (CADIC), Consejo Nacional de Investigaciones Científicas y Técnicas (CONICET), Ushuaia, Tierra del Fuego, Argentina

L. Huertas · E. Vettese
Instituto Nacional de Tecnología Agropecuaria (INTA), Río Gallegos, Santa Cruz, Argentina

G. Martínez Pastur
Consejo Nacional de Investigaciones Científicas y Técnicas, Centro Austral de Investigaciones Científicas y Técnicas, Ushuaia, Tierra del Fuego, Argentina

P. L. Peri et al. (eds.), *Ecosystem Services in Patagonia*, Natural and Social Sciences of Patagonia, https://doi.org/10.1007/978-3-030-69166-0_14

being similar for both locals and visitors. Our results showed that social perception of values can substantially contribute to identify ES by focusing on the conflicts that emerge among different stakeholder groups. The sociocultural information of the present study can provide important inputs into negotiations in a decision-making process, allowing participants to compare positive and negative impacts of various options for ES management.

Keywords Decision-making process · Landscape planning · Human wellbeing · Stakeholder perception

1 Introduction

There is international recognition that ecosystem services (ES) are directly or indirectly related to ecosystem structures, functions or processes that contribute to human wellbeing (MEA 2005). In Southern Patagonia (Santa Cruz Province), previous studies at regional level reported results of provisioning ES such as timber from native forests (Peri et al. 2019a) and livestock and firewood from silvopastoral systems (Peri and Ormaechea 2013) and regulating ES such as soil carbon (Peri et al. 2018) and nitrogen content (Peri et al. 2019b) and some studies that analysed cultural ES at landscape level (Martínez Pastur et al. 2016; Rosas et al. 2020). However, in areas like Patagonia, sociocultural values such as social needs and perceptions of the stakeholders towards ES usually are poorly investigated in ES assessments (Bryan et al. 2010; Chan et al. 2012). The perceptions of the stakeholders about ES depend on their type of knowledge, place attachment and how they interact with the natural ecosystems surroundings (Russell et al. 2013), e.g. local stakeholders with longer time of residency near protected areas placed more value on the ES provided by their ecosystems (Sodhi et al. 2010).

Using sociocultural valuation enables the assessment of a broad range of ES by making explicit the stakeholders' interests (Chan et al. 2012). A good understanding of the social perception of values is required when designing agricultural and environmental policies to promote multifunctionality taking into account the views of stakeholders with different roles and interests (van Oudenhoven et al. 2012). Geijzendorffer et al. (2015) highlighted the need to analyse the role of multiple stakeholder groups and their relationships with the provision, demand and management of ecosystem to improve sustainable management. Furthermore, the importance of feeling connected to nature or the connections between ecosystems and people facilitate a decision-making process of benefits that ecosystems provide to societies (de Groot et al. 2010). In particular, the cultural ES represent intangible dimensions of the links between people and ecosystems that determine human preferences and values from psychological, social and spiritual aspects (Satz et al. 2013). In this context, human wellbeing surveys can be used to evaluate the importance of ES from stakeholders whose wellbeing is more directly dependent on ES or external people (tourists or visitors) with less dependency on ES (Reed et al. 2009). This chapter presents results from an analysis of stakeholders' perceptions of actual ecosystem services, wellbeing and connectedness to nature at regional level in Southern Patagonia (Santa Cruz Province, Argentina). For this, based on participant observation and semi-structured interviews, we designed a questionnaire and conducted 451 interviews.

2 Study Area and Methodology

The study area of the present work was the whole province of Santa Cruz (243,943 km^2) located between latitudes 46° 00′ and 52° 30′ S (South Patagonia, Argentina) (Fig. 14.1). The main economic activities have been related to mining (e.g. coal, gold, silver), livestock (e.g. notably sheep), agriculture (e.g. crops and fruit production near the Andes Mountains and valleys) as well as oil and gas after its discovery near Comodoro Rivadavia in 1907. Natural steppe grasslands, characterized by the presence of tussock (*Festuca*, *Stipa*), short grasses (*Poa*, *Carex*) and shrubs, occupy near 85% of the land and contribute as a main feed resource for sheep rearing for meat and wool production (Peri et al. 2013). The Andean native forests

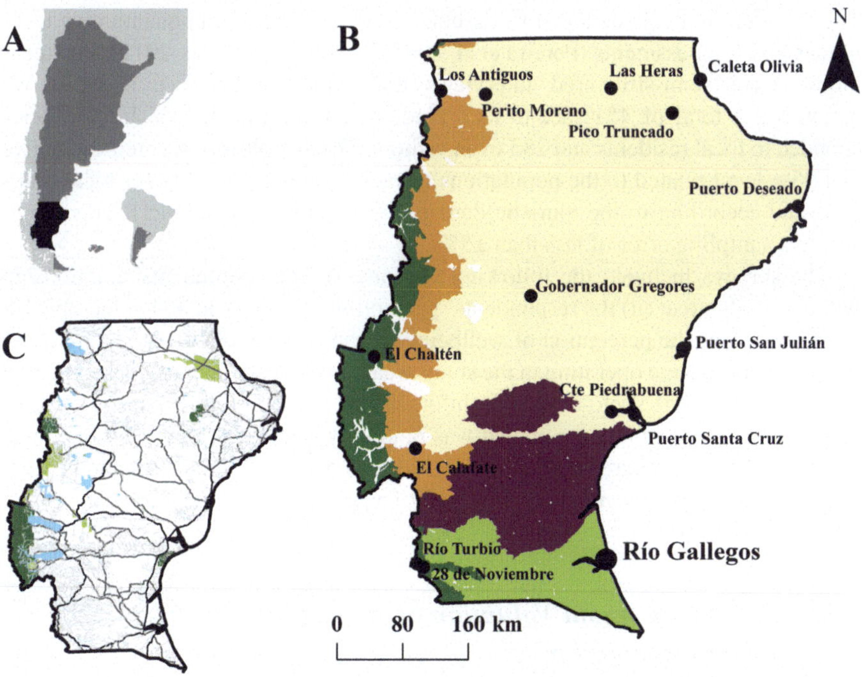

Fig. 14.1 Characterization of the study area: (**a**) location of Argentina (dark grey) and Santa Cruz Province (black); (**b**) provincial cities and main ecological areas (light pink = dry steppe, light green = humid steppe, purple = shrublands, orange = sub-Andean grasslands, dark green = forests and alpine vegetation); (**c**) main roads, rivers and lakes and natural reserve network (dark green = national parks, light green = provincial reserves)

cover a narrow (near 100 km) and long (near 1000 km) strip of land. The southern beeches, lenga (*Nothofagus pumilio*) and ñire (*N. antarctica*), are the most common forest species, covering 335,450 ha. In the irrigated valleys, there are approximately 3600 ha protected by windbreak (1500 km) to allow establishment of fruit trees, pastures and horticultural crops and to protect agricultural crops, livestock and rural houses (Peri and Bloomberg 2002). Instead sheep farming introduced in the late nineteenth century has been the main economic activity, where tourism became an ever more important part of Patagonia's income. The climate in this region is dry, cold and windy. Rainfall decreases from 800 to 1000 mm to 200 mm yr^{-1} from west to east across the Andes Mountains, which act as an orographic barrier to moist winds coming from the west. Temperatures are highest from December to February and present their minimum during June–July. The predominant wind direction is from the south-southwest quarter. Severe and frequent windstorms occur in spring and summer, with wind speeds over 100 km h^{-1}. For distances and climate, Patagonia is sparsely populated with a density of 1.9 inhabitants km^{-2}.

2.1 Survey Design

The research methods included a combination of qualitative and quantitative techniques on ES assessments (Pereira et al. 2005; Martín-López et al. 2012). The techniques used semi-structured interviews and direct face-to-face surveys. We conducted a total of 451 face-to-face randomized surveys, in which 168 corresponded to local residents and 283 from visitors. The sample size was representative at a 95% level related to the populations in each provincial locality for local's surveys and according to the touristic demands for visitors (Table 14.1). This represented a sampling error of less than ±5%.

The surveys included the following sections: (i) the respondents' relationship with the study area; (ii) the respondents' perception of important and vulnerable ES in the area; (iii) the perception of wellbeing by the residents in the study area; (iv) the drivers of change operating in the study area; (v) the respondents' environmental behaviour; and (vi) socio-economic information. For both questionnaires, locals and visitors, the population sampled was randomly selected to cover a wide range of respondents' backgrounds (age, sex, type of work). More participants were female (57.4%) and the average age was 39.1 years.

2.2 Identification and Valuation of Important and Vulnerable ES

The first part of the questionnaire used in the survey was designed to explore the knowledge of local people with the study area and their existing knowledge about ES delivery. In the second part, each respondent selected the three main ES (provisioning, regulating and cultural) and determined which are the most important and

Table 14.1 Distribution of surveys carried out in Santa Cruz Province sorted by cities and towns for locals according to total population (234,132 people) and for visitor related to touristic demand

City/town	Population %	Surveys for locals	% touristic demand	Surveys for visitors
Río Gallegos	35	52	18	48
Caleta Olivia	19	29	8	22
Pico Truncado and Las Heras	15	23	1	3
El Calafate and El Chaltén	8	18	42	118
Puerto Deseado	5	11	9	28
Río Turbio and 28 de Noviembre	6	12	5	12
Puerto San Julián	3	6	6	15
Piedra Buena and Pto. Santa Cruz	4	8	6	16
Perito Moreno and Los Antiguos	3	5	5	14
Gobernador Gregores	2	4	2	7
Total	**100**	**168**	**100**	**283**

vulnerable for wellbeing from a panel designed with examples and pictures of the potential ES provided by the studied area. The list of ES (Table 14.2) was derived from interviews to experts, bibliography and classifications used in previous studies (MEA 2005; CICES 2013). The panels with images (pictures) were chosen as a means to facilitate respondents' comprehension of ES.

Then, ES were classified into four types using an importance-vulnerability matrix: critically perceived as both important for wellbeing and vulnerable (score 4), important but not vulnerable (score 3), vulnerable but not important (score 2) and less relevant neither are perceived as important for wellbeing nor as vulnerable (score 1) (Palomo et al. 2011). The aim of the importance-vulnerability matrix was to prioritize ES in the study area according to how they are perceived by the stakeholders. We calculated the median number of respondents, expressed in percentages, who perceived the ES' importance and vulnerability; we then used those figures as cut values to decide which ES were highly perceived as important or vulnerable.

2.3 Local Perceptions of Wellbeing

A section of the questionnaire explored the local respondents' wellbeing through a set of 20 items related to the 5 components (basic materials for a good life, health, good social relations, security and freedom of choice and action of human wellbeing) identified in the Millennium Ecosystem Assessment (MEA 2005). These items were also measured on a Likert scale, ranging from 1 (completely disagree) to 4 (completely agree). Although wellbeing was measured at an individual level, some items were related to the perceptions on the community performance since wellbeing is a multidimensional concept. To examine the responses regarding wellbeing,

Table 14.2 Potential ecosystem services detected as provided in Southern Patagonia and included in the direct face-to-face questionnaires conducted. Images (pictures) were included for each service type to facilitate respondents' comprehension of ES

Category	Service division	Service group	Service type	Example in Southern Patagonia
Provisioning	Nutrition	Biomass	Traditional agriculture	Fruit trees, berries, lucerne
			Intensive agriculture (greenhouse)	Tomato, lettuce, strawberry, chard
			Livestock	Sheep, cow
			Fishing and shellfish	Trout, snook, spider crab, mussels
			Forest harvesting	Mushrooms, berries
			Medicinal, therapeutic products	Honey, infusion for tea
		Water	Fresh water	Water for agriculture and human consumption
	Materials	Biomass	Timber	*Nothofagus* and poplar wood
			Construction materials	Stones, boulders, sands
	Energy	Renewable abiotic energy sources	Clean energy	Wind power and solar energy
Regulating	Maintenance of physical, chemical, biological conditions	Atmospheric composition and climate regulation	Climate regulation	CO_2 sequestration from vegetation
		Pest and biological invasions control	Reduction in incidence, risk	Invasive alien species (*Hieracium praealtum*, *Taraxacum officinale*)
		Soil formation and composition	Soil fertility	Water courses and riversides, litter
		Lifecycle maintenance, habitat and gene pool protection	Habitat for species	Natural protected areas for huemul (*Hippocamelus bisulcus*), carpenter woodpecker (*Colaptes pitius* and *Picoides lignarius*), chinchillón anaranjado (*Lagidium wolffsohni*)
			Pollination	Pollinating insects

(continued)

Table 14.2 (continued)

Category	Service division	Service group	Service type	Example in Southern Patagonia
	Mediation of flows	Liquid flows	Water regulation	Riparian vegetation, water infiltrations
		Mass flows	Erosion control	Desertification, deforestation, vegetation cover threshold
Cultural	Spiritual and symbolic interactions with ecosystems	Other cultural outputs	Existence	Satisfaction for species conservation: huemul (*Hippocamelus bisulcus*)
			Tranquility and relaxation	Water, snow, forest and mountainous landscapes
	Physical and intellectual interactions with ecosystems	Intellectual and representative interactions	Traditional knowledge	Traditional *boleadoras* for hunting animals, ethnographic museums, animal herding
			Environmental education and scientific knowledge	Books, research and activities about the environment and traditions in Patagonia
			Aesthetic enjoyment	Beautiful landscapes
			Local identity	Cook an entire lamb across an iron cross over an open fire
		Physical and experiential interactions	Recreational hunting and fishing	Small-game and big-game hunting (hare, fox, goose, guanaco) and fishing (trout)
			Nature tourism	Hiking, horse riding, mountain activities
			Rural tourism	Related to traditional sheep stations, gastronomy and agro-tourism

we first used Cronbach's alpha test (Cronbach 1951) to analyse the internal consistency of the 20 wellbeing items. Then, we performed a hierarchical cluster analysis (HCA) to explore how the different components of human wellbeing were perceived and identified.

Respondents were sorted into three stakeholder categories to determine which social actors were affected by changes to ES delivery. These were (i) high degree of influence in decision-making group involved in ES decision-making processes (people with the capacity to affect policies or sustainable development plans such as local mayor, nature protection agents, scientists); (ii) locals dependent on provisioning ES (tour operators, fishermen, farmers, adventure enterprises); and (iii) locals not directly dependent on ES (public workers, residents, local teachers, students, technicians, retired residents). We explored differences in the perceived importance of ES for wellbeing among the stakeholder groups by using the non-parametric Kruskal-Wallis test. Also, to test the links between vulnerability of ES and their effect on human wellbeing, a principal component analysis (PCA) was used. Vulnerability of ES was measured on a Likert scale, ranging from 1 (most vulnerable) to 5 (no vulnerable).

2.4 *Connectedness to Nature*

For both local and visitors, we used the Inclusion of Nature in Self (INS) scale as 'the extent to which an individual includes nature within his/her cognitive representation of self' (Schultz 2001). For this, participants were asked to select from a series of five overlapping circles labelled 'self' and 'nature'. The item read 'Please circle the picture that best describes your relationship with the natural environment. How interconnected are you with nature?' Scores ranged from 1 (where the circles touched but did not overlap) to 5 (where the circles were nearly entirely overlapping). For visitors, the Nature Relatedness Short Version (NR-6) was chosen because it is a widely used measure of the subjective wellbeing and environmental variables as a self-nature connection construct (Nisbet and Zelenski 2013). It displays a similar pattern of correlations as the full 21-item scale (Nisbet and Zelenski 2011). Four items of the NR-6 scale assess the self-identification with nature, a sense of connectedness that may be reflected in spirituality, awareness or subjective knowledge about the environment (e.g. 'My relationship to nature is an important part of who I am'), and two items capture individual differences in the need for nature and comfort with wilderness (e.g. 'I take notice of wildlife wherever I am'). Participants respond to statements using a five-point Likert scale (1 = strongly disagree, 5 = strongly agree), and items are averaged with higher scores indicating stronger connectedness.

3 Results

3.1 *Identification and Valuation of Important and Vulnerable ES*

Ecosystem services were classified depending on the degree of perceived importance and vulnerability for wellbeing. Twelve ES (5 provisioning, 6 cultural and 1 regulating) were perceived as important for wellbeing. Clean energy, fresh water, aesthetic values, tranquillity and relaxation, livestock, traditional knowledge, nature tourism, environmental education, existence, fishing and shellfish, intensive

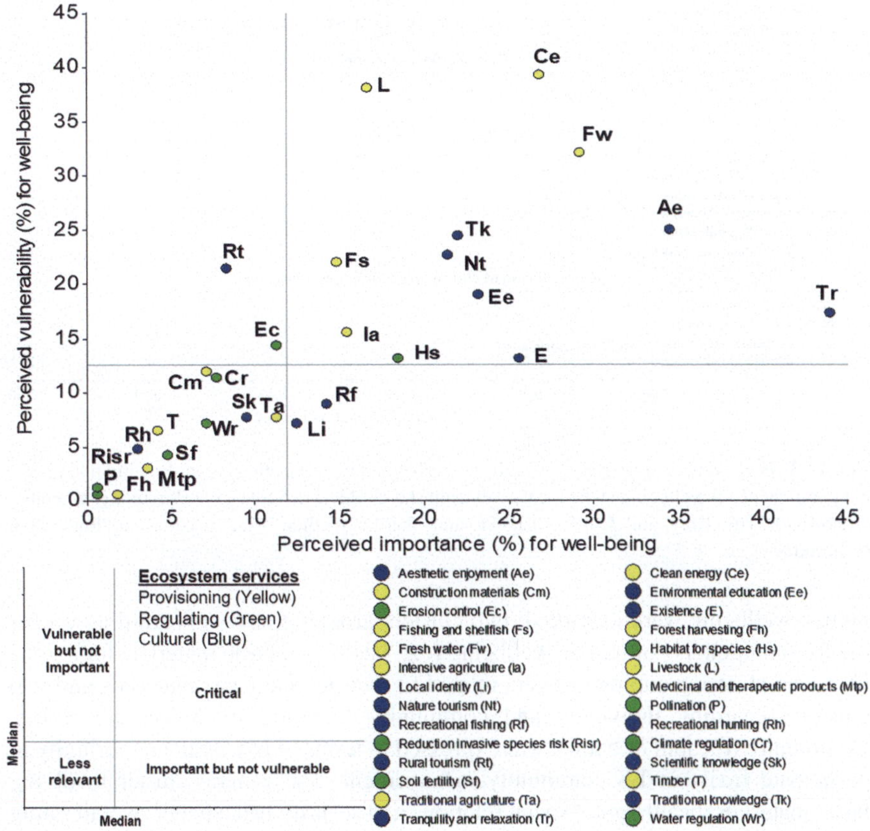

Fig. 14.2 Scatter plots representing the perceived importance of ecosystem services for wellbeing (x-axis) and the perceived vulnerability (y-axis) in Santa Cruz Province, Patagonia, Argentina

agriculture and habitat for species were the critical ES. The important but not vulnerable category was characterized by cultural ES including recreational fishing and local identity (Fig. 14.2). The category of vulnerable but not important ES included rural tourism and erosion control. Lastly, in the category of less relevant services, we found some regulating (e.g. pollination), some provisioning (e.g. forest harvesting) and some cultural services (e.g. recreational hunting).

3.2 *Local Perceptions of Wellbeing*

The reliability for the human wellbeing for the 20 items determined by Cronbach's alpha was 0.857, suggesting that the different dimensions of human wellbeing were highly intercorrelated. The HCA shows how different components of wellbeing relate to each other (Fig. 14.3). The following five main groups of dimensions of

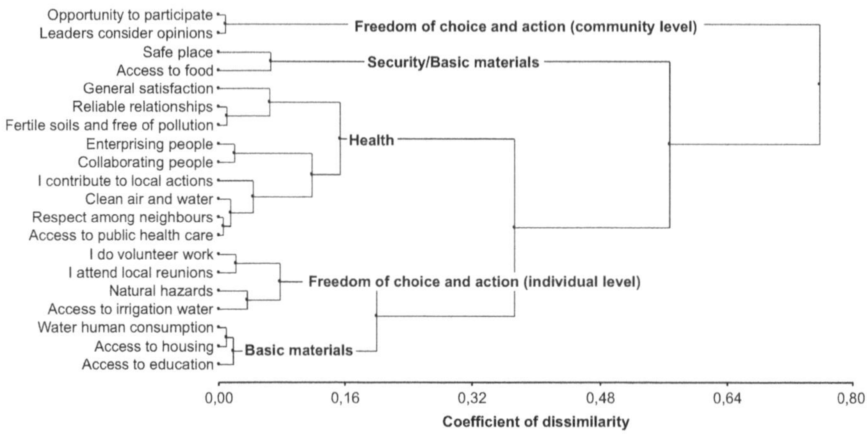

Fig. 14.3 Hierarchical cluster analysis (HCA) performed with questions regarding the five different components of wellbeing at the local level from stakeholders in Santa Cruz Province (Patagonia, Argentina). The Bray and Curtis distance and Ward's method were used as agglomerative techniques

human wellbeing were identified: two clusters grouping answers regarding one of the five components of human wellbeing (e.g. health and basic materials for a good life), one cluster grouping answers related to security and basic materials and two clusters regarding individual and community freedom of choice and action. Community freedom of choice was expressed in terms of having the opportunity to participate freely in ES community management. The general cluster involving basic materials, health, good social relations and security had a better overall rating than those regarding freedom of choice and action (Table 14.3).

Table 14.4 shows the results of the Kruskal-Wallis tests where the perceptions of vulnerable ES by each stakeholder group were compared. The local stakeholder groups comprised local dependent on provisioning ES and locals not directly dependent on ES. Both groups consisted of people who resided in the study area, and their level of knowledge and familiarity with the study area was high. The third group is related to a high degree of ES decision-making processes. Both groups of locals and decision-makers perceived provisioning services (mainly livestock, fresh water, timber and fishing and shellfish) and regulating (erosion control, habitat for species and climate regulation) as important (Table 14.4). Although all groups perceived the cultural services of existence and environmental education as important, traditional knowledge was significantly more relevant for local dependent on provisioning ES (Table 14.4).

The links between vulnerability of ES and their effect on human wellbeing among stakeholders are presented in Fig. 14.4. The first PCA axis (74.9% of the total variance) represented in the positive loadings mainly the perceptions of vulnerable cultural services (particularly aesthetic values and traditional knowledge) associated with locals dependent on provisioning ES and locals not directly dependent on provisioning ES (Fig. 14.4). The second PCA axis (15.9% of the total variance)

Table 14.3 Mean and standard deviation (S.D.) of the set of 20 items related to the 5 components of human wellbeing identified in the Millennium Ecosystem Assessment (MEA 2005): the basic materials for a good life, health, good social relations, security and freedom of choice and action. Items are ordered by the preference score (1–4) obtained in the questionnaires

Item statement	Dimension of wellbeing	Mean	S.D.
I have access to food	Basic materials	3.67	0.62
It is a safe place to live	Security	3.48	0.73
I have access to housing	Basic materials	3.14	1.06
I have access to fresh water for consumption	Basic materials	3.08	1.07
I have everything to live happily	Basic materials	3.07	0.84
I have access to education	Basic materials	2.99	1.04
I have access to fresh water for irrigation	Basic materials	2.86	1.17
Water and air are clean and unpolluted	Health	2.75	0.96
Soils are fertile and free of pollution	Health	2.73	0.86
I contribute to local causes or charity actions in my community	Freedom of choice and action (individual level)	2.69	1.06
I have access to the public health system	Health	2.68	0.97
There are good relations among the neighbours in town	Good social relations	2.66	0.86
Neighbours respect each other	Good social relations	2.64	0.98
It is probable that a natural accident could happen in the future (landslides, fires, floods)	Security	2.51	1.14
Neighbours take initiative	Freedom of choice and action (community level)	2.48	0.93
We collaborate to improve the village	Good social relations	2.42	0.90
I volunteer in activities for the benefit of the town	Freedom of choice and action (individual level)	2.29	1.13
I participate in meetings about town issues	Freedom of choice and action (individual level)	2.12	1.12
I have the opportunity to participate in the decision-making process	Freedom of choice and action (community level)	1.67	0.82
The municipality leaders take into account my opinion	Freedom of choice and action (community level)	1.65	0.82

represented in the positive loadings the group of decision-makers associated with perceptions of regulating ES (particularly climate regulation) as vulnerable ES (Fig. 14.4).

3.3 Connectedness to Nature

Survey respondents generally indicated a high level of connectivity with nature being similar for both locals and visitors (Fig. 14.5). For the self-and-nature circles, only about <2% of the respondents indicated that they felt separate from nature,

Table 14.4 Perceived importance of ecosystem services for wellbeing considered by stakeholders, in percentage (%), and differences among stakeholders calculated by the Kruskal-Wallis test

Ecosystem services	Stakeholders			
	High degree of influence in decision-making	Locals dependent on provisioning ES	Locals not directly dependent on ES	Kruskal-Wallis
Provisioning				
Traditional agriculture	12.5	7.8	13.6	$X^2 = 0.39$
Intensive agriculture	18.8	14.1	10.2	$X^2 = 0.35$
Livestock	18.8	20.3	21.6	$X^2 = 0.20$
Fishing and shellfish	18.8	29.7	14.8	$X^2 = 2.58$
Forest harvesting	6.3	3.1	10.2	$X^2 = 0.56$
Construction materials	0.0	1.6	5.7	$X^2 = 0.25$
Fresh water	31.3	37.5	39.8	$X^2 = 0.18$
Clean energy	12.5	10.9	15.9	$X^2 = 0.31$
Timber	31.3	25.0	22.7	$X^2 = 0.22$
Medicinal and therapeutic products	12.5	7.8	8.0	$X^2 = 0.06$
Regulating				
Climate regulation	37.5	23.4	19.3	$X^2 = 1.64$
Habitat for species	37.5	31.3	30.7	$X^2 = 0.32$
Water regulation	6.3	18.8	27.3	$X^2 = 2.18$
Erosion control	37.5	26.6	38.6	$X^2 = 1.07$
Soil fertility	43.8	23.4	23.9	$X^2 = 1.46$
Reduction invasive species risk	0.0	4.7	8.0	$X^2 = 0.31$
Pollination	0.0	1.6	4.5	$X^2 = 0.15$
Cultural				
Existence	31.3	29.7	23.9	$X^2 = 0.33$
Traditional knowledge	12.5	42.2	26.1	$X^2 = 4.47$**
Tranquility and relaxation	25.0	14.1	18.2	$X^2 = 0.58$
Local identity	0.0	12.5	11.4	$X^2 = 0.64$
Environmental education	37.5	17.2	23.9	$X^2 = 1.45$
Scientific knowledge	12.5	12.5	13.6	$X^2 = 0.01$
Nature tourism	0.0	7.8	5.7	$X^2 = 0.24$
Recreational hunting	12.5	20.3	15.9	$X^2 = 0.37$
Recreational fishing	18.8	18.8	10.2	$X^2 = 1.01$
Rural tourism	18.8	14.1	6.8	$X^2 = 0.92$
Aesthetic enjoyment	6.3	21.9	29.5	$X^2 = 2.62$

**Indicates statistical significance at the $p < 0.05$

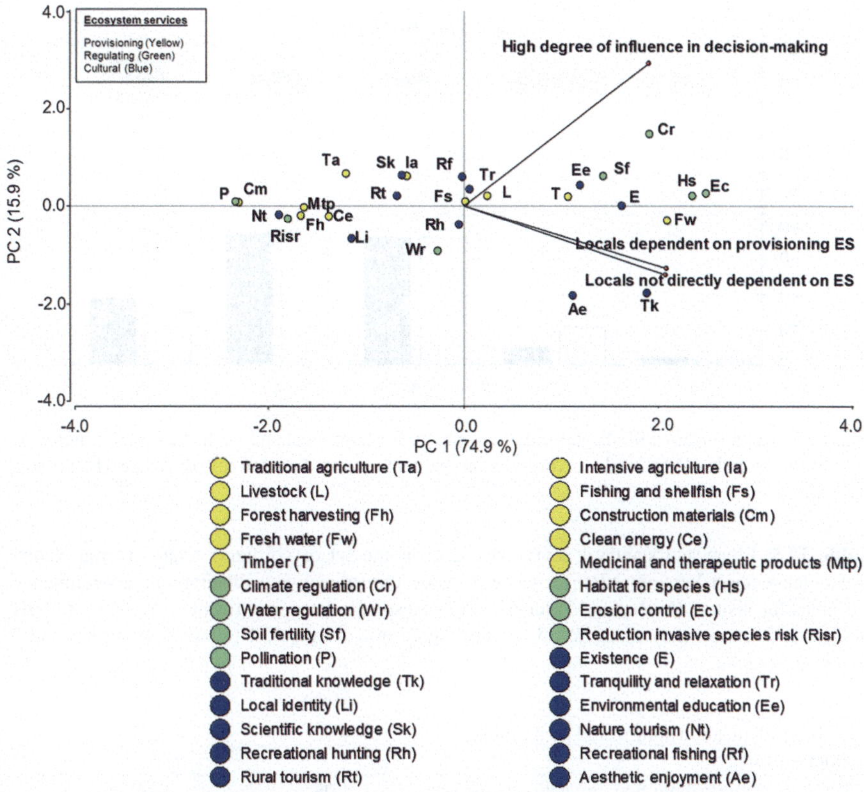

Fig. 14.4 Principal component analysis (PCA) plot of stakeholders and vulnerable ecosystem services in Santa Cruz Province (Patagonia, Argentina)

whereas more than 54% (52.4% for locals and 56.5% for visitors) identified with the circles depicting self and nature as mostly the same (scores 4 and 5, Fig. 14.5).

The Nature Relatedness Short Version (NR-6) analysis determined that visitors scored a high self-nature connection (Table 14.5). From this the highest score was related to individual differences in the need for nature and comfort with wilderness, as well as awareness of local wildlife (e.g. 'my ideal vacation spot would be a remote, wilderness area').

4 Discussion

We found that 12 ES (5 provisioning, 6 cultural and 1 regulating) were perceived among stakeholders as important for wellbeing. Usually, people tend to identify ES that can be perceived by the senses or more directly linked to the human-made components of landscapes such as agriculture and other extractive activities detected in the present work (Lewan and Söderqvist 2002; Lamarque et al. 2011). However,

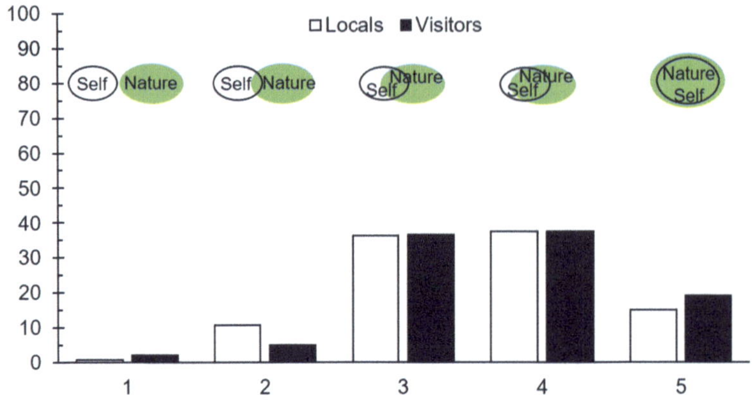

Fig. 14.5 Relationship with the natural environment for both local and visitors using the Inclusion of Nature in Self (INS) scale (expressed in percentage) in Santa Cruz Province (Patagonia, Argentina)

Table 14.5 Mean and standard deviation (S.D.) of the set of six items related to the Nature Relatedness Short Version (NR-6) used to measure the subjective wellbeing and environmental variables for visitors (Nisbet and Zelenski 2013). Response categories for items were 1 (strongly disagree), 2 (moderately disagree), 3 (neither agree nor disagree), 4 (moderately agree) and 5 (strongly agree)

Item	Mean	S.D.
(i) My ideal vacation spot would be a remote, wilderness area	4.02	1.03
(ii) I always think about how my actions affect the environment	4.08	0.99
(iii) My connection to nature and the environment is a part of my spirituality	3.76	1.06
(iv) I take notice of wildlife wherever I am	4.24	0.87
(v) My relationship to nature is an important part of who I am	4.02	0.91
(vi) I feel very connected to all living things and the earth	3.76	1.10

regulating and cultural ES, associated with less tangible components of landscapes, had been also identified by stakeholders in rural systems, as it was the case here for traditional knowledge (Hauck et al. 2013; Martín-López et al. 2012). From these, the provisioning ES of livestock production, fishing, fresh water, clean energy and intensive agriculture were the critical ES and highly related to the semi-arid characteristics of the study area. Regarding livestock production, natural grasslands occupy most of Santa Cruz Province and are the principal food resource for sheep, reared for meat and wool (Peri et al. 2013, 2016). However, in Patagonia over the last 70 years, we have witnessed extensive degradation of once productive steppe ecosystems (desertification) (Golluscio et al. 1998). Thus, heavy and unsustainable grazing conditions threaten the future of livestock productivity, therefore threatening the long-term wellbeing of the local economy (Aguiar and Sala 1998; Bertiller and Bisigato 1998). Regarding fishing as critical, mainly in coastal and marine areas in Patagonia, it may be reflected the unsustainable practices that generate anthropogenic drivers related to food production and marine biodiversity (Rocha et al. 2014). However, there is limited information about ES provided by marine and coastal habitats and ecosystems, creating knowledge gaps about the importance that people assign to these areas (Martin et al. 2016). Provisioning fresh water also was perceived as critical. In the study areas, most important watersheds are located at the Andes where main rivers fed and cross the plateau steppe and outflow to the Atlantic Ocean. People may perceive that livestock, farming, energy and urban and rural populations will be impacted by climate change-induced changes in glacier runoff and therefore less available water coming from glaciers and mountain forests (Cuesta et al. 2019). The glaciers of the Southern Andes showed the highest glacier mass loss rates worldwide with more than 40 m water equivalent over the period 1961–2016 (Zemp et al. 2019). Meier et al. (2018), covering the area between 41 and 56°S, reported an absolute glacierized area loss of 5455 km^2 (19.4%) in the last ~150 years, where the annual area reduction increased by 0.25% for the periods 2005–2016. According to Aylward et al. (2005), this ES provides a great contribution to human wellbeing if society improves the design and management of water resource infrastructure, establishes more inclusive governance and integrated approaches to water management and adopts water conservation technologies and demand management that increase water productivity. Another reason that may explain why provisioning ES were highly identified as critical in the study area relates to the contribution of traditional activities and knowledge not only to food provision but to the delivery of other ES, such as landscape aesthetic values or tranquillity and relaxation as contributions to wellbeing. This is consistent with rural areas suffering from depopulation where traditional agriculture was highly related to the maintenance of local identity and to the contribution of social capital and enhancement of wellbeing (Pereira et al. 2005).

Habitat for species was the only regulating ES perceived as critical. This is consistent with the idea that habitat for rare or endangered species decreases due to several factors related to human activities (e.g. forestry, ranching, mining) or climate change (Newbold 2010; Badiane et al. 2017; Godsoe et al. 2017). For example, for huemul (*Hippocamelus bisulcus*), the most threatened flag species of Southern Patagonia, Rosas et al. (2017) found that habitat losses occurred in the extreme potential distribution areas (northern and southern areas of Santa Cruz Province), related to the increasing ranch activities. Also, Rosas et al. (2018) showed hotspots of lizard biodiversity in the north-east area as related to conditions of desertification due to livestock breeding production. The knowledge of habitat requirement for a target species is a key issue in the management and conservation planning (Villero et al. 2017).

The important but not vulnerable category was characterized by cultural ES including recreational fishing and local identity. Martínez Pastur et al. (2016) reported that local identity in Santa Cruz was mainly related to small cities (e.g. El Chaltén and El Calafate) and areas with special cultural interest (e.g. Cueva de las

Manos UNESCO World Heritage) associated with the presence of flora, terrestrial native fauna, water (e.g. sea coast, lakes and rivers) and human buildings. Recreational fisheries (mainly fishing in fresh water in lakes and rivers by locals and international fly fishermen) are developed throughout Patagonia, both in Chile and Argentina, with a significant local and regional economic impact (Vigliano and Alonso 2000). Although recreational fishing was classified as important but not vulnerable, there are evidences of a rudimentary fresh water stock assessment in Patagonia, a declining quality in several trout recreational fisheries (both in catch rate and size of the fish caught) and the introduction of exotic fishes (Pascual et al. 2007).

When the reliability of human wellbeing on nature was analysed, different dimensions of human wellbeing were highly intercorrelated, e.g. we found that items regarding the basic materials for good life, security, health and good social relations had higher appraisals among stakeholder groups. In contrast, issues regarding freedom of choice and action received lower scores and differences among stakeholder groups. This suggests that differences in beliefs and preferences are also often linked to differences in the power to pursue goals (McShane et al. 2011).

We disaggregated ES values at a stakeholder group level to analyse if perceptions of wellbeing relate to sociocultural values (Table 14.4). We found that both locals and decision-maker groups perceived provisioning services (e.g. livestock, fresh water, timber and fishing) and regulating (e.g. erosion control, habitat for species and climate regulation) as important. We found also that while all groups perceived the cultural services of existence and environmental education as important, traditional knowledge was significantly more relevant only for locals dependent on provisioning ES. Thus, while the local development professionals who resided in the study area (locals not directly dependent on ES) tended to acknowledge mostly the cultural dimensions of land use relating the endangerment of traditional and aesthetic values, the local's dependent on provisioning ES tended to relate it to the degradation of their livelihoods. Relationships between vulnerability of ES and their effect on human wellbeing among stakeholders determined they the perceptions of vulnerable cultural services (particularly aesthetic values and traditional knowledge) associated with locals dependent on provisioning ES and locals not directly dependent on provisioning ES. The group of decision-makers presented perceptions of regulating ES (particularly climate regulation) as vulnerable ES. These divergent stakeholder priorities can be used to visualize possible trade-offs between different ES, mainly because people's willingness to conserve one ES might be at the expense of another (Martín-López et al. 2012).

Both respondents, locals and visitors, indicated a high level of connectivity with nature by using the Inclusion of Nature in Self (INS) scale. Also, the Nature Relatedness Short Version (NR-6) analysis determined that visitors scored a high self-nature connection mainly related to individual differences in the need for nature and comfort with wilderness, as well as awareness of local wildlife. This highlights that environmental values derive from a sense of connectivity with nature and measure a value orientation that underlies environmental concern and behaviour (Dutcher et al. 2007). Nature relatedness contributes to potentially important implications by contributing to human wellbeing as well as environmental sustainability and protection (Saunders 2003). However, cultivating a greater sense of connectivity in the effort to achieve ecological and economic sustainability may require working through and across existing belief systems.

5 Recommendations for Policy Makers

One of the most important challenges in ES is managing the emerging trade-offs for making decisions (Bennett et al. 2009). In this context, our results showed that social perception of values can substantially contribute to identify ES by focusing on the conflicts that emerge among different stakeholder groups. We found that perceptions of the relationships between ES and individual wellbeing varied at stakeholder group level. Divergent sociocultural values among stakeholder priorities can be used to visualize possible trade-offs between different ES. This was because people's willingness to conserve one ES might be at the expense of another. However, value conflicts do not only arise from perceiving different ES, but they can also arise from different content in valuing the same ES, implying contrasting actions or policies (Trainor 2006). Although there were differences in perceptions between stakeholders according to their particular interests, they also shared common views for many ES. Also, we confirmed that sociocultural valuation is case sensitive by detecting differences in perceptions in different areas and stakeholder-sensitive tool by detecting differences in perceptions among stakeholder groups.

Decisions should follow concrete goal or objectives in ways that are meaningful to local residents and to stakeholders, such as promoting sustainability, promoting human wellbeing or achieving better management of a resource, but also the procedures involved in the decision-making process itself, focusing on public processes of inclusion and deliberation (Norton 2005). Although there is a plurality of values associated with the complex and multifaceted services, the sociocultural information of the present study can provide important inputs into negotiations, allowing participants to compare positive and negative impacts of various options for ES management. Since decision-making processes are based on value systems usually derived from particular scientific disciplines, our results became important because they represented the diversity of perspectives from broader segments of society, which are directly affected by the outcomes of such decisions (Lynam et al. 2007). Traditional environmental management strategies that do not take into account stakeholders' and local communities' perceptions of ES on wellbeing can fail by social resistance (Menzel and Teng 2009). Therefore, encouraging experience exchange through participatory mechanisms is very relevant to guarantee that multiple stakeholders with contrasting perceptions regularly interact and discuss about their interests, needs and management of ecosystem services (Kenward et al. 2011). This participatory system could be implemented by local governments and mediated by third-party institutions at different organizational levels and adapted to the cultural and geographical characteristics of each social-ecological system.

References

Aguiar MR, Sala OE (1998) Interactions among grasses, shrubs, and herbivores in Patagonian grass-shrub steppes. Ecol Austral 8:201–210

Aylward B, Bandyopadhyay J, Belausteguigotia JC (2005) Freshwater ecosystem services. Ecosyst Hum Well Being Policy Responses 3:213–254

Badiane A, Matos C, Santos X (2017) Uncovering environmental, land-use and fire effects on the distribution of a low-dispersal species, the Hermann's tortoise Testudo hermanni. Amphib-Reptil 38:67–77

Bennett EM, Peterson GD, Gordon LJ (2009) Understanding relationships among multiple ecosystem services. Ecol Lett 12:1–11

Bertiller MB, Bisigato A (1998) Vegetation dynamics under grazing disturbance. The state-and-transition model for the Patagonian steppes. Ecol Austral 8:191–199

Bryan BA, Raymond CM, Crossman ND, Macdonald DH (2010) Targeting the management of ecosystem services based on social values: where, what, and how? Landsc Urban Plan 97:111–122

Chan KMA, Guerry AD, Balvanera P, Klain S, Satterfield T, Basurto X, Bostrom A, Chuenpagdee R, Gould R, Halpern BS, Levine J, Norton B, Ruckelshaus M, Russell R, Tam J, Chan KM, Guerry D (2012) Where are cultural and social in ecosystem services? A framework for constructive engagement. Bioscience 62:744–756

CICES (2013) The common international classification of ecosystem services. Retrieved from www.cices.eu

Cronbach LJ (1951) Coefficient alpha and the internal structure of tests. Psychometrika 16:297–334

Cuesta F, Llambí LD, Huggel C, Drenkhan F, Gosling WD, Muriel P, Jaramillo R, Tovar C (2019) New land in the Neotropics: a review of biotic community, ecosystem, and landscape transformations in the face of climate and glacier change. Reg Environ Chang 19:1623–1642

de Groot RS, Alkemade R, Braat L, Hein L, Willemen L (2010) Challenges in integrating the concept of ecosystem services and values in landscape planning, management and decision making. Ecol Complex 7:260–272

Dutcher DD, Finley JC, Luloff AE, Buttolph Johnson J (2007) Connectivity with nature as a measure of environmental values. Environ Behav 39:474–493

Geijzendorffer IR, Martín-López B, Roche PK (2015) Improving the identification of mismatches in ecosystem services assessments. Ecol Indic 52:320–331

Godsoe W, Franklin J, Blanchet FG (2017) Effects of biotic interactions on modeled species' distribution can be masked by environmental gradients. Ecol Evol 7:654–664

Golluscio RA, Deregibus V, Paruelo JM (1998) Sustainability and range management in the Patagonian steppe. Ecol Austral 8:265–284

Hauck J, Görg C, Varjopuro R, Ratamäki O, Jax K (2013) Benefits and limitations of the ecosystem services concept in environmental policy and decision making: some stakeholder perspectives. Environ Sci Pol 25:13–21

Kenward RE, Whittingham MJ, Arampatzis S, Manos BD, Hahn T, Terry A, Simoncini R, Alcorn J, Bastian O, Donlan M, Elowe K, Franzén F, Karacsonyi Z, Larsson M, Manou D, Navodaru I, Papadopoulou O, Papathanasiou J, von Raggamby A, Sharp RJA, Söderqvist T, Soutukorva Å, Vavrova L, Aebischer NJ, Leader-Williams N, Rutz C (2011) Identifying governance strategies that effectively support ecosystem services, resource sustainability, and biodiversity. PNAS 108:5308–5312

Lamarque P, Tappeiner U, Turner C, Steinbacher M, Bardgett RD, Szukics U, Schermer M, Lavorel S (2011) Stakeholder perceptions of grassland ecosystem services in relation to knowledge on soil fertility and biodiversity. Reg Environ Chang 11:791–804

Lewan L, Söderqvist T (2002) Knowledge and recognition of ecosystem services among the general public in a drainage basin in Scania, Southern Sweden. Ecol Econ 42:459–467

Lynam T, de Jong W, Sheil D, Kusumanto T, Evans K (2007) A review of tools for incorporating community knowledge, preferences, and values into decision making in natural resources management. Ecol Soc 12:5

Martin C, Momtaz S, Gaston T, Moltschaniwskyj N (2016) A systematic quantitative review of coastal and marine cultural ecosystem services: current status and future research. Mar Policy 74:25–32

Martínez Pastur G, Peri PL, Lencinas MV, García-Llorente M, Martín-López B (2016) Spatial patterns of cultural ecosystem services provision in Southern Patagonia. Landsc Ecol 31(2):383–399

Martín-López B, Iniesta-Arandia I, García-Llorente M, Palomo I, Casado-Arzuaga I, García del Amo D, Gómez-Baggethun E, Oteros-Rozas E, Palacios-Agundez I, Willarts B, González JA, Santos-Martín F, Onaindia M, López-Santiago CA, Montes C (2012) Uncovering ecosystem services bundles through social preferences: experimental evidence from Spain. PLoS One 7:e38970

McShane TO, Hirsch PD, Trung TC, Songorwa AN, Kinzig A, Monteferri B, Mutekanga D, Thang H, Van Dammert JL, Pulgar-Vidal M, Welch-Devine M, Brosius JP, Coppolillo P, O'Connor S (2011) Hard choices: making trade-offs between biodiversity conservation and human wellbeing. Biol Conserv 144:966–972

MEA (Millennium Ecosystem Assessment) (2005) Ecosystems and human wellbeing: current state and trends. Island Press, Washington, DC

Meier WJH, Grießinger J, Hochreuther P, Braun MH (2018) An updated multi-temporal glacier inventory for the Patagonian Andes with changes between the Little Ice Age and 2016. Front Earth Sci 6(62)

Menzel S, Teng J (2009) Ecosystem services as a stakeholder-driven concept for conservation science. Conserv Biol 24:907–909

Newbold T (2010) Applications and limitations of museum data for conservation and ecology, with particular attention to species distribution models. Prog Phys Geogr 34:3–22

Nisbet EK, Zelenski, JM (2011) Bridging sustainability and wellbeing with the NR-6, a brief new measure of nature relatedness. Proceedings 72nd Annual Convention of Canadian Psychological Association, Toronto, Ontario, Canada

Nisbet EK, Zelenski JM (2013) The NR-6: a new brief measure of nature relatedness. Front Psychol 4:813

Norton B (2005) Sustainability: a philosophy of adaptive ecosystem management. University of Chicago Press, Chicago

Palomo I, Martín-López B, López-Santiago C, Montes C (2011) Participatory scenario planning for protected areas management under the ecosystem services framework: the Doñana social-ecological system in southwestern Spain. Ecol Soc 16(1):23

Pascual MA, Cussac V, Dyer B, Soto D, Vigliano P, Ortubay S, Macchi P (2007) Freshwater fishes of Patagonia in the 21st century after a hundred years of human settlement, species introductions, and environmental change. Aquat Ecosyst Health Manag 10:212–227

Pereira E, Queiroz C, Pereira HM, Vicente L (2005) Ecosystem services and human wellbeing: a participatory study in a mountain community in Portugal. Ecol Soc 10:14

Peri PL, Bloomberg M (2002) Windbreaks in South Patagonia: growth models, windspeed reduction and effects of shelter on crops. Agrofor Syst 56:129–144

Peri PL, Ormaechea S (2013) Relevamiento de los bosques nativos de ñire (*Nothofagus antarctica*) en Santa Cruz: base para su conservación y manejo. Ed INTA, Río Gallegos

Peri PL, Lencinas MV, Martinez Pastur G, Wardell-Johnson GW, Lasagno R (2013) Chapter 4: Diversity patterns in the steppe of Argentinean Southern Patagonia: environmental drivers and impact of grazing. In: Morales Prieto MB, Traba Díaz J (eds) Steppe ecosystems: biological diversity, management and restoration. NOVA Science Publishers Inc, New York, pp 73–95

Peri PL, Lencinas MV, Bousson J, Lasagno R, Soler R, Bahamonde H, Martínez Pastur G (2016) Biodiversity and ecological long-term plots in southern Patagonia to support sustainable land management: the case of PEBANPA network. J Nat Conserv 34:51–64

Peri PL, Rosas YM, Ladd B, Toledo S, Lasagno R, Martínez Pastur G (2018) Modelling soil carbon content in South Patagonia and evaluating changes according to climate, vegetation, desertification and grazing. Sustainability 10(2):438

Peri PL, Monelos L, Díaz B, Mattenet F, Huertas L, Bahamonde H, Rosas YM, Lencinas MV, Cellini JM, Martínez Pastur G (2019a) Estado y usos de los bosques nativos de lenga, siempreverdes y mixtos en Santa Cruz: Base para su conservación y manejo. 108 pp. Ed INTA-Consejo Agrario Provincial (CAP), Río Gallegos

Peri PL, Rosas YM, Ladd B, Toledo S, Lasagno RG, Martínez Pastur G (2019b) Modeling soil nitrogen content in South Patagonia across a climate gradient, vegetation type, and grazing. Sustainability 11:2707

Reed MS, Graves A, Dandy N, Posthumus H, Hubacek K, Morris J, Prell C, Quinn CH, Stringer LC (2009) Who's in and why? A typology of stakeholder analysis methods for natural resource management. J Environ Manag 90:1933–1949

Rocha J, Yletyinen J, Biggs R, Blenckner T, Peterson G (2014) Marine regime shifts: drivers and impacts on ecosystems services. Philos Trans R Soc B Biol Sci 370:20130273

Rosas YM, Peri PL, Huertas Herrera A, Pastore H, Martínez Pastur G (2017) Modelling of potential habitat suitability of *Hippocamelus bisulcus* in Southern Patagonia: how much protect the Natural Reserve network of Argentina? Ecol Process 6:28

Rosas YM, Peri PL, Martínez Pastur G (2018) Potential biodiversity map of lizard species in Southern Patagonia: environmental characterization, desertification influence and analyses of protection areas. Amphibia-Reptilia 39:289–301

Rosas YM, Peri PL, Bahamonde H, Cellini JM, Barrera MD, Huertas Herrera A, Lencinas MV, Martínez Pastur G (2020) Chapter 6: Trade-offs between management and conservation for the provision of ecosystem services in the southern Patagonian forests. In: Stanturf J (ed) Achieving sustainable management of Boreal and temperate forests, Burleigh Dodds Series in Agricultural Science, Cambridge, UK, 33 pp

Russell R, Guerry AD, Balvanera P, Gould RK, Basurto X, Chan KMA, Klain S, Levine J, Tam J (2013) Humans and nature: how knowing and experiencing nature affect wellbeing. Annu Rev Environ Res 38:473–502

Satz D, Gould RK, Chan KMA, Guerry A, Norton B, Satterfield T, Halpern BS, Levine J, Woodside U, Hannahs N, Basurto X, Klain S (2013) The challenges of incorporating cultural ecosystem services into environmental assessment. Ambio 42:675–684

Saunders CD (2003) The emerging field of conservation psychology. Hum Ecol Rev 10:137–149

Schultz PW (2001) The structure of environmental concern: concern for self, other people, and the biosphere. J Environ Psychol 21(4):327–339

Sodhi NS, Ming T, Cagan L (2010) Local people value environmental services provided by forested parks. Biodivers Conserv 19:1175–1188

Trainor SF (2006) Realms of value: conflicting natural resource values and incommensurability. Environ Values 15:3–29

van Oudenhoven APE, Petz K, Alkemade R, Hein L, de Groot RS (2012) Framework for systematic indicator selection to assess effects of land management on ecosystem services. Ecol Indic 21:110–122

Vigliano PH, Alonso M (2000) Potencial económico de la pesca recreacional en la Argentina: Una forma de pesca artesanal poco conocida y su posible impacto en economías regionales de países no desarrollados. Gayana Zool 64:109–114

Villero D, Pla M, Camps D, Ruiz-Olmo J, Brotons L (2017) Integrating species distribution modelling into decision-making to inform conservation actions. Biodivers Conserv 26:251–271

Zemp M, Huss M, Thibert E, Eckert N, McNabb R, Huber J, Barandun M, Machguth H, Nussbaumer SU, Gärtner-Roer I, Thomson L, Paul F, Maussion F, Kutuzov S, Cogley JG (2019) Global glacier mass changes and their contributions to sea-level rise from 1961 to 2016. Nature 568(7752):382

Chapter 15
Looking Beyond Ecosystem Services Supply: Co-production and Access Barriers in Marine Ecosystems of the Chilean Patagonia

Ximena Vergara, Alejandra Carmona, and Laura Nahuelhual

Abstract In this chapter, we propose a framework of analysis based on the ecosystem service cascade model, to describe and explore the distribution mechanisms of ecosystem services (ES) and their benefits and to inquire into the processes of co-production, capture and access to benefits. As a case study, we selected the Magallanes region in the southern Chilean Patagonia. We mapped five ES: sense of place, food from aquaculture, recreational opportunities, food from artisanal fisheries, and education and knowledge opportunities. For each ES, we determined the number and location of direct and indirect beneficiaries using the Final Ecosystem Goods and Services Classification System (FEGS-CS). Each ES showed a distinctive pattern of supply distribution, and most of the marine space presented low to very low values in all indicators. The areas where ES indicators increased corresponded to small areas and did not necessarily overlap, suggesting few spatial positive and negative synergies between ES. This dispersed distribution of the ES did not coincide with the highly concentrated locations of direct and indirect beneficiaries in the four cities of the region. Moreover, a large part of the ES supply were captured by foreign entities, either by foreign companies in the case of food from artisanal fisheries and aquaculture or by extra-regional beneficiaries in the case of

X. Vergara (✉)
Centro de Investigación en Dinámica de Ecosistemas Marinos de Altas Latitudes, Universidad Austral de Chile, Valdivia, Chile
e-mail: ximena.vergara@uach.cl

A. Carmona
Centro de Educación Continua, Universidad Austral de Chile, Valdivia, Chile

L. Nahuelhual
Instituto de Economía, Universidad Austral de Chile Centro de Investigación: Dinámica de Ecosistemas Marinos de Altas Latitudes Instituto Milenio en Socio-Ecología Costera, Valdívia, Chile

P. L. Peri et al. (eds.), *Ecosystem Services in Patagonia*, Natural and Social Sciences of Patagonia, https://doi.org/10.1007/978-3-030-69166-0_15

recreation and education and knowledge opportunities. The analysis of co-production and access mechanisms evidences that past policies have favored a pattern where the regional population is deprived of accessing the direct benefits of these ES through consumption, enjoyment, or learning. Instead, the benefits that they receive are restricted to employment (sometimes low-quality employment such as at fish processing plants) and income. A better distribution of benefits requires deep transformations of preceding policies and institutions, which not only defined the concentration of people in certain territories but also defined the access mechanisms and power relations that determine who access ES benefits and who does not.

Keywords Marine ecosystem services · Off-site effects · Ecosystem services flows · Access barriers · Distributive inequality · Chilean Patagonia

1 Introduction

Marine systems provide multiple and varied ecosystem services (ES) on which much of the world's population depends (Worm et al. 2006). Although the ES concept has proliferated in the scientific and practical fields, it has had little application to marine ecosystems (Arkema et al. 2015; Townsend et al. 2018; Martino et al. 2019). In assessing marine ES, difficulties are encountered in linking ecological functioning and the generation of services (Jobstvogt et al. 2014; Townsend et al. 2018), which can be summarized as (i) integration of ocean depth as another spatial axis of assessment; (ii) complex spatial and temporal dynamics; (iii) nonlinearity of marine ecological processes; (iv) high mobility of resources and people (Hargreaves-Allen 2020); and (v) shortage of ecological, social, and economic data across the vast majority of the world's oceans.

In this context, most marine ES research has focused on mapping service supply (Nahuelhual et al. 2020) as the first level of the ES cascade conceptual framework (Potschin-Young et al. 2018). Although understanding the factors determining supply is an important contribution to marine ES research, the usefulness of the ES concept in decision-making requires going beyond supply, towards understanding the interactions between supply and demand, their spatial relationships, and the relationships between the natural and social system.

In this sense, little has been studied about the role that societies and institutions have in the production, distribution, and governance of ES and the benefits they provide (Bennett et al. 2015; Rova and Pranovi 2017; Martino et al. 2019). Thus, large gaps in the assessment of ES remain, especially in marine systems. As explained by Bennett et al. (2015), we still do not have a deep understanding of (i) the co-production of ES, understood as the specific social and ecological interactions that enable service generation (Fischer and Eastwood 2016); (ii) the factors driving preference distribution and access to ES benefits; (iii) stakeholder diversity, potential social conflicts, and inequities arising from access to specific ES; and (iv) issues such as how and when existing governance structures prevent or enhance access to ES benefits (Ruckelshaus et al. 2015).

In this chapter, we propose a framework of analysis to describe and explore the distribution mechanisms of ES and their benefits. Using the framework, we investigate the processes of co-production, capture, and access to benefits, accounting for the spatial relationships between ES supply and capture, based on existing frameworks.

The proposed framework of analysis can serve as an input for (i) identifying the diversity of beneficiaries and asymmetries among them (Felipe-Lucia et al. 2015; Laterra et al. 2019a); (ii) raising awareness among groups based on knowledge of the direct and indirect benefits generated by marine ecosystems (Schirpke et al. 2014); and (iii) identifying the barriers to access the benefits from ecosystems mainly by the most vulnerable groups.

In order to apply this framework, we selected five ES, namely, sense of place, food from aquaculture, recreation opportunities, food from artisanal fisheries, and education and knowledge opportunities. We based our analysis on the Magallanes region in the southern Chilean Patagonia, since this study area is a particularly relevant case for understanding access to ES. Firstly, its naturalness and vastness inspired naturalists and adventurers (Rozzi et al. 2019) and more recently an exclusive mass tourism with special interests in nature (Rozzi et al. 2010). Secondly, it is a region of long-standing conflicts in the use of the marine and coastal space, such as the conflict between canoe peoples who inhabited the marine and coastal areas (Kaweskar and Yaganes) and the foreign powers that plied the Strait of Magallanes (Gleisner and Montt 2014). Finally, given its geopolitical location, the development of the area has been permanently driven by foreign capital to the detriment of local capital (Harmabour 2019).

2 Methods and Data

2.1 Study Area

The Magallanes region and Chilean Antarctica is the most extensive and the second least populated of the Chilean territory. It extends between 53°09 and 70°55 SL and marks the southern limit of Patagonia and the American continent. According to the 2017 population census, it has 164,661 inhabitants who are concentrated in the cities of Punta Arenas, Puerto Natales, Porvenir, and Puerto Williams. Its creation as a region dates back to 1978, and its final boundaries were drawn in the twentieth century (Supreme Decree 2339/YEAR). The territory was historically inhabited by the Kaweskar, Yagan, Tehuelches, and Selknam peoples, for thousands of years. However, the occupation of Patagonia and the advent of sheep farming in the years 1880–1920 from Europeans led to the genocide and acculturation of these people almost leading to their extermination (Harambour 2017).

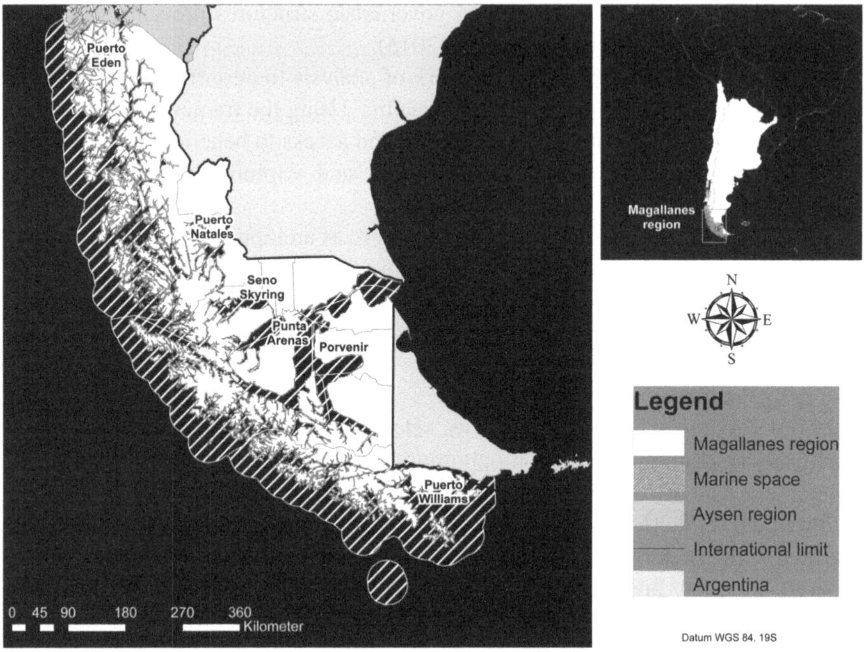

Fig. 15.1 Study area in the Magallanes region, Chilean Patagonia

Currently the region has an economy focused on commodity exports. The main activities are cattle raising, the exploitation of hydrocarbons, the salmon industry, fishing (industrial and artisanal), and tourism (Fig. 15.1).

2.2 *Framework of Analysis to Explore Mechanisms of Appropriation, Co-production, and Access to Ecosystem Services Benefits*

We built our analyses on the ES cascade framework (Potschin-Young et al. 2018), where a "service production chain" is represented. We start with the biophysical structures and processes of the ecosystems and end with the social benefits that society obtains (Liquete et al. 2013). The cascade framework has been enriched by incorporating new interactions mainly allusive to the relationships between social and biophysical systems (Spangenberg et al. 2014; (Felipe-Lucia et al. 2015). The adapted framework (Fig. 15.2) allowed us to explore the relationships of co-production of ES, the mechanisms of access to benefits, the migration of ES flows, and the impacts of these relationships and flows.

We defined each of the elements in Fig. 15.2 as follows:

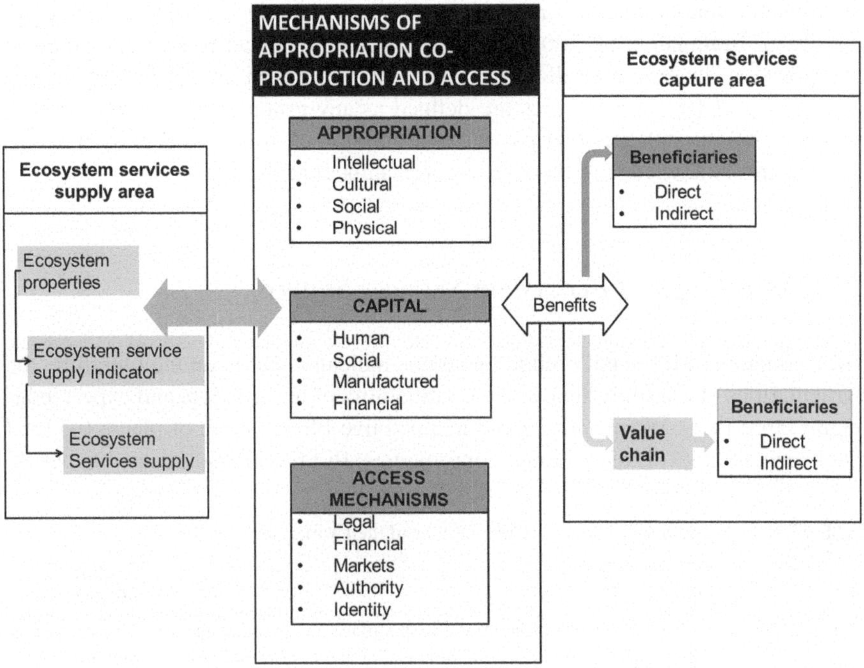

Fig. 15.2 Analysis framework to explore mechanisms of appropriation, co-production, and access to ecosystem services benefits. (Modified from Felipe-Lucia et al. 2015)

(i) *Ecosystem services supply areas*: Ecosystem service supply is defined as "the ability to provide a specific package of ecosystem goods and services within a specific spatial extent and temporal duration" (Burkhard et al. 2012). A supply area is defined as the space in which a certain supply accrues and is usually represented by ES supply hotspots.

(ii) *Mechanisms of co-production and access to ecosystems*: They allow us to understand the contributions of different forms of capital (human, social, manufactured, and financial) to the supply of ES and the capture of benefits by human beneficiaries (Cook et al. 2020). Ownership mechanisms, for example, are understood as forms of capturing, transferring, transforming, processing, and/or providing the services to generate ecosystem benefits, which require investments of time, labor, resources, technology, and money as a means to make them available (Braat and De Groot 2012). In addition, goods and services from ecosystems enter value chains of economic activities that are linked at the local level (Laterra et al. 2019b). On the other hand, access to benefits from ES is affected by a network of formal and informal institutions and policies (Scoones and Wolmer 2003).

(iii) *Area of capture of ES and benefits*: A capture area is where ES supply (totally or partially) reaches a human beneficiary and generates a direct or indirect benefit. Supply and demand for ES are often spatially uncoupled (Hein et al. 2016; Brauman et al. 2007; Vergara et al. 2020). Benefits are understood as positive changes in well-being given the fulfillment of needs and desires (TEEB 2010). Beneficiaries are defined as any group of people who benefit from ES through active or passive consumption or through recognition resulting from knowledge of these services (Nahlik et al. 2012).

2.3 Assessment of Ecosystem Services Supply

The evaluation of ES supply relied on ad hoc indicators based on the literature, the particularities of the study area, and the availability of information, and expert judgment (Table 15.1). We quantified and mapped five ES: (i) sense of place; (ii) food from aquaculture; (iii) recreational opportunities; (iv) food from artisanal fisheries;

Table 15.1 Ecosystem services supply indicators and their variables

ES	Description	Spatial attributes
Sense of place	It reflects the emotional bond with a place, created through direct interaction between humans and places (Kaltenborn 1998)	Iconic marine species
		Special lasting places
Food from aquaculture	Quantity of fish extracted from mariculture units, measured as biomass (kg km^2). The main species cultivated in the region are coho salmon, salmon, and rainbow trout, which represented 100% of the regional harvest in 2018	Annual productivity
Recreation opportunities	Capacity of marine and coastal areas to support tourism and recreational activities	Accessibility
		Scenic beauty
		Capacity for tourist use
		Unique natural resources
		Cultural sites
Food supply from artisanal fisheries	It reflects the amount of fish extracted from the marine area of the artisanal fishery, expressed as biomass (kg km^2). Information is included on five artisanal species of regional importance, which together represent 67% of landings in 2018 (SERNAPESCA 2019): spider crab (*Lithodes santolla*), sea asparagus (*Ensis macha*), scallop (*Austrochlamys natans*), conger eel (*Genypterus chilensis*), and sea urchin (*Loxechinus albus*)	Fishing area of each species
		Landings
Education and knowledge opportunities	It reflects the amount of research carried out in the region	Number of publications
		Research effort

and (v) education and knowledge opportunities. The selection and weighting of variables to be included in each indicator were made based on a bibliographic review and expert judgment. The variables selected for each indicator were spatially processed using Geographic Information Systems (GIS) (see Vergara et al. 2020 for details on indicator construction and supplementary material within).

2.4 Benefit Assessment

We adapted the Final Ecosystem Goods and Services Classification System (FEGS-CS) developed by the US-EPA (Landers and Nahlik 2013), which is a two-part classification system composed of (i) the identification of ES and (ii) the identification of an explicit human beneficiary of these specific goods and services. Direct beneficiaries are defined as those who directly use, consume, and/or enjoy the benefits of the ES, while indirect beneficiaries are those that benefit through the generation of employment and wealth (Daw et al. 2011).

A review of information was carried out for the representation of beneficiaries. The main source of information was the databases of the Internal Revenue Service (SII 2019), where workers are reported by company and economic category. In many cases, direct data is not available, and potential beneficiaries can only be estimated using spatial information and statistical information available at the municipal level, such as population censuses (Schirpke et al. 2014). A description of each type of beneficiary for each ES can be found in Vergara et al. (2020).

2.5 Co-production and Access to Ecosystem Services

Human intervention in ecosystems has been the fundamental factor driving the supply and distribution of ES in the Anthropocene (MEA 2005). People protect, conserve, use, challenge, alter, exploit, destroy, change, and rehabilitate ecosystems, consciously and unconsciously, for their own benefit or that of another person, with implications for ecosystem functions and services (Bennett et al. 2015). The analysis of co-production and access mechanisms relied on the questions in Table 15.2. This description was based on a literature review, expert knowledge, and government newsletters and information.

Table 15.2 Key questions to explore the co-production of and access to ecosystem services

Co-production mechanisms	
What are the mechanisms governing the co-production of the ecosystem service?	Appropriation When does nature become an ecosystem service?
	How are ecosystem services captured, transformed, transported, and processed in order to generate ecosystem benefits?
	Entering the value chain How does the benefit enter the network of relationships or links that aim to bring the greatest possible value to the benefit?
Access mechanisms	
How do access barriers to capture ecosystem service benefits operate?	Rights-based access (legal) What legal mechanisms operate to restrict access to beneficiaries?
	Access to financial capital What financial mechanisms operate to restrict access to beneficiaries?
	Access to market What market mechanisms operate to restrict access to beneficiaries?
	Access to authority What governance mechanisms operate to restrict access to beneficiaries?
	Access through social identity What cultural and identity mechanisms operate to restrict access to beneficiaries?

3 Results

3.1 Ecosystem Service Supply Areas

Figure 15.3 and Table 15.3 show the distribution of ES supply in Magallanes region and the percentage and area (km^2) by value category. Sense of place showed low to medium values of the indicators (measured in normalized values from 0 to 100) across most of the region (Fig. 15.3, panel A). The areas of higher indicator values concentrated in Cape Horn, in Navarino Island. This is due to the concentration of attributes like the presence of emblematic species such as leopard seal *(Hydrurga leptonyx*), southern elephant seal (*Mirounga leonine*), and black-browed albatross (*Thalassarche melanophris*) and areas with indigenous toponymy. The areas of provision of food from aquaculture concentrated in Puerto Natales and in Skyring Sound. Both areas hold the most suitable hydrodynamic characteristics, but they have been declared saturated areas for aquaculture (Res. n° 2189). Most parts of the region (97.8%) do not have any supply (Fig. 15.3, panel B).

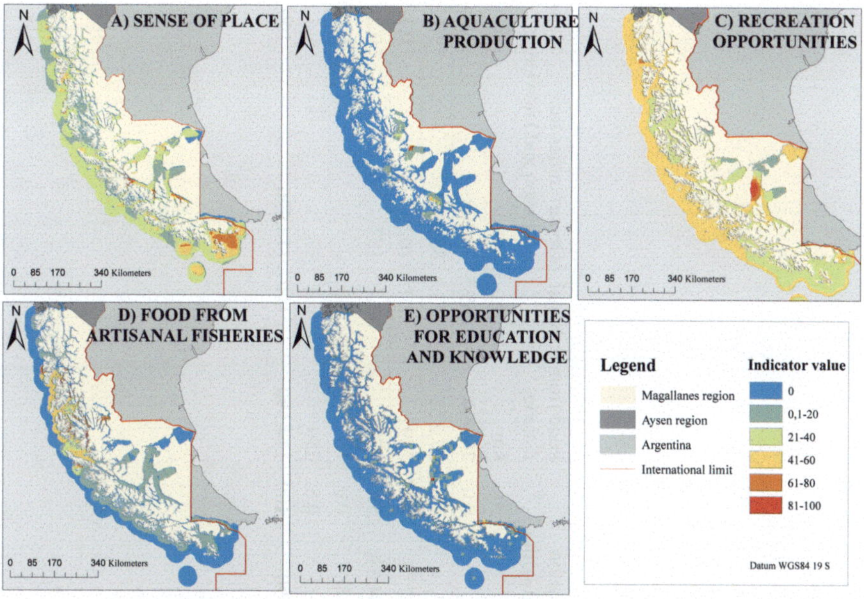

Fig. 15.3 Distribution of the ecosystem services supply indicators in the Magallanes region

The highest values of the recreation opportunities indicator concentrated in the Strait of Magallanes, an iconic place that has high values of accessibility attributes (with consolidated marine routes of international importance), unique natural resources (due to the presence of southern right whale, *Eubalaena australis*, and sei whale, *Balaenoptera borealis*), and cultural heritage. More than 50% of the region's marine area had medium values of the indicator, which makes it the indicator with the largest regional distribution (Fig. 15.3, panel C).

In the case of food from artisanal fisheries, medium, high, and very high values of the indicator (covering 8.5% of the marine area) concentrated in the north of the region, near the towns of Puerto Natales and Puerto Eden. This is explained by the presence of the urchin (*Loxechinus albus*) fishery, which by itself represents 43% of regional landings (Fig. 15.3, panel D).

Education and knowledge opportunities were widely distributed across the region but with only two small areas with high and very high indicator values. The first area is the central section of the Strait of Magallanes, which due to its proximity to the city of Punta Arenas (which groups universities and research centers) becomes an area of easy management and access. The second area was located in the Beagle Channel (Fig. 15.3, panel E).

Table 15.3 Percentage and area of provision of ES by category of indicator value

Indicator value category (ES supply)	Indicator normalized value	Sense of place		Food from aquaculture		Recreation opportunities		Food from artisanal fisheries		Opportunities for education and knowledge	
		km^2	%	km^2	%	km^2	%	km^2	%	km^2	%
No ES flow	0	5.8	0.0	112,020	97.8	5.17	0.0	75,094	65.5	76,108	66.4
Very low	0.1–2.0	23,484.8	20.5	1995	1.7	5551	4.8	27,272	23.8	38,096	33.2
Low	2.1–40.0	73,186.6	63.9	431	0.4	46,647	40.7	2397	2.1	102	0.1
Medium	40.1–60.0	13,476.5	11.8	96	0.1	60,444.63	52.7	7702	6.7	187	0.2
High	60.1–80.0	4307.3	3.8	19	0.0	1064.2	0.9	1866	1.6	0	0.0
Ver high	80.1–100	129	0.1	29	0.0	878	0.8	259	0.2	97	0.1
Total		**114,590**	**100**	**114,590**	**100**	**114,590**	**100**	**114,590**	**100**	**114,590**	**100**

3.2 *Benefit Assessment*

3.2.1 Beneficiaries in the Magallanes Region (Local Beneficiaries)

Table 15.4 shows the number of direct beneficiaries (DB) and indirect beneficiaries (IB) of ES in Magallanes region and their percentage over the regional population.

In the case of sense of place, the totality of the regional population corresponded to direct beneficiaries. Indirect beneficiaries (508) represent 0.3% of the regional population. The greater percentage of indirect beneficiaries relates to activities of publication of books linked to the marine and coastal territory of the region, followed by the workers of regional museums. Although this ES is distributed throughout the region and the entire population is the direct beneficiary of it, it does not generate significant production chains in the region.

For food from aquaculture, direct beneficiaries (salmon consumers) are 10,656 people representing 6.7% of the regional population. The indirect beneficiaries are 420 people who live in the region and work in the salmon production chain, representing only 0.3% of the population. For recreational opportunities, the beneficiaries are visitors from outside the region who register their visit to national parks. In 2018, 544,523 visits were registered, of which 280,755 were Chilean visitors and 263,768 were foreigners. The number of indirect beneficiaries employed by industries related to tourism and recreation was 7888, which represents 4.9% of the regional population. Food from fisheries had 4441 direct beneficiaries (fish consumers) and 9368 indirect beneficiaries who work as artisanal fishermen or in processing plants and represent 5.6% of the regional population.

Finally, the ES of education and knowledge opportunities had 29,892 direct beneficiaries who were mainly students involved in activities related to the knowledge of marine and coastal areas. The indirect beneficiaries were 2667 people who work in educational institutions. They have access to teaching activities related to marine and coastal areas and represent 1.7% of the regional population.

3.2.2 External Beneficiaries

Figure 15.4 shows the distribution of direct beneficiaries of ES that are exported as products outside the country and/or attract beneficiaries from other parts of the world. The availability of data only allows us to present the external beneficiaries of the ES of food from fisheries, food from aquaculture, and recreation opportunities. Panel A shows the distribution of direct beneficiaries of food from artisanal fisheries. In this case, 39.6% of the ES flow is exported to Japan, where 245,480 people can meet their annual consumption demand for marine products only with what is imported from the Magallanes region. Another 25% goes to China where 163,800 people can satisfy their annual fish demand with these imports. Spain, Taiwan, and the United States are the next recipients of the region's marine and fishing products with exports representing 9.3%, 7.5%, and 3.0% of the region's ES flow, respectively.

Table 15.4 Local beneficiaries of the Magellan region

	Number	Percentage of total population
Population (number of persons[a])	166,533	100.0
DB sense of place		
Native population	1261	0.8
Municipality population[a]	165,272	99.2
Total DB sense of place	166,533	100.0
IB sense of place		
Artists whose work is related to the identity of marine ecosystems	79	0.0
Film producers related to the landscape of seas and fjords	50	0.0
Other cultural activities inspired by the marine and coastal area such as book publishing and other publishing activities	282	0.2
Workers of museums and cultural activities	97	0.1
Total IB sense of place	508	0.3
DB food from aquaculture		
People who can feed with the availability of fish given consumption recommendation	10,656	6.4
Total DB food from aquaculture	10,656	6.4
IB food from aquaculture		
Workers of related aquaculture services	420	0.3
Total IB food from aquaculture	420	0.3
DB recreation opportunities		
Total visitors to national parks[b]	458,447	
Total DB recreation opportunities	458,447	
IB recreation opportunities		
Workers in travel agencies, tour operators, and tour guides	239	0.1
Workers in transportation activities	300	0.2
Workers in tourist hosteling	7349	4.6
Total IB recreation opportunities	7888	4.9
DB food from fisheries		
People who can feed with the availability of fish given consumption recommendation	4441	2.7
Total DB food from fisheries	4441	2.7
IB food from fisheries		
Artisanal fishers	9136	5.5
Workers of related fisheries service	232	0.1
Total IB food from fisheries	9368	5.6
DB education and knowledge opportunities		
Teachers of university and technical careers that can relate to marine and coastal themes	101	0.1
Schoolchildren that are related to marine and coastal knowledge activities	28,880	17.3

(continued)

Table 15.4 (continued)

	Number	Percentage of total population
Researchers who publish on marine and coastal issues located in the Magellan region	91	0.1
Secondary technical education coastal and marine specialties	820	0.5
Total DB education and knowledge opportunities	29,892	18.0
IB education and knowledge opportunities		
People who work at state educational establishments that have access to marine and coastal educational activities	2667	1.7
Total IB education and knowledge opportunities	2667	1.7

[a]Represents the total of the municipalities population subtracting the native population
[b]They are foreign visitors who register entry into national parks; the national tourism service does not record domestic tourism

Panel B shows the destinations of the food from aquaculture service: 27% of the service's flow is exported to the United States where 475,168 people can supply their annual fish consumption demands with the Magallanes region's salmon, followed by Russia with 22% of the exports (395,171 beneficiaries), Japan with 12.0% (207,777 beneficiaries), and China with 8.9% (155,289 beneficiaries). The total number of external beneficiaries of food from aquaculture is 1,726,413 people.

Panel C shows the origin of foreign visitors to the Magallanes region. The largest number of beneficiaries comes from Germany with 9.2% of visitors (21,811 recreationists), France 8.3% (19,696 recreationists), Brazil 7% (16,693 recreationists), and Argentina 6.9% (16,453 visitors). The total number of foreign visitors to Magallanes Patagonia is 237,338 (SERNATUR 2018).

3.3 Co-production Mechanisms

In the case of sense of place, the external factors that contribute to the generation of the service are the social and cultural production and reproduction that influence in the intellectual and affective act of the places that are inhabited (Mora 2012; Gustafson 2001). Food from aquaculture is generated by taking advantage of the ecosystem conditions for the cultivation of aquaculture species, which is achieved by modifying the marine environment through the construction of cage rafts for fish farming (Landers and Nahlik 2013), in this case salmonids, a species introduced into the territory. Recreation opportunities depend on the natural attributes of the region but are modeulated by external factors in the generation of the service, which respond to the information dissemination, ideas, and opinions with the intention of positioning a geographic place that meets the desired attributes for recreation and tourism (Stokowski 2002). Particularly in the Magallanes region, the ES for recreation opportunities is linked to the nature tourism market where its remote location and extreme weather conditions (presence of land and sea ice) have represented

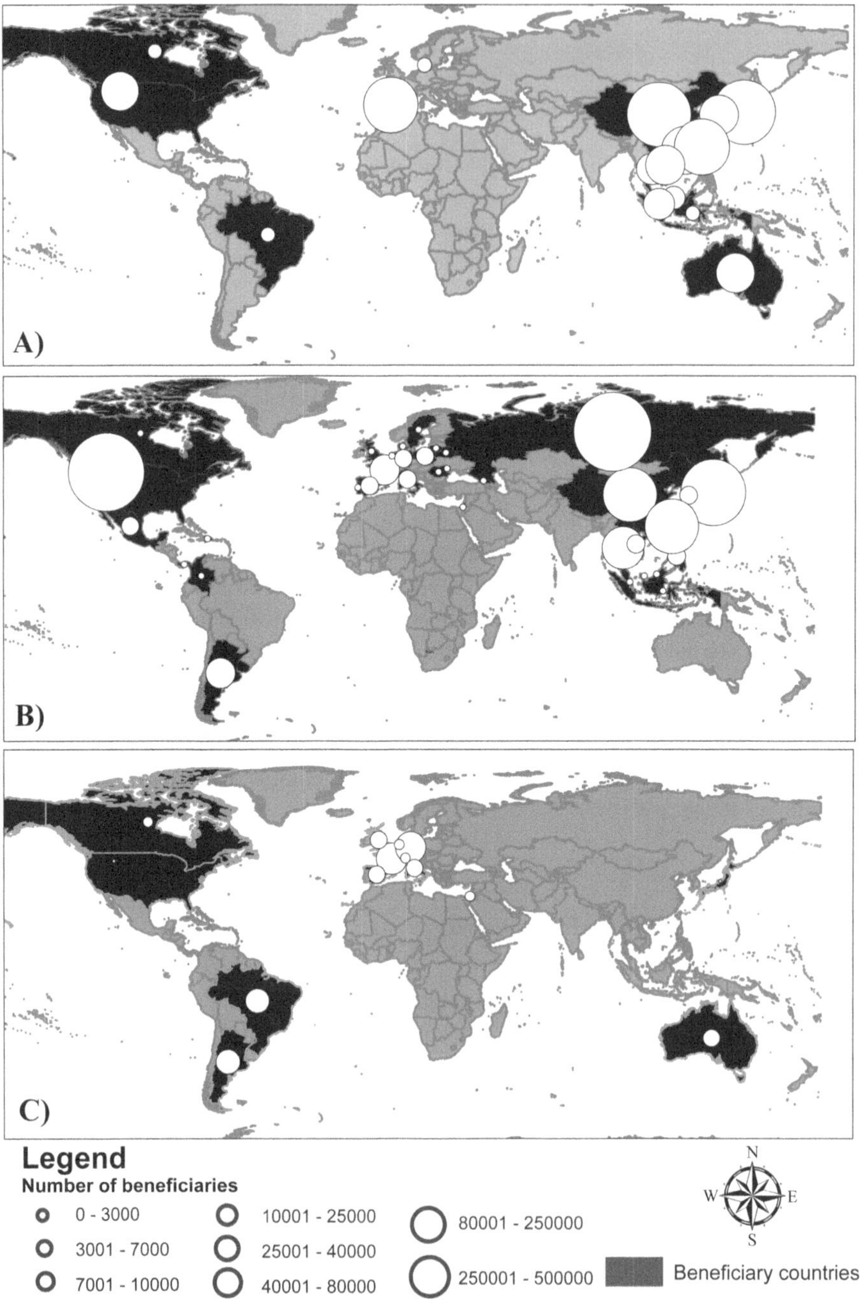

Fig. 15.4 External beneficiaries of food from artisanal fisheries (**A**), food from aquaculture (**B**), and recreation opportunities (**C**)

prohibitive conditions and important limitations for human activity (Lamers et al. 2008). However, the tourism industry began to overcome these limitations, positioning the region through market products using concepts such as nature tourism among others (Hartwell et al. 2012). In the case of food from artisanal fisheries, it is the fishing activity that generates the ES, which is reflected into the capture of fish for human consumption (Anticamara et al. 2011). For the ES of education and knowledge opportunities, nature becomes a service through learning as a result of the interaction with the social and biophysical environment (Alexander et al. 2009).

Direct beneficiaries capture the ES of sense of place, based on the links with identity and belonging to the community (Gallina and Williams 2015), and it is transformed into benefits through artistic expressions for example (Spencer and Werness-Rude 2015). For the capture, transformation, transport, and processing of food from aquaculture, a variety of processes are required such as breeding, fattening, feeding, and processing that depend on specific ecosystem conditions. These processes culminate in the production of food that constitutes the service (García Moreno 2005). In the case of recreational opportunities, the capture of the service is almost exclusively through the tourist market. The tourism companies in the area, based on complex logistics that can reach the Antarctic Region (Lamers et al. 2008), transport the recreationists (direct beneficiaries) to the places where the service is provided. For their part, the indirect beneficiaries (owners, suppliers, and employees of these industries) capture the service from the operation of these companies (Laterra et al. 2019b). In the case of food from artisanal fisheries, the capture is carried out by the fishermen (indirect beneficiaries) using fishing gear that is specific to each species. Fishing takes place in fishing grounds located in fjords and channels, and fish is transferred in fishing boats to the coves. From the coves, the fish is distributed to the direct beneficiaries, which are local and foreign consumers (Castillo 2011). The capture of education and knowledge opportunities is carried out through exploratory and descriptive research of the components of the ecosystem and their relationships, applying different methods. The indirect beneficiaries, on the other hand, make use of the captured knowledge for educational purposes.

For the ES of sense of place, the attributes of marine and coastal locations are catalysts for individual creativity and creative industries (Drake 2003; Clare 2013). The creative industries are integrated in place, where the importance of aesthetics and social networks leads to a narrow geographical clustering (Clare 2013; Gertler 2004). For food from aquaculture, the benefit enters the value-adding network through the global food industry. In the case of recreation opportunities, the ES is transformed into tourist products. Tourist products are characterized by a set of tangible and intangible elements that allow the development of specific activities in certain destinations (Nasimba and Cejas 2015). For ES of food from fisheries, the service enters the network of value addition through the local sale of products and the global food industry. The knowledge coming from the ES of education opportunities suffers a process of value addition, from public and private institutions that are working on programmatic axes of research and formal and nonformal education to improve understanding of the value of ecosystems for the well-being of humanity (MEA 2005; TEEB 2010).

3.4 Access Mechanisms Acting as Barriers to Capture ES Benefits

For the place-based ES, there are at least three mechanisms that restrict or favor access to beneficiaries, all of which are linked to the zoning of marine coastal areas:

(i) Indigenous Peoples' Marine Coastal Spaces. They are a form of spatial property right that favors access to one type of direct beneficiaries of ES and excludes another. It aims to protect ecosystems and native cultures (Zelada Muñoz and Park Key 2013).
(ii) Marine Protected Areas. There are two main categories of marine areas: the first, marine parks, restricts almost all economic uses and access to beneficiaries, and the second, multiple-use marine protected areas, has a flexible framework to design that activities can be developed within your area (such as artisanal fishing).
(iii) Salmonidae Concessions: They function as concessions of marine space where rafts-cages for salmonids are installed. A concession is restrictive in its entry to people not related to the particular salmon farming company.

In the case of the indirect beneficiaries of sense of place service, there is a legal framework that regulates the creative industries. These aspects condition the access to benefits, as they are necessary requirements for its formal development. Within these regulations are employment regulations, health, safety, product standards, antitrust laws, and patent laws, among others. In addition, the creative industries integrate the sectors in which the value of goods and services offered is based on intellectual property (Erickson 2018), a limitation that must be respected by the rest of the indirect beneficiaries.

There are no legal access barriers for the direct beneficiaries of food from aquaculture (salmon consumers). In the case of the indirect beneficiaries of the ES for food from aquaculture, sectorial permits are required for the installation and operation of the plants (SERNAPESCA, Navy, Environment Superintendence, Environmental Impact Evaluation Service). There are no legal restrictions on access to tourism products for the direct beneficiaries of the ES of recreation opportunities. However, for the IB, there is a need for formalization, which implies patent rights, navigation rights, and sectorial permits for access to areas for recreation, among others (Subsecretaria de Turismo 2020).

In the case of the ES for the food from fisheries, the law that regulates the inspection allows sanctioning consumers (direct beneficiaries) that do not respect the control mechanisms of the artisanal fishing specific to each fishery (biological closures, size regulation, and noncompliance with quotas, among others) (Law 21.132, Ministerio de Economía, Fomento y Turismo 2019). Artisanal fishermen as indirect beneficiaries of this service have a legal regulation of access to benefits, through (i) the limitation of fishing licenses (closure of the list of fishermen authorized to operate in a given fishery); (ii) the establishment of fixed fishing quotas; and (iii) regulation by size and closures (SUBPESCA 2020). For their part, the processing plants must comply with the same standards of reception of fishing products described for the direct beneficiaries, in addition to being regulated in terms of health and trade.

In the case of researchers (direct beneficiaries of education and knowledge opportunities), the activity is restricted by the need to have legal documents such as sectorial permits that allow, for example, navigation, research fishing, and access to certain places, among others (Artigas and Escobar 2001; CONAF 2013). On the other hand, secondary technical education establishments with a coastal and marine specialty (direct and indirect beneficiaries) require legal and administrative procedures to validate their formal educational programs. However, these mechanisms do not restrict access to the benefit, but rather give it a degree of recognition and allow them the necessary funding for its operation (MINEDUC 2011).

No financial mechanisms were identified that restrict or facilitate access to the ES direct beneficiaries of sense of place. However, for some indirect beneficiaries (those linked to the creative industries), access to funding sources is limited, even in the case of activities and projects that have an attractive "expected" private profitability. For food from aquaculture, the direct benefit of the ES is mediated by the capacity to pay for salmonid products. One single kilogram of salmon costs approximately US$15 (FAO 2019), and salmon sales points are concentrated in two cities in the region. As for the workers (IB) of this service, there are no restrictions mediated by financial capital. However, the owners and investors need financial capital for the installation of breeding, fattening, feeding, and processing centers for salmonid products (SERNAPESCA 2001). It should be noted that with the exception of Agrosuper (Chilean holding of food companies), the important capitals of fishing farming are foreign and associated with large investment holding companies such as Legend Holdings Corporation (Lenovo) or Mitsubishi (Irarrázaval and Bustos 2020).

For the direct beneficiaries of recreation opportunities, there is a financial restriction in the access since marine and coastal Patagonian tourism is planned towards foreign markets (Mac-Lean 2010). In Magallanes, 50% of the population has a salary equal to or less than US$620 per month. The value of a tourist cruise in Patagonia starts at US$2000, and a 1-day kayak trip starts at US$150 (EPAUSTRAL 2019). Additionally, from the perspective of the indirect beneficiaries, it is necessary to emphasize that tourism linked to the SE for recreation can have high installation costs.

The tourism promotion programs and the financing lines of the Chilean Development Corporation and the National Tourism Service do not offer capital access programs focused on marine tourism (Subsecretaría de Turismo 2016).

Access to food from fisheries is mediated by the capacity to pay. The average prices of the first transaction of the fish products that make up the indicator are crab, 9 US$ per kg; urchin, 1 US$ per kg; golden conger, 3 US$ per kg; huepo, 0.7 US$ per kg; and scallop, 1 US$ per kg (SERNAPESCA 2019). The prices of sale to the consumer in the established commerce are crab, 50 US$ per kg; urchin, 33 US$ per kg; golden conger, 16 US$ per kg; huepo, 10 US$ per kg; and scallop, 17 US$ per kg. Beyond price restrictions, it is important to emphasize that practically there are no points of sale of fishing products in the region and they can only be found in the municipalities of Puerto Natales and Punta Arenas.

On the other hand, artisanal fishermen need access to financial capital for the acquisition and maintenance of boats and for their operation. Access to such financial capital is limited by the informality of the activity, whereas state promotion instruments are sporadic and insufficient. This situation means that fishing boat owners usually end up being financed by the processing plants through an authorization contract (Poblete et al. 2013). Beneficiaries who are processing plant workers do not need access to financial capital.

For direct beneficiaries of education and knowledge opportunities, specifically for scientific researchers, financial capacity is a determining factor in conducting research (e.g., fees, sailing campaigns, laboratories, etc.). In general, financial capital is obtained by applying for scientific development funds through the National Agency for Research and Development and is mostly restricted to those with PhD degrees, being a restriction on access to benefits. The existing secondary technical education establishments of coastal and marine specialty in Magallanes region are mostly public, so there is no restriction of access by financial mechanisms to students and teachers related to them.

In the case of the direct beneficiaries of the ES sense of place, there is no market associated that can restrict access to beneficiaries; however, in the case of the indirect beneficiaries, it is the market that regulates who can access the benefits of the service. In this sense, market gaps are evident, which can be summarized in the centralization of purchasing powers, platforms to promote creative industries, and spaces for training, exhibition, and marketing. For the ES of food from aquaculture, the food coming from the aquaculture in Magallanes (salmonids) is oriented towards international markets. This generates two mechanisms of exclusion towards the local consumers: lacking salmonid products in the region and acting synergistically with the financial access barriers through the local prices of the salmon. Regarding indirect beneficiaries of this service, the employees of the salmon plants access the market through their employment, and there are no mechanisms for excluding beneficiaries. On the other hand, investors in processing plants access the market through ProChile's positioning programs (Vera Garnica 2009), the promotion instruments of Fundación Chile, and their own business networks.

In the case of recreation opportunities, there are no market mechanisms that restrict access to the direct beneficiaries; restrictions are rather of price, as already noted in the financial mechanisms. On the contrary, in the case of the indirect beneficiaries, it is important to point out that tourism planning in Magallanes is oriented towards the investment of foreign capital or consolidated national capital, leaving smaller local tourism operators outside the objective image of regional tourism (Collipal Pichicona 2017). In the case of food from artisanal fisheries, artisanal fishermen (indirect beneficiaries) access the formal market only through the sale of marine products to intermediaries or processing plants. Their access is limited by contracts that are previously made through the authorization of boats (Poblete et al. 2013). Processing plant workers access the market through their work. Process plant investors access international markets, thanks to the support of ProChile (state agency of the Ministry of Foreign Affairs which is in charge of promoting the exportable supply of Chilean goods and services) (Aldunate Wegner 2013).

There are no market mechanisms restricting access to ES beneficiaries of education and knowledge opportunities. In the region, research with a high I + D component is favored through the valorization of the latter in the market. It should be noted that the region's science policy recognizes as an objective "Promoting the application of knowledge and technology in productive activities to increase the competitiveness of the regional productive structure" (Conicyt 2009).

For the ES of sense of place, the existing governance mechanisms do not determine access. However, there has been a demand (recently recognized) from the Selknam people towards the authority for their recognition as a living people, which due to their delay has resulted in a lack of protection of their places of cultural importance. Furthermore, between 2004 and 2010, under Article 23 of Convention 169 of the International Labour Office (ILO) and Article 1 of the Indigenous Law, the Fishing Sub-secretariat of Chile granted quotas for the capture of 60 specimens per year, exclusively aimed to Kawesqar Indigenous Community (Cruz-Rueda 2008).

For the food from aquaculture, the access of direct beneficiaries and indirect beneficiaries is not mediated by access to authority. Nevertheless, the salmon industry after the ISA virus crisis (2016) formed a public-private coalition with the State where the general guidelines of the industry were established. After this the salmon companies that were summoned to the coalition positioned their preferences in the new regulations (Irarrázaval and Bustos 2020).

For recreation opportunities, direct beneficiaries do not face identified barriers to access the authority. However, in the case of the indirect beneficiaries, it is necessary to clarify that the municipalities are administrative nodes with power that in Chile have a central role in the development of tourism. Nevertheless, in Magallanes, the municipalities and their authorities concentrate an unequal political power, and, with some exceptions, they do not have the capacity to be an articulating agent of the development of the marine and coastal tourism in their territory (Valenzuela 2015).

Researchers and research centers (indirect beneficiaries of education and knowledge opportunities) need to position their research topics before local and regional authorities to obtain public funding for their field campaigns; such positioning is done through lobby (Guinovart 2009).

In the case of the ES of sense of place, the access of the direct beneficiaries is mediated by social and cultural identity (Gustafson 2001). For the ES of recreation opportunities, certain groups, such as artisanal fishermen, can access capture areas of this ES through the practice of their occupation, exceptionally being an access not mediated by the tourism industry. In this case, being part of this particular social group facilitates access to the benefits of the service. Likewise, there is a line of financing aimed at strengthening the supply of indigenous tourism products (indirect beneficiaries); however, no native people of Magallanes region has been awarded such financing (Subsecretaría de Turismo 2020). Salmon farming is not culturally rooted in Magallanes, and there is no consumption linked to social identity (Chávez Zúñiga & Milahuichún Mayorga 2011). There are no mechanisms that restrict access through cultural identity.

In the ES of education and knowledge, opportunities and researchers (direct beneficiaries) who belong to groups that have historically been related to the sea (canoeing or fishing villages) may have greater facility to generate knowledge that links different forms of knowledge; an example of this is the scientific work of Jose Tonko (2008), researcher and representative of the Kawésqar people. A similar case occurs with the relation of students and teachers (direct beneficiaries and indirect beneficiaries). Activities related to the sea through family inheritance (such as work in fishing or aquaculture) can favor the learning process (Ávila-Ruiz 2005). In the case of ES of provision of food from fishery, the informal consumption of fishery products may be mediated by the proximity of the consumer to fishermen (Castillo 2011). The access of indirect beneficiaries is not mediated by access to authority.

4 Discussion and Recommendations

In the collective imagination, the southern seas are drawn pristine and devoid of human activity. From the end of the nineteenth century and the beginning of the twentieth century, this romantic idea of Patagonia as the last empty border where everything is possible (Harambour 2019) was spread throughout the global north. This idea of spaces without sovereignties (neither of the nation states nor of other peoples) promoted a logic of free capital flows throughout Patagonia based on the activities of European companies.

The truth is that the southern seas have been populated for at least 11 thousand years (Méndez 2011) and since then humans have used, transformed, managed, and adapted to the sea to co-produce goods and services for their long-term benefits. This co-production has intensified in recent decades because of technology that makes possible the "materialization" of many ES, which would, otherwise, be impossible to capture (e.g., opportunities for education and knowledge). On the other hand, technology has allowed the generation of ES under "semi-natural" or highly artificial conditions, such as the production of food from aquaculture in marine farms.

Understanding the mechanisms that determine how ES are co-produced and how they are distributed goes beyond observing the ecosystem variables that determine ES supply (first stage in the cascade model). Certainly, ES as waste assimilation are represented exclusively by biophysical variables. But others, such as those described in this chapter, need different degrees of socio-ecosystem interaction for nature to become a service. On the other hand, the ways in which ES are distributed and captured have also changed. In ancient times, many ES (both marine and terrestrial) were potentially "open access." However, the institutions that have contributed to ES protection against increasing pressures on the marine space have also generated access barriers, causing great asymmetries and inequalities in the capture of benefits derived from ES (Laterra et al. 2019a; Vergara et al. 2020).

For the marine ES studied in this chapter, we observe a gradient in co-production in terms of the "human effort" delivered so that nature becomes a service and society can appropriate its benefits. At the beginning of this gradient, we have the ES of sense of place, for which there is an almost straight appropriation of benefits by local inhabitants. This means that the capture of sense of place is not mediated by other factors. Then, food from artisanal fisheries requires a fishing effort for the service to be appropriated. The ES of education and knowledge opportunities requires close interaction between people and nature so that questions are produced that motivate research, knowledge production and learning. Towards the end of the gradient, there are recreation opportunities, which need built capital and managerial actions to position places in the tourism markets and advanced logistics and infrastructure to make the tourism activity possible, particularly given the conditions of the southern seas.

Finally, there is the production of food from aquaculture, which is the ES with the greatest need for human intervention in order to become available. This gradient is aligned with the nature of each ES, ranging from pure public goods (sense of place) to private goods traded in markets (aquaculture products). In the latter case, salmon farming is based on the existence of secure and individual property rights (granted to each firm through an aquaculture concession) that make this provision ES perfectly exclusive and rival. Both the co-production and the public/private good gradient are associated with the level of appropriation of each ES. While sense of place can be freely appropriated, salmon consumption is affordable only by wealthy consumers in foreign markets.

It is also possible to order the ES according to the type and number of barriers they have. For this, it is necessary to differentiate the barriers that direct versus indirect beneficiaries must overcome. Although in this study there are four types of beneficiaries according to type (direct versus indirect) and location (local within the region and extra-regional or international), here we will especially discuss the barriers that local inhabitants have to access benefits. Food from artisanal fisheries, food from aquaculture, and recreation opportunities have few barriers to legal access and higher financial barriers. On the other hand, sense of place and opportunities for education and knowledge are mediated (to some degree) by legal barriers in terms of the existence of areas with different restrictions and the need to have different permits to carry out marine research.

Market mechanisms, which are reflected in the structure of value chains, generate significant access barriers for the direct local beneficiaries of food from artisanal fisheries and aquaculture and recreation opportunities. Since the fish and tourism industry target global markets, they have restricted the emergence of local market niches or other forms of access not based on markets. This access barrier works synergistically with the financial access barrier, while access to authority does not appear to be a significant barrier for any local direct beneficiary.

Certain beneficiaries access various services through identity, e.g., people who have historically been linked to fishing activities can access marine and coastal areas with cultural significance through their boats. Likewise, certain beneficiaries can access the consumption of marine and coastal products informally, without being mediated by a market. This also shows that an individual or groups of individuals can be direct beneficiaries for one ES and indirect beneficiaries for another.

In this context, ES valuations must consider these dynamics when quantifying benefits.

For indirect beneficiaries, the access barriers identified are generally more significant than those identified for direct beneficiaries. Legal access barriers are binding for food from aquaculture and recreation opportunities. Who can overcome these barriers to be able to insert themselves into the value chains that produce these services? Almost exclusively companies and capitals with knowledge of the laws that regulate activities and with access to legal advice. The creative industries go through the same filters. The ES of food from fisheries has multiple legal barriers that exclude indirect beneficiaries. Who can access? In general, artisanal fishers who have been in the activity for a longer time and who fish even before the current fishing legislation was generated. In this case there is a synergy between access through identity and legal access.

Financial barriers are relevant in the case of food from aquaculture and recreation opportunities since the costs of installing an aquaculture concession or a marine and coastal tourism service in Patagonia are high. Who can overcome these barriers to be able to insert themselves into the value chains that produce these services? Almost exclusively foreign or national companies, but as global chains. The provision of food from fisheries encounters financial access barriers related to investment in vessels and the costs of the fishing operation itself. For the creative industries, access barriers are high, but they are at least partially overcome through state financial instruments.

In the case of ES inserted in global distribution chains, access to markets requires consolidated business networks that are supported by governments to position Chilean products abroad (e.g., ProChile). Who has access? Consolidated national or global financial capitals that exclude smaller local capitals. These same indirect beneficiaries are the ones who more easily access authority through public-private alliances, which allows them to better position their interests on the public agenda.

Any recommendation arising from our results requires answering two normative questions: Is it desirable that the benefits of the ES are captured primarily by local inhabitants? Is it securing their access over foreign beneficiaries a value in itself? These questions are relevant, and answering them is not in the hands of researchers, but they should emerge from the social agreements that the different actors must make within a democratic governance system. If the agreed-upon answers are affirmative, then it would be necessary to make institutional adjustments that allow the removal of several barriers. For example, the tourism policy of Magallanes is oriented towards the positioning of Patagonia as a global destination; this is reinforced because of the public-private alliances that articulate the policy, which are made up of the same actors that currently provide tourism services, do not include potential local beneficiaries. Regarding tourism, the focus has been continuously placed on the generation of economic profits rather than on allowing local inhabitants to get to know their region or on generating small-scale enterprises with a local identity. Directing this type of policy towards local markets can also help the region to gain resilience in the face of events such as the coronavirus pandemic experienced during 2020–2021 and that has severely affected the Magallanes region.

A major territorial transformation is the generation of the necessary infrastructure to promote local capture of ES benefits, which includes infrastructure for roads, tourism, ports, and spaces for the exchange of products. In the same way, a profound review is necessary of the way in which extractive use rights are granted and the privileges that these grant. These rights, in the figure of a marine concession (aquaculture, mining), are at the center of territorial disputes in southern Chile (Tecklin 2016). Although these rights embody the privatization of certain marine spaces, they have remained unchanged since their enactment. In summary, for the ES approach to have a true option of influencing marine planning, it must go beyond merely evaluating ES supply. The way in which ES are co-produced tells us much about who may or may not benefit, while their current distribution and the barriers preventing equitable distribution can teach us what must be overcome for fair and sustainable marine governance.

Acknowledgments This study was funded by FONDAP Grant 15150003.

References

Aldunate Wegner P (2013) Imagen país: Los casos del vino y el salmón chileno (2007–2011) [Master's thesis, Universidad de Chile]. http://repositorio.uchile.cl

Alexander PA, Schallert DL, Reynolds RE (2009) What is learning anyway? A topographical perspective considered. Educ Psychol 44(3):176–192

Anticamara JA, Watson R, Gelchu A, Pauly D (2011) Global fishing effort (1950–2010): trends, gaps, and implications. Fish Res 107(1–3):131–136. https://www.sciencedirect.com

Arkema KK, Verutes GM, Wood SA, Clarke-Samuels C, Rosado S, Canto M, Rosenthal A, Ruckelshaus M, Guannel G, Toft J, Faries J, Silver JM, Griffin R, Guerry D (2015) Embedding ecosystem services in coastal planning leads to better outcomes for people and nature. Proc Natl Acad Sci 112(24):7390–7395. https://www.pnas.org

Artigas C, Escobar J (2001) Perfil de la cooperación para la investigación científica marina en América Latina y el Caribe. CEPAL

Ávila Ruiz RM (2005) Reflexiones sobre la enseñanza y el aprendizaje del patrimonio integrado: una experiencia en la formación de maestros. Revista de Investigación en la Escuela 56:43–53. https://idus.us.es

Bennett EM, Cramer W, Begossi A, Cundill G, Díaz S, Egoh BN, Geijzendorffer IR, Krug CB, Lavorel S, Lazos E, Lebel L, Martín-López B, Meyfroidt P, Mooney HA, Nel JL, Pascual U, Payet K, Pérez-Harguindeguy N, Peterson GD, Prieur-Richard AH, Reyers B, Roebeling P, Seppelt R, Solan M, Tschakert P, Tscharntke T, Turner BL II, Verbug PH, Vigilizzo EF, White PCL, Woodward G (2015) Linking biodiversity, ecosystem services, and human well-being: three challenges for designing research for sustainability. Curr Opin Environ Sustain 14:76–85. https://www.sciencedirect.com

Braat LC, De Groot R (2012) The ecosystem services agenda: bridging the worlds of natural science and economics, conservation and development, and public and private policy. Ecosyst Serv 1(1):4–15. https://www.sciencedirect.com

Brauman KA, Daily GC, Duarte TKE, Mooney HA (2007) The nature and value of ecosystem services: an overview highlighting hydrologic services. Annu Rev Environ Resour 32:67–98. https://www.annualreviews.org

Burkhard B, Kroll F, Nedkov S, Müller F (2012) Mapping ecosystem service supply, demand and budgets. Ecol Indic 21:17–29. https://www.sciencedirect.com

Castillo C (2011) Informe Final Proyecto FIP 2008–45
Chávez Zúñiga E, Milahuichún Mayorga C (2011) Los efectos socioeconómicos de la salmonicultura en la región de Magallanes. [tesis de pregrado, Universidad de Magallanes]. http://www.bibliotecadigital.umag.cl
Clare K (2013) The essential role of place within the creative industries: boundaries, networks and play. Cities 34:52–57. https://www.sciencedirect.com
Collipal Pichicona C (2017) Profesionalización de la actividad turística en Chile. Evolución y avances en el capital humano de la actividad turística chilena. Trilogía Ciencia-Tecnología-Sociedad 29(40):46–59. https://repositorio.utem.cl
CONAF (2013) Catastro y Evaluaci´on de Recursos Vegetacionales Nativos de Chile, Informe Regional. Uso del suelo de la XII Region
CONICYT PR (2009) Diseño de la Política Regional de Ciencia, Tecnología e Innovación de la Región de Magallanes
Cook D, Malinauskaite L, Davíðsdóttir B, Ögmundardóttir H, Roman J (2020) Reflections on the ecosystem services of whales and valuing their contribution to human well-being. Ocean Coast Manag 186(15) https://www.sciencedirect.com
Cruz Rueda E (2008) Mecanismos de consulta a los pueblos indígenas en el marco del convenio 169 de la OIT: el caso mexicano. Revista Pueblos y fronteras digital 3(5):74–111
Daw T, Brown K, Rosendo S, Pomeroy R (2011) Applying the ecosystem services concept to poverty alleviation: the need to disaggregate human Well-being. Environ Conserv 38(4):370–379. https://www.cambridge.org
Drake G (2003) 'This place gives me space': place and creativity in the creative industries. Geoforum 34(4):511–524. https://www.sciencedirect.com
EPAUSTRAL (2019) Memoria 2019. https://www.epaustral.cl
Erickson K (2018) Intellectual property and creative industries policy in the UK. In: Abbe E, Brown L (eds) Research handbook on intellectual property and creative industries. Edward Elgar Publishing, Cheltenham
FAO (2019) Fishery and Aquaculture Statistics. Global Fisheries commodities production and trade 1976–2017 (FishstatJ). In: FAO Fisheries and Aquaculture Department [online]. Rome. Updated 2019. https://www.fao.org/fishery/statistics/software/fishstatj/en
Felipe-Lucia MR, Martín-López B, Lavorel S, Berraquero-Díaz L, Escalera-Reyes J, Comín FA (2015) Ecosystem services flows: why stakeholders' power relationships matter. PLoS One 10(7):e0132232. https://journals.plos.org
Fischer A, Eastwood A (2016) Coproduction of ecosystem services as human–nature interactions—an analytical framework. Land Use Policy 52:41–50. https://www.sciencedirect.com
Gallina M, Williams A (2015) Variations in sense of place across immigrant status and gender in Hamilton, Ontario; Saskatoon, Saskatchewan; and, Charlottetown, Prince Edward Island, Canada. Soc Indic Res 121(1):241–252. https://pubmed.gov
Garcia SM, De Leiva Moreno JI (2005) Evolution of the state of fish stocks in the Northeast Atlantic within a precautionary framework, 1970–2003: a synoptic evaluation. ICES J Mar Sci, 62(8):1603–1608
Gertler MS (2004) Creative cities: what are they for, how do they work, and how do we build them? (pp. 1–21). Ottawa: Canadian Policy Research Network, Family Network
Guinovart JJ (2009) Mind the gap: bringing scientists and society together. Cell 137(5):793–795. https://europepmc.org
Gleisner C, Montt S (2014) Yagán: serie introducción histórica y relatos de los pueblos de Chile
Gustafson P (2001) Roots and routes: exploring the relationship between place attachment and mobility. Environ Behav 33(5):667–686. https://journals.sagepub.com
Harambour AR (2017) Soberanía y corrupción. La construcción del Estado y la propiedad en Patagonia austral (Argentina y Chile, 1840–1920) 1. Historia (Santiago), 50(2):555–596
Harambour A (2019) Soberanías fronterizas: Estados y capital en la colonización de Patagonia (Argentina y Chile, 1830–1922). Ediciones Universidad Austral de Chile

Hargreaves-Allen VA (2020) The economics of marine reserves. In Oxford Research Encyclopedia of Environmental Science. https://oxfordre.com

Hartwell H, Hemingway A, Fyall A, Filimonau V, Wall S (2012) Tourism engaging with the public health agenda: can we promote 'wellville' as a destination of choice? Public Health 126(12):1072–1074. https://pubmed.gov

Hein L, van Koppen CK, van Ierland EC, Leidekker J (2016) Temporal scales, ecosystem dynamics, stakeholders and the valuation of ecosystems services. Ecosyst Serv 21:109–119. https://www.sciencedirect.com

Irarrázaval F, Bustos B (2020) ¿Peces Gordos y Peces pequeños?: Los empresarios salmoneros, cambio institucional y conflicto sectorial en Chile. Polis (Santiago) 19(56):158–184. https://scielo.conicyt.cl

Jobstvogt N, Watson V, Kenter JO (2014) Looking below the surface: the cultural ecosystem service values of UK marine protected areas (MPAs). Ecosyst Serv 10:97–110. https://www.sciencedirect.com

Kaltenborn BP (1998) Effects of sense of place on responses to environmental impacts: a study among residents in Svalbard in the Norwegian high Arctic. Appl Geogr 18(2):169–189. https://www.sciencedirect.com

Lamers M, Haase D, Amelung B (2008) Facing the elements: analysing trends in Antarctic tourism. Tour Rev 63(1):15–27. https://www.emerald.com

Landers D, Nahlik A (2013) Final ecosystem goods and services classification system (FEGS-CS). U.S. Environmental Protection Agency, Washington, DC, EPA/600/R-13/ORD-004914, 2013. https://www.epa.gov

Laterra P, Nahuelhual L, Vallejos M, Berrouet L, Arroyo Pérez E, Enrico L, Jiménez-Sierra C, Mejía K, Meli P, Rincón-Ruíz A, Salas D, Špirić J, Villegas JC, Villegas-Palacios C (2019a) Linking inequalities and ecosystem services in Latin America. Ecosyst Serv 36:100875. https://www.sciencedirect.com

Laterra P, Nahuelhual L, Gluch M, Sirimarco X, Bravo G, Monjeau A (2019b) How are jobs and ecosystem services linked at the local scale? Ecosyst Serv 35:207–218

Liquete C, Piroddi C, Drakou EG, Gurney L, Katsanevakis S, Charef A, Egoh B (2013) Current status and future prospects for the assessment of marine and coastal ecosystem services: a systematic review. PLoS One 8(7):e67737. https://journals.plos.org

Mac-Lean Bravo CA (2010) Plan de Negocios para un Hotel Boutique Ubicado en la Patagonia

Martino S, Tett P, Kenter J (2019) The interplay between economics, legislative power and social influence examined through a social-ecological framework for marine ecosystems services. Sci Total Environ 651:1388–1404. https://www.sciencedirect.com

MEA (2005) Ecosystems and human Well-being: current state and trends, volume 1. Island Press, Washington, DC

Méndez C, Reyes O, NUEVO DELAUNAY, A. M. A. L. I. A., Trejo V, Barberena R, Velásquez H (2011) Ocupaciones humanas en la margen occidental de Patagonia Central: eventos de poblamiento en Alto Río Cisnes. Magallania (Punta Arenas) 39(2):223–242

MINEDUC, Ley N. (2011) 20.529 que crea el Sistema Nacional de Aseguramiento de la Calidad de la Educación Parvularia. Básica y Media y su Fiscalización, año.2011

Mora M (2012) Territorialidad nasa en Bogotá: apropiación, percepción y sentido de lugar. Cuadernos de Geografía: Revista Colombiana de Geografía 21(1):83–98. https://revistas.unal.edu.co

Nahlik AM, Kentula ME, Fennessy MS, Landers DH (2012) Where is the consensus? A proposed foundation for moving ecosystem service concepts into practice. Ecol Econ 77:27–35. https://www.sciencedirect.com

Nahuelhual L, Vergara X, Bozzeda F, Campos G, Subida MD, Outeiro L, Villasante S, Fernández M (2020) Exploring gaps in mapping marine ecosystem services: a benchmark analysis. Ocean Coast Manag 192:105193. https://www.sciencedirect.com

Nasimba C, Cejas M (2015) Diseño de productos turísticos y sus facilidades. Qualitas 10:22–39. https://www.unibe.edu.ec

Poblete EG, Olivares JQ, Experto T, González GM, Araya EL, Álvarez FT, Díaz JM (2013) Política Pública de Desarrollo Productivo para la Pesca Artesanal
Potschin-Young M, Haines-Young R, Görg C, Heink U, Jax K, Schleyer C (2018) Understanding the role of conceptual frameworks: Reading the ecosystem service cascade. Ecosyst Serv 29:428–440. https://www.sciencedirect.com
Rova S, Pranovi F (2017) Analysis and management of multiple ecosystem services within a social-ecological context. Ecol Indic 72:436–443
Rozzi R, Massardo F, Cruz F, Grenier C, Muñoz A, Mueller E (2010) Galapagos and Cape Horn: ecotourism or greenwashing in two iconic Latin American archipelagoes? Environ Philos 7(2):1–32. www.umag.cl
Rozzi R, Heidinger K, Massardo F (2019) Tracing Darwin's path in Cape Horn. University of North Texas Press, Denton
Ruckelshaus M, McKenzie E, Tallis H, Guerry A, Daily G, Kareiva P, Polasky S, Ricketts T, Bhagabati N, Wood SA, Bernhardt J (2015) Notes from the field: lessons learned from using ecosystem service approaches to inform real-world decisions. Ecol Econ 115:11–21. https://www.sciencedirect.com
Ruiz RMÁ (2005) Repensando la Enseñanza del Arte desde una perspectiva multicultural. Implicaciones didácticas. Enseñar ciencias sociales en una sociedad multicultural. Una mirada desde el mediterráneo, 60:33
Schirpke U, Scolozzi R, De Marco C, Tappeiner U (2014) Mapping beneficiaries of ecosystem services flows from Natura 2000 sites. Ecosyst Serv 9:170–179. https://www.sciencedirect.com
Scoones I, Wolmer W (2003) Introduction: livelihoods in crisis: challenges for rural development in Southern Africa. IDS Bull 34(3):1–14. https://onlinelibrary.wiley.com
SERNAPESCA (2001) R. A. Decreto Supremo 320. Ministerio de Economía Fomento y Reconstrucción, Santiago, Chile. 2019
SERNAPESCA (2019) Listado de precios promedio primera transacción. Información obtenida por ley de transparencia. Ministerio de economía, Fomento y reconstrucción, Santiago, Chile. 2019
Sernatur (Servicio Nacional de Turismo) (2018) Plan de Accion Region de Magallanes y la Antartica Chilena. Sector Turismo. 2014–2018. Gobierno de Chile.
SII, Servicio de impuestos (2019) Estadísticas de empresas por región y comuna. Departamento de estadísticas.
Spangenberg JH, von Haaren C, Settele J (2014) The ecosystem service cascade: further developing the metaphor. Integrating societal processes to accommodate social processes and planning, and the case of bioenergy. Ecol Econ 104:22–32. https://www.sciencedirect.com
Spencer KR, Werness-Rude MD (eds) (2015) Maya imagery, architecture, and activity: space and spatial analysis in art history. University of New Mexico Press, Albuquerque
Stokowski PA (2002) Languages of place and discourses of power: constructing new senses of place. J Leis Res 34(4):368–382. https://www.tandfonline.com
SUBPESCA (2020) Estado de Situación de Las Principales Pesquerías Chilenas, Año 2019. Departamento de pesquerías, División de Administración Pesquera. http://www.subpesca.cl
Subsecretaría de Turismo S (2016) Herramientas para emprendedores y empresarios turisticos. http://www.biblioteca.digital.gob.cl
Subsecretaría de Turismo (2019) Informe de cruceros internacionales en los puertos de Chile. Ministerio de Economía, Fomento y Reconstrucción, Santiago, Chile, 2019. 7
Subsecretaria de Turismo S (2020) Lineamientos para la comercialización de experiencias auténticas de turismo indígena en Chile. http://www.biblioteca.digital.gob.cl
Tecklin D (2016) Sensing the limits of fixed marine property rights in changing coastal ecosystems: salmon aquaculture concessions, crises, and governance challenges in southern Chile. J Int Wildl Law Policy 19(4):284–300. https://www.tandfonline.com
TEEB (2010) In: Kumar P (ed) The economics of ecosystems and biodiversity ecological and economic foundations. London and Washington, Earthscan

Tonko J (2008) Relatos de viaje kawésqar. Onomázein: Revista de lingüística, filología y traducción de la Pontificia Universidad Católica de Chile 18(2):11–47. http://onomazein.letras.uc.cl
Townsend M, Davies K, Hanley N, Hewitt JE, Lundquist CJ, Lohrer AM (2018) The challenge of implementing the marine ecosystem service concept. Front Mar Sci 5:359
Valenzuela JR (2015) Gestión local del turismo: el municipio y su rol en el desarrollo turístico en Chile. Gestión Turística (23):34–56. https://revistas.uach.cl
Vera Garnica J (2009) Cluster del Salmón en Chile: análisis de los factores de competitividad a escala internacional. Revista venezolana de gerencia 14(47):343–370. http://ve.scielo.org
Vergara X, Carmona A, Nahuelhual L (2020) Spatial coupling and decoupling between ecosystem services provisioning and benefiting areas: implications for marine spatial planning. Ocean Coast Manag 105455. https://www.sciencedirect.com
Worm B, Barbier EB, Beaumont N, Duffy JE, Folke C, Halpern BS, Jackson J, Heike KL, Micheli F, Palumbi SR, Sala E, Selkoe KA, Stachowicz J, Watson R (2006) Impacts of biodiversity loss on ocean ecosystem services. Science 314(5800):787–790. https://science.sciencemag.org
Zelada Muñoz S, Park Key J (2013) Análisis crítico de la Ley Lafkenche (N 20.249): El complejo contexto ideológico, jurídico, administrativo y social que dificulta su aplicación. Universum (Talca) 28(1):47–72. https://scielo.conicyt.cl

Chapter 16
Ecosystem Services and Human Well-Being: A Comparison of Two Patagonian Social-Ecological Systems

Luisa E. Delgado, Ignacio A. Marín, and Víctor H. Marín

Abstract The literature shows several possible relationships between ecosystem degradation, ecosystem services, and human well-being. In this chapter, we compare two Patagonian social-ecological systems (the Aysén watershed and Chiloé Island) regarding the use of ecosystem services by rural people and the relationships with ecosystem degradation. Results showed that people living in isolated, less modified systems have higher use of ecosystem services and material well-being. However, they have a lower quality of life. We discuss these issues and propose a conceptual model.

Keywords Degradation · Ecosystem services · Human well-being · The environmentalists' expectation

L. E. Delgado (✉)
Escuela de Ciencias, Facultad de Ciencias, Universidad de Chile, Santiago, Chile

Fundación Centro Transdisciplinario de Estudios FES-Sistémicos, Santiago, Chile
e-mail: ldelgado@uchile.cl

I. A. Marín
Ingeniería Comercial, Universidad Adolfo Ibañez, Santiago, Chile
e-mail: imarin@alumnos.uai.cl

V. H. Marín
Departamento de Ciencias Ecológicas, Facultad de Ciencias, Universidad de Chile, Santiago, Chile
e-mail: vmarin@uchile.cl

P. L. Peri et al. (eds.), *Ecosystem Services in Patagonia*, Natural and Social Sciences of Patagonia, https://doi.org/10.1007/978-3-030-69166-0_16

1 Introduction

The Millennium Ecosystem Assessment (MEA 2005) states that ecosystem services (ES) are essential for human well-being (HWB). Although the literature shows an average of 130 articles per year published during the last 15 years,[1] the relationships between ES and HWB are still far from being solved (Costanza et al. 2017; Cruz-García et al. 2017). When the MEA proposed a relationship between ES and HWB, it generated a new complexity into well-being assessments, requiring rethinking the efficiency and effectiveness of existing HWB indices such as the United Nations Index for Human Development (Delgado and Marín 2017). Current academic theories on HWB center on (a) subjective or psychological arguments, (b) economic ideas, and (c) sociological reasons or the normative ideal (Aguado et al. 2012). However, HWB indicators do not consider the benefits that people get from nature, often due to the lack of regional databases (Delgado and Marín 2017).

Sen (2009) defines HWB as the extent to which people can live the kind of life they value and develop their potentials under opportunities. So, many instrumental types of freedom determine the people's capacity to generate the kind of life they value. Today HWB implies personal and environmental security, access to material goods, good health, and social relationships related to decision-making freedom.

From this perspective, it is vital to re-evaluate the public policies related to the ecosystems' sustainability, since they define the opportunities for the access and distribution of the benefits ecosystems provide to the people. They also determine the people's capacity for ecosystems' management and, consequently, their well-being. Favorable public policies' perceptions by rural people (or ecosystems' people) would increase their opportunity to maximize their well-being through sustainable ecosystems (Delgado et al. In press).

Cruz-García et al. (2017) and Delgado and Marín (2016), working on ES and HWB in rural areas, show the benefits (economic, psychological, sociological, and cultural) of living in social-ecological systems (SES) with low anthropization levels. The authors propose that people feel happy living in such ecosystems in developing countries even when lacking essential services such as potable water and sewage systems. Nevertheless, the lack of health systems and the low presence of governmental institutions affect their well-being. Bachmann-Vargas and van Kopppen (2020) mention that in remote regions, such as Patagonia, low anthropization levels and high nature conservation values have a high cost for human beings given their distance to urban areas, generating unequal access to opportunities. Nevertheless, people consider that living in these systems is beneficial in terms of ES provision (Sangha 2019; Delgado et al. 2013; Zorondo-Rodríguez et al. 2019). For example, Delgado et al. (2013) show that free access to firewood, fungi, fishes, and macroalgae, among others, improves people's material well-being, an issue also discussed by other authors (Forest Trends, The Katoomba Group, and UNEP 2008).

[1] https://webofknowledge.com

ES provision changes due to anthropogenic causes, affecting HWB (Montes and Salas 2007). From this perspective, HWB acquires a multidimensional meaning, beyond the economy, to include health, security, social interactions, and recreation (MEA 2005). Quétier et al. (2007) state that local ES are defined by the social context, determining if an ecosystem component or function will bring concrete benefits to human life. However, the effective use of services is also conditioned by the relationships between social actors and their appropriation schemes. So, as Schmitz (2010) proposes for ecosystems, the links between anthropization, ES, and HWB in social-ecological systems seem contextual. What are their relationships in Patagonia? In this chapter, we explored this question comparing two social-ecological systems of Chilean Patagonia, with different degrees of anthropization: the Aysén watershed (Aysén Region of the General Carlos Ibañez del Campo) and Chiloé Island (Los Lagos Region).

2 Methods

2.1 Study Areas

2.1.1 The Aysén Watershed

The Aysén watershed is an extensive (surface area = 11,456 km^2) exhoreic catchment system located between 45°S and 46°S in the head of the Aysén fjord in the Chilean Patagonia (Fig. 16.1A). We have studied it from several social-ecological perspectives, including social participation (Bachmann et al. 2007), governance (Delgado et al. 2007), socioeconomic impacts (Yarrow et al. 2008), landscape structure (Torres-Gómez et al. 2009), public perception on socioeconomic development (Ianni et al. 2009), and ES and HWB (Delgado et al. 2013; Delgado and Marín 2016). Although nearly 60% of its forests were burned at the beginning of the twentieth century to produce grazing land for cattle (Torres-Gómez et al. 2009), the watershed is a low anthropization ecosystem given its low human population density (Delgado et al. 2013). Economic development in the basin comprises seven sectors (Yarrow et al. 2008), mining, aquaculture, forestry, industries, agriculture, tourism, and livestock, with aquaculture and mining as the main activities. Provisioning ES (i.e., water and firewood) used by the rural population represents an average economic contribution of 148 USD per month (Delgado et al. 2013).

2.1.2 Chiloé Island

Isla Grande de Chiloé (or Chiloé Island) is an insular space located in the northern part of the Chilean Patagonia (Fig. 16.1B). With a surface area of 8394 km^2 is among the ten largest islands on the continent. The current island's culture (Chilota culture) is a syncretism between local Mapuche-Huilliche peoples and Spaniards

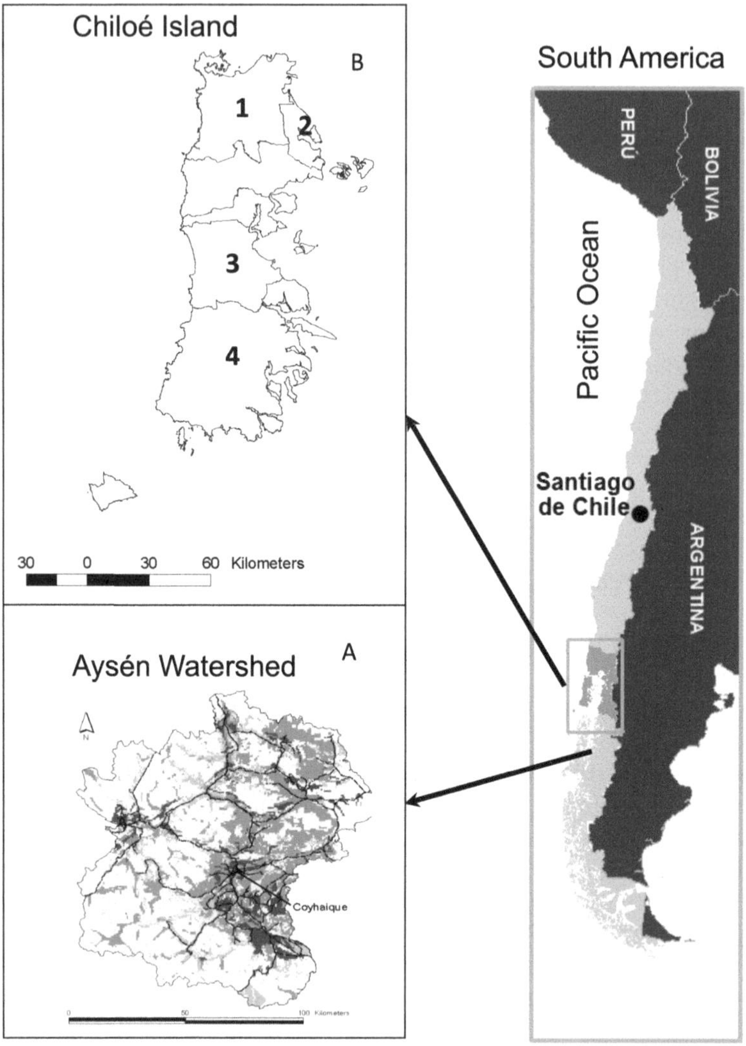

Fig. 16.1 Study areas of Chilean Patagonia. (A) Aysén watershed and (B) Chiloé Island. Numbers in the Chiloé Island correspond to surveyed communes. (1) Ancud and, (2) Quemchi, (3) Chonchi, and (4) Quellón

that arrived during the sixteenth century. Delgado et al. (2019) describe their social-economic and social-ecological conflicts. The island's main economic activities are salmon and mussel farming, potatoes agriculture, and cultural tourism (Pérez-Orellana et al. 2020). During the year 2017, the Chilean National Agency for Research and Development (ANID, former CONICYT) funded a 4-year research project[2] oriented to study its rural social-ecological relationships. In this chapter, we use part of its results.

2.2 Databases

Delgado et al. (2013) and Delgado and Marín (2016) describe all the social-ecological data for the Aysén watershed. In Aysén and also in Chiloé, the primary research tool was a social-ecological household survey. We developed the Aysén watershed survey in October 2009 (Delgado et al. 2013) and almost 9 years later (January–February 2019) on Chiloé Island. Although the surveys' structure was different (e.g., number of questions), the elements used to calculate the ES and well-being indices were the same.

Delgado et al. (2013) describe the survey structure for the Aysén watershed. The survey universe corresponded to the number of households in the Aysén watershed (5400). Thus, the sample size (83 households) generated 95% confidence and an 11% error. Chiloé's survey included four communes (Fig. 16.1B) with a total sample size of 228 households that, considering the survey universe (12,563 households), provided results with 95% confidence and 7% error. In both cases, we coded and analyzed the data using the IBM Software Statistical Package for the Social Sciences (SPSS Version 26). In the next section, we describe the information used to calculate ES and HWB indices. The full structure of the Chiloé survey is available from the first author of this chapter upon request.

Rural family size (3 ± 1.5) and the percentage of families owning the property where they live (85%) are similar in both areas (Aysén and Chiloé). However, while 100% of Aysén rural households use firewood as their primary house energy source, only 21.9% of Chiloé households use it exclusively, while 76.5% use a mixture of firewood and gas. Another difference between areas relates to the temporality of the households' head jobs. In the case of Aysén, 77% have permanent jobs, while for Chiloé, it decreases to 48.4%. Finally, although agriculture and cattle raising are the two most important activities in both areas, in Aysén, it represents 56% of the households, while in Chiloé Island, it is less than half (23.7%) of Aysén.

2.3 Indices

The question posed in this chapter required calculating three types of indices: (1) an anthropization index, (2) ES indices, and (3) HWB indices. According to Delgado and Marín (2020), ecosystem degradation is a concept that many people use, but few define. Here we define it as the changes in the ecosystem's structure and

[2] Proyecto Fondecyt Regular No. 1170532 awarded to Luisa E. Delgado

functioning due to human activities or anthropization. We used a modified version of Sanderson et al.'s (2002) "Human Influence Index" (HII) as a proxy to evaluate ecosystems' anthropization conditions (Delgado and Marín 2020):

$$\mathrm{HII}_{\mathrm{mod}} = \mathrm{Pop.density} + \mathrm{Urban}(\%) * \mathrm{Sc}_2 + \mathrm{agr} - \mathrm{for}(\%) * \mathrm{Sc}_3 \quad (1)$$

where Pop. density = influence scores of the human population per unit area (1/km^2) extracted from WCS-CIES (2005), using population data from the 2017 Chilean population census (INE 2017); Urban (%) = percentage of urban areas, extracted from CONAF/UACH (2014); Sc_2 = influence score for urban areas extracted from WCS-CIES (2005): agr-for (%) = percentage covered by agricultural or forestry sectors, obtained from CONAF/UACH (2014); and Sc_3 = average influence score for agricultural-forestry areas (6), modified from WCS-CIES (2005).

We also calculated the percentage of land use covered with native forests as an independent variable. We obtained the data from governmental databases (National Forest Corporation; CONAF[3]).

In the ES's case, we used provisioning, regulation, and maintenance services indices, using critical elements for Patagonia's rural people (Delgado et al. 2013). We calculated the provisioning services index *(PSI)* using one biotic provisioning service (wild plants as energy sources; wood) and one abiotic service (surface water or groundwater for drinking; water) according to the formula:

$$(PSI) = Wood + Water \quad (2)$$

where *Wood* = fraction of households (between 0 and 1) obtaining native wood from their properties and *Water* = fraction of households (between 0 and 1) getting water from sources other than the private (paid) service. In both cases, 0 = none of the interviewees obtain native wood from their properties, and 1 = 100% obtained it. Thus, if all rural households obtain wood from their properties and use water without paying a private service, *(PSI)* has a maximum value of 2. We chose these two variables since they represent the most basic provisioning service in rural areas of southern Chile (Delgado and Marín 2016).

In the case of the regulation and maintenance index *(RSI)*, we used four elements: cattle raising (other maintenance services by living processes according to CICES V5.1[4]); subsistence agriculture without using chemical fertilizers (biotic regulation of soil quality) and the services generated from ecosystems to maintain the chemical conditions of freshwaters using the formula:

$$(RSI) = Cattle + (Agriculture * Fertility) + Sewage \quad (3)$$

where *Cattle* = fraction of households raising cattle in their property, *Agriculture* = fraction of households developing subsistence (non-commercial)

[3] http://www.conaf.cl

[4] http://www.cices.eu

agriculture, *Fertility* = fraction of households not adding fertilizers, and *Sewage* = fraction of housing not using the public (paid) sewage system. All variables had a 0 to 1 rank. If value is 0, none use the service, and 1 means that all interviewees use them. If all rural households use these four services (e.g., regulation of biological conditions providing food for cattle, without using fertilizers, regulation of water, sewage, conditions[4]), *(RSI)* = 3.

Thus, *(PSI)* and *(RSI)* reflect the extent of the current use of ES by the rural populations of both areas, based on the services researchers found are the most important in Patagonia. So, the composition of *(PSI)* and *(RSI)* is contextual; their usage in other areas should include a preliminary analysis of whether these elements are also the most basic for the local peoples.

We calculated two HWB indices material conditions index *(MCI)* and quality of life index *(QLI)* using a modified version of the equations discussed by Delgado and Marín (2016):

$$\left(MCI\right) = Housing + income + Jobs \tag{4}$$

where *Housing* = fraction of property owners, *Income* = fraction of households over the poverty line based on the definitions of the Chilean Ministry for Social Development and Family,[5] and *Jobs* = fraction of household's heads with permanent jobs (either with an employer or independent). All variables had a 0 to 1 rank, where 0 means that no household has the analyzed attribute and 1 that all households have them.

$$\left(QLI\right) = Social\,connections + Education + Health\,status \tag{5}$$

where *Social connections* = fraction of households participating in at least one social organization, *Education* = fraction of households whose head had more than the 8 years of primary education, and *Health status* = fraction of households having a health program (either private or public). The maximum theoretical value for both indices is 3.

3 Results and Discussion

The degree of anthropization was three times higher for Chiloé Island than for the Aysén watershed (Table 16.1). Human population density explains over 70% of the difference. Still, native forest land cover is more extensive in Chiloé Island than in the Aysén watershed. The lower percentage of Aysén native forest seems to be related to the colonization process during the nineteenth and twentieth centuries that burned nearly 60% of the forests, transforming them into pastureland (Torres-Gómez et al. 2009). Even so, less than 1% of both systems' total area corresponds

[5] http:// www.desarrollosocialyfamilia.gob.cl

to forestry, agriculture, and urban land. Thus, Patagonia has a smaller forestry development than other southern Chilean terrestrial ecosystems, located outside Patagonia, such as the Río Cruces watershed in Valdivia, with a 38.5% forestry land (Delgado and Marín 2016).

Results showed higher ES indices associated with a lower degree of anthropization (Table 16.2). Thus, a larger percentage of households use the analyzed services in the Aysén watershed than in Chiloé Island. The sum of both indices (**ES**, Table 16.2) was nearly 25% lower than its maximum theoretical value (5) for Aysén and 50% lower for Chiloé. Hence, even in this area with low anthropization conditions such as Patagonia, rural households depend on other sources for basic provisioning (e.g., wood) and regulation and maintenance (e.g., soil fertility) services. How much do rural people save using services directly from the ecosystem? Table 16.2 shows an example of wood, the principal heating source in Patagonian households. A cubic meter of wood in Chile cost 48 USD in 2018 (INFOR 2018). Aysén and Chiloé surveys showed that families use 5 m^3 per month in the former and 2.1 m^3 in the latter (Table 16.2). Thus, if we use INFOR (2018) price, families using the wood ES save between 101 and 240 USD per month, contributing to their economic well-being.

Table 16.1 Value and components of the modified Human Influence Index (HII_{mod}) for the Aysén watershed (Aysén) and Chiloé Island (Chiloé)

Components	Aysén	Chiloé
Population density per km^2	0.80	11.70
Forestry land (%)	0.53	0.85
Agriculture land (%)	0.10	0.01
Urban land (%)	0.11	0.32
$\mathbf{HII_{mod}}$ (dimensionless)	5.88	18.39
Native forest (%)	46	61

Table 16.2 Ecosystem services indices and their components for Aysén watershed (Aysén) and Chiloé Island (Chiloé)

Indices and components	Aysén	Chiloé	Percent change
(PSI)	*1.58 ± 0.13*	*1.32 ± 0.14*	*−16*
Wood	0.77 ± 0.04	0.39 ± 0.13	−49
Water	0.81 ± 0.10	0.99 ± 0.02	+16
(RSI)	*2.16 ± 0.13*	*1.20 ± 0.06*	*−44*
Cattle	0.57 ± 0.06	0.16 ± 0.08	−72
Agriculture	0.69 ± 0.05	0.32 ± 0.09	−54
Fertility	0.83 ± 0.04	0.36 ± 0.36	−57
Sewage	1.00 ± 0.08	0.94 ± 0.08	−6
ES	**3.74**	**2.52**	**−33**
Monthly wood use (m^3)	5.00 ± 2.00	2.10 ± 1.40	
Savings (USD)	240	101	

PSI provisioning services index, *RSI* regulation and maintenance services index. **ES** = *(PSI)* + *(RSI)*. Savings = monthly money savings, in US dollars, due to the use of wood ES

Although both ES indices had the same trend (i.e., higher values in the system with lower anthropization), regulation and maintenance services showed the most substantial difference. The component with the most significant decrease (−72%) in the more anthropized system (Chiloé) was cattle raising. Indeed, the percentage of households raising cattle in Chiloé (16%) was closer to that of ecosystems outside Patagonia with higher anthropization levels (Delgado and Marín 2016). Although with a smaller percent change (−57%), a similar difference relates to not using fertilizers for subsistence agriculture. In this case, 83% of Aysén households do not use them and 36% in Chiloé.

On the other hand, it is notorious that rural people from both areas "trust" the ecosystem to clean their sewage. Indeed, results show that almost 100% of the households do not use public or private sewage services, evacuating their liquid and solid residues into the soil through cesspits. The surveys did not provide information to know whether this is a conscient trust or the easiest, less expensive thing to do. However, since more than 80% of households use rural water, untreated by private companies (Table 16.2), they trust that its quality is suitable for human consumption.

HWB, considering the sum of both indices (*(MCI)* and *(QLI)*), was higher at Aysén watershed than at Chiloé Island (Table 16.3). However, Chiloé Island showed higher social connections and health status than Aysén. So, quality of life tends to be better in more anthropized than in less anthropized systems. The *(QLI)* value for Chiloé (2.00; Table 16.3) was almost the same (1.99) as that calculated for the Río Cruces' watershed by Delgado and Marín (2016). So, remoteness and isolation seem to affect life quality (see also Bachmann-Vargas and van Kopppen 2020).

Quality of life can be assessed in terms of "objective measurements" (i.e., the elements included in the *(QLI)*), but it can also include "subjective elements" such as life satisfaction (Delgado and Marín 2016; Costanza et al. 2017). In the Chiloé survey, we included a question to assess that element of rural quality of life: Do you have negative comments on the meaning of rural life? We then estimated life satisfaction as the fraction (between 0 and 1) of households without negative comments.

Table 16.3 Human well-being indices and their components for Aysén watershed (Aysén) and Chiloé Island (Chiloé)

Indices and components	Aysén	Chiloé	Percent change
MCI	*2.36 ± 0.08*	*1.71 ± 0.27*	*−28*
Housing	0.86 ± 0.05	0.86 ± 0.04	0
Income	0.72 ± 0.05	0.36 ± 0.11	−50
Jobs	0.77 ± 0.07	0.49 ± 0.15	−36
QLI	*1.63 ± 0.07*	*2.00 ± 0.20*	*+23*
Social connect.	0.58 ± 0.08	0.77 ± 0.06	+33
Education	0.48 ± 0.08	0.43 ± 0.09	−10
Health status	0.58 ± 0.05	0.79 ± 0.11	+36
WELL-BEING	**3.99**	**3.71**	**−7**

MCI material conditions index, *QLI* quality of life index. **WELL-BEING** = *(MCI)* + *(QLI)*

Since we did not measure this HWB component in Aysén, we can only compare it with previous results obtained in other rural areas 250 km north of Patagonia (Río Cruces, Valdivia) by Delgado and Marín (2016). Results from Chiloé (0.69 ± 0.05) and Río Cruces (0.75 ± 0.09) did not differ significantly, suggesting that nearly 72% of the people living in these rural areas are satisfied with their way of life. Also, we would like to emphasize that the most frequent negative comment about rural life, in both areas, was "isolation in case of emergencies." Thus, living in areas with low anthropization levels, such as Patagonia, seems to benefit people in terms of ES and material conditions, but with high costs in terms of their quality of life (e.g., social connections, health systems, fast solution of emergencies). We discuss these issues in the following paragraphs.

Pristine ecosystems are almost nonexistent. Moreover, the few remaining are better described as having a low anthropization level (Delgado and Marín 2020), mostly found in areas far from large urban cities. Current studies show that even remote areas, such as the Aysén watershed in Patagonia (Torres-Gómez et al. 2009), are no longer pristine (e.g., Kruse 2016). Still, our results show that its anthropization degree is lower than that of Chiloé Island (Table 16.1). They further show that this decrease in anthropization increases ES use by rural people and their HWB (Tables 16.1 and 16.2). So, they agree with "the environmentalists' perspective" (Raudsepp-Hearne et al. 2010), but not for all well-being components. Two components of the quality of life index (*QLI*; Table 16.3) were smaller in the less degraded system, which relates to the lack of infrastructure and connectivity for Aysén (Bachmann-Vargas and van Kopppen 2020). Still, Aysén isolation may appear positive under the current health conditions (i.e., COVID-19 pandemic).[6] But one of Aysén's commercial ventures is nature-based tourism (Aliste et al. 2018). For example, a Google search using the phrase "Aysén tourism"[7] produced 339 thousand pages. One consequence was that the first COVID-19 case in the Aysén Region on March 13, 2020,[8] was a British tourist arriving on a cruise. Still, Aysén is, so far, the Chilean region with the lower contagion rate.[9] However, local people fear that in case of emergencies, they are isolated. So, the benefits of isolation seem to be only partial and, at times, problematic for rural people. Furthermore, studies of Chiloé Island show that rural people perceive a low governmental presence (Delgado et al. submitted) and a lack of services such as sewage systems, trusting the ecosystem for the quality of the water they consume.

The appropriation and use of ES in rural areas may have different forms, modifying their relationships with HWB. This multiplicity of forms requires transdisciplinary studies, especially if the goal is to solve real problems such as degradation and poverty (Balvanera et al. 2020). ES in Patagonia contribute to economic and material HWB (Table 16.2). Nevertheless, there is an increase in social-ecological

[6] https://www.emol.com/especiales/2020/internacional/coronavirus/casos-chile.asp

[7] Conducted on August 4, 2020

[8] http://www.eldivisadero.cl/noticia-56048

[9] https://www.emol.com/especiales/2020/internacional/coronavirus/casos-chile.asp

conflicts, mostly due to low social participation, lack of empowerment (Pérez-Orellana et al. 2019), and the strong relationship between economic and cultural activities and ES (Bachmann-Vargas and van Kopppen 2020) and also a lack of feedback mechanisms between local collective management groups and the government (Delgado et al. In press).

Hobbs et al. (2006) and Hobbs et al. (2009) proposed two concepts incorporating the effects of human beings on earth ecosystems: "novel ecosystems" and "historical ecosystems." The first type corresponds to ecosystems that have been modified by human actions. The second type corresponds to those that "retain the biota and ecosystem properties that were prevalent in the past" (Hobbs et al. 2009: page 600). However, since finding historical ecosystems is unlikely, even in isolated areas such as Patagonia, we propose to replace the terms by "novel social-ecological systems" and "historical social-ecological systems."

Novel social-ecological systems (novel SES) have been modified in their structure and processes by human actions. They are the most common type on earth today. Historical social-ecological systems (historical SES) are those with low anthropization levels, mainly in remote, isolated areas. Both systems generate services to human beings, but relationships are contextual (Fig. 16.2). Historical SES produce more regulation and maintenance services than their novel counterparts, but provisioning services may go in different directions (Raudsepp-Hearne et al. 2010). For well-being, humans living in novel SES have lower material conditions but a higher quality of life than those living in historical SES. However, the decrease in quality of life will depend on the degree of isolation, the active or passive role of

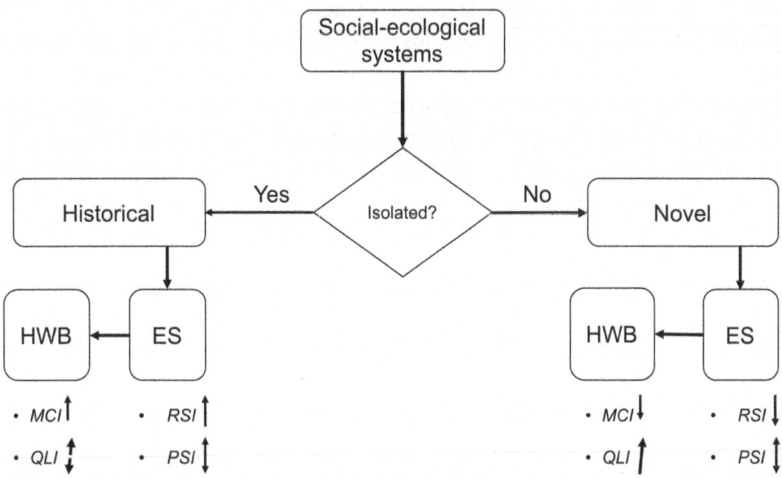

Fig. 16.2 Conceptual model on the relationships between the anthropization of historical and novel social-ecological systems, ecosystem services (ES), and human well-being (HWB). RSI maintenance and regulation services index, PSI provisioning services index, MCI material conditions well-being index, QLI quality of life well-being index. Arrows next to the indices show the expected trends for each type of service and well-being component. ↓ = decrease, ↑ = increase, ↕ = both trends. See Indices in section 2 for further details

the regional/local government, and their development pattern (Aliste et al. 2018). In summary, isolated areas such as Patagonia, with high ecological value and equally high nature' contributions to the material conditions of HWB, do not necessarily correspond to areas with high quality of life for the people living in its rural environments.

Ecosystem services are synergic to the development of other conditions that, together, satisfy essential human needs and the components of well-being. The MEA (2005) framework proposes four constituents: (1) security, (2) health, (3) basic material goods, and (4) good social relationships. Living in areas with low anthropization levels, such as Aysén in the Chilean Patagonia, contributes directly to the people's health (e.g., adequate oxygen levels, direct nature recreation, etc.). However, it also means living in an "isolation paradox" (Shihipar 2020), where remoteness may go against security and health in emergency conditions. So, remote areas require integrated ecosystem management approaches that should include ecosystems sustainability and human well-being (objective and subjective) as goals.

4 Recommendations for Policy Makers

Currently, the Chilean government has community development programs (PLADECO; Valdivieso 2020) in the studied areas. However, they do not include ecosystem services and their contribution to local households' economy, despite available information (Delgado et al. 2013; Nahuelhual et al. 2015; De la Barrera et al. 2015; Delgado and Marín 2016). Therefore, current territorial development plans have a vision of no relationships between economic development, human well-being, and ecosystem structures and functions. Bachmann-Vargas and van Kopppen (2020) propose that these partial territorial views result from the centralism of public investments in Chile and the geographic distribution of human capital. Thus, the management of territories with low anthropization requires local participation (social and governmental actors) and co-learning, given the contextual relationships between anthropization, ecosystem services, and human well-being (Fig. 16.2), to include the different social-ecological perspectives.

Acknowledgments This work for this chapter was financed by CONICYT-Chile FONDECYT Grant No, 1170532 awarded to L. E. Delgado.

References

Aliste E, Folchi M, Nuñez A (2018) Discourses of nature in new perceptions of the natural landscape in southern Chile. Front Psychol 9:1177

Bachmann-Vargas P, van Kopppen CSA (2020) Disentangling environmental and development discourses in a peripheral spatial context: the case of the Aysén regional, Patagonia, Chile. J Environ Dev. https://doi.org/10.1177/1070496520937041

Bachmann P, Delgado L, Marín V (eds) (2007) Toward the integrated ecological management of coastal zones in Chile: contributions of the ECOMANAGE project. Salesianos Impresores SA, Santiago, Chile
Balvanera P, Pérez-Harguindeguy N, Perevochtchikova M et al (2020) Ecosystem services research in Latin America 2.0: expanding collaboration across countries, dsiciplines, and sectors. Ecosyst Serv 42:10186
Aguado M, Calvo D, Dessal C, Riechmann et al (2012) La necesidad de repensar el bienestar humano en un mundo cambiante. Papeles de relaciones ecosociales y cambio global. No.119, pp. 49–76
CONAF/UACH (2014) Monitoreo de cambios, corrección cartográfica y actualización del catastro de recursos Vegetacionales Nativos de la Región de los Lagos. Valdivia, Chile. Corporación Nacional Forestal/Universidad Austral de Chile, Valdivia, Chile
Costanza R, de Groot R, Braat L et al (2017) Twenty years of ecosystem services: how far have we come and how far do we still need to go? Ecosyst Serv 28:1–16
Cruz-García G, Sachet E, Blundo-Canto G et al (2017) To what extent have the links between ecosystem services and human well-being been researched in Africa, Asia, and Latin America? Ecosyst Serv 25:201–212
De la Barrera F, Bachmann-Vargas P, Tironi A (2015) La investigación de servicios ecosistémicos en Chile: una revisión sistemática. Investigaciones Geográficas 50:3–18
Delgado LE, De Ríos R, Perevochtchikova M et al (In press) Water governance in rural communities of Chiloé Island, southern Chile: a multi-level analysis. J Rural Stud
Delgado LE, Marín VH (2020) Ecosystem services and ecosystem degradation: environmentalist's expectation? Ecosyst Serv 45:101177
Delgado LE, Marín VH, Asún R et al (2019) Environmental governance for the coastal marine ecosystem services of Chiloé Island (southern Chile). In: Delgado LE, Marín VH (eds) Social-ecological systems of Latin America. Springer Nature Switzerland AG
Delgado LE, Marín VH (2017) Human well-being and historical ecosystems: the environmentalist's paradox revisited. Bioscience 67:5–6
Delgado LE, Marín VH (2016) Well-being and the use of ecosystem services by rural households of the Río cruces watershed, southern Chile. Ecosyst Serv 21:81–91
Delgado LE, Sepúlveda MB, Marín VH (2013) Provision of ecosystem services by the Aysén watershed, Chilean Patagonia, to rural households. Ecosyst Serv 5:102–109
Delgado LE, Bachmann PL, Oñate B (2007) Gobernanza ambiental: una estrategia orientada al desarrollo sustentable a través de la participación local. Revista Ambeinte y Desarrollo de CIPMA 23:68–73
Forest Trends, The Katoomba Group & UNEP (2008) Payments for ecosystem services: getting started: a primer. ISBN: 978-92-807-2925-2
Hobbs RJ, Higgs E, Harris JA (2009) Novel ecosystems: implications for conservation and restoration. Trends Ecol Evol 24:599–605
Hobbs RJ, Arico S, Aronson J et al (2006) Novel ecosystems: theoretical and management aspects of the new world order. Glob Ecol Biogeogr 15:1–7
Ianni E, Feoli E, Marín VH et al (2009) Quantifying public perceptions on socio-economic development- an example from the Aysén watershed, Chile. Int J Sustain Dev 12:303–320
INE (2017) Resultados CENSO 2017. Instituto Nacional de Estadísticas Chile. http://resultados.censo2017.cl/
INFOR (2018) Bletín de precios forestales 2018. N° 165 Junio. Instituto Forestal. https://wef.infor.cl/publicaciones/precios/2018/06/precios201806.pdf
Kruse F (2016) Is Svalbard a pristine ecosystem? Reconstructing 420 years of human presence in an Arctic Archipelago. Polar Record 52:518–534
MEA (2005) Ecosystems and human well-being: synthesis report. In: Millennium ecosystem asessment. Island Press, Washington, DC
Montes C, Salas O (2007) La evaluación de los ecosistemas del milenio. Las relaciones entre el funcionamiento de los ecosistemas y el bienestar humano Ecosistema 16:1–11

Nahuelhual L, Laterra P, Villarino S et al (2015) Mapping of ecosystem services: missing links between purposes and procedures. Ecosyst Serv 13:162–172
Pérez-Orellana DC, Delgado LE, Marín VH (2020) Panarchy and ecosystem services: a social-ecological analysis of Chiloé Island, southern Chile. Ecol Soc 25(4):34
Quétier F, Tapella E, Conti G et al (2007) Servicios Ecosistémicos y actores sociales. Aspectos conceptuales y metodológicos para un estudio interdisciplinario. Instituto Nacional de Ecología México. Gaceta Ecológica 84–85:17–26
Raudsepp-Hearne C, Peterson GD, Tengö M et al (2010) Untangling the environmentalist's paradox: why is human well-being increasing as ecosystem services degrade? Bioscience 60:576–589
Sanderson EW, Jaiteh M, Levy MA, Redford KH, Wannebo AV, Woolwer G (2002) The human footprint and the last of the wild. Bioscience 52:891–904
Sangha KK (2019) Failure of mainstream well-being measures to appropriately reflect the well-being of Indigenous and local communities and its implications for welfare policies. Indig Policy J 3:1–14
Schmitz OJ (2010) Resolving ecosystem complexity. Princeton University Press, Princeton, NJ
Sen A (2009) Desarrollo y Libertad. Planeta, Bogotá
Shihipar, A (2020) Coronavirus and the isolation paradox. The New York Times, March 13, 2020. https://www.nytimes.com/2020/03/13/opinion/coronavirus-social-distancing.html
Torres-Gómez M, Delgado LE, Marín VH et al (2009) Estructura del paisaje a lo largo de gradientes urbano-rurales en la cuenca del río Aisén (Región de Aisén, Chile). Rev Chil Hist Nat 87:73–82
Valdivieso, G (2020) ¿Para qué sirve el PLADECO? Revista PLANEO. Espacio para Territorios Urbanos y Regionales. http://revistaplaneo.cl/2012/10/09/para-que-sirve-el-pladeco/
WCS-CIES (2005) Wildlife Conservation Society and Center for International Earth Science Information Network Columbia University (2005). Last of the wild project, version 2, 2005 (LWP-2): global human footprint dataset (geographic). Palisades: NASA Socioeconomic Data and Applications Center. Available Online at: http://sedac.ciesin.columbia.edu/data/set/wildareas-v2-human-footprint-geographic
Yarrow MM, Tironi A, Ramírez A et al (2008) An applied assessment model to evaluate the socioeconomic impact of water quality regulations in Chile. Water Resour Manag 22:1531–1543
Zorondo-Rodríguez F, Carrasco-Oliva G, Alfonso A et al (2019) Vinculando bienestar humano y servicios ecosistémicos: Ganancias y pérdidas de bienestar de comunidades rurales por cambios ecosistémicos. In: Cerda C, Silva-Rodríguez E, Briceño C (eds) Naturaleza en sociedad. Una mirada a la dimensión humana de la conservación de la biodiversidad. Ocholibros Editores, Santiago, Chile, pp 207–239

Chapter 17
Urban Planning in Arid Northern Patagonia Cities to Maximize Local Ecosystem Services Provision

Luciano Boyero, Leonardo Datri, Micaela Lopez, Clara Rodríguez Morata, Mario Robertazzi, Hernán Lopez, Maira Kraser, Tamara Canay, Juan Valle Robles, and Silvia Matteucci

Abstract The cities of arid North Patagonia are located mainly on valleys framed by slopes. In recent decades they have had exponential demographic growth. The unplanned expansion of the urban area occurs over sensitive and potentially attenuating landscape areas of extreme climatic and hydrological events. In this chapter we present through three case studies the general characteristics of the cities of the region, which should serve as the basis for their planning. We identified that both peri-urban natural areas and green infrastructure provide locally appropriated ecosystem services, with broad potential to ensure good conditions for urban sustainability.

Keywords Urban ecosystem · Risk · Fluvial · Valley · Urban planning

L. Boyero (✉) · L. Datri · C. Rodríguez Morata · M. Robertazzi · H. Lopez
T. Canay · J. Valle Robles
Laboratorio de Ecología de Bordes (LEB), Universidad de Flores (UFLO),
Cipolletti, Rio Negro, Argentina
e-mail: luciano.boyero@uflouniversidad.edu.ar

M. Lopez
Laboratorio de Ecología de Bordes (LEB), Universidad de Flores (UFLO),
Cipolletti, Rio Negro, Argentina

Instituto Patagónico de Estudios de Humanidades y Ciencias Sociales
(IPEHCS – CONICET), Neuquén, Argentina

M. Kraser
Laboratorio de Ecología de Bordes (LEB), Universidad de Flores (UFLO),
Cipolletti, Rio Negro, Argentina

Facultad de Ciencias del Ambiente y la Salud (FACIAS) – Universidad Nacional del
Comahue, Neuquén, Argentina

S. Matteucci
Consejo Nacional de Investigación, Ciencia y Tecnología (CONICET),
Buenos Aires, Argentina

P. L. Peri et al. (eds.), *Ecosystem Services in Patagonia*, Natural and Social
Sciences of Patagonia, https://doi.org/10.1007/978-3-030-69166-0_17

1 Introduction

Most of the ecosystem services consumed in cities are generated by ecosystems located outside of the cities themselves (Rees 1992; Folke et al. 1997; Rees and Wackernagel 1996; Deutsch and Folke 2005). Cities represent radically altered environments locally and have effects on a regional and global scale (Grimm et al. 2008). The effects they have on different biogeochemical and climatic processes at a global level are disproportionate towards the surface they occupy (3% of the earth's surface, 78% of the emissions of carbon, and 60% of water use, Brown 2001). Economic growth and demographic changes increase with the growth of urban settlements that increment not only the transformed local surface but also the demands of nearby and distant ecosystem services (Grimm et al. 2008). Despite this, the well-being of local populations is highly dependent on the ecosystem services provided in situ. Given that the proportion of the urban population has already exceeded the rural population in much of the world – with a sustained increasing trend – the total benefit of locally perceived urban ecosystem services is significant.

Among the ecosystem services provided and perceived in the urban landscapes themselves, we find provision, regulation, cultural, and support services. Although in most cases cities produce only a small proportion of the food they consume (Folke et al. 1997; Ernstson et al. 2010), there are examples where the production of urban agriculture plays important roles for food security, especially during economic and social crises (Page 2002; Buchmann 2009; Barthel et al. 2010; Barthel and Isendahl 2013); this represents a provision ecosystem service. The green infrastructure of cities regulates the local temperature and buffers the effects of the Urban Heat Island (Moreno-García 1994), which provides climate comfort and reduces GHG emissions (Grimmond and Oke 1999; Arnfield 2003; Pataki et al. 2011). Runoff mitigation is another regulation service provided by urban green elements such as urban forestry, which has a function of delaying the contribution of rainfall (Pataki et al. 2011), and public and private permeable green spaces which increase infiltration (Hood et al. 2007; Shuster et al. 2008). The high population density of urban ecosystems makes them, on the one hand, stressful environments for people and at the same time sites of cultural innovation par excellence. For this reason, cultural services take on particular importance in these ecosystems (Bolund and Hunhammar 1999; Chiesura 2004; Konijnendijk et al. 2013). These values are derived mainly from green infrastructures and urban nature reserves that have implications for recreation (Rall and Haase 2011) and aesthetic benefits on health (Maas et al. 2006; van den Berg et al. 2010), give a sense of place and social cohesion (Gotham and Brumley 2002), and promote cognitive development that enhances the understanding of ecological dynamics and strengthens social commitment towards the conservation of other services (Colding et al. 2006; Barthel et al 2010). In this sense, various cities around the world are recognizing the ecosystem services associated with green infrastructures and reconsidering the design and planning of their cities. Some examples of them are the city of Curitiba (Brazil) with the consolidation of an ecological planning model (Montaner 1999); the city of Madrid (Spain) that has prepared a Green Infrastructure and Biodiversity Plan with guidelines for strategic planning based on urban trees and green space networks (Madrid City Council 2018); and the "Room for the River" program undertaken by the Dutch government which aims to restore the floodplain of the Rhine river basin as a tool for integrated water and water risk management (Rijke et al. 2012).

Cities are complex and changing systems, where human action has modified the bio-geoecological patterns of ecosystems, water, and the quality of air. Cities could be defined as human ecosystems or socio-ecological systems differentiated from other ecosystems because of their high human density (Frank et al. 2017). According to the Population Division of the United Nations Department of Economic and Social Affairs (2018), about half of the global urban population lives in cities and settlements with fewer than 500,000 inhabitants, while approximately 1 out of 8 persons lives in one of the 33 megacities with more than ten million inhabitants. At present, the sustained increase in urban population and its population density ratio (United Nations Department of Economic and Social Affairs 2018) and the scientific evidence of changes in weather global patterns (Allen et al. 2018) make matters about urban planning necessary and pertinent. Urban and territorial planning (or their absence) are the main drivers of the structure and socio-ecological processes that develop in a city and its peripheries. The systemic functioning of a city and the relationship of its inhabitants with the system can be regulated through four axes: (i) urban density, (ii) design of public space, (iii) types of land use (mixed vs. specialized), and (iv) social cohesion. Each of these axes, in turn, is conditioned by the environmental context in which a city is located and is traversed by the historic context.

Some of the first and most relevant settlements and urban centers were founded in arid areas. Most of them were developed near the coast or in the margins of some river or lagoon and its river valleys (Brown 2001). The geographic location was a decisive settlement factor, comprising the survival of its inhabitants and human activities, mainly associated with the use of water for agriculture (Evenari et al. 1982). This river condition is also the case for most of the urban settlements in the arid region of northern Patagonia. However, the technological and socioeconomic contexts of the development of modern cities in arid regions combine new factors and scenarios, usually without urban planning or strategy, with territorial, socioeconomic, and climatic risks produced by urbanization. Somehow, the distribution of the world's deserts matches the location of less privileged populations, particularly in relation to the availability of arable land, natural resources, population structure, and socioeconomic potential (Alshuwaikhat and Nkwenti 2002). In most settlements within deserts or semi-desert areas, challenges arise from extreme temperatures and large daily thermal amplitudes. In addition, these areas are often far from supply centers of tangible and intangible resources and especially far from potential markets which makes industrial development difficult. However, the arid region of northern Patagonia has an important hydrocarbon reservoir in much of its territory which has been exploited in recent decades creating skewed effects and unbalanced regional development. About 85% of the non-tax revenues of the Province of

Neuquén represent hydrocarbon royalties, which makes almost 40% of the total revenues depend on this economic activity. The province's gross geographic product (GGP) is comprised in order of importance by mining and the industry associated with services and real estate and commercial activities, which in turn are highly dependent on the socioeconomic dynamics of the former. These three sectors by 2017 represented 63.42% of the province's GDP (Provincial Directorate of Statistics and Censuses). This economic growth and its projections drove population growth, given by a high rate of people settling in the region (Perrén 2010; Landriscini et al. 2014; Roca and Manacorda 2017; Pérez 2018). On the other hand, the inequities generated by price distortions and the unsteadiness of the activity have increased social and residential segmentation (Perrén et al. 2019, Perrén and Pérez 2020). The lack of territorial planning and the absence of productive diversification have generated an anarchic occupation of the territory which results in large irregular settlements in areas of high environmental risk.

Urban river floods associated with extreme precipitation events have the potential to cause significant economic and personal damage since they often impact residential, commercial, and industrial areas (Leitao et al. 2013). This represents the greatest urban environmental threat in the region of northern Patagonia. Furthermore, it is expected that the impact of climate change increasing the frequency and magnitude of extreme events magnifies the risk of hydroclimatic disasters as floods, storms, landslides, and droughts in arid zones (IPCC 2007; Ugarelli et al. 2010). When talking about risk, it is often expressed in terms of danger, vulnerability, and exposure (Barroca et al. 2006). The identification of the local factors that control these components is the first step towards a correct approach to disaster risk reduction in any location (Sudmeier-Rieux et al. 2019). In the area of the Alto Valle del Río Negro and the lower basins of its Neuquén and Limay tributaries, the greatest threat associated with the water factor has been the floods derived from riverine floodings that threaten coastal developments. More recently, and associated with trends of increment of rainfall levels that various authors predict for the arid region of northern Patagonia (Barros and Mattio 1978; Barros and Rodriguez Sero 1979; Castañeda and González 2008, in Romero et al. 2014), an increasing probability is expected of avalanches and mass movements in areas of plateaus and slopes. Then, as Maskrey (1993) states, disasters are not "natural"; rather, it is essential to incorporate the social dimension into risk studies. From a contextual and historical approach (García Acosta 2005), the social construction of the risk refers to the different material conditions (demographic, socioeconomic, morphological, and functional factors) and immaterial (ideological, cultural and political-territorial) that entail a differential vulnerability within the same territory (Rebotier 2009).

In recent decades, the city of Neuquén has established itself as the main logistics, commercial, financial, and administrative center of northern Patagonia and represents the node of a region that is experiencing strong population growth accompanied by a dispersed and fragmented occupation of the territory (Perrén 2010; Landriscini et al. 2014; Roca and Manacorda 2017; Pérez 2018). The urbanization process in the region has dramatically degraded the natural protection offered by ecosystems against these threats, exacerbating urban risk (United Nations 2018). In this sense, during the last 20 years, the urban sprawl has expanded both on plateau and slope areas, as well as along the banks of the Limay and Neuquén rivers. This situation has modified land uses, impacting on the natural infiltration capability of the land, speeding up surface water runoff, and causing flooding. On the plateau area, the process of urbanization has reduced vegetation cover and biodiversity with the consequent degradation of the soil structure and its infiltration capacity due to the increment of the waterproofed surface. In the context of river catchments/riverine, urban development projects have been authorized on riparian forests, hindering their function as flood mitigators, modifying its ecosystem structure, and decreasing the infiltration rate of the soils and, therefore, the capacity of the forest to mitigate flooding. In other words, the urbanization process has diminished the functionality of ecosystems, limiting considerably the provision of ecosystem services (ES) and thus reducing the ecological resilience of the city environment.

In this chapter we will describe the potentialities of urban planning to maximize the provision of urban ecosystem services in the arid northern Patagonia. In addition, the relevance of the urban ecosystem services approach for disaster risk management will be evaluated based on case studies. We will present three case studies that are located close to the Alto Valle del Río Negro but whose characteristics are repeated in other urban enclaves of the arid northern Patagonia region. In the first place, the results of a series of investigations that address the links between biodiversity conservation, the provision of cultural services, and flood risk management will be presented. Secondly, the relationship between regulation services and alluvial risk will be evaluated, through an urban natural protected area undergoing transformation which is located on a site between the runoff area and an urbanized floodplain. Finally, results of a study of urban trees effect on thermal regulation and its variation on topographic gradients will be presented. From all the cases, proposals for urban planning from the ecosystem services approach are derived and presented.

2 Historical and Ecological Context of Arid North Patagonia

The geographic area that is the subject of study in this chapter is composed of two dimensions, defined with different degrees of subjectivity. We define the term "arid" from the Aridity Index (AI) proposed by Martonne (1926), which is calculated based on the mean annual rainfall and the mean annual temperature ($AI = P/T + 10$). In this chapter we will address a region that encompasses areas classified as semi-arid, arid, and hyperarid according to this index. On the other hand, the term "Patagonia" is a concept crossed by geographical, geological, biological (botanical and fauna), social, cultural, political, and historical perspectives. Its northern limit is diffuse and variable depending on the perspective. On this occasion, we will consider the Colorado river and its tributary, the Barrancas river, as the northern limit of the region. Since in this chapter we only address the northern sector of Patagonia, it is necessary to determine a southern limit of this subregion. We will consider the political limit between the provinces of Río Negro and Chubut (42nd parallel south) as the southern limit of the northern Patagonian subregion. From that border on, towards the south, progressive changes in climatic terms are evident (mainly lower temperatures). This conditions other biological and social parameters (lower population density and connectivity) and hardens living standards. Patagonia's Limay and Neuquén rivers drain in a west-east direction and converge to form the Negro river, which eventually flows into the Atlantic Ocean. The study area includes a total population of 1.180.000 inhabitants, mostly residing in urban conglomerates located on the river valleys named above (784.000 inhabitants approximately according to the census 2010).

The occupation of arid northern Patagonia was historically and spatially associated with the river valleys that cross it from west to east and towards the sea coast (Prates 2008; Mange 2019). According to Álvarez Palau (2012), from the pre-Columbian stages to the present, five phases of development and use of natural resources can be counted (Table 17.1): (1) a pre-Columbian period, with nomadic hunter-gatherer societies; (2) an equestrian stage, with an increase in commercial exchanges (cattle, textiles, salt) with other regions; (3) a period of Euro-American colonization, construction of the railroad, and the irrigation system (from 1879 to 1960); (4) a boom stage of intensive fruit growing (from 1960 to 1980), and (5) a post-crisis stage of fruit growing that is synchronous with the increase in the exploitation of energy resources (hydroelectric and hydrocarbon).

Different phases can be recognized within the period that begins with the late nineteenth-century colonization which determine the subdivision and distribution of land (Blanco 2007; Muñoz 1996). This stage of development began with the arrival of the railway at the confluence of the Limay and Neuquén rivers in 1902, built by the British company Ferrocarril del Sud, connecting what we now know as the Alto Valle with the Bahía Blanca Port. Irrigation works were being built, as the railroad developed, starting at the Ballester dam on the Neuquén river. This system of hydraulic works and canals configures the irrigated oasis of the Alto Valle, from which the colonization process and distribution of land to small farmers (mostly European) became the norm.

With the development of agriculture and urbanization, the heterogeneity and complexity of the valley's landscape increased (Datri and Maddio 2010, Datri et al. 2016). Due to the regulation of the rivers and the development of the irrigated oasis, new ecosystems of swamps, hydrophilic grasslands, and riparian forests of exotic Salicaceae originated, while the forest matrix was reduced to discrete patches of Zygophyllaceae and saline grasslands. At the end of the 1980s, the global context of the fruit market started a crisis in the activity that, together with the economic development based on the extraction of hydrocarbons, motivated the abandonment of productive lands. This abandonment is encouraged in turn by urban policies deployed since the beginning of this century that favor the current urban expansion over the valley and the irrigated oasis to the very riverbeds of the Limay, Neuquén, and Negro rivers. The data provided by the Regional Agricultural Census and the National Agricultural Censuses show the reduction in production and increased land abandonment (Chiementon and Cogliatti 2011; Urraza and

Table 17.1 Phases of development and use of natural resources in river valleys of arid northern Patagonia

	Pre-Columbian stage	Equestrian stage	Colonization, railroad, and irrigation system stage		Fruit growing and hydroelectric energy booming stage	Post-crisis of fruit growing and exploitation of fossil energy stage
	Until ~1700	~1700 to 1879	1879 to 1960		1960 to 1980	1980 to present
Milestones and historical processes	Nomadic hunter gatherers, close to river valleys. The end of the stage begins with the arrival of Europeans to the continent	Commercial exchange to the east and west of the Andes (cimarrón cattle, textiles, and salt). Semipermanent strategic settlements near rivers and marshes	1879: Military conquest, subjugation of native peoples, and state control of the territory. Occupation and livestock agricultural expansion in valleys 1882: First agricultural colonies 1889: Construction of a railway line	1899: Hydrological surveys. 1916: Dam on the Neuquén river and start of the main irrigation canal (130 km) 1918: Finding of oil in Plaza Huincul 1930: Ending of the works of the irrigation system in the Alto Valle del Río Negro	Continuation of the water regulation plan (4 dams in the Neuquén river and 4 in the Limay river) Expansion of fruit growing upriver and towards clogged secondary channels and consolidated bars. 1975: Extraordinary rain (deaths, evacuees, and material damage)	2006: Flow river record registered to Neuquén river (10,300 m^3/s) 2011: Announcement of the existence of a hydrocarbon reserve estimated at 27,000 M 2014: Extraordinary rainfall (historical record of intensity) 2016: Extraordinary rainfall (urban floodings in Neuquén city) Increased population growth rate Growth of the urban sprawl on riverbank and riverbed areas, and towards arid plateau

Table 17.1 (continued)

	Pre-Columbian stage	Equestrian stage	Colonization, railroad, and irrigation system stage		Fruit growing and hydroelectric energy booming stage	Post-crisis of fruit growing and exploitation of fossil energy stage
	Until ~1700	~1700 to 1879	1879 to 1960		1960 **to** 1980	1980 to present
Use of natural resources	Varied diet (small fish, mollusks, some fruits) Chiselling of local boulders (silicon and basalts)	Diet with high energy return (large wild vertebrates and feral cattle). The consumption of fish and shellfish is abandoned Use of pastures and wetlands for cattle and horse grazing	Incipient transformation of the natural surface of the valley for production of cereals and then alfalfa First implantations of exotic forest specie. Hydrological risk (Neuquén river floods of up to 9000 m3, every 5–10 years)	Accelerated increase of vine and fruit trees (apple and pear), in valleys and floodplains Expansion of network of forest curtains for shelter, firewood, and wood (Salicáceas) Livestock exploitation on the semi-arid plateau (mainly sheep) Energy supply with power plants based on coal or oil.	Homogenization of river flows (max. 1100 m3 / s, TR 100 years) Transformation of fluvial geomorphology (from flanged rivers to meandering rivers with consolidated islands) Leveling and clearing of banks and clogged channels. Formation of agricultural soils Growth of urban nodes, constituting a dispersed metropolis on the river valleys. Displacement of the native species *Salix humboldtiana* by invasion of exotics (*Salix* spp. and others)	Increase in the frequency of convective rainfall in the central-eastern zone Annual rainfalls decrease in west (high basins or great rivers) Increase in agricultural area abandoned or converted to urban uses (housing, parties, sports) Increased exposure of the urban population to threats such as avalanches and floods Hydraulic fracturing for hydrocarbon extraction

Muñiz 2017). This situation occurs simultaneously with a process of increased population growth rate of the main urban center, the city of Neuquén, and the expansion of the urban area over the irrigated valley which includes all the cities in the current conglomerate.

The new urban agents modeling the valley landscape give place to a new configuration of the rural landscape and society-nature relations. At the interface of the city with the irrigated valleys, floodplains, rivers, and the valley's slope, new areas of very dynamic changes are established, among which there is a new exchange of materials, energy flow, and opportunistic species, which give rise to a whole riparian neo-ecosystem (Datri et al. 2016). This new condition on the urban edges represents new direct and indirect threats against the well-being of the population. On the one hand, the risk is exacerbated by the occupation of areas on the valley's slope – with steep slopes subject to erosive processes – and of the floodplains of the Neuquén and Limay rivers. On the other hand, the same process affects the biodiversity of riparian wetlands and the vegetation cover of the highest part of the micro-basins that drain the Patagonian plateau towards the river valleys. Thus, a two-way map of vulnerabilities is configured, which affects the security of the urban population at the same time that the resilience of the ecosystem is weakened to absorb the changes imposed by the ecohydrological dynamics of the system.

The riparian forest allows, in the proposed regional context, to link the conservation of biodiversity, the provision of cultural ecosystem services, and the management of flood risk, favoring both the natural and social resilience of the urban ecosystem. In the same way, although in a more limited way, the xerophilous vegetation of the woods constitutes one of the few variables of the ecosystem of the arid slope of the valley. At the two extremes of the arid-riverine gradient, the Metropolitan Region leads to changes in the natural landscapes of geo-biophysical support. For these reasons, its approach implies two perspectives to the variables that define the ecohydrological dynamics that configure the functioning dynamics of both ecosystems and consequently the services they offer or are affected by.

The Metropolitan Region of Confluence (MRC) has unique characteristics, given that few regions of the world have a similar structural complexity. The urbanization model of the MRC – fast and expansive in the territory – increasingly distant from the compact and sustainable city model has not attended or planned a symbiosis between the landscape, the climate, and the city. Instead, the city of the MRC, understood as everything that has urbanized land, seems to have sought urban and territorial models where the economy of energy resources has had and has great economic success. Among these urban models, we could mention the city of Dubai, with major architectural and infrastructure milestones, as symbols of strength and economic success, or the city of Riyadh where the lack of comprehensive and sustainable urban and territorial planning has urbanized and completely transformed the landscape in which the city settles.

One peculiarity of urban landscapes is that transformations tend to homogenize ecosystems, erasing original differences and replacing them with structures and functions that are common for cities in different parts of the world (Pickett et al. 2011; McKinney 2006; Kühn and Klotz 2006; Ignatieva 2011). However, the topographic and water subsystems that underlie and contextualize cities geographically can impose on them unique imprints. One of those singularities in the landscape of arid northern Patagonia is the establishment of cities in river valleys delimited by high slopes and foothills ("bardas"). This topographic difference configures the plateau-valley slope that determines most of the biophysical processes that shape the urban landscape.

River valleys contain the highest proportion of local biodiversity, and on a regional scale, they compose and shape large biogeographic islands and biological corridors between the Andean region and the Atlantic sea coast. On the profile of the valley slope and the riverbeds, there is an energy gradient that results from the precipitation water runoff on the slope. In the longitudinal direction, there is a cyclical flood contribution that deposits silts and other fine sediments, which give origin to soils suitable for agriculture (Morello 1995). For these reasons, according to the geomorphology of each valley, there are different instances of retention and cycling of nutrients and water availability in phreatic levels close to the soil surface that give unique edapho-ecological conditions to the arid regional context (Datri et al. 2016). The natural processes of sediment deposition, added to anthropic processes of leveling and clearing of the banks, favored the generation of soils with high agricultural aptitude, which due to their productive capacity are identified as providers of provision services.

But this context establishes a set of natural risks that are accentuated with climate change and with deficiencies or absence of territorial planning. On the one hand, occupation of river floodplains involves risks of flooding and social displacement. On the other hand, settlements at the foot of slopes induce risks due to avalanches and mass movements. Finally, rising temperatures create the possibility of heat waves in cities in arid regions. Within the management of environmental risk, ecosystem services offer a conceptual and practical framework, where the provision of regulation services of the modeling effects of energy in the landscape can be perceived by the local population with an increase in security conditions (Munang et al. 2013). Ecosystem services associated with river landscapes can include protection against flooding of wetlands and more favorable habitability conditions due to the regulatory factor of climatic extremes in an arid context. Urban planning that contemplates the regulatory services provided by both riparian environments and ecological infrastructure would reduce the adverse effects of the natural dynamics of the landscape and climate change.

3 Study Cases: Methods and Results

In the framework of a research program, different lines of work were approached to explore, in general terms, urban strategies to plan cities that maintain beneficial ecosystem functions for their inhabitants. Within the program, data derived from each independent line are gathered and interrelated in a single Geographic Information System (GIS) associated with the study area (Metropolitan Region of Neuquén). The hypotheses of each project constitute the theoretical set that seeks to

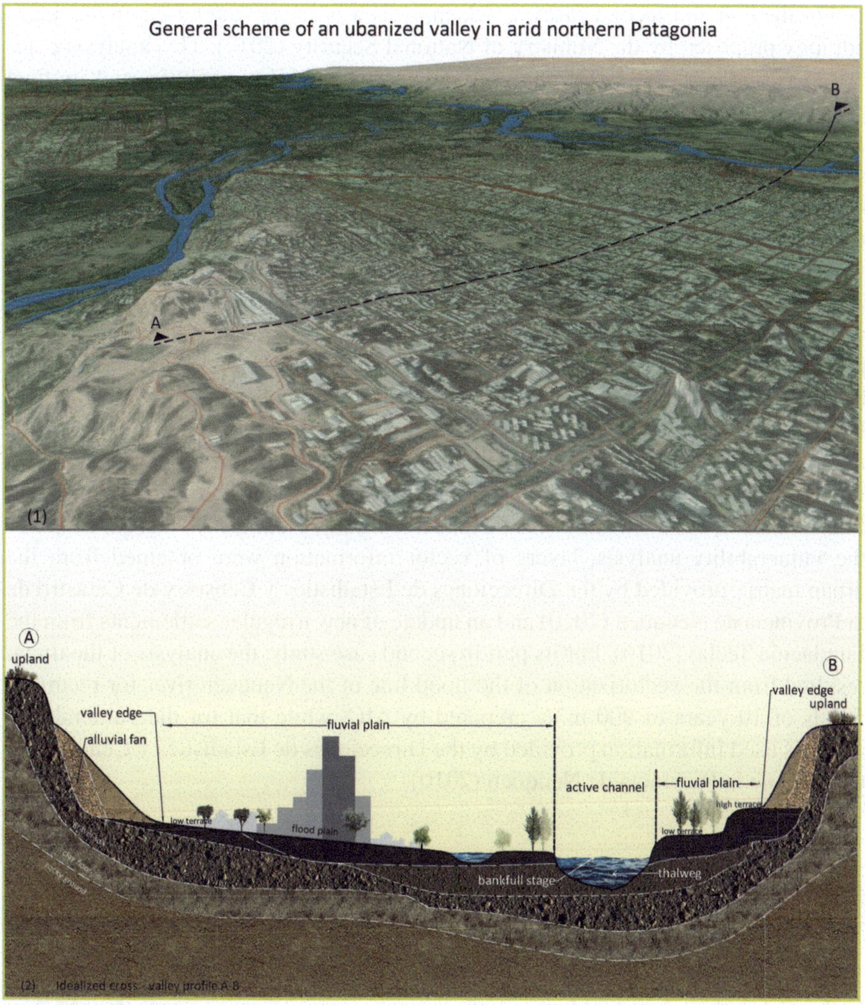

Fig. 17.1 General scheme of the valleys, with characteristic topographic profiles of the cities of the arid northern Patagonia. Profile AB: cut of the valley of the confluence of the Limay and Neuquén rivers. (Source: own elaboration based on a DEM ALOS PALSAR)

define and test a more operative landscape concept for the studies of the valleys and their processes of change in northern Patagonia (Fig. 17.1). Using descriptive and experimental approaches, a set of results from three cases was analyzed from a perspective directed by the concept of urban ecosystem services. The cases address the following issues:

(a) Riverbanks for flood risk management
(b) The urban transformation of an urban natural area on high runoff basins
(c) The spatial variations of the effect of urban trees as a thermal regulator

For the first and second cases, a risk analysis was carried out based on the methodology proposed by the Ministry of National Security (2017). This analysis comprised two stages, the threat analysis and the vulnerability analysis, and its final product was a cartographic representation (risk maps) that identified various risks associated with water threats in the territory. For this purpose, time series of SPOT-5 and Sentinel-2 scenes, Digital Elevation Models (DEM) provided by the National Geographic Institute (IGN) at 1: 5000 scale, and Landsat 8 OLI/TIRS images were used. These scenes form the basis of the plot for digital processing and obtaining green indexes. The data were superimposed on the data and made it possible to analyze the characteristics of urban processes and their relationship with ecosystem services. Specifically for the first case study, the analysis of the threat consisted of the vectorization of an extraordinary flood of the Limay river that occurred in 2001, by means of a combination of RGB of two infrared bands and one visible from a Landsat image July 5, 2001 belonging to a flood scene of July 17, 2001, corresponding to a flood of 1700 m^3/s. This flood line was taken as a reference for a maximum flood of the Liam river, close to the equation line established by the Autoridad Interjurisdiccional de las Cuencas de los Ríos Limay, Neuquén y Negro (AIC). For the vulnerability analysis, layers of vector information were obtained from the urban matrix provided by the Direcciones de Estadística y Censos y de Catastro de la Provincia de Neuquén (2010) and an update of new irregular settlements from the Fundación Techo (2016). For its part in second case study, the analysis of the threat resulted from the vectorization of the flood line of the Neuquén river for recurrent floods of 10 years of 900 m3/s prepared by AIC, while that for the vulnerability analysis used information provided by the Direcciones de Estadística y Censos y de Catastro de la Provincia de Neuquén (2010).

3.1 Case 1: Multipurpose Riverbanks: Risk Management and Biodiversity Conservation

In the region there is a double trend of urban expansion on river coasts (Landriscini 2017): on the one hand, the advance of irregular settlements on the peri-urban and, on the other hand, associated with an ideal of "green imaginary" (Irarrázaval Irarrázaval 2012, p. 79) of tranquility and purity, the emergence of real estate ventures. The neighborhoods Valentina Sur Rural (north bank of the Limay river, in the city of Neuquén) and Las Perlas (south bank of the Limay river, in the city of Cipolletti) are clear examples of this. In the sector, there are six popular neighborhoods, La Costa (Valentina Sur Rural neighborhood), Puente Santa Mónica, Costa Esperanza, Primeros Pobladores, Parque, and Vista del Valle (Las Perlas neighborhood). A few meters away are six closed neighborhoods, Costa Nogal, Ferri, Lomas del Limay, Sauces del Limay, La Castellana, and Los Notros. While the former are born in response to a process of exclusion based on the commercialization of urban land, the latter arise associated with state projects for urban renewal and revaluation of the riparian coast. However, both types of growth imply the advance of urban land over wetlands and riparian forests, which constitute biologically unique environments in the region with reduced spatial occupancy. By means of an analysis of exceptional floods, the risk of flooding on the banks of the Limay river was evaluated. This analysis sought to contrast the hydrological dynamics resulting from current development as opposed to natural dynamics and to investigate urbanization alternatives that can prevent the occurrence of disasters while promoting the conservation of ecosystem functions useful to local people.

Figure 17.2a shows the result of a flood risk estimation for an event similar to that of July 2001, with $Q = 1740$ m^3/s. For an event of these characteristics, with an expected recurrence period of 14 years, most of the areas immediately adjacent to the main river channel present a high risk of flooding. This result becomes relevant when we analyze the process of occupation of the Limay river coasts in the last decades. In this sense, it can be observed that most of the latest developments built since 2016 are in many cases within the limits of the 2001 flood (Fig. 17.2a).

In fact, when we observe the potential evacuation lines of a flood flow associated with an event such as the one in 2001 (Fig. 17.2b), we see that they overlap recently urbanized areas in these sectors. On the other hand, associated with the phenomenon of urbanization, on the map with a continuous black line (Fig. 17.2a, b), the raised linear infrastructures corresponding to embankments, artificial slopes, and raised streets are indicated. In the case of the structures present in the area of South Valentina (Fig. 17.2b), it is observed that in many cases these interrupt or modify the natural lines of flow evacuation in the event of flooding. These infrastructures have been built both to facilitate access to these new developments and to protect them from the flooding of the Limay river. However, the interruption of natural flood flow lines can generate impacts in several ways (SEPA 2011). On the one hand, the modification of the flow direction can channel the water to specific points of the coast, generating problems of erosion, sedimentation, and impoundment downstream (Arnaud-Fassetta et al. 2005). On the other hand, in the event of a

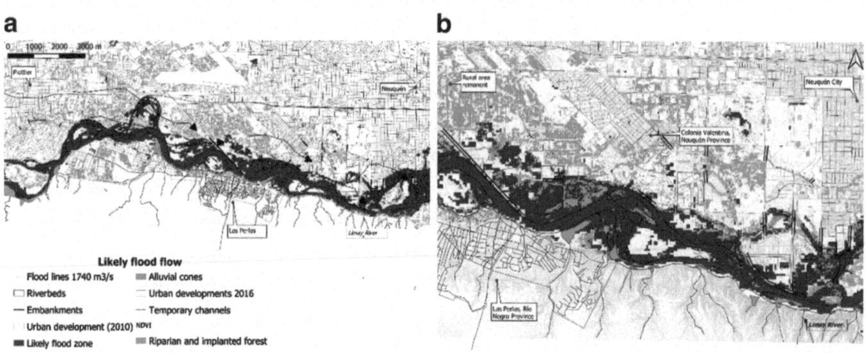

Fig. 17.2 (**a**) Flood risk map of the Limay river for a recurrence flow of 14 years ($Q = 1740$ m^3/s). (**b**) Possible channels for channeling a flood ($Q = 1740$ m^3/s) in an area occupied by embankments, artificial slopes, and elevated streets

possible breakdown of these infrastructures, urbanizations may be isolated without communication routes that would allow the evacuation of the population or the entry of aid. In line with this, until recently, most risk management plans have resorted to options based on conventional engineering measures, also called "gray" infrastructure. However, the role that ecosystems play in providing essential services to reduce and mitigate flood risk has been widely recognized in recent years (Nel et al. 2014). Thus, under the right circumstances, healthy riparian ecosystems that replace or complement gray structures can play a critical role in flood risk management, mitigating the impacts of disasters, reducing physical vulnerability, and strengthening resilience (Ilieva et al. 2018). It is essential, in this sense, to develop capacities and knowledge for the correct management and planning of an urban ecosystem as heterogeneous as that of the region.

At the same time, the riparian forest represents a local and community setting for learning through various teaching channels (Novo 2012) and for environmental action through the development of conservation and restoration activities open to the community. In this sense, environmental education strengthens social empowerment (Telias 2010) and contributes to the development of individual and collective capacities of a society, through strategies that increase adaptation, risk management, and community resilience (González-Gaudiano and Maldonado-González 2017). With this, the riparian forest allows, in the regional context, to link the conservation of biodiversity, the provision of regulatory and cultural ecosystem services, as well as the management of flood risk, favoring both the natural and social resilience of the urban ecosystem.

3.2 Case 2: Alluvial Risk

On the west bank of the Neuquén river, another urbanization process takes place on the floodplain. In this sector, the buildings were protected from the eventual flooding of the river by means of an artificial slope. On the opposite side of the urbanization, there is a larger natural slope or foothill: *bardas*. This area has been declared Urban Nature Reserve although, in recent decades, different activities and disturbances have taken place transforming around 20% of its surface. This natural area currently offers a recreational space that is used by people from all over the region for recreation and environmental education (Cultural Services). Presently, the residential settlement in the lower zone gives the area new functionalities in relation to human well-being. Figure 17.3 shows that the natural network that drains water from the upper area in this sector is configured by a series of micro-basins that currently drain into the urbanization sector. The inhabitants of the urbanization formed on the floodplain could perceive benefits or damages depending on the topographic, hydrological, and vegetation characteristics of the micro-basins of the higher levels. Both the input flows and the concentration peaks and delay times of the precipitations respond to geobiological factors such as slope, soil structure (infiltration rate), and vegetation (cover and structure). Any modification made to any of these factors

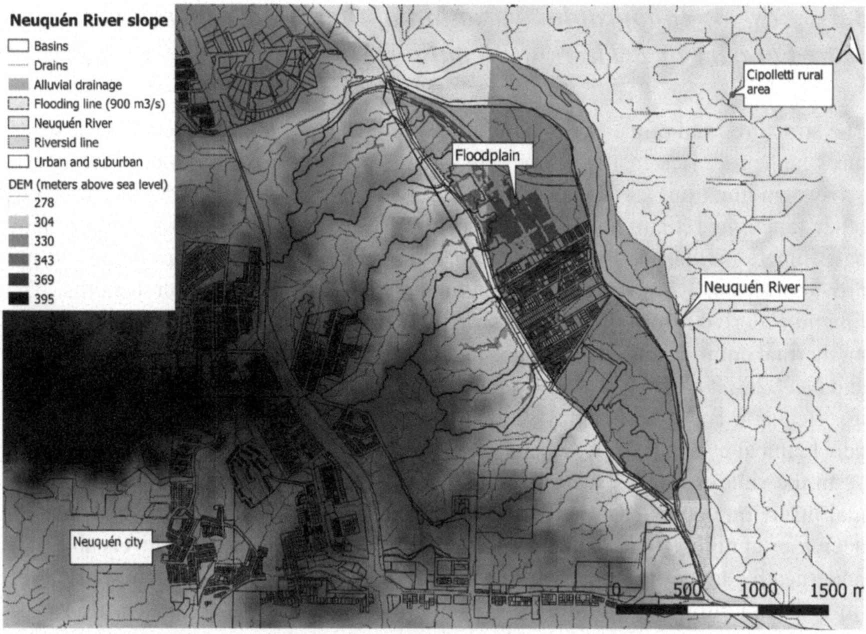

Fig. 17.3 Alluvial risk on the "bardas" in an urbanized area on the floodplain of the Neuquén river ($Q = 900$ m^3/s and design precipitation TR = 100 years)

could alter the hydrological dynamics, either increasing or decreasing the alluvial risk. Flood events that have recently occurred could be associated with the interruption of the natural drainage from the *bardas* to the river. As a consequence, the water has accumulated in the urbanized areas where a large amount of sediment has also been deposited due to the erosion of the bare soil of the slope. Based on forecasts of increased rainfall intensity and the frequency of extreme events in the region, an increasing probability of avalanches and mass movements is expected (Barros and Mattio 1978; Barros and Rodriguez Sero 1979; Castañeda and González 2008, in Romero et al. 2014). It is necessary to deepen the understanding of the ecohydrological functioning of these areas, including the study of alternative urban settings in the protected natural area. The realization of spatial models would allow us to evaluate accurately the benefits or damages derived from different land uses, thereby contributing to the landscape design. From disaster risk management, it is essential to develop capacities and knowledge for the planning and management of a heterogeneous urban ecosystem that includes the barda, the valley, and the associated topographic gradient. This gradient favors natural drainage and represents a regulatory ecosystem service promoting the evacuation of water and erosion control. Thus, natural drainage (and each of the subsystems that integrate it) must be interpreted with the ecosystem services approach and included in urban planning and management, representing a key element in risk reduction.

3.3 Case 3: Topographic Variation of the Effect of Urban Forestry as a Thermal Regulator

The structures and processes of natural ecosystems are often replaced by new structures and processes in the urban ecosystem. In this transformation, some of the ecosystem functions are lost and, therefore, also some ecosystem services. However, other functions remain and even new natural structures (green infrastructure) may have functions that lead to ecosystem services that the natural ecosystem did not provide. Urban forestry regulates thermal dynamics, among other benefits. The functions derived from this natural structure can have direct positive consequences on thermal comfort and the health of local inhabitants (mitigating heat waves, Heidt and Neef 2008; Chen et al. 2014; Jenerette et al. 2016; Saaroni et al. 2018), but also its effects on energy efficiency scale up and affect global phenomena such as biogeochemical cycles.

In the valleys, as well as in the cities, a meso climatic phenomenon called thermal inversion occurs (Kuttler et al. 1996; Rendón et al. 2014). As the temperature decreases at night (by radiative cooling), the cold air drains to lower levels and goes underneath a layer of warmer air. The inversion is maximized during sunrise, at the time of minimum temperature. This phenomenon, which involves energy flows and mass movements, could be altered by the presence and characteristics of urban trees.

Through a quasi-experimental study, we studied the topographic variation of the effect of urban trees on the regulation of the Urban Heat Island (UHI), in summer. The experimental units corresponded to Urban Lane Channels (ULCH) of similar geometry, width, materiality, and orientation (N-S), but with variations in the levels of two combined variables: urban tree cover (%) and topographic position (masl) (Table 17.2). Using Arduino sensors, we measured the atmospheric temperature at 2 m from the ground for at least 3 days in each of the ULCH, taking data every 15 minutes. In addition, the temperature was simultaneously measured at a site used as a natural reference (non-urbanized) located in a high topographic position with sparse shrub vegetation. The temperature difference between each ULCH and the reference site is considered as an index of the UHI, for which we obtained UHI values for each of the measurement days at times of minimum temperature (UHI min) and maximum temperature (UHI max).

At the time of maximum daily temperature (Fig. 17.4a), the UHI max value varied between positive (Heat Island) and negative (Cold Island) depending on the degree of urban tree cover. In the low cover ULCH, the atmospheric temperature of the city was up to 2 °C higher than the temperature in the reference site. When increasing urban tree cover, the UHI is reduced, even registering a lower temperature in ULCH with high coverage than in the reference site. However, this effect of the urban trees didn't result the same in the entire road network, as it was affected by the topographic position of each ULCH. The variations of the UHI given by tree cover were more intense in the ULCH of low topographic position (valley), while in the high position (plateau), the effect of tree cover was less, although they followed the same trend. These results suggest that in cities located in valleys, the

Table 17.2 Characteristics of Urban Lane Channels (ULCH) used as experimental units

	Coverage		Altitude		Coordinates	
ULCH	Level	%	Topo.	m.a.s.l.	Lat.	Long.
La Pampa 1950	Low	8.7	Valle	264	−38.976372	−68.062322
Leguizamon 1900	Medium	29.9	Valle	265	−38.976035	−68.267943
Winter 150	High	45.1	Valle	265	−38.956671	−68.048004
Fotteringam 150	Low	10.5	Central	267	−38.953401	−68.069102
Sgo. Del Estero 450	Medium	26.9	Central	281	−38.950272	−68.064916
La Rioja 350	High	50.6	Central	276	−38.951221	−68.063531
La Rioja 950	Low	17.8	Meseta	311	−38.944494	−68.063424
Las Violetas 1100	Medium	29.6	Meseta	308	−38.942883	−68.05059
Hirigoyen 850	High	41.1	Meseta	299	−38.945858	−68.060742

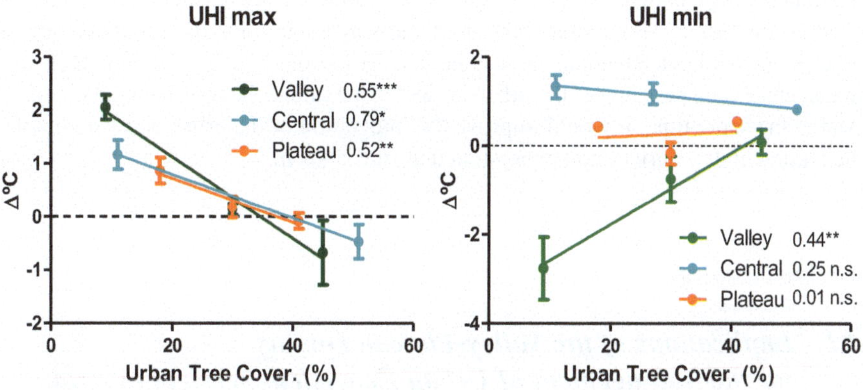

Fig. 17.4 Urban Heat Island (UHI, Δ°C) at times of maximum daily temperature (UHI max) and minimum daily temperature (UHI min) as a function of urban tree cover (%), for ULCH grouped by topographic position. The R2 and the significance of each linear regression are indicated. *$p < 0.05$; **$p < 0.01$; ***$p < 0.001$

topographic position is more important than the urban infrastructure in determining the atmospheric temperature.

At the time of minimum daily temperature (Fig. 17.4b), the UHI min value was also affected by an interaction between topography and urban tree cover. In intermediate and high topographic positions (central and plateau), the UHI min was positive or null, respectively, regardless of the urban trees. However, in the lower positions of the valley, the UHI min remains zero (the same urban temperature as in a natural site) only in the ULCH with high cover of urban trees, while when the cover is lower, the UHI becomes strongly negative (more than 2 °C colder in the ULCH than in the natural site). Again a strong influence of the topographic position is appreciated, in this case perhaps explained by the thermal inversion. However, the urban trees seem to interfere in the thermal inversion process, keeping the atmospheric temperature warmer even in the lower positions of the valley.

The network of ULCH can occupy up to 25% of the surface of a city. Together with public green spaces, they make up an urban ecosystem management unit, from which the provision of a series of locally perceived ecosystem services can be regulated. The type and structure of the vegetation implanted in cities affects temperature, air quality, biodiversity (Sandström et al. 2006), aesthetic comfort, and people's sense of identity. At the local level, these variables affect people's physical and mental health (Grahn and Stigsdotter 2003; White et al. 2013) and can modulate social dynamics (Dige 2011). At the same time, thermal regulation improves energy efficiency and consequently reduces the urban carbon footprint (Matteucci 2005) which implies global benefits. However, the provision of these ecosystem services could be heterogeneous across the entire surface of cities, showing interaction between the effects of each structure and its context (geographical, morphological, climatic, social, etc.). For the case study presented here, it was observed that the effect of urban trees on thermal regulation varies along a topographic gradient. Among the current challenges of urban ecology, an inquiry is required about the factors that determine the heterogeneity of other services. Understanding these patterns, on the one hand, will strengthen the management and planning of the urban project to maximize the well-being of citizens. And at the same time, it would improve the integration of the urban ecosystem with the natural and/or agricultural ecosystem that surrounds it.

4 Discussion

4.1 Implications of the Valley-Plateau Duality for the Management of Urban Ecosystem Services in Arid North Patagonia

The fragmented and dispersed matrix of the northern Patagonia cities responds to population and economic dynamics associated with extractivism. With the development boom of shale gas and oil production in the last decade, the city of Neuquén became an administratively hierarchical node within the territory of Neuquén province, but spatially dispersed. This is due to poor planning and low compliance with current urban development regulatory standards. In this way, cities of the *Alto Valle* such as Plottier, Centenario, and Cipolletti were absorbed into the matrix of a diffuse urban conglomerate, which includes four municipal and two provincial jurisdictions. Due to this political fragmentation, cities such as Cipolletti or General Roca acquire certain administrative importance, within the limits of the urban conglomerate that belongs to the province of Río Negro. In this way, territorial units are intertwined (urban, suburban, rural and riverside, administrative, landscape, and urbanly heterogeneous and complex). The urban development of the region is led by the city of Neuquén, while other cities grow demographically without getting to constitute administrative, political, and socioeconomic hierarchies to allow their inhabitants to have job opportunities and adequate service provision. This shapes a high dependence of the region on the urban center of Neuquén and conditions the urban policy of neighboring municipalities to yield agricultural and riverside lands in response to the demand for urban land for residential purposes. In terms of urban development trends, the city of Neuquén also promotes urban occupation of the floodplains of the Limay and Neuquén rivers, ignoring the technical standards that regulate the limits of flood lines. At the same time, rivers constitute the interprovincial boundary between the provinces of Neuquén and Río Negro, generating different speculative expectations on the shores corresponding to both municipal and provincial jurisdictions. This political-administrative particularity constitutes a frame of reference to understand the nature of the problem in the regulation of urban ecosystem services. In general, all the cities in the arid northern Patagonia developed on river valleys. This configures cities that coexist with an ecosystem matrix formed by two well-differentiated ecosystems: on the one hand, the slopes of the valleys and the Patagonian plateau, with attributes regulated by the regularity of the drought, low vegetation cover, poor and skeletal soils, and a generalized slope on the foothills strip and, on the other hand, the floodplains of the Limay, Neuquén, and Negro river valleys, accompanied by an interconnected network of riparian wetlands and the most fertile soils in the region under an irrigation system. In this way, the arid northern Patagonia cities have a great spatial heterogeneity both at an urban and ecological level. Urbanistically, heterogeneity manifiests itself with public policies aimed towards the urbanization of the valley, both in wetlands and irrigated oases, while spontaneous urban developments spread over the *bardas* with little planning. As a consequence, urban ecosystem services on the *bardas* and wetlands are restricted to the form of occupation and use of the land but also to their own nature. In this way, an urban asymmetry becomes established in the socioeconomic aspect between *bardas* and riverbanks but also in the ecological aspect due to the functions and services that natural remnants can offer according to the singularities of one and another ecosystems.

If there is something in common between the *bardas* and wetland ecosystems, it is that the environmental risks are given by the dynamics of water. While the *bardas* landscape is formed by the shaping energy of rainwater runoff along the slopes of a vast network of micro-basins collected by rivers, the valleys are depositories of storm drains and floods with different levels of recurrence that overflow on the floodplain. This means that any surface susceptible of being flooded or to channel alluvial processes already in itself constitutes an ecosystem service. If we add that vegetation plays a fundamental role in delimiting wetlands, fixing soils, or affecting the architecture of temporary or permanent riverbeds, the ecological functions of these spaces begin to acquire the capacity to provide services in the urban context. For this reason, for each urban and ecological situation, there is a potential risk regulator depending on the urban configuration and the ecosystem services associated with the particular location. Thus, at the *bardas*, the upper basins are areas that determine a good part of the stability of the foothills and the city's cost of investment in drainage infrastructure. At the wetlands, their conservation in addition to promoting biodiversity also constitutes a city's source of financial savings when they can function efficiently as water gardens, providing green spaces capable not only of cushioning the effect of flooding but also recycling and channeling storm drains. In the same way, adequate urban planning delimited by a harmonious separation between urbanization and the wetlands based on the flood lines would result in a reduction of the proliferation of zoonotic diseases and avian risk for urban settlements.

4.2 *Urban Ecological Infrastructure: Planning from the Ecosystem Services Perspective*

Ecosystem services depend on the maintenance and proper functioning of urban ecosystems (Szumacher and Malinowska 2013) and, consequently, are influenced by urban management practices and policies (Andersson et al. 2007 in Karis et al. 2019). The valley's region can be understood as a hydrosocial territory, insofar as it articulates three territorial spaces: physical, social, and legal (Damonte-Valencia 2015). In the first place, the physical space, associated with a complex water system, connects the plateau area in a north-south direction with the floodplain of the Limay, Neuquén, and Negro rivers, through surface and underground drainage channels. Secondly, social spaces make material and symbolic, diverse, and differential use of the xeric and coastal physical space. In this sense, the fluvial bodies have represented and represent structuring axes for the cities (Bendini and Steimbreger 2007), while the areas of slopes and plateaus are consolidated as settlements of relegated sectors of society (Perrén 2010). This socioeconomic segregation closely related to the symbolic idea of the river as a space of nature and well-being, versus the plateau as "no place" (Nogué and Albert 2004), results in the disconnection of the hydrosocial territory and threatens the ecosystem services that the cities could offer. This is manifested in the last territorial space, the legal one. Regarding this, the region has a profuse and historical institutional regulatory framework in terms of territorial ordering of coastal areas, currently focused on urban capital gains, with the creation of new residential, work, and recreational locations. Regulated management objectives, ambiguous water planning, and the exclusion of riparian forest from the Native Forest Territorial Planning (Lopez and Gentili 2020) reduce the coastal space to provide cultural and scenic ecosystem services to the detriment of regulation, provision, and support services. At the other extreme, the plateau as Landscape Heritage (Neuquén city ordinance 502/68) also provides cultural ecosystem services, but lacks adequate management to maximize other types of urban ecosystem services. This is the consequence of a disorderly urban expansion and territorial ordering, which, guided by the spontaneous emergence of settlements, does not consider water, ecological, or geomorphic planning criteria (Pérez 2018). The city of Neuquén represents an emblematic case, because since 1998 it has had an urban planning code, which clearly establishes characteristics of peri-urban areas of *bardas* and riverbanks. In fact, in recent years the creation of urban protected natural areas on *bardas*, but not on wetlands, has proliferated. Shortly after it came into force, however, the code was modified by regulations that favored the urbanization of peri-urban, productive, and riparian zones and even the intensive use of land in protected natural areas. By 2003, urban developments were already being authorized in floodplains, islands, and wetlands. With the authorization of provincial technical bodies regulating water resources, a process of filling the wetlands began that brought the ground level above flood levels corresponding to regular floods. As a consequence, there is a riparian development axis that connects Neuquén and Plottier on the coast of the Limay river. Many of these developments on *bardas* and wetland environments have financial support from multilateral entities such as the Inter-American Development Bank (IDB). In the case of urban development on floodplains of the Neuquén river, this puts a strong pressure on the conversion of productive land into urban on the shore belonging to the city of Cipolletti.

Below we summarize the ecological infrastructure in the metropolitan area of northern Patagonia for each class of ecosystem service, and we make proposals to incorporate into future urban planning.

4.3 Regulation Ecosystem Services

The urban epicenters in the arid northern Patagonia are mainly located in riparian areas. Many of the recent developments occur in areas at risk of flooding which naturally absorb the extraordinary floods of the rivers. As noted in one of the case studies in this chapter, incorporating the concept of ecosystem services into urban planning would be a viable strategy for disaster risk reduction. The identification of ecosystem services associated with risk mitigation coincides in these areas with riparian forests of invasive and native Salicaceae, wetlands, quarries, and abandoned farms. Therefore, the conservation of these places would serve multiple purposes. In fact, these purposes are also served by reclaimed artificial wetlands such as quarries and drainage channels or disturbed surfaces that are restored, such as paleo-channels or ponds. Another characteristic of the arid urbanizations of northern Patagonia is their location in valleys with geomorphological variations. These variations determine a heterogeneity in terms of the provision of ecosystem services of the same element or structure of the urban ecosystem. *Bardas* configure the headwaters of micro-basins that develop on urbanized and highly waterproofed areas. Their vegetation cover, depending on the topographic position, does not exceed 40%. That is why they are vulnerable and very dynamic ecosystems. For this reason, its conservation and even more so regulation with bioengineering techniques such as the one developed in the 1970s in the city of Neuquén already comprise an abandoned tradition in a way of mitigating alluvial risk through the consolidation of community forests and protectors. Urban trees make up a structure that provides direct (local, through climate comfort) and indirect ecosystem services (global, by mitigating heat from urban islands and reducing energy consumption). The arrangement of these in the valleys is not homogeneous because it varies in a gradient from plateau (arid) to valley (wetlands). Considering these variations in planning offers the double benefit of increasing the efficiency of the use of resources and the design of the structure (green infrastructure). One strategy in this sense is to give continuity to the principles of promoting protective forests in high watersheds and adapting the selection of native tree species and shrubs, following the experience accumulated in Parque Norte, while in the valley floors, riparian wetlands can be incorporated into a functional conservation system aimed at keeping the architecture of rivers stable and ensuring the provision of water. Another strategy is to restore the old drainage and irrigation channels of the old farms and incorporate them with their green hedges of poplars and willows, managing them and replacing them with species more suitable for the urban environment.

4.4 Cultural Ecosystem Services

In the last decades, the society of northern Patagonia has started to appreciate the *bardas* as a place for contact with nature and the practice of outdoor sports. A systematic and objective study of the perception and valuation of these peri-urban environments is necessary, which allows this variable to be included in territorial planning. Likewise, in the case studies presented in this chapter, it is evident that since the arid condition of the region leads to a prominent valuation of river environments, the conservation of these areas as natural green spaces (avoiding their urbanization) offers not only regulation services but also cultural services that cannot be obtained in other areas of the urban landscape.

The vertiginous growth of the cities of the MRC, which has led to the doubling of their population between two intercensal periods, has the main consequence of the scarcity of green spaces. This is manifested in the extensive recreational use of informal spaces and without adequate infrastructure on the shores of rivers and bardas. In addition to the previous proposal, the same spaces for the conservation, restoration, or promotion of protective forests, even as spaces destined to the conservation of an ecological function, must also provide for recreational and cultural uses.

4.5 Provisioning Ecosystem Services

The urban nodes of northern Patagonia have suffered abrupt demographic increases in the last decade. Urban growth has been and continues to be, in part at the expense of the loss of agricultural land. In the world, the population migrates from rural areas to urban areas and modifies their lifestyle by increasing consumption. In the study region, this phenomenon was underpinned by the exploitation of hydrocarbons, which generated a decade of economic prosperity. Notwithstanding, the fruit crisis deepened. This problem becomes more relevant as the productive area is limited to the irrigated valley; then its loss would lead to the need to transport food from remote regions (Deutsch and Folke 2005). The COVID-19 pandemic and the global crisis that is predicted as a result have accelerated the search for alternatives to the current food production system (Lal 2020; Pulighe and Lupia 2020; Altieri and Nicholls 2020; Loker and Francis 2020). Faced with scenarios of socioeconomic fragility, urban agriculture is shown as a viable option to achieve food security in times of crisis (Santandreu et al. 2009; Delgado 2017; Kutiwa et al. 2010). Incorporating food production into new land uses that replace productive units would ensure the preservation of provision services but would also favor the provision of other services such as biodiversity (Orsini et al. 2014; Clucas et al. 2018). Urban planning, and especially the planning of the peri-urban edges of the cities of northern Patagonia, should be directed from the concept of multifunctional landscapes (Lovell 2010), promoting a design and management of the territory that maximizes the benefits of the population and above all anticipates the changes that are forecast.

There are still enough spaces in production or abandoned in the urban matrix of the MRC, which can be converted into urban agriculture spaces. These spaces should not only be conceived as mere production units but also green, educational, and innovation spaces in urban agriculture and sustainability with the goal of reducing the ecological footprint of the cities of the MRC.

4.6 Of Support

The concentration of the population in productive valleys inevitably generates a compromise between urban and agricultural uses, resulting in a landscape where both uses intertwine becoming increasingly difficult to separate and sectorize them. The urban ecosystem overlaps spatially with the agricultural ecosystem including in a fragmented way some remnants of natural (or low intervention) ecosystems. In this context, the ecosystem processes in both ecosystems interact adding complexity to the socio-ecological system. Both for a scenario where the predominance of fruit growing activity is maintained and for a scenario of productive diversification, the conservation and management of urban biodiversity will have repercussions on the conservation of peri-urban biodiversity and local food production. On the one hand, riparian areas have suffered in the last century an enrichment of exotic plant species, which in part come from urban trees. The selection of urban tree species with low invasive potential for the new urbanized areas could prevent or mitigate the occurrence of invasions in areas with still low transformation. On the other hand, pollinators represent a determining link in food production in the region. It would be advisable to consolidate local studies and the application of known tools to conserve pollinator communities in urban areas (Threlfall et al. 2015; Hicks et al. 2016; Hall et al. 2017).

Only a comprehensive planning perspective, having as its axis water and the challenges it implies for risk and its availability due to climate change, allows complementing the parts of a complex urban and environmental system. The parts and their functions as services that articulate this system in the arid valleys of Patagonia are the regulatory functions of the high micro-basins that drain into rivers, urban and rural artificial drainage, secondary channels, paleo-channels and riparian wetlands, and riparian forests. The four variables configure the structure of drainage and regulation of water distribution in valleys of an extremely arid region. And these variables comprise the basic and functional structure for architectural and urban design on which any intervention that seeks to obtain adequate urban trees, regulate the urban thermal island, and mitigate environmental risk must be supported. From the measurements of the streets and avenues, their orientation, the line of retreat of buildings, and their heights largely depend on how the city relates to its trees and the urban atmosphere. Factors such as permeability but also recreation and comfort indirectly depend on the distances and average times with which citizens move towards green spaces. For this reason, support services can only be conceived in a comprehensive and articulated strategy. An ecosystem service component cannot be seen without its counterpart of how the city is built at the same time that a safe and resilient city is only so if it articulates and adjusts the operating dynamics of the natural processes that govern its environment.

5 Conclusions

The urban ecosystems of the arid north of Patagonia are eminently hydric. The cities and societies that develop in the region are riverside in nature, and, furthermore, a large proportion of the population is located along the valleys of a unique river system (Río Negro, with its tributaries Limay and Neuquén). Although another important part of these cities is located in the arid portion of the valley and even on the Patagonian plateau, their ecosystems, by their very nature, are shaped by the dynamics of the runoff of rainwater towards the river floodplains. As has been observed in the case studies presented, considering this condition in urban planning would result in the reduction of the risk of disasters due to floods and mass removal movements, through the conservation of biodiversity and the functional aspects of vegetation cover, its spatial heterogeneity, and the geomorphology of the temporary and permanent channels that drain the valleys and cities. Planning with a regional ecological perspective, based on the management of micro-basins at urban scales and valleys, would also allow an adequate strategy to safeguard the provision and quality of water in a context of reduction of river flows at the same time as intense rains are increasing as a consequence of climate change. This chapter lays the foundations and indicates the basic guidelines for incorporating the ecosystem services approach both in urban planning in the arid north of Patagonia and in local lines of research. These principles are based on the restoration of natural drainage, implementation of sustainable urban drainage, promoting the connectivity of the natural drainage network, the establishment of riverine lines, creation of conservation areas for the upper micro-basins of the valley and wetlands, and configuration of agrarian or recreational parks as buffer areas between floodplains and cities.

Acknowledgments We thank José Gatica for guiding us in the risk analysis and Christian Kreber for valuable comments on the earlier version of the manuscript. We are grateful to Luciana Campos for translating to English. The study was funded by UFLO.

References

Allen M, Babiker M, Chen Y, de Coninck H, Connors S, van Diemen R et al (2018) Summary for policymakers global warming of 1.5 C: an IPCC special report on the impacts of global warming of 1.5 C above pre-industrial levels and related global greenhouse gas emissions pathways, in the context of strengthening the global response to the threat of climate change. World meteorological. Organization:1–24

Alshuwaikhat HM, Nkwenti DI (2002) Developing sustainable cities in arid regions. Cities 19(2):85–94

Altieri MA, Nicholls CI (2020) Agroecology and the reconstruction of a post-COVID-19 agriculture. J Peasant Stud 47(5):881–898

Álvarez Palau E (2012) La colonización del Alto Valle del Rio Negro (Neuquen) en Argentina: ferrocarril, obras hidráulicas y electricidad y consolidar del poblamiento. In *Simposio internacional globalización, innovación y construcción de redes técnicas urbanas en América y Europa, 1890-1930. Brazilian Traction, Barcelona Traction y otros conglomerados financieros y técnicos* (pp 1–20)

Andersson E, Barthel S, Ahrné K (2007) Measuring social–ecological dynamics behind the generation of ecosystem services. Ecol Appl 17(5):1267–1278

Arnaud-Fassetta G, Cossart E, Fort M (2005) Hydro-geomorphic hazards and impact of man-made structures during the catastrophic flood of June 2000 in the Upper Guil catchment (Queyras, Southern French Alps). Geomorphology 66:41–67

Arnfield AJ (2003) Two decades of urban climate research: a review of turbulence, exchanges of energy and water, and the urban heat island. Int J Climatol 23(1):1–26

Ayuntamiento de Madrid (2018) Bases y directrices generales del plan de infraestructura verde y biodiversidad de la ciudad de Madrid. España, Madrid

Barros VR, Mattio HF (1978) Tendencias y fluctuaciones en las precipitaciones de la región patagónica. Meteorológica; VIII-IX:237–248

Barros VR, Rodriguez Sero JA (1979) Estudio de las fluctuaciones y tendencias de la precipitación en el Chubut utilizando funciones ortogonales empíricas. Geoacta 10(1):1979–1204

Barroca B, Bernardara P, Mouchel JM, Hubert G (2006) Indicators for identification of urban flooding vulnerability. Nat Hazards Earth Syst Sci 6(4):553–561

Barthel S, Folke C, Colding J (2010) Social–ecological memory in urban gardens—retaining the capacity for management of ecosystem services. Glob Environ Chang 20(2):255–265

Barthel S, Isendahl C (2013) Urban gardens, agriculture, and water management: sources of resilience for long-term food security in cities. Ecol Econ 86:224–234

Bendini M, Steimbreger N (2007) Nuevos espacios productivos en la Patagonia: reestructuración social de una cadena tradicional agrícola. Pampa:145–164

Blanco G (2007) La ocupación de la tierra en Neuquén: política distributiva, formas de tenencia y uso del suelo (1880-1920). Mundo agrario 7

Bolund P, Hunhammar S (1999) Ecosystem services in urban areas. Ecol Econ 29(2):293–301

Brown LR (2001) Eco-economy: building an economy for the earth. Routledge

Buchmann C (2009) Cuban home gardens and their role in social–ecological resilience. Hum Ecol 37(6):705–721

Castañeda ME, González MH (2008) Some aspects related to precipitation variability in the Patagonia region in Southern South America. Atmósfera 21(3):303–317

Chen D, Wang X, Thatcher M, Barnett G, Kachenko A, Prince R (2014) Urban vegetation for reducing heat related mortality. Environ Pollut 192:275–284
Chiesura A (2004) The role of urban parks for the sustainable city. Landsc Urban Plan 68(1):129–138
Chiementon ME, Cogliati MG (2011) Variaciones del uso del suelo en Cipolletti, Provincia de Río Negro, Argentina. Contribuciones Científicas GAEA 23:51–60
Clucas B, Parker ID, Feldpausch-Parker AM (2018) A systematic review of the relationship between urban agriculture and biodiversity. Urban Ecosyst 21(4):635–643
Colding J, Lundberg J, Folke C (2006) Incorporating green-area user groups in urban ecosystem management. AMBIO J Hum Environ 35(5):237–244
Damonte-Valencia GH (2015) Redefiniendo territorios hidrosociales: control hídrico en el valle de Ica, Perú (1993-2013). Cuadernos de Desarrollo Rural 12(76):109–133
Datri LA, Maddio R (2010) Geometrías fluviales en la definición del paisaje de la Colonia Valentina Sur (Neuquén) y Balsa Las Perlas (Río negro). Calidad de Vida y Salud 3(2)
Datri LA, Faggi AM, Gallo LA, Carmona F (2016) Half a century of changes in the riverine landscape of Limay River: the origin of a riparian neoecosystem in Patagonia (Argentina). Biol Invasions 18(6):1713–1722
Delgado C (2017) Mapping urban agriculture in Portugal: lessons from practice and their relevance for European post-crisis contexts. Moravian Geograph Rep 25(3):139–153
Deutsch L, Folke C (2005) Ecosystem subsidies to Swedish food consumption from 1962 to 1994. Ecosystems 8(5):512–528
Dige G (2011) Green infrastructure and territorial cohesion. The concept of green infrastructure and its integration into policies using monitoring systems. European Environment Agency, Copenhagen
Direcciones de Estadística y Censos y de Catastro de la Provincia de Neuquén (2010). Censo Nacional de Población, Hogares y Viviendas 2010. Neuquén, Argentina
Ernstson H, Barthel S, Andersson E, Borgström ST (2010) Scale-crossing brokers and network governance of urban ecosystem services: the case of Stockholm. Ecol Soc 15(4)
Evenari M, Shanan L, Tadmor N, Shkolnik A (1982) The Negev: the challenge of a desert. Harvard University Press
Folke C, Jansson Aa, Larsson J, Costanza R (1997) Ecosystem appropriation of cities. Ambio 26(3):167–172
Frank B, Delano D, Caniglia BS (2017) Urban systems: a socio-ecological system perspective. Sociol Int J 1(1):1–8
Fundación Techo (2016) Relevamiento de asentamientos informales 2016, Buenos Aires
García Acosta V (2005) El riesgo como construcción social y la construcción social de riesgos. Desacatos 19:11–24
Gotham KF, Brumley K (2002) Using space: agency and identity in a public–housing development. City Community 1(3):267–289
González-Gaudiano EJ, Maldonado-González AL (2017) Amenzas y riesgos climáticos en poblaciones vulnerables. El papel de la educación en la resiliencia comunitaria. Teor educ 29(1):273–294
Grahn P, Stigsdotter UA (2003) Landscape planning and stress. Urban For Urban Green 2(1):1–18
Grimmond CSB, Oke TR (1999) Heat storage in urban areas: local-scale observations and evaluation of a simple model. J Appl Meteorol 38(7):922–940
Grimm NB, Faeth SH, Golubiewski NE, Redman CL, Wu J, Bai X, Briggs JM (2008) Global change and the ecology of cities. Science 319(5864):756–760
Hall DM, Camilo GR, Tonietto RK, Ollerton J, Ahrné K, Arduser M, Goulson D (2017) The city as a refuge for insect pollinators. Conserv Biol 31(1):24–29
Heidt V, Neef M (2008) Benefits of urban green space for improving urban climate. In: Ecology, planning, and management of urban forests. Springer, New York, pp 84–96
Hicks DM, Ouvrard P, Baldock KC, Baude M, Goddard MA, Kunin WE, Mitschunas N, Memmot J, Morse H, Nikolitsi M, Osgathorpe LM, Potts S, Robertson KM, Scott AV, Sinclair F, Westbury

DB, Stone GN (2016) Food for pollinators: quantifying the nectar and pollen resources of urban flower meadows. PLoS One 11(6):e0158117

Hood MJ, Clausen JC, Warner GS (2007) Comparison of Stormwater lag times for low impact and traditional residential development 1. JAWRA J Am Water Resour Assoc 43(4):1036–1046

Ilieva L, McQuistan C, van Breda A, Rodriguez AV, Guevara O, Cordero D, Podvin K, Renaud F (2018) Adoptando soluciones basadas en la naturaleza para la reducción del riesgo de inundación en América Latina. Practical Action / WWF / IUCN CEM

Ignatieva M, Stewart GH, Meurk C (2011) Planning and design of ecological networks in urban areas. Landsc Ecol Eng 7(1):17–25

Irarrázaval Irarrázaval F (2012) El imaginario 'verde' y el verde urbano como instrumento de consumo inmobiliario: configurando las condiciones ambientales del área metropolitana de Santiago. Revista INVI 27(75):73–103

IPCC (2007) Climate change 2007: impacts, adaptation and vulnerability. In: Parry ML, Canziani OF, Palutikof JP, van der Linden PJ, Hanson CE (eds) Contribution of working group II to the fourth assessment report of the intergovernmental panel on climate change. Cambridge Univ Press, Cambridge, UK

Jenerette GD, Harlan SL, Buyantuev A, Stefanov WL, Declet-Barreto J, Ruddell BL, Myint SW, Kaplan S, Li X (2016) Micro-scale urban surface temperatures are related to land-cover features and residential heat related health impacts in Phoenix, AZ USA. Landsc Ecol 31(4):745–760

Karis CM, Mujical CM, Ferraro R (2019) Indicadores ambientales y gestión urbana. Relaciones entre servicios ecosistémicos urbanos y sustentabilidad. CUADERNO URBANO. Espacio, cultura, sociedad 27(27):9–30

Konijnendijk CC, Annerstedt M, Nielsen AB, Maruthaveeran S (2013) Benefits of urban parks. A systematic review. A report for IFPRA, Copenhagen & Alnarp

Kühn, I., & Klotz, S. (2006). Urbanization and homogenization–comparing the floras of urban and rural areas in Germany. Biol Conserv 127(3):292–300

Kutiwa S, Boon E, Devuyst D (2010) Urban agriculture in low income households of Harare: an adaptive response to economic crisis. J Hum Ecol 32(2):85–96

Kuttler W, Barlag AB, Robmann F (1996) Study of the thermal structure of a town in a narrow valley. Atmos Environ 30(3):365–378

Lal R (2020) Home gardening and urban agriculture for advancing food and nutritional security in response to the COVID-19 pandemic. Food Security:1–6

Landriscini G, Domeett G, Rivero I (2014) Neuquén, ciudad intermedia y nodo regional. Algunas evidencias de la dinámica económica y la centralidad urbana. Revista Pilquen-Sección Ciencias Sociales 1(17):1–15

Landriscini SG (2017) Transformaciones territoriales y producción del hábitat en la Norpatagonia. Neuquén de los tiempos de la provincialización a los del neoliberalismo

Leitao JP, Almeida MDC, Simões NE, Martins A (2013) Methodology for qualitative urban flooding risk assessment. Water Sci Technol 68(4):829–838

Loker A, Francis C (2020) Urban food sovereignty: urgent need for agroecology and systems thinking in a post-COVID-19 future. Agroecol Sustain Food Syst:1–6

Lopez M, Gentili J (2020) Análisis normativo-institucional de los espacios verdes públicos ribereños en el Área Metropolitana de Neuquén (Argentina). Entorno Geográfico, N°20, 42–67.

Lovell ST (2010) Multifunctional urban agriculture for sustainable land use planning in the United States. Sustainability 2(8):2499–2522

Maskrey A, Cardona O, García V, Lavell A, Macías JM, Romero G, Chaux GW (1993) Los desastres no son naturales. Red de Estudios Sociales en Prevención de Desastres en América Latina

Maas J, Verheij RA, Groenewegen PP, De Vries S, Spreeuwenberg P (2006) Green space, urbanity and health: how strong is the relation? J Epidemiol Community Health 60(7):587–592

Mange E (2019) Investigaciones arqueológicas en la margen sur del valle medio-superior del río Negro (provincia de Río Negro). Universidad de La Plata

Martonne ED (1926) L'indice d'aridité. Bulletin de l'Association de géographes français 3(9):3–5

Matteucci SD (2005) De la ecología a la urbanoecologìa. Fronteras 4(4):18–30

McKinney ML (2006) Urbanization as a major cause of biotic homogenization. Biol Conserv 127(3):247–260
Ministerio de Seguridad de la Nación (2017) Manual para la elaboración de mapas de riesgo, Buenos Aires
Montaner J (1999) El modelo Curitiba: Movilidad y espacios verdes. Ecología Política 17:69–71
Moreno-Garcia MC (1994) Intensity and form of the urban heat island in Barcelona. Int J Climatol 14(6):705–710
Morello J (1995) Grandes Ecosistemas de Suramérica. En: Gallopin, G. (comp.) El futuro ecológico de un continente. Editorial de la Universidad de las Naciones Unidas y Fondo de Cultura Económica. México
Muñoz SF (1996) Colonización, regadío y dinámica fluvial en la Patagonia argentina. El Alto Valle del Río Negro. Ería. Revista Cuatrimestral de Geografía, (39–40), 33–49
Munang R, Thiaw I, Alverson K, Liu J, Han Z (2013) The role of ecosystem services in climate change adaptation and disaster risk reduction. Curr Opin Environ Sustain 5(1):47–52
Nel JL, Le Maitre DC, Nel DC, Reyers B, Archibald S, van Wilgen BW, Engelbrecht FA (2014) Natural hazards in a changing world: a case for ecosystem- based management. PLoS One 9(5)
Nogué y Albert (2004) Cartografía de los cambios sociales y culturales En: Romero, J, (coord.) Geografía Humana. (pp. 159–202) Editorial Ariel. Buenos Aires
Novo M (2012) La Educación Ambiental: Bases éticas, conceptuales y metodológicas. Editorial Universitas S.A, Madrid
Orsini F, Gasperi D, Marchetti L, Piovene C, Draghetti S, Ramazzotti S, Gianquinto G (2014) Exploring the production capacity of rooftop gardens (RTGs) in urban agriculture: the potential impact on food and nutrition security, biodiversity and other ecosystem services in the city of Bologna. Food Secur 6(6):781–792
Page B (2002) Urban agriculture in Cameroon: an anti-politics machine in the making? Geoforum 33(1):41–54
Pataki DE, Carreiro MM, Cherrier J, Grulke NE, Jennings V, Pincetl S, Zipperer WC (2011) Coupling biogeochemical cycles in urban environments: ecosystem services, green solutions, and misconceptions. Front Ecol Environ 9(1):27–36
Pérez GG (2018) La conurbación en torno a la ciudad de Neuquén. Perspectiva regional y aportes para el ordenamiento territorial. La Plata, Buenos Aires
Perrén J (2010) Estructura urbana, mercado laboral y migraciones. In: Una aproximación al fenómeno de la segregación en una ciudad de la Patagonia (Neuquén: 1960–1990). Movimiento, Miradas en
Perrén J, Pérez G (2020) La segregación urbana desde el tiempo y el espacio. Una aproximación en una ciudad patagónica (Neuquén, 1904-2010). Revista Digital de la Escuela de Historia 12(28):9
Perrén J, Pérez G, Cabezas S (2019) Crecimiento en altura en una ciudad intermedia argentina. Neuquén hacia comienzos del siglo XXI. Revista Pilquen. Sección Ciencias Sociales 22(3):54–69
Pickett ST, Cadenasso ML, Grove JM, Boone CG, Groffman PM, Irwin E et al (2011) Urban ecological systems: scientific foundations and a decade of progress. J Environ Manag 92(3):331–362
Prates L (2008) Los Indígenas del río Negro. Un Enfoque Arqueológico. SAA, Buenos Aires
Pulighe G, Lupia F (2020) Food first: COVID-19 outbreak and cities lockdown a booster for a wider vision on urban agriculture. Sustainability 12(12):5012
Rall EL, Haase D (2011) Creative intervention in a dynamic city: a sustainability assessment of an interim use strategy for brownfields in Leipzig. Germany landscape and urban Planning 100(3):189–201
Rebotier J (2009) La dimensión territorial del riesgo urbano en Caracas: características y alcances. Una propuesta integradora para pensar el riesgo en una realidad socio-espacial compleja. Trace. Travaux et recherches dans les Amériques du Centre, (56):11–25

Rees W (1992) Ecological footprints and appropriated carrying capacity: what urban economics leaves out. Environ Urban 4(2):121–130

Rees W, Wackernagel M (1996) Urban ecological footprints: why cities cannot be sustainable—and why they are a key to sustainability. Environ Impact Assess Rev 16(4–6):223–248

Rendón AM, Salazar JF, Palacio CA, Wirth V, Brötz B (2014) Effects of urbanization on the temperature inversion breakup in a mountain valley with implications for air quality. J Appl Meteorol Climatol 53(4):840–858

Rijke J, Van Herk S, Zevenbergen C, Ashley R (2012) Room for the river: delivering integrated river basin management in the Netherlands. Int J River Basin Manag 10(4):369–382

Roca SA, Manacorda A (2017) Ambiente y Salud en un área periurbana de Neuquén: condiciones del hábitat en asentamientos cercanos al área natural protegida "Parque de los Dinosaurios". Educo, Buenos Aires

Romero PE, González EM, González MH (2014) Características hídricas y climáticas del norte Patagónico. Tagliavini et al., eds. II Encuentro de investigadores en formación en Recursos Hídricos, 9–10

Saaroni H, Amorim JH, Hiemstra JA, Pearlmutter D (2018) Urban Green Infrastructure as a tool for urban heat mitigation: survey of research methodologies and findings across different climatic regions. Urban Clim 24:94–110

Sandström UG, Angelstam P, Mikusiński G (2006) Ecological diversity of birds in relation to the structure of urban green space. Landsc Urban Plan 77(1–2):39–53

Santandreu A, Perazzoli AG, Terrile R, Ponce M (2009) Urban agriculture in Montevideo and Rosario: a response to crisis or a stable component of the urban landscape. Urban Agriculture Magazine 22:12–13

SEPA (2011) Controlled activities regulations (CAR) flood risk standing advice for engineering, discharge and impoundment activities

Shuster WD, Pappas E, Zhang Y (2008) Laboratory-scale simulation of runoff response from pervious-impervious systems. J Hydrol Eng 13(9):886–893

Sudmeier-Rieux K, Nehren U, Sandholz S, Doswald N (2019) Disasters and Ecosystems: Resilience in a Changing Climate-Source Book. Geneva: UNEP and Cologne: TH Köln - University of Applied Sciences

Szumacher I, Malinowska E (2013) Servicios ecosistémicos urbanos según el modelo de varsovia. Revista del CESLA (16):81–108

Telias A (2010) Educación para la ciudadanía y educación ambiental: una articulación necesaria. Congreso Iberoamericano de Educación, Buenos Aires

Threlfall CG, Walker K, Williams NS, Hahs AK, Mata L, Stork N, Livesley SJ (2015) The conservation value of urban green space habitats for Australian native bee communities. Biol Conserv 187:240–248

Ugarelli R, Leitão JP, do Céu Almeida M, Bruaset S (2010) Overview of climate change effects which may impact the urban water cycle. Water Supply 4(4):47–54

United Nations (2018) World urbanization prospects: the 2018 revision. Population Division of the UN Department of Economic and Social Affairs, New York

Urraza MS, Muñiz JA (2017) Crecimiento urbano en el área de la Agencia de Extensión Rural Cipolletti. INTA

Van den Berg AE, Maas J, Verheij RA, Groenewegen PP (2010) Green space as a buffer between stressful life events and health. Soc Sci Med 70(8):1203–1210

White MP, Alcock I, Wheeler BW, Depledge MH (2013) Would you be happier living in a greener urban area? A fixed-effects analysis of panel data. Psychol Sci 24(6):920–928

Chapter 18
Land Size, Native Forests, and Ecosystem Service Inequalities in the Rural Chilean Patagonia

Cristobal Jullian and Laura Nahuelhual

Abstract In this chapter we pose two questions: (i) Does land size relates to forests and ecosystem service (ES) provision concentration? (ii) Do these spatial concentration patterns relate to expressions of vulnerability such as poverty, low human development, and natural capital loss at the municipality level? For the entire Chilean Patagonia and neighboring Biobío region, we found strong positive correlations (>0.92) between land holding size, native forest, and ES provision (as measured by the Ecosystem Service Provision Index). At the municipality level (n = 147), spatial cluster analysis identified five groups. Group 1 comprised 12 municipalities with high Gini coefficients (high inequality) for land size, native forest area, ES provision, and income, high poverty rates, low levels of human development, medium level of rural and indigenous population, high deforestation and afforestation rates, and high ES loss rates. In the other extreme, group 3 comprised 13 municipalities characterized by the coincidence of low Gini coefficients for land size, native forest area, ES provision, and income, lower poverty rates, higher levels of human development, higher levels of rural population, lower deforestation and low afforestation rates, and low ES loss rates. The results corroborate (i) a trilogy of inequalities, where larger landholders concentrate on land, native forests, and ES provision and (ii) a "trap"-like pattern in some municipalities, where inequalities are linked to poverty and natural capital loss, which is more prevalent in municipalities with a medium proportion of indigenous population. The results corroborate the great inequalities that prevail in the Chilean Patagonia, which demand a shift in

C. Jullian (✉)
Centro de Investigación en Dinámica de Ecosistemas Marinos de Altas Latitudes (IDEAL); Universidad Austral de Chile, Valdivia, Chile

L. Nahuelhual
Instituto de Economía, Universidad Austral de Chile Centro de Investigación: Dinámica de Ecosistemas Marinos de Altas Latitudes Instituto Milenio en Socio-Ecología Costera, Valdívia, Chile

P. L. Peri et al. (eds.), *Ecosystem Services in Patagonia*, Natural and Social Sciences of Patagonia, https://doi.org/10.1007/978-3-030-69166-0_18

land, rural development, and natural resources policies. Any conservation or management instrument based on ES provision must consider this context, with a view to avoiding further inequalities.

Keywords Geographic information systems (GIS) · Cluster analysis · Ecosystem services · Inequality · Development policies · Southern Chile

1 Introduction

Distributional issues and inequality (the disparity of distribution or opportunity) are important considerations for rural economic and social development and have gained a renowned interest in the Sustainable Development Agenda (United Nations 2020). Distributional issues in rural areas and agriculture-dependent societies have a prominent position within economic and rural development literature, particularly in relation to farm income inequality (Jayne et al. 2003; Loughrey and Donnellan 2017). In addition to a number of other well-established factors, the distribution of farm income is highly influenced by the distribution of land size (Severini and Tantari 2015; Lowder et al. 2016). For this reason, the literature on land size is increasingly concerned with distributional issues remarkably in the work of Olper (2007), Kay (2006, 2007), and Lowder et al. (2016), among others. Frankema (2010) describes land inequality as "one of the crucial underpinnings of long-run persistent wealth and asset inequality."

Land inequality is multidimensional involving, at a minimum, landlessness, insufficient land (land size inequality), and land concentration (Kay 2002; von Bennewitz 2017). A more general concept of rural inequality would add other factors to these dimensions, such as income inequality, the security of tenancy and paid labor arrangements, and restriction on access to essential natural resources and cultural places for rural and indigenous peasants (Bengoa 2007; Lowder et al. 2016).

Land as a functional ecological structure provides support for the growth and development of biodiversity and multiple ecosystems, such as grasslands and native forests, which support a range of ecosystem services (ES) that contribute to human well-being directly and indirectly (MEA 2005; IPBES 2019). Therefore, land size matters for nature conservation, sustainability, and justice in a series of relevant ways (Coomes et al. 2016). First, land size conditions how people use resources. For example, landholders with large parcels may use their land and forests more extensively, whereas those with small land holdings may use both more intensively, with relevant implications for biodiversity and ES provision. Second, land as a productive asset equates to wealth, so landholders with more land have a greater asset base to diversify production, manage risk, and self-insure as compared to small landholders (Byerlee and Deininger 2013; Yamauchi 2016). Third, the income derived from land can be reinvested in improving access to new resources and thus wealth, thereby changing the social distribution of resource access (Yamauchi 2016).

Latin America, with a land Gini coefficient of 0.79, is the world's most unequal region in terms of land distribution (von Bennewitz 2017). Chile occupies the second place in land inequality (after Paraguay) within Latin America. The most relevant components of this inequality are insufficient land (land size inequality) and extreme land concentration with a Gini coefficient of 0.91, representing near perfect inequality (OXFAM 2016). The phenomenon of concentration of land in large land holdings, which began in Latin America in the colonial period, was reinforced after the independence (Chonchol 2003) when colonial power was replaced by landed oligarchies that concentrated the best land (Wiener 2011). The *hacienda* system (latifundio-minifundio complex) expanded from the 1850s to the 1930s and achieved a dominant position within Latin America's agrarian structure (Kay 2002). This system was highly profitable for the landed elites who controlled political and economic power since colonial times and allowed them to shape agrarian institutions in their own interests (Barraclough 1999). By 1960, "latifundistas" ("landholders") owned roughly 5% of the land holdings and about four-fifths of the land area, while minifundistas (smaller estates) owned four-fifths of land holdings but only 5% of the land (Griffin et al. 2002). The middle-sized farm sector was relatively insignificant (Barraclough 1999). It is estimated that approximately one-third of the agricultural labor force was landless at the time (Griffin et al. 2002).

The establishment of land holdings and rural inhabitants in the southern Patagonia took place later, as compared to the settlement of the northern Patagonia. The colonizing occupation process began in 1878, through a cycle of auctions of fiscal properties that concentrated between 1903 and 1906 (Martinic 2006). This process allowed the constitution of the agrarian latifundio linked to sheep farming (Sasso 2006).

The land policies established since then, and particularly the neoliberal policies enacted in the early 1970s, consolidated a highly unequal development model that also promoted strong levels of poverty and marginalization (Davis-Hamel 2012; Hojman 1996).

A recent literature has started to associate spatial land inequalities to the concentration of other natural resources and ES (e.g., Benra and Nahuelhual 2019; Wegerif and Guereña 2020). This chapter contributes to this emerging literature by posing two questions: (i) Does land size relates to forests and ES provision concentration? (ii) Do these spatial concentration patterns relate to expressions of vulnerability such as poverty, low human development, and natural capital loss at the municipality level? To answer these questions, we relied on land holding and municipal spatial data, covering the Chilean Patagonia (36–55° S; 67–73° O) and the neighboring Biobío region to the north.

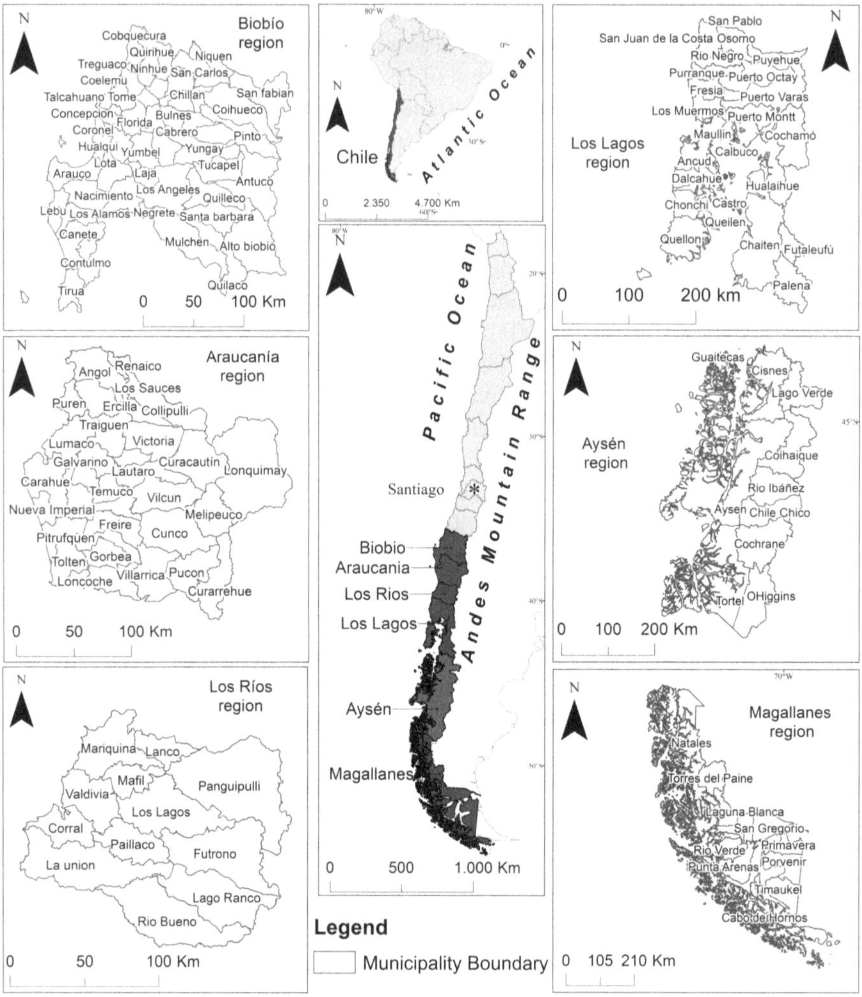

Fig. 18.1 Study area, regions and municipalities

2 Study Area

The study area (Fig. 18.1) comprises the Chilean Patagonia (Araucanía, Los Ríos, Los Lagos, Aysén, and Magallanes regions) and the neighboring Biobío region, from 36°46'S – 73° 03'W to 55°70'S – 67°06'W, excluding the Chilean Antarctic. These six regions represent 49% of the country's area (369,938 km^2) and 26% of the country's population (4,477,874 persons), of which 24% live in rural areas (INE 2018). The regions comprise 147 municipalities that vary in size from 72 km^2 (Chiguayante in the Biobío region) to 47,543 km^2 (Natales in the Magallanes region) (Table 18.1).

Table 18.1 Summary of landholding data per region within the study area

Region	Regions area (km^2)	N° landholdings	Landholding size (ha)			Std. deviation
			Maximum	Average	Minimum	
Biobío	37,112	104,670	75,868	30.9	0.02	420.5
Araucanía	31,812	136,401	54,464	20.4	0.01	306.2
Los Ríos	18,390	19,969	47,962	66.5	0.02	587.3
Los Lagos	48,424	41,758	71,973	50.5	0.01	706.5
Aysén	105,629	6751	44,031	430.9	0.02	1438.6
Magallanes	128,572	2071	96,800	2291.9	0.10	5620.4
Total	369,938	311,595				

3 Methods and Data

3.1 Spatial Correlations Between Land Size, Native Forest Area, and Ecosystem Services Provision

We explored the spatial correlation between land size, native forest area, and ES provision for a universe of 311,595 land holdings within the study area. Land size data came from two sources: (i) the Farm Cadastral Map, which is a digital cartography of rural properties at scale 1:20,000, that provides information on landholdings area and boundaries (CIREN–CORFO 1999), and (ii) the Servicio de Impuestos Internos (SII) farm database which is a digital cartography of properties at scale 1:10,000 for the year 2016 that shows the spatial distribution of each landholder with its respective boundary (SII 2016). Our assembled database comprised private farm properties, private protected areas, and forest companies' landholdings. We excluded public protected areas and other state-owned lands. We present the regional summary of landholdings in Table 18.1. It is important to mention that there are no community lands or forests in the study area, as in other Latin American countries. Thus, the conservation of nature and ES outside the protected areas of the state is entirely in the hands of private individuals.

Based on the vegetation surveys elaborated by the Corporación Nacional Forestal (CONAF 1998, 2013) and using GIS techniques, we obtained the old-growth and secondary forest area of each land holding.

To assess ES provision, we relied on a synthetic index called Ecosystem Services Provision Index (ESPI) (Paruelo et al. 2016). ESPI is based on two attributes of the seasonal dynamics of the Normalized Difference Vegetation Index (NDVI) (Rouse et al. 1974), namely, the annual mean (NDVI mean) and the intra-annual coefficient of variation of the NDVI, a descriptor of canopy seasonality. Derived from the Advanced High Resolution Radiometer, NDVI has been used, among other applications, to monitor desertification, land use and cover changes, and effects of global warming at high latitudes and as input in ES indicators (Paruelo et al. 2016; Stoms and Hargrove 2000). To the best of our knowledge, ESPI has not been applied to the Chilean Patagonia, and it is likely to have some limitations to represent some ES, particularly cultural ES. However, it has the advantage of being less costly than type-specific ES indicators in terms of construction time. We assigned the native forest area and the accumulated ESPI (sum of all pixel values within a property) to each landholding through the intersect tool using ArcGIS 10.8.

3.2 Land Size, Native Forest, and Ecosystem Services Inequalities

Based on land size, native forest area, and ESPI data, we built three Gini coefficients for each municipality in the study area in order to explore inequality patterns. The standard Gini coefficient (Gini 1909) is mathematically defined based on the Lorenz curve, which plots the cumulative proportions of a variable, sorted in an increasing order (y axis), against corresponding cumulative proportions of a second variable (x axis). The 45° line represents perfect equality. The area between the Lorenz curve and the line of perfect equality represents the degree of concentration. We adapted a version of the Gini coefficient for land developed by Sun et al. (2010) according to the following equation:

$$G = 1 - \sum_{k=1}^{k=n-1} \left(X_{k+1} - X_k\right)\left(Y_{k+1} - Y_k\right) \tag{18.1}$$

where G is the final Gini coefficient value, X_{k+1} is the cumulative number of landholdings, and Y_{k+1} is either the cumulative percentage of land, native forest area, or ESPI of each landholding. When $k = 1$, $Xi - 1$ and $Yi - 1$ are both equal to 0. We used the ranges proposed by Zheng et al. (2013) to classify the resulting Gini values in "absolutely equal" (<0.2), "relatively equal" (0.2–0.3), "reasonable" (0.3–0.4), "relatively unequal" (0.4–0.5), and "absolutely unequal" (>0.5).

3.3 Cluster Analysis

To explore the relationship between inequality and other variables associated with vulnerability, we used a cluster analysis. We selected the variables (Table 18.2) based on the three dimensions of a socio-ecological trap of inequality-poverty-degradation of natural resources (Nahuelhual et al. 2020). The relationships between these three dimensions have been studied by various authors in sub-regional case studies in Chile and other countries. Examples can be found in Agostini et al. (2008), Agostini and Brown (2007), Clarke (1975), FAO (2006), Janvry and Sadoulet (2001), Keswell and Carter (2014), Laterra et al. (2019), and Leonard et al. (2020). However, the same relationships have not been explored at a regional scale, for an area as extensive as Patagonia.

Table 18.2 Variables used in the cluster analysis

Variable	Data source	Brief description	Code	Range
Gini land	CIREN-CORFO (1999), SII (2016)	Based on landholding size distribution	G_LAND	0–1
Gini forest	CONAF (2013)	Based on native forest distribution across landholdings	G_FOREST	0–1
Gini ESPI	Paruelo et al. (2016)	Based on ESPI provision across landholdings	G_ESPI	0–1
Poverty rate	Agostini et al. (2008)	Based on three information layers: Demographic (number of household members, preschool-age household members) Head of household (sex, educational level, and ethnicity to which they belong) Assets at home (e.g., heating system, cell phone)	POV	0–1
Gini income	Agostini and Brown (2007)	A model of household income or consumption is estimated using survey data, restricting the explanatory variables to those also available in both the survey and a census undertaken at a similar point in time	G_INCOME	0–1
Human development index	PNUD (2000)	Based on: Health (life expectancy) Education (average years of education for adults and children) Wealth (GDP per person)	HDI	0–1
Rural population	INE (2018)	Proportion of the inhabitants of a municipality that live in rural areas	P_RUR	0–1
Indigenous population	INE (2018)	Proportion of the inhabitants of a municipality that belongs to an indigenous ethnic group	P_INDIG	0–1
Deforestation rate ($ha^{1}year^{-1}$)	CONAF (1998, 2013), Puyravaud (2003)	Deforestation rate ($ha^{1}year^{-1}$) was calculated as* $rd = \left(\left(\frac{1}{(t2-t1)}\right) * \left(\ln\left(\frac{A2}{A1}\right)\right)\right)$	NF_RATE	0–10.5
Non-native tree plantation afforestation rate ($ha^{1}year^{-1}$)	CONAF (1998, 2013), Puyravaud (2003)	Afforestation rate ($ha^{1}year^{-1}$) was calculated as: $rd = \left(\left(\frac{1}{(t2-t1)}\right) * \left(\ln\left(\frac{A2}{A1}\right)\right)\right)$	NNTP_RATE	0.001–0.08
ESPI annual change rate	Paruelo et al. (2016), Puyravaud (2003)	ESPI change rate was calculated as $rd = \left(\left(\frac{1}{(t2-t1)}\right) * \left(\ln\left(\frac{A2}{A1}\right)\right)\right)$	ESPI_RATE	0.03–0.06

[a] *rd* = Deforestation, afforestation, or annual ESPI change rate; *t2* = year at the time 2; *t1* = year at the time 1; *A2* = native forest, non-native tree plantation area, or ESPI (accumulated) at the time 2; *A1* = native forest, non-native tree plantation area, or ESPI (accumulated) at the time 1

We carried on the cluster analysis using the "grouping analysis" tool available from the mapping cluster tools of ArcGIS 10.8. This tool assembles features based on features' attributes and optional spatial or temporal constraints. This analysis shows how each variable behaves in a spatial way with respect to the others. The evaluation of the optimal number of groups suggested three groups (pseud F-statistic = 26), but given the extension of the study area, the cluster was forced to five groups, which yielded a reasonably high value of pseudo-F statistic equal to 23. The analysis requires the variables to be normalized because those that have large variations (where the data values are widely dispersed around the mean value) tend to influence the clusters more than the variables that have small variations. The standardization involves a Z transformation in which the mean value is subtracted from each value and divided by the standard deviation of all values.

4 Results

4.1 *Spatial Correlations: Land Size, Native Forest Area, and Ecosystem Service Provision*

Our results show a high spatial correlation between land size, native forest area, and ESPI provision in all the regions of the study area (Table 18.3).

Although we do not attempt to demonstrate causal relationships, the scatterplots in Fig. 18.2 show more clearly the association pattern between land size, native forest area, and ESPI provision. As expected, the proportion of forest and the accumulated ESPI increase with land size.

4.2 *Land Size, Native Forest Area, and Ecosystem Services Inequalities*

In the case of land size distribution, 27% of municipalities fell within the category of absolutely equal (Gini land <0.2), 29% within the category of relatively equal (0.2–0.3), 15% within the category of reasonable (0.3–0.4), 0.8% within the relatively unequal category (0.4–0.5), and 28% within the absolutely unequal category (>0.5).

In the case of native forest distribution, 5% of the municipalities fell within the category of absolutely equal (Gini forest <0.2), 46% within the category of relatively equal (0.2–0.3), 37% within the reasonable category (0.3–0.4), 7% within the relatively unequal category (0.4–0.5), and 5% within the absolutely unequal category (>0.5).

In the case of ESPI distribution, 37% of the municipalities fell within the category of absolutely equal (Gini ESPI <0.2), 19% within the category of relatively

Table 18.3 Pairwise correlations between land sizes, native forest are, and ESPI provision in each region

	Mean correlation coefficients		
Region	Land size-forest area	Land size-ESPI provision	Forest area-EPSI provision
Biobío	0.94	0.88	0.89
Araucanía	0.96	0.83	0.87
Los Ríos	0.97	0.99	0.97
Los Lagos	0.98	0.99	0.97
Aysén	0.78	0.86	0.92
Magallanes	0.93	0.94	0.94
Study area	0.93	0.92	0.93

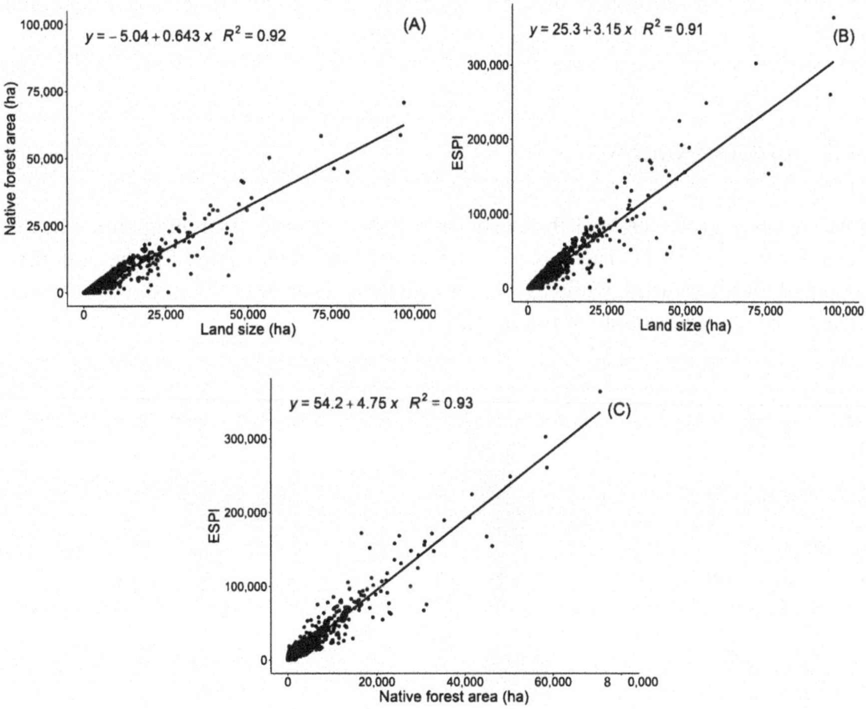

Fig. 18.2 Scatterplots between land size and native forest area (panel **a**), land size (axis *x*) and ESPI accumulated (panel **b**), and native forest area and ESPI accumulated (panel **c**) for a stratified sample on 311,595 landholdings

equal (0.2–0.3), 15% within the category of reasonable (0.3–0.4), 0.4% within the relatively unequal category (0.4–0.5), and 29% within the absolutely unequal category (> 0.5).

The three municipalities with the greatest inequality in the distribution of land (absolutely unequal) were Chaitén (0.85) in the Los Lagos region, followed by Antuco (0.78) and Lota (0.75), both in the Biobío region. The three municipalities with the lowest land inequality (absolutely equal) were San Gregorio (0.012) in the Biobío region and Puqueldón (0.013) and Castro (0.02) in the Los Lagos region (Fig. 18.3a).

The three municipalities with the greatest inequality in the distribution of the native forest (absolutely unequal) were Yungay (0.82) in the Biobío region and Chaitén (0.81) and Los Muermos (0.80), both in the Los Lagos region. In turn, the municipalities with the greatest equality (absolutely equal) were Purranque (0.08) in the Los Lagos region and San Carlos (0.07) and Portezuelo (0.03), both in the Biobío region (Fig. 18.3b).

Regarding the distribution of ESPI, the municipalities with the greatest inequality (absolutely unequal) were Chile Chico (ESPI Gini 0.99) in the Aysén region, Quemchi (0.95) in the Los Lagos region, and Quilaco (0.93) in the Biobío region. In turn, the three municipalities with the highest equality (absolutely equal) were Aysén (0.02) in the region of the same name and Castro (0.01) and Puqueldón (0.01), both in the Los Lagos region (Fig. 18.3c).

4.3 Cluster Analysis

Our group analysis yielded the clustering of five groups of municipalities in the study area (Fig. 18.4). The box plot shows the statistical values for the complete group of municipalities, while each point and their respective colors show the statistical values for the sample of municipalities.

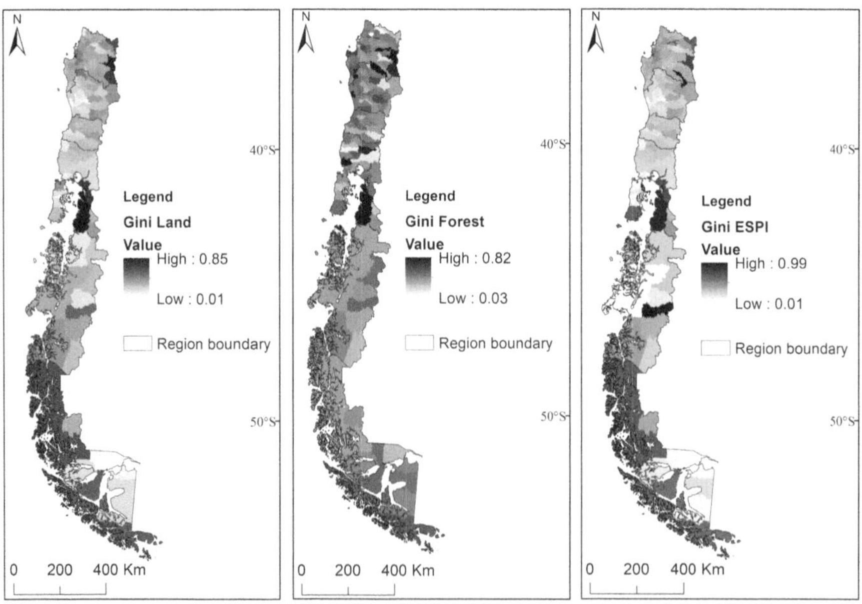

Fig. 18.3 Spatial distribution of land Gini (**a**), forest Gini (**b**), and ESPI Gini (**c**) at the municipality level. The value 1 is perfect inequality while the value 0 is perfect equality

The first group (blue color) comprised 12 municipalities distributed across the Biobío (6), Araucanía, Los Lagos (2), Aysén (1), and Magallanes (3) regions. These municipalities gathered 3756 (65.3%) small landholdings (<60 ha), 1423 (24.8%) medium landholdings (60–1000 ha), and 571 (9.9%) large properties (>1000 ha). These municipalities showed Gini coefficients in the range absolutely unequal for ESPI ($\bar{x}$= 0.69), land distribution ($\bar{x}$ = 0.60), and income ($\bar{x}$ = 0.51) and in the relatively unequal range for native forest ($\bar{x}$ = 0.40). Furthermore, they showed the highest poverty rates ($\bar{x}$ = 0.43) and the lowest levels of the human development index ($\bar{x}$ = 0.60). The proportion of the rural population was medium to low ($\bar{x}$ = 0.30), while the proportion of the indigenous population was below the mean of the study area ($\bar{x}$ = 0.20). It was the second group with the highest rates of deforestation ($\bar{x}$ = 2.68), while the rates of afforestation with exotic species were above the average ($\bar{x}$ = 0.12). They were the third group of municipalities with the highest rates of annual loss of ESPI ($\bar{x}$ = 0.04).

The second group (red color) comprised 64 municipalities distributed across the Biobío (1), Araucanía (21), Los Ríos (11), Los Lagos (26), Aysén (4), and Magallanes (1) regions. These municipalities concentrated a total of 134,065 (88.4%) small properties, 16,479 (10.9%) medium properties, and 1115 (0.7%) large landholdings. These municipalities showed Gini coefficients in the ranges absolutely equal for land distribution ($\bar{x}$ = 0.19), reasonable for native forest and ESPI ($\bar{x}$ = 0.32, $\bar{x}$ =

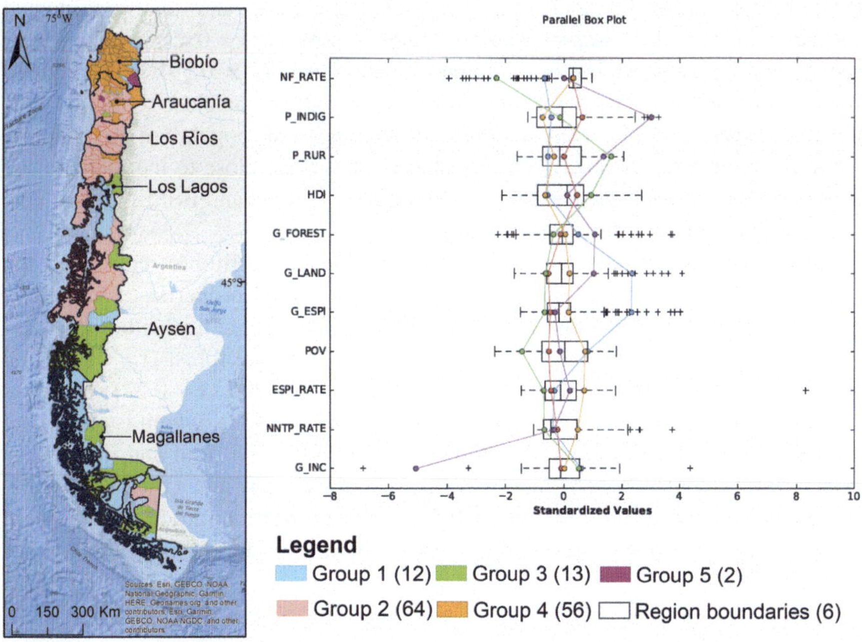

Fig. 18.4 Spatial distribution of cluster-group analysis and box plot

0.30), and relatively unequal for income ($\bar{x}$ = 0.49). Poverty rates in these municipalities were lower than the mean for the study area ($\bar{x}$ = 0.26), which is consistent with levels of human development, rural population, and indigenous peoples above the mean ($\bar{x}$ = 0.69,$\bar{x}$ = 0.45 and($\bar{x}$ = 0.38, respectively). In addition, the rates of deforestation ($\bar{x}$ = 0.25), afforestation ($\bar{x}$ = 0.12), and annual loss of ESPI ($\bar{x}$ = 0.04) were close to the mean of the study area.

The third group (green color) was made up of 13 municipalities distributed along the Araucanía (1), Los Lagos (1), Aysén (5), and Magallanes (6) regions. They comprised 3651 (61%) small properties, 1445 (24.2%) medium landholdings, and 885 (14.8%) large landholdings. They showed Gini coefficients in the range of absolutely equal for land distribution ($\bar{x}$ = 0.17) and ESPI ($\bar{x}$ = 0.16), relatively equal for native forest ($\bar{x}$ = 0.28), and relatively unequal for income inequality ($\bar{x}$ = 0.50). They also showed the lowest poverty rates of all groups ($\bar{x}$ = 0.14), which is consistent with the highest human development indices ($\bar{x}$ = 0.72). The rural population ($\bar{x}$ = 0.89) and indigenous population were slightly below the mean ($\bar{x}$ = 0.15). On the other hand, these municipalities showed the highest deforestation rates ($\bar{x}$ = 6.86), the lowest afforestation rates ($\bar{x}$ = 0.07), and the lowest rates of annual loss of ESPI ($\bar{x}$ = 0.02).

The fourth group (brown color) included 56 municipalities distributed across the Biobío (45), Araucanía (9), Los Ríos (1), and Los Lagos (1) regions. This group of municipalities comprised 129,422 (91.4%) small landholdings, 11,491 (8.1%) medium-size holdings, and 622 (0.4%) large properties. They were characterized by Gini coefficients in the range of relatively equal for land distribution ($\bar{x}$ = 0.29), reasonable for native forest ($\bar{x}$ = 0.34) and ESPI ($\bar{x}$ = 0.31), and relatively unequal for income ($\bar{x}$ = 0.48). They showed high rates of poverty ($\bar{x}$ = 0.43), which is consistent with the lowest levels of human development ($\bar{x}$ = 0.59) which ranged between 0.5 and 0.7. The percentage of rural population was close to the mean of the study area ($\bar{x}$ = 0.35), while the presence of indigenous population was the lowest in the study area ($\bar{x}$ = 0.13). Deforestation rates were close to the mean of the study area ($\bar{x}$ = 0.18), whereas afforestation rate ($\bar{x}$ = 0.26) and ESPI loss rate ($\bar{x}$ = 0.07) were the highest of the entire study area.

Group 5 included only two municipalities, one in the Biobío region and the other in Araucanía. They comprised 6377 (95.6%) small properties, 255 (3.8%) medium, and 38 (0.6%) large and showed degrees of inequality within the relatively unequal range for land distribution ($\bar{x}$ = 0.41), native forest ($\bar{x}$ = 0.47), and income ($\bar{x}$ = 0.48), although relatively equal levels for ESPI ($\bar{x}$ = 0.22). The poverty rates were close to the mean of the study area ($\bar{x}$ = 0.31), which is consistent with average levels of human development ($\bar{x}$ = 0.66). The rural and indigenous population percentages were among the highest in the study area ($\bar{x}$ = 0.8, $\bar{x}$= 0.82, respectively). Deforestation rate ($\bar{x}$ = 0.99) and afforestation rate ($\bar{x}$ = 0.12) were close to the mean of the study area, although with high levels of annual loss of ESPI ($\bar{x}$ = 0.05).

5 Discussion

5.1 Landscapes of Inequality and Implications for Planning

Three principal findings emerge from the results. First, we corroborate that land size is associated with the proportion of native forest at the property level. Large properties, usually classified as those over 1000 ha, represent only 26% of the total area in the study area, but they concentrate 55% of the land and 62% of the native forest area. These properties include three different types of holdings: farms dedicated to cattle ranching and dairy; properties dedicated to forestry operations, which include single landholdings and forest companies; and privately protected areas (PPA) dedicated to nature conservation, which include individual owners, corporations, foundations, and NGO. Their average proportion of native forest is 68% as compared to small landholdings with 34%. They are mostly located near the Andes range, and therefore they hold other outstanding attributes (e.g., lakes, volcanoes, waterfalls) that can sustain diverse ES such as recreation opportunities (Nahuelhual et al. 2016).

In 2016, the area of forest plantations within the study area reached 2,114,975 ha (INFOR 2018), which represented 69% of the country's total in the same year. Of this amount, 59% were located in the Biobío region, while 24% in the Araucanía region and 8% in Los Ríos, and the remaining 9% were divided between the Lagos, Aysén, and Magallanes regions. Most of them belonged to 16 forest companies (<30,000 ha) (mostly from the Matte and Angelini groups). On the other hand, there are 338 medium-sized owners and 19,579 small ones with forest plantations. Most of this plantation area has been established through the Decree Law 701, legal body enacted in 1974 with the aim of promoting forestry development in Chile, which established incentives (subsidies) for the forestry activity of exotic species (Labbé 2017). As for PPA, there are different types of owners. For the year 2011 (year of the most updated cadaster), there were 266 APP in the study area with surfaces ranging from 1.4 ha to 291,773 ha. An emblematic case is the Tantauco Park, owned by the current president, Sebastian Piñera, who acquired 118,000 ha (15% of the territory that comprises the Big Island of Chiloé) in the commune of Quellón (cluster group 2), Los Ríos region. These lands were bought from an offshore company created in Panama in order to evade taxes in and had been historically claimed by Hulliche communities. Tantauco Park is the third largest PPA in the study area, comprising 100,000 ha of native forest, which include patches of the Ciprés de las Guaitécas and Tepuales. In addition, within the park there are 1600 ha of lakes, lagoons, and rivers, which considers 107 km of rivers for recreational use and 7600 ha of wetlands.

Second, we corroborate that lager properties are associated with the highest accumulated values of ESPI, which can be explained by the fact that land area directly determines the amount of accumulated ESPI and because larger landholdings have well-conserved native vegetation, which is closely related to high ESPI values. Larger properties (> 1000 ha) account for 49% of ESPI accumulated provision. On the other extreme, we find subsistence farms with less than 30 ha that represent 49% of land but only the 10% of the forest area and 17% of ESPI accumulated provision. Additionally, they exhibit high rates of deforestation (0.06 $ha^{1}year^{-1}$) due to unsustainable logging and forest degradation (Reyes et al. 2016).

Third, we find a "trap"-like pattern in 8% of the municipalities, mainly in northern Patagonia. Social-ecological traps are path-dependent processes, which are causally produced through a conjunction of events over time (Boonstra and Boer 2014). A path-dependent process refers to the reproduction, or persistence, of certain phenomena in the absence of the forces that were responsible for the original production of these phenomena (Mahoney 2000). The present state of the trap is one where inequalities (land, forests, ES, and income), poverty, and natural resources loss and degradation are spatially coupled, as we can observe in cluster group 1. Yet, this present picture is the result of a series of complex interactions among policies (land, forest, and rural development), external drivers (e.g., timber demand), and people responses (e.g., resignation or confrontation) over time.

Today, Patagonia reflects what Coomes et al. (2016) denominate "landscapes of social inequality" referring to land size inequality and inequality in land use. The clearest expression of this denomination is that larger properties have conserved their forests through a more extensive use or conservation practices, whereas small and medium properties have reduced their forest area along with forest degradation (Bopp et al. 2020), which translate in a smaller capacity to provide ES (Benra and Nahuelhual 2019). These inequalities have also been analyzed from the lens of land grabbing. As with other land grabs, land acquisition for conservation is often considered problematic, undermining local sovereignty, allowing benefits of resources to be captured by outsiders, and causing harm to local people (Benjaminsen and Bryceson 2012). Southern Chile has been cited as an example of conservation land grabs (Pearce 2012; Holmes 2014).

These findings have important implications for conservation and development planning alike. Conservation policies usually highlight the role of private landholders in achieving biodiversity and ES conservation goals (e.g., Kamal et al. 2015). However, the results suggest that smallholders in Patagonia, regardless of their motivations, are in a disadvantaged position to become part of traditional conservation programs such as native forest subsidies, private protected areas, or conservation easements. For the Aysén and Magallanes regions, since the startup of the fund for the conservation, recovery and sustainable management of the native forest in 2008 (Law 20,283), USD $ 1,701,750 have been transferred to small owners. Of this amount, only 14% has been allocated to native forest preservation plans, while 67% has been allocated to management plans for logging. It could be expected that small landholders face the same limitations with regard to ES-based incentives. Studies conducted in countries that have implemented payments for ecosystem services (PES), for example, show that an increasing number of payment contracts have started to target medium to large properties (Alix-Garcia and Wolff 2014; Arriagada et al. 2012), which has led to significant critiques on equity and environmental justice grounds (He and Sikor 2015; Sikor 2013). In response to this equity claims, there is an ongoing effort to develop a grounded, institutionally oriented model of PES that relaxes some economic emphasis that defined de earliest PES versions.

Much of this recent work is place- and actor-centered (i.e., agency, local contingencies, and individuals' knowledge and subjectivities are taken seriously) (Rommel and Anggraini 2018; Wang and Wolf 2019). Additionally, the organizational and cultural aspects of institutional innovation are increasingly recognized (Potter and Wolf 2014; Schröter et al. 2017). These types of considerations could apply more widely to other conservation instruments such that smaller landholders are not excluded on the grounds of land and forest size and quality, but they become recognized for their important role in ensuring landscape connectivity.

In terms of development policies, the results confirm the need for territorial emphasis since Chilean Patagonia is far from being a homogeneous territory. Particular attention deserve those territories with persistent inequality, poverty, and loss of natural resources (e.g., cluster group 1), which are also the center of socio-environmental conflicts related to ownership of land and other natural resources such as water (Bengoa 2007; Pinto 2012).

In summary, the results corroborate the great inequalities that persist in the Chilean Patagonia after more than a century, which hamper inclusive development in the present. The results also highlight the need to think about the development and conservation of nature and ES as a whole and not as divergent objectives. An inclusive development requires, among other things, promoting ES policies that, instead of increasing current inequalities, aim at minimizing them.

Acknowledgments This work was sponsored and funded by FONDECYT Grant N° 1190207. The authors wish to thank Felipe Benra, Claudio Agostini, and Pedro Laterra whose comments and data helped improve the chapter.

References

Agostini CA, Brown PH (2007) Desigualdad Geográfica en Chile. Revista de Análisis Económicos 21:3–33

Agostini CA, Brown PH, Góngora DP (2008) Distribución Espacial de la Pobreza en Chile. Estudios Económicos 35:79–110

Alix-Garcia J, Wolff H (2014) Payment for ecosystem services from forests. Ann Rev Resour Econ 6:361–380. https://doi.org/10.1146/annurev-resource-100913-2524.

Arriagada RA, Ferraro PJ, Sills EO, Pattanayak SK, Cordero-sancho S (2012) Do payments for environmental services affect Forest cover? A farm-level evaluation from Costa Rica. Land Econ 88:382–399

Barraclough S (1999) 'Land Reform in Developing Countries'. UNRISD Discussion Paper 101. Geneva: United Nations Research Institute for Social Development.

Bengoa J (2007) Historia de un conflicto: los mapuches y el estado nacional durante el siglo XX. 242 pp.

Benjaminsen TA, Bryceson I (2012) Conservation, green/blue grabbing and accumulation by dispossession in Tanzania. J Peasant Stud 39:335–355. https://doi.org/10.1080/03066150.2012.667405

Benra F, Nahuelhual L (2019) A trilogy of inequalities: land ownership, forest cover and ecosystem services distribution. Land Use Policy 82:247–257. https://doi.org/10.1016/j.landusepol.2018.12.020

Boonstra W, Boer F (2014) The historical dynamics of social-ecological traps. Ambio 43:260–274
Bopp C, Engler A, Jara-Rojas R, Arriagada R (2020) Are forest plantation subsidies affecting land use change and off-farm income? A farm-level analysis of Chilean small forest landowners. Land Use Policy 91. https://doi.org/10.1016/j.landusepol.2019.104308.
Byerlee D, Deininger K (2013) The rise of large farms in land-abundant countries: do they have a future? L Tenure Reform Asia Africa:333–353. https://doi.org/10.1057/9781137343819_14.
Chonchol J (2003) La Reforma Agraria en América Latina. Proceso Agrar en Boliv y América Lat:205–221
CIREN-CORFO (1999) Digital cartography of rural properties. https://www.ciren.cl/tipo/propiedades/
Clarke D (1975) Land inequality and income distribution in Rhodesia. Afr Stud Rev 18(1):1–7. https://doi.org/10.2307/523705
CONAF (1998). Cadaster and evaluation of native vegetation resources of Chile. http://sit.conaf.cl/
CONAF (2013). Cadaster and evaluation of native vegetation resources of Chile. http://sit.conaf.cl/
Coomes OT, Takasaki Y, Rhemtulla JM (2016) Forests as landscapes of social inequality. Ecol Soc 21:1–20. https://doi.org/10.5751/ES-08684-210320
Davis-Hamel A (2012) Successful neoliberalism?: state policy, poverty, and income inequality in Chile. Int Soc Sci Rev 87:79–101
FAO (2006) Hacia una política pertinente para el desarrollo integral de los pueblos indígneas. 112 pp. Available on http://www.fao.org/3/i3001s/i3001s.pdf
Frankema E (2010) The colonial roots of land inequality: geography, factor endowments, or institutions? Econ Hist Rev 63:418–451. https://doi.org/10.1111/j.1468-0289.2009.00479.x
Gini C (1909) Il diverso accrescimento delle classi sociali e la concentrazione della ricchezza. G degli Econ 38:27–83
Griffin K, Khan AR, Ickowitz AMY et al (2002) Poverty Distrib Land 2:279–330
He J, Sikor T (2015) Notions of justice in payments for ecosystem services: insights from China's Sloping Land Conversion Program in Yunnan Province. Land Use Policy 43:207–216. https://doi.org/10.1016/j.landusepol.2014.11.011
Hojman D (1996) Poverty and inequality in Chile: are democratic politics and neoliberal economics good for you? J Interam Studies World Aff 38:73–96. https://doi.org/10.2307/166361
Holmes G (2014) What is a land grab? Exploring green grabs, conservation, and private protected areas in southern Chile. J Peasant Stud 41:547–567. https://doi.org/10.1080/03066150.2014.919266
INE (2018) Síntesis de resultados CENSO 2017. Available on https://www.censo2017.cl/descargas/home/sintesis-de-resultados-censo2017.pdf
INFOR (2018) Anuario forestal 2018. Available on https://wef.infor.cl/publicaciones/anuario/2018/Anuario2018.pdf.
IPBES (2019) Regional Assessment Report on Biodiversity and Ecosystem Services for the Americas. Chapter 4. Plausible futures of nature, its contributions to people and their good quality of life. Available on https://ipbes.net/sites/default/files/ipbes_global_assessment_chapter_4_unedited_31may.pdf.
Janvry A, Sadoulet E (2001) Income strategies among rural households in Mexico: the role of off-farm activities. World Dev 29(3):467–480. https://doi.org/10.1016/S0305-750X(00)00113-3.
Jayne TS, Yamano T, Weber MT et al (2003) Smallholder income and land distribution in Africa: implications for poverty reduction strategies. Food Policy 28:253–275. https://doi.org/10.1016/S0306-9192(03)00046-0
Kamal S, Grodzinska-Jurczak M, Kaszynska AP (2015) Challenges and opportunities in biodiversity conservation on private land: an institutional perspective from Central Europe and North America. Biodivers Conserv 24:1271–1292. https://doi.org/10.1007/s10531-014-0857-5
Kay C (2002) Chile's neoliberal agrarian transformation and the peasantry. J Agrar Chang 2:464–501. https://doi.org/10.1111/1471-0366.00043
Kay C (2006) Rural poverty and development strategies in Latin America. J Agrar Chang 6:455–508. https://doi.org/10.1111/j.1471-0366.2006.00132.x.

Kay C (2007) Land, conflict, and violence in Latin America. Peace Rev 19:5–14. https://doi.org/10.1080/10402650601181808

Keswell M, Carter MR (2014) Poverty and land redistribution. J Dev Econ 110:250–261. https://doi.org/10.1016/j.jdeveco.2013.10.003

Labbé D (2017) El lucrativo negocio del acaparamiento de tierras. El Ciudadano. https://www.elciudadano.com/economia/el-lucrativo-negocio-del-acaparamiento-de-tierras/11/10/

Laterra P, Nahuelhual L, Vallejos M et al (2019) Linking inequalities and ecosystem services in Latin America. Ecosyst Serv 36:1–14. https://doi.org/10.1016/j.ecoser.2018.12.001

Leonard B, Parker DP, Anderson TL (2020) Land quality, land rights, and indigenous poverty. J Dev Econ 143:102435. https://doi.org/10.1016/j.jdeveco.2019.102435

Loughrey JJ, Donnellan T (2017) Inequality and Concentration in Farmland Size: A Regional Analysis for Western Europe. 15th EAAE Congr "Towards Sustain Agri-food Syst Balanc Between Mark Soc 20.

Lowder SK, Skoet J, Raney T (2016) The number, size, and distribution of farms, smallholder farms, and family farms worldwide. World Dev 87:16–29. https://doi.org/10.1016/j.worlddev.2015.10.041

Mahoney J (2000) Path dependence in historical sociology. Theor Soc 29(4):507–548

Martinic M (2006) El poblamiento rural en magallanes durante el siglo xx.: realidad y utopia. Magallania (Punta Arenas), 34(1):5–20

Millennium Ecosystem Assessment (2005) Ecosystems and Human Well-being: Synthesis. Island Press, Washington, DC.

Nahuelhual L, Benra F, Rojas F, Ignacio Díaz G, Carmona A (2016) Mapping social values of ecosystem services: what is behind the map? Ecol Soc 21. https://doi.org/10.5751/ES-08676-210324

Nahuelhual L, Saavedra G, Mellado MA et al (2020) A social-ecological trap perspective to explain the emergence and persistence of illegal fishing in small-scale fisheries. Marit Stud 19:105–117. https://doi.org/10.1007/s40152-019-00154-1

Olper A (2007) Land inequality, government ideology and agricultural protection. Food Policy 32:67–83. https://doi.org/10.1016/j.foodpol.2006.03.009

OXFAM (2016) Unearthed: land, power and inequality in latin America. Available on https://www-cdn.oxfam.org/s3fs-public/file_attachments/bp-land-power-inequality-latin-america-301116-en.pdf

Paruelo JM, Texeira M, Staiano L et al (2016) An integrative index of Ecosystem Services provision based on remotely sensed data. Ecol Indic 71:145–154. https://doi.org/10.1016/j.ecolind.2016.06.054

Pearce F (2012) The landgrabbers: the new fight over who owns the Earth. Eden Project Books, Boston

Pinto J (2012) El conflicto Estado-Pueblo Mapuche 1900-1960. Universum 1:167–189

PNUD (2000) Desarrollo humano en las comunas de Chile. 102 pp. Available on http://bibliotecadigital.ciren.cl/bitstream/handle/123456789/15748/Desarrollofinal.pdf?sequence=1&isAllowed=y

Potter CA, Wolf SA (2014) Payments for ecosystem services in relation to US and UK agri-environmental policy: disruptive neoliberal innovation or hybrid policy adaptation? Agric Human Values 31:397–408. https://doi.org/10.1007/s10460-014-9518-2

Puyravaud JP (2003) Standardizing the calculation of the annual rate of deforestation. For Ecol Manag 177:593–596. https://doi.org/10.1016/S0378-1127(02)00335-3

Reyes R, Blanco, G, Lagarrigue A, Rojas F (2016) Ley de Bosque Nativo: Desafíos Socioculturales para su Implementación. Instituto Forestal y Universidad Austral de Chile. 82 p.

Rommel J, Anggraini E (2018) Spatially explicit framed field experiments on ecosystem services governance. Ecosyst Serv 34:201–205. https://doi.org/10.1016/j.ecoser.2018.10.002

Rouse JW, Haas RH, Schell JA, Deering, DW (1974) Monitoring vegetation systems in the great plains with erts. In Third earth resources technology satellite- 1 symposium, Greenbelt, NASA SP-351; pp. 301–317

Sasso M (2006) Remate de tierras fiscales en el terriotiro de Magallanes (1903). Magallania 34:157–160. https://doi.org/10.4067/S0718-22442006000100010

Schröter M, Stumpf KH, Loos J, Van Oudenhoven APE, Böhnke-henrichs A, Abson DJ (2017) Refocusing ecosystem services towards sustainability. Ecosyst Serv 25:35–43

Severini S, Tantari A (2015) The distributional impact of agricultural policy tools on Italian farm household incomes. J Policy Model 37:124–135. https://doi.org/10.1016/j.jpolmod.2015.01.004

SII (2016) Digital cartography of rural properties. Available on https://www4.sii.cl/mapasui/internet/#/contenido/index.html.

Sikor T (2013) The justices and injustices of ecosystem services.

Stoms DM, Hargrove WW (2000) Potential NDVI as a baseline for monitoring ecosystem functioning. Int J Remote Sens 21:401–407. https://doi.org/10.1080/014311600210920

Sun T, Zhang H, Wang Y, Meng X, Wang C (2010) The application of environmental Gini coefficient (EGC) in allocating wastewater discharge permit: the case study of watershed total mass control in Tianjin. China Resour Conserv Recycl 54:601–608. https://doi.org/10.1016/j.resconrec.2009.10.017

United Nations (2020). World Social Report 2020. Inequality in a rapidly changing world. 216 p. https://www.unilibrary.org/content/books/9789210043670#chapters

von Bennewitz E (2017) Land tenure in Latin America: from land reforms to counter-movement to neoliberalism. Acta Univ Agric Silvic Mendelianae Brun 65:1793–1798. https://doi.org/10.11118/actaun201765051793

Wang P, Wolf SA (2019) A targeted approach to payments for ecosystem services. Glob Ecol Conserv 17:e00577. https://doi.org/10.1016/j.gecco.2019.e00577

Wegerif MCA, Guereña A (2020) Land inequality trends and drivers. Land 9:1–23. https://doi.org/10.3390/land9040101

Wiener E (2011) La concentración de la propiedad de la tierra en América Latina: una aproximación a la problemática actual. Coalición Internacional para el Acceso a la Tierra:64

Yamauchi F (2016) Rising real wages, mechanization and growing advantage of large farms: evidence from Indonesia. Food Policy 58:62–69. https://doi.org/10.1016/j.foodpol.2015.11.004

Zheng X, Xia T, Yang X, Yuan T, Hu Y (2013) The land Gini coefficient and its application for land use structure analysis in China. PLoS One 8:1–10. https://doi.org/10.1371/journal.pone.0076165

Chapter 19
Imaginaries, Transformations, and Resistances in Patagonian Territories from a Socio-Ecological Perspective

Pedro Laterra , Laura Nahuelhual, Mariana Gluch, Pablo L. Peri, and Guillermo Martínez-Pastur

> *Nowadays myths are generally short-lived, but that of Patagonia is among the more resilient ones. It's true: the myth seems to be on sale, just like the land itself. Power has been fighting for the land of Patagonia for centuries and hasn't stopped yet*
>
> *—Aliaga 2019*

Abstract Socially shared expectations about people's behavior and worldviews, both in terms of how they work and how they should work, are part of the concept of social imaginaries. Beyond representing the idiosyncrasy of a society, imaginaries can play cohesive, critical, or transformative roles over societies and their

P. Laterra (✉)
Fundación Bariloche, San Carlos de Bariloche, Río Negro, Argentina

Consejo Nacional de Investigaciones Científicas y Técnicas (CONICET), CCT Patagonia Norte, San Carlos de Bariloche, Río Negro, Argentina
e-mail: pedro.laterra@conicet.gov.ar

L. Nahuelhual
Instituto de Economía, Universidad Austral de Chile Centro de Investigación: Dinámica de Ecosistemas Marinos de Altas Latitudes Instituto Milenio en Socio-Ecología Costera, Valdívia, Chile

M. Gluch
Fundación Bariloche, San Carlos de Bariloche, Río Negro, Argentina

P. L. Peri
National Agricultural Technology Institute, Río Gallegos, Argentina

G. Martínez-Pastur
Consejo Nacional de Investigaciones Científicas y Técnicas, Centro Austral de Investigaciones Científicas y Técnicas, Ushuaia, Tierra del Fuego, Argentina

P. L. Peri et al. (eds.), *Ecosystem Services in Patagonia*, Natural and Social Sciences of Patagonia, https://doi.org/10.1007/978-3-030-69166-0_19

supporting systems. Therefore, we postulate that by modulating nature-society relationships, social imaginaries can represent transformative or resistance forces affecting the provision, capture, and distribution of ecosystem services (ES). We analyze nature-society social imaginaries (NSI) based on previously published information, with particular attention to their influences on the provision, capture, and social distribution of ES in Patagonian landscapes of Argentina and Chile. Firstly, we built and used a conceptual NSI framework that integrates the concepts of ES, social imaginaries, and the governance of natural capital. Secondly, we applied the NSI framework to selected case studies addressing four main drivers of change in Patagonian socio-ecological systems (SES): (i) land dispossession, (ii) industrial forestry expansion, (iii) touristification, and (iv) damming of rivers for hydroelectric uses. According to our analyses, Patagonian SES are being affected by typical transforming forces of the modern imaginary (e.g., short term, productivity, reductionist-mechanistic vision, individualism) as well as forces of resistance that characterize postmodern (e.g., intergenerational concern, ecocentrism, holism) and indigenous (e.g., territorial ancestral rights, living well in harmony with nature) imaginaries. These three main NSI underlie controversies around public policies and governance of ES, by influencing the shared perception of nature contributions to well-being, the shared expectations about human behaviors (governance institutions), and/or the final decisions affecting natural capital. We discuss the variation of the relative importance of these three mechanisms across NSI types.

Keywords Land dispossession · Forestry Expansion · Touristification · Hydroelectric uses · Nature-society relationships

1 Introduction

The Patagonian territory we know today was shaped hierarchically, not by the mythological beings and beliefs of its native peoples but through the visions and decisions of central powers from northern latitudes. The great national interests and the profitability of large enterprises based on the exploitation of natural resources sustained the early colonization of Patagonia and the well-being of their original and new inhabitants. Since then, the ways of accessing the benefits derived from nature have diversified, and antagonisms between alternative demands over nature have emerged. From a sociocultural point of view, Patagonia is currently a place where diverse cultures and their different relationships with nature meet. However, influences of extraterritorial cultures on the access to ecosystem services (ES) and the social distribution of their benefits prevail over local cultures. The nineteenth-century imaginary of Patagonia as a vast land of opportunities open to big business with ancient inhabitants decimated or "pacified" created resistance expressions that have emerged during the last decades from local communities, including the descendants of first-nation's genocide survivors (Blanco and Mendes 2006).

Most of these resistances hold "nature" and the emerging disputes over the different views on nature-society relationships as their emblem. While some actors value Patagonia's pristine environment, others defend the need to use natural resources under a new extractive logic (neo-extractivism). In this context, local people appropriate academic and political concepts to support their claims.

In contrast to previous fragmentary views of natural resources and the environment, the concept of ES integrates the diversity of demands and values of a society in relation to nature, either as a source of primary products or of less tangible contributions to well-being (MEA 2005; Díaz et al. 2015). More importantly, the ES approach allows to visualize the synergistic or antagonistic relationships (trade-offs) that arise between the opportunities of supply, capture, and distribution of ES benefits, disaggregating beneficiaries as "winners" and "losers" of the possible decisions of nature use.

In comparison with the research on ES values, and despite their relevance for the achievement of more sustainable trajectories, the role of beliefs (Raymond and Kenter 2016), shared intentions (Tomasello et al. 2005), and imaginaries (Archibald et al. 2020) that influence values, norms, and behaviors has been much less studied. A strong influence of psychosocial factors affecting values and intentions or decisions of management has been illustrated for different SES (e.g., Mastrangelo et al. 2014), but we still have a poor understanding about the dynamics of those factors on long-term nature conservation. Against the tendency to understand and influence the behavior of complex systems by focusing on proximate and dominant factors, Meadows (1999) highlighted that factors acting at higher hierarchical levels, like paradigms ("the shared ideas in the minds of society" or "deepest set of beliefs about how the world works"), provide significant, sustained, and beneficial effects on the redesign of the system or governing policy. In other terms, it is not just how much people value a particular good or service but the expectations that people have about what is real/unreal and normal/abnormal around their social existence, the social imaginaries – what shapes the high leverage points capable to move SES into more sustainable trajectories. A challenging question is to what extent those imaginaries can or should be changed and into which directions (Lotz-Sisitka 2010; Stephenson Jr 2011).

In this chapter we offer a theoretical and hermeneutical exploration of the main nature-society imaginaries (NSI) that coexist in Patagonia, with emphasis in their consequences for biodiversity and the provision of ES, nature governance, and nature contributions to people's well-being. We analyze the interplay between different NSI and their transformative vs resistance roles on nature's governance in the context of four main drivers of change in Patagonian SES, namely, land dispossession, industrial forestry expansion, touristification, and damming of rivers for hydroelectric uses. In the next sections, we firstly provide some insights about social imaginaries and their role on the relationship between societies and nature (Sect. 2), and then we present a conceptual framework for the analysis of the roles of collective imaginaries within the context of SES (Sect. 3). Next, we apply the conceptual framework to understand the role of imaginaries on land-use decisions and the application of alternative management practices, for selected study cases within the Patagonian region (Sect. 4). Finally, we synthesize main results and discuss the possible contribution of this conceptual framework to a better understanding of SES functioning (Sect. 5).

2 Social Imaginaries

Conceptual basis and definitions: Social imaginaries are depicted as social institutions that create a shared universe of meaning to which a society owes its unity and coherence. They constitute "socially constructed schemes that allow us to perceive something as real, explain it and intervene operatively" (Pintos 2005). They are historically given constructions of meaning that are naturalized and that have an autonomous existence, independent of social subjects and allow us to make reality intelligible, and ultimately give a specific orientation to society over a determined period (Díaz 1996; Agudelo 2011).

The concept of social imaginaries contains a tension between what "should be," established in a top-down manner, and a more dynamic "what is," as emerging from social self-organization. Social imaginaries act as homogenizing agreements with a social cohesive function and constitute the necessary innovation to adjust these agreements to variable local challenges. Thus, social imaginaries have elements of both moral structure (what is right) and moral agency (what is worth striving for); they are a blend of "how things usually go, but this is interwoven with an idea of how they ought to go" (Taylor 2003). This tension between the normative and the factual understandings implies conflicts, but, eventually, it also generates changes, risks, and progress.

Archetypal narratives like the "myth of progress" or "foundational myths" are good examples about how social imaginaries may support "normal" social behaviors. However, it is central to this chapter to note that imaginaries do not necessarily constitute constructs that identify and determine a certain society in a homogeneous and exclusive way. For example, social imaginaries embodied by minorities offer a pressure on the status quo, leading to conflicts with dominant imaginaries, with different social and environmental consequences in the long term. Rather than fixed and formal normative institutions, here we understand the social imaginaries as the semi-unconscious expectations that people have about what is real/unreal and normal/abnormal around their social existence, including stabilizing and/or transformative aspirations.

Social imaginaries about nature-society relationships (NSI) Imaginaries of each social group change over time due to multiple factors (socioeconomic, sociopolitical, marketing factors) ranging from value transformations to the influence of social networks.

The fundamental characteristics of the main imaginaries manifested in Patagonia, along two main imaginary axes, are: intergenerational and territorial significance (Fig. 19.1). Despite its schematic nature, this exercise provides a starting point to support subsequent analyses. Based on the emergence and domain of individualism and equality principles, modern social imaginaries overcome the community and hierarchy principles leading to a "vertical world of mediated access" which characterized the premodern social imaginaries of European societies (Taylor 2003). But, this moral agency of modern social imaginaries contributed to the rise of individual initiatives, innovations, wealth, more needs, and also, the big environmental challenges that humanity faces today. Environmental or nature-society imaginaries were examined by Buckles (2018), who described the two extreme perspectives of a continuum, from Taylor's modern to postmodern social imaginaries, by opposing different time scales (contemporary vs. intergenerational-oriented focuses), technological styles (high-input technological control of natural resources vs. low-input demanding of process management), and complexity conscience (reductionist vs. holistic understanding), among other aspects (Fig. 19.1). Within the range of greatest intergenerational significance, the sustainability criterion is only the starting point of a multiplicity of environmental imaginaries that, in the direction of a growing ecocentrism, finally meet with the deep ecology (e.g., McGregor 2004).

The western and dominant NSI normally excluded the visions of indigenous cultures. However, indigenous knowledge and worldview make up the conceptual framework of the Intergovernmental Platform on Biodiversity and Ecosystem

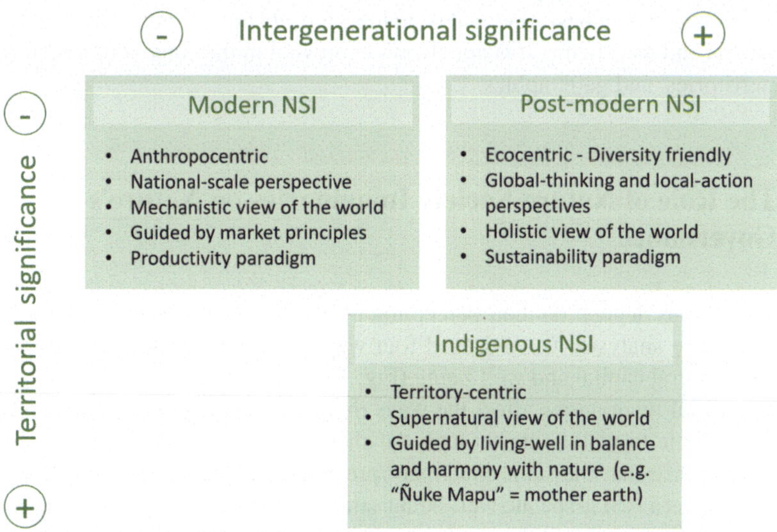

Fig. 19.1 Simplified imaginaries about nature-society relationships (NSI) and their preliminary characteristic features for the analysis of Patagonian cases

Services (IPBES), in recognition of its practical and symbolic relevance for peoples and communities around the world (Díaz et al. 2015). Of course, we continue talking about imaginaries whose expression in actions and lifestyles can be highly variable in the Patagonian territories. After all, imaginaries are not always aligned with individual and collective behaviors and actions toward the environment. A main distinctive aspect of indigenous worldview usually consists in a nonmechanistic but supranatural and holistic conscience about human-nature interconnectedness, leading to environmental accountability and stewardship.

Although the holistic vision and awareness of the interconnections between the human and the natural world is not absent in postmodern NSI, in indigenous NSI this interconnection is mainly associated with the supernatural dimension, spirituality, and symbolism (Skewes et al. 2012; Rojas et al. 2020). Another distinctive feature of indigenous NSI is the place occupied by the concept of territoriality (Montalba et al. 2005). Under the indigenous vision, the territory is not merely a geographical area, soil, subsoil, water, animals, plants, or even the people that inhabit the place; the territory is all these dimensions together, plus culture, memory, the right to decide on natural resources as well as to exercise its own organizational forms, and people's identities. However, indigenous communities are not homogeneous in their beliefs, depending on their livelihoods (rural versus urban), level of identity (self-recognition as a part of an indigenous collectivity), possession of land, and belonging to indigenous communities, among others. Therefore we cannot assume a single imaginary that represents all indigenous people in Patagonia.

One of the best-known indigenous cultures of Patagonia is the Mapuche culture, the prevailing ethnic group in the region today, in terms of population size and visibility of claims. In the Mapuche worldview, human well-being, history, culture, and the territory are deeply and inextricably linked to each other (Tricot 2020). While this vision may have emerged from long-term/historical interactions between ecosystems and well-being, it is not always expressed in the context of transformed SES, territories, and geographies.

3 The Role of Nature-Society Imaginaries on Nature's Governance

Humans actions depend on their perception of nature and environment. In order to guide our case analysis, we envisaged four ways in which social imaginaries may influence natural capital and well-being (Fig. 19.2).

First, social imaginaries affect the ways in which manufactured goods and services contribute to well-being through the shared values or shared perceptions of their importance that the same imaginary presupposes (Fig. 19.2, path 1). At one extreme, modern NSI associate well-being satisfaction to consumption of numerous goods, some of them involving high-impact technology (e.g., dams). At the other extreme, postmodern NSI question consumption based on those technologies.

Implicitly, the former NSI suppose a weak sustainability and therefore a broad substitution capacity between natural and manufactured capital, while the latter assume a strong sustainability and therefore a relatively low substitution capacity.

Second, the imaginaries also influence the ways in which natural capital mediates well-being (Fig. 19.2, path 2), from a productivity perspective under the modern imaginary to an intergenerationally responsible use, which privileges sustainability over economic growth (postmodern imaginary). Therefore, the contribution of goods and services from nature to well-being not only depends on their flow rates, as usually acknowledged in previous frameworks (Díaz et al. 2015; Potschin-Young et al. 2018), but it is also modulated by NSI that influence the extent to which nature contributes to well-being.

Third, taking into account how natural capital and its services contribute to well-being, NSI are able to affect that capital through shared expectations about what should or should not be done in terms of the practices and management decisions (Fig. 19.2, path 3).

Finally, NSI may affect the natural capital by social self-organization in opposition to top-down expectations about "what should be done" (Fig. 19.2, path 4). Governance is related to the perception of future scenarios linked to risk analysis and actions from each stakeholder. The notion of power between involved parties is part of the social imaginary and helps to guide the actions undertaken by the

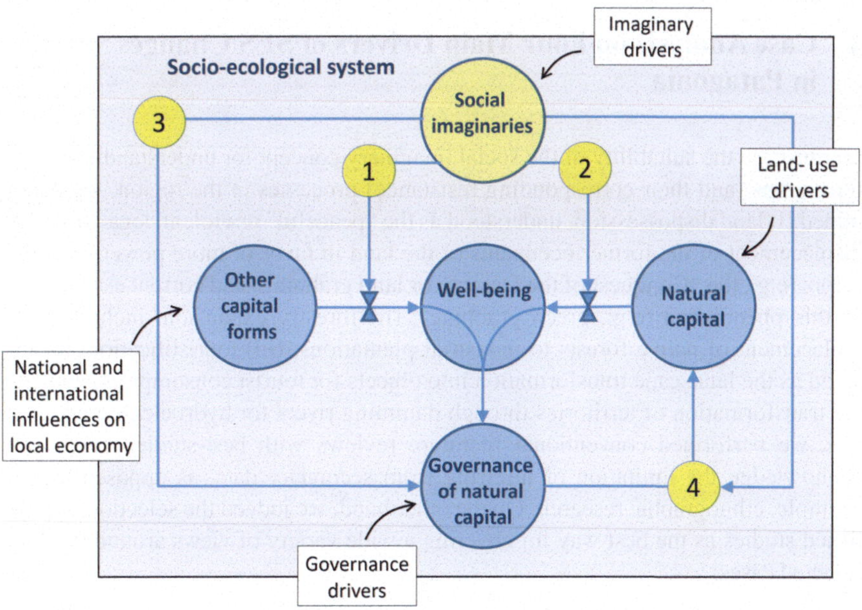

Fig. 19.2 A conceptual framework for the analysis of collective imaginaries' roles within the context of socio-ecological systems. The numbers identify four ways in which social imaginaries influence natural capital and well-being from ES: shared values and/or shared perceptions (1 and 2), shared expectations (3), and shared behaviors (4)

stakeholders based on risk-benefit calculations and the possibilities of cooperation or conflict that may occur between them.

Power relationships emerge as a key factor influencing the course of events. For example, some stakeholders' actions may drive ES conservation by preventing opposing actors with conflicting views on land use. Other stakeholders may try to use their political or economic influence to impose their own imaginaries of development with a particular perception of the future, the risk involved, and the management of environmental resources. This lack of consensus among actors (e.g., distrust between opposition groups and government) may create conflicts for management of natural capital. Particular imaginaries around management and conservation of natural capital and land-use planning that influence ecosystems and ES may lead to a confrontation between different development and governance models.

The NSI are characterized not only by the values, beliefs, norms, and practices that they promote but also by the magnitude of the social forces that they exert on SES. Thus, NSI can promote transformations (transforming forces), or they can endorse resistance against other drivers. The coexistence of these forces underlies different distributive conflicts or access to available ES. Furthermore, here we emphasize that behind the main drivers of SES changes examined here, opposing or antagonistic NSI can be recognized and characterized by forces of cohesion, transformation, or resistance they can exert on SES.

4 Case Analysis on Four Main Drivers of SES Changes in Patagonia

To illustrate the suitability of the social imaginary concept for understanding transformations (and their corresponding resistance) processes in the region, we determined (i) land dispossession, understood as the "peaceful" or violent, legal or illegal displacement of the former occupants of the land in favor of more powerful social actors (e.g., the "Conquest of the Desert," or land grabbing) and current expressions of this phenomena (e.g., green grabbing); (ii) forestry expansion, including the replacement of native forests to industrial plantations; (iii) touristification, understood as the landscape transformation into objects for tourist consumption; and (iv) the transformation of territories through damming rivers for hydroelectric use. For this, we performed conventional literature reviews with best-studied cases. We acknowledge the limitation of inferring from secondary data, as opposed to, for example, ethnographic research. On the other hand, we judged the selection of published studies as the best way for covering a wide variety of views around the four selected cases.

4.1 Land Dispossession

The colonization of the Patagonian territories involved the eradication of most indigenous people, by the military campaigns known, ironically, as the "Pacification of Araucanía" (1861–1881) in Chile and the "Conquest of the Desert" (1869–1888) in Argentina (Hasbrouck 1935). Subsequently, the national governments promoted the establishment of military settlements in strategic areas and the sale and concession of public lands to private landowners, mostly foreign settlers. Once these territories were administratively incorporated into the national territories of Argentina and Chile, the objective was to develop Patagonia based on the agricultural model, in correspondence to the national economic ideal based on European development.

Biographic, traveling, and historical literature provide great testimonies that sustain social imaginaries about Patagonia nature. One of the most clear expressions of such imaginary is reflected in the psychological and ideological profile of Domingo F. Sarmiento, an educator, thinker, journalist, writer, military, politician, and finally, one of the most influential presidents of Argentina (1868–1874) during the "Conquest of the Desert." Pastor and Mora (2013) analyzed the imaginary of Sarmiento in four axes: (i) representations of nature, where the idea of the "desert" only populated by Indians contrasts with the idea of productive lands, and urban development (as stated in the title of his famous book *Facundo: Civilization and Barbarism*, 1845); (ii) the differentiation and racial discrimination – the foreign colonist dominating technology against the lazy Indian and the speed and dynamism against slowness and immutability; (iii) ideal societies, where the United States was a development model to follow; and (iv) exemplary lives and professions of scientists and technologists. Sarmiento's imaginary about the "South" was externally influenced by the vision of Argentina of the English explorer and writer Francis Bond Head published in 1926 ("Rough Notes Taken During Some Rapid Journeys Across the Pampas and Among the Andes") and other European writers (Jagoe 2008).

Land dispossession during the Patagonian settlement stage Land grabbing by actors inside or outside the national territories usually involved the displacement of original inhabitants. This shaped the Patagonian territories since the late eighteenth century, when the settlement policies of Argentinean and Chilean governments displaced native peoples and promoted the occupation by foreign and domestic settlers. Through different mechanisms, European immigrants enjoyed privileged access to land, giving place to extensive sheep production systems in Argentina and Chile, originally in the hands of English settlers and English capitals, and German farmers in Chile. Numerous historical testimonies show that these colonization policies depicted indigenous cultures as an obstacle to development. Briefly, "Argentine oligarchy detested the reality that it imagined as a threat, and sought to become independent from the American past by building a new country through expansion and replacement: the territory is desert, its vastness is a scourge, its inhabitants are savages, their past is shame, their future, if any, is in the image and likeness of Europe" (Harambour 2019). Following the conceptual framework (Fig. 19.2), during the period of colonization of Patagonia, the management of the territories and their natural resources was shaped under the imaginaries of modernity, leading to a clear division between the winners and losers of this process.

The National Parks systems or precluding settlements for the intergenerational well-being A second mechanism of occupation was the creation of national parks, partially driven by border policies of each country and partially supported by the imaginary of nature conservation inspired by pioneers from the northern hemisphere. These actions, promoted from imaginaries aligned with the intrinsic value of biodiversity, were confronted with imaginaries of sovereignty and identity of the first nations still present today. The national parks systems of Argentina and Chile were mainly developed on empty or previously eradicated territories. These displacements for conservation reasons in Patagonia created severe conflicts some of which persist until the present date (Serenari et al. 2015, 2017a, b).

In Argentina, the creation of the national parks system in Patagonia was originally linked to tourism development (Aizen and Tam Muro 1992). These socioeconomic and cultural changes occurred as a result of specific actions taken by the National Argentine Government through the Directorate of National Parks enacted in 1934, which shaped a particular productive model that gradually became the single economic engine of the region (Bessera 2006). The Directorate had two main objectives: the conservation of the natural resources and the consolidation of territorial sovereignty near to international borderlines. These actions were in line with a particular social imaginary that conceived nature as pristine, understood as "free from human interference" (Carpinetti 2006). This implied, in practice, the eviction of rural settlers that resided in the areas. The preservation of scenic beauty and the desire to populate Patagonia with European descendants imposing elitist land-use models resulted in the marginalization and disregard of the rights of these rural communities (Carpinetti 2006).

Successive development models, mainly from the nation's capital, continued delineating the area's trajectory. During the Peronist government (1946–1955), tourism in Argentina went through a deep transformation that favored "social tourism" by promoting popular leisure and recreational activities for the working class (Méndez 2016). This completely changed the physiognomy of towns like the emblematic San Carlos de Bariloche, which constituted one of the "targets" of this new social tourism policy.

Land grabbing nowadays Current expressions of land grabbing in Patagonia can be classified into two basic types: (i) the acquisition of large areas, frequently by foreign investors, and (ii) the growing territorial claims of the descendants of formerly displaced, native peoples.

During the last decades, the acquisition of large land tracts by foreign investors has reached a prominent public status in Argentina, where the economic and legal conditions of the 1990s favored acquisitions and at the same time the dispossession of rural sectors with precarious land tenure (Murmis and Murmis 2012; Zoomers

2011). In Chile, new land acquisitions have been motivated by conservation objectives, not only by foreign buyers but also by local NGOs and other philanthropist individuals and organizations. The nature and dimension of this phenomenon has been reviewed by different authors (Di Giminiani and Fonck 2018; Holmes 2014, 2015). For example, the acquisitions made by foreign investors like Douglas Tompkins, Joe Lewis, Ted Turner, and Luciano Benneton on the Argentine side (Gorenstein and Ortiz 2016; Murmis and Murmis 2012; Smit 2017) are worth highlighting. Part of these acquisitions, particularly those by the Conservation Land Trust Foundation (Tompkins Conservation), pursue conservation objectives, either for their use value (e.g., ecotourism) or for the intrinsic value of biodiversity. Particularly in Chile, private protected areas are rapidly expanding and represent an important portion of the national territory (Rivera and Vallejos-Romero 2015).

While private conservation initiatives mostly fit into the postmodern imaginary (e.g., intergenerational responsibility, holistic integration, respect for diversity), modern and development visions of this phenomena have accused private protection initiatives of blocking productive uses of natural resources (Tecklin and Sepúlveda 2014). Thus, setting aside productive lands for conservation purposes has been qualified as a case of capital accumulation using politically correct motives or green grabbing (Holmes 2015). The indigenous imaginary generally rejects these two visions, by considering them forms of market-driven instruments that ultimately harm their ancestral rights in the territory (Meza 2009; Serenari et al. 2017b).

These new forms of conservation are highly contested, and their societal acceptance is still to be determined. Nonetheless "this imaginary has helped forge a hegemonic front among resident, corporate, and state actors supporting eco-regionalism: Southern Andean Patagonia as a space committed to green development" (Mendoza et al. 2017).

The Argentina´s "Land Law" (Law 26,737) was enacted in 2011 setting the maximum surface by province that can be owned by foreigners, but it was made more flexible by the successor government in 2017. This law was initially promoted by an organization of small and medium farmers (Federación Agraria Argentina), in response to the impact exerted on this sector by the demand from foreign investors with high purchasing power. In Chile, while the attempt to enact a law limiting the extension of private reserves (presumably in favor of forestry enterprises) failed, the imaginaries behind private conservation have been successfully incorporated within the Chilean legislation, through, for example, the "Derecho Real de Conservación" (conservation easement) (Tecklin and Sepúlveda 2014) enacted in 2017.

On the other hand, the indigenous actions against these forms of appropriation of land in detriment of their own territorial claims have been the object of negative public reactions, violent repressions, and judicial persecution. In response to indigenous claims, the governments have promoted new institutions such as the "conservation communities" (Tecklin and Sepúlveda 2014), the "comanagement of protected areas" (Sepúlveda and Guyot 2016), and the "Espacios Costeros Marinos para Pueblos Originarios (ECMPO)" for the protection of customary uses practiced by indigenous communities settled in the coastal zone of Chile. They represent different syntheses between a certain recovery of indigenous territorial rights and the postmodern imaginary of relationship with nature (Araos et al. 2020; Candia 2013).

Far from an apparent "pacification," the present decades are also the scene of new territorial disputes arising from conflicting imaginaries. In relative terms, the most recent stage of this long-lasting process of land grabbing is characterized by the following: (i) the gravitation of claims for land restitution than for further dispossessions, (ii) claims are featured by descendants of the many ethnic groups (e.g., Mapuche, Kaweskar, Yagán) who, even though no longer live in communities, are still organized around self-recognition of their cultural legacies and their ancestral rights over the territory (Zorondo-Rodríguez et al. 2019), and (iii) after several generations of apparent silence and invisibility, new demands include expressions of power, such as roadblocks, arson attacks, and the occupation of public and private lands. These shared ideas and expectations, from self-recognition to public interventions, clearly constitute a new imaginary, a type of neo-ethnicity (Bergesen 1977; Moynihan and Glazer 1975). This neo-ethnicity revives, for example, in the Wallmapu ancestral notion of the Mapuche's territory at both sides of the mountain range (Barrera et al. 2019; Vitar 2010). Faced with the imaginaries of a desert, first, and of a land free of Indians, later, Mapuche demands for the restitution of their territories have awakened Argentines to the existence of other imaginaries that are not willing to disappear (Aliaga 2019).

4.2 Forestry Expansion

Whether as sources of natural resources, water and climate regulation, scenic beauty, biodiversity, or cultural identity, the Andean-Patagonian forests are an inescapable component of the narratives and social imaginaries about this region. Both modern and postmodern NSI about forests and forestry have contributed and still contribute to explain the current gravitation on Patagonian landscape.

Collisions between modern and postmodern NSI are clearly illustrated by the trajectory of Patagonian forestry. The next cases are aimed at illustrating the interplay between different NSI, scientific knowledge, and different stakeholders, around the access to which are perceived as antagonistic goods and services provided by forest ecosystems.

Transformations and deforestation of native forest lands during the early settlement period Since the settlement times (early twentieth century), Argentine and Chilean governments promoted visions of state territoriality based on export-oriented agrarian capitalism, where landowners (mainly Europeans) established oligopolistic control over production and distribution networks to foreign markets (Bandieri 2009). These policies led to the increase in livestock (high stocking rates), mainly sheep, and promoted the overuse of forests by affecting natural regeneration dynamics (Gea-Izquierdo et al. 2004).

In the Chilean zone of Aysén as well as other Chilean forested zones in Southern Patagonia, the scarcity of land for ranching purposes and the directives from the central government led to extensive forest burnings that cleared large portions of the original forest cover (Robinson 2013). For example, Bizama et al. (2011) determined a loss of approximately 23% of native forests and also an increase in the number of forest fragments (<100 ha) as a result of the settlement processes in the Aysén River area during the twentieth century.

A quite different situation occurred in the Argentinean side of southern continental Patagonia, where the ranchers had a positive perception of most of the native forests that occupy less than 2% of the total area by providing shelter and good forage for animals (Peri et al. 2013), timber for rural construction, fences, and firewood (Peri et al. 2019). In Tierra del Fuego, the first settlements occurred in the grasslands and ecotone areas with *Nothofagus antarctica* forests. Through the years, and when the displacement of the native people was effective, the ranching moved to the southern areas where native forests were the dominant land cover. The girdling (capados) and burning of trees in large areas of forests were a common practice to increase the pasture allowance for livestock, leading to the degradation of most of the *N. antarctica* forests in the Chilean side of the Tierra del Fuego island and more than 30,000 ha (representing 4% of total forest original cover) of *N. pumilio* in the Argentinean side (Collado 2001).

The imaginary of the forest industry in Tierra del Fuego Island An early perception of the native forest as the main obstacle for regional development allowed certain land conversion of the island's forests to other uses, including grasslands and commercial plantations. However, starting a few decades ago, the public perception of foreign forest companies reduced forestry investments. This was because of society's concern about large-scale forestry environmental impacts and potential antagonisms between forestry and nature-based tourism and the belief that forestry benefits are better secured by relatively small local companies as compared to big foreign enterprises. On the Argentine side, the discrediting of foreign companies increased after the war between Argentina and Great Britain for the Malvinas Islands in 1982.

The above sketched tension between modern and postmodern imaginary about forests and forestry can be illustrated by the case of the US-based Trillium Corporation, who acquired 625,000 ha on the Chilean and 185,000 ha on the Argentinean side of Tierra del Fuego (near 70% of the forests available for timber purposes) and sought to implement a large-scale logging project in Tierra del Fuego during the 1990s. Then a conflict started between Trillium Co. and environmentalist organizations in Argentina and Chile (Klepeis and Laris 2006), when NGOs and scientists suspected that logging rates exceeded forest resilience, generating losses in ES provision for the next generation.

The environmental movement in the case of Argentina was based on the number of hectares to be harvested each year, which overlooked the sustainable silvicultural proposal of Trillium. Environmental groups defended that big companies were unable to sustainably manage forests, as compared to smaller ones (Gamondes

Moyano et al. 2016), and that large foreign companies were less prone to protect the natural resources as compared to local companies, against available information on the contrary (Martínez Pastur et al. 2007; Gea Izquierdo et al. 2004).

On the Chilean side, Trillium proposed the innovative Rio Condor sustainable forestry project (272,729 ha) to extract timber and enhance socioeconomic development in Tierra del Fuego. The imaginary of modernity associated with timber extraction found opposition from environmental groups, forest scientists, but also from within the government and the forest industry. After a 13-year dispute, the project was dismissed in 2004. Goldman Sachs, a global investment banking firm, acquired the loans and land and finally donated the property to the Wildlife Conservation Society (WCS) which created Karukinka Park. This conservation ensured the long-term protection of Nothofagus forests but also had negative economic impacts for local communities (e.g., Cameroon and El Porvenir on the Chilean side of Tierra del Fuego).

On the Argentinean side of the Tierra del Fuego Island, the company stopped their operations during several years and finally started again in 2019, but the conflict limited the development of the forestry sector due to the negative social perception and the native forestland availability. Nowadays, the company continues their operations at a lower scale (harvesting, silvicultural management, removal of cattle from forests, sawmill and secondary industry, and export of dry timber wood).

Chile's industrial forestry expansion and Mapuche struggles for land The expansion of large-scale industrial plantations based on nonnative tree species (*Pinus radiata* and *Eucalyptus* sp.) was imposed as a development model in Chile, and it was assisted by the controversial Law Decree 701 subsidy, which covered up to 70% of plantation costs. This process led to the replacement of thousands of hectares of native forest between the early 1970s and 1990s, with well-documented negative effects on biodiversity and ES (e.g., Heilmayr et al. 2016; Braun et al. 2017), which became to be known as the "Chilean native forest tragedy" (Hoffmann 1998). Despite the environmental and social impacts of native forest replacement, the plantation expansion model continued without much resistance until the 1990s when a great national awareness of the transformative power of nonnative tree plantations occurred, channeled through the actions of eminent scientists, civil society organized groups, and national and global environmental NGO.

Native forest replacement concentrated in central-southern Chile and northern Patagonia, where Mapuche communities are mainly located (Araucanía region). Industrial plantations were aligned with the modern imaginary against the imaginary of those who defended indigenous territorial identity and became the center of the Mapuche conflict (Latorre and Pedemonte 2016). The "Chilean forest model" based on plantations became internationally known for its economic success (Clapp 1995; Salas et al. 2016; Mora 2018) but at present suffers a legitimation crisis that derives from deep social discontent due to its socio-ecological effects at the local level (Mora 2018).

The recent stage of forest expansion in northern Chilean Patagonia has been problematized as a (neo) extractive process set in Chile's turn toward a "green economy" which provides a framework within which forest policy is restructured, seeking an apparent more sustainable model (Mora 2018). On the one hand, the discourse of governments and the forest companies considers plantation as the only alternative for "sustainable development"in these areas, which reflects the renewal of the modern social imaginary currently framed in the green development model. Forest plantations become, specifically, the representative element in the construction of the territory as a productive cluster (Farris and Martínez-Royo 2019). On the other hand, opposing perspectives from organized local communities consider forest expansion as an imposition of extractive models that cause serious effects on their economic, environmental, and cultural system (Montalba et al. 2005). A central element of indigenous resistance to forest expansion is the imaginary of water as a common good rather than a human right vision which has been co-opted by neoliberalism. Among other effects, the plantations generate water scarcity, depriving the communities of this common good, while the native forest ensures the water supply: "*Las forestales fuera de secar las aguas de las vertientes, esteros o ríos, también impiden que llueva. Ya que al exterminarse el bosque nativo huyen también los poderes o energías de las aguas, por eso los hermanos de más edad dicen: Los animalitos del agua se van, el Mowelfe wigkul (concepto que refiere a las fuerzas que posee la montaña) está dentro de la forestal*" (Montalba and Carrasco 2003; Montalba et al. 2005), which can be translated like "The forestry companies outside of drying the waters of the springs, estuaries or rivers, also prevent rain, because when the native forest is exterminated, the powers or energies of the waters also flee. That is why the older brothers say: The animals are leaving the water, the Mowelfe wigkul (a concept that refers to the forces that the mountain possesses) is within the forest."

The commons' imaginary (Appadurai 1986) can be understood as a response to recent changes in our social, economic, and political lives, particularly those associated with economic globalization, decline of place-based communities and social identities, and global environmental and economic crises (Wagner et al. 2012).

4.3 Touristification

Patagonia is identified with notions of pristine, exotic, adventurous, uncontaminated, wild, majestic, and freedom, which feed the touristic stereotype that becomes its symbolic capital (Dimitriu 2002; Rodríguez et al. 2014; Decasper and Servalli 2016). For these reasons, this region was early perceived as having the potential to be sold as a destination and then promoted as such via market mechanisms, a process typified for other regions and localities of the world as "touristification."

During a touristification process, the landscape per se does not immediately transform a place into a touristic destination, it must be conceptualized as such by particular actors (Dimitriu 2002; Bertoncello 2006; Merlos and Otero 2013; Nuñez 2018). Certain territorial aspects are privileged over others and highlighted to prospective customers, and these characteristics end up forming part of the social imaginary these regions are associated with. In this sense, territories become commodified, which means that the landscape is reconfigured and acquires certain economic value for people that can live or not inside the territories. In other words, places become products (Dimitriu 2002). Returning to our conceptual frameworks (Figs. 19.1 and 19.2), touristification processes can be seen as an expression of the modern imaginary about the relationship between society and nature, a shared behavior which influences the way in which natural capital is transformed into "consumer products" (shared expectations, governance) and the way it contributes to well-being (shared values and perceptions).

A brief history of Patagonian touristification Some decades after the emergence of "social tourism" in charge of the Argentine government (see Sect. 4.1), the touristification process received a new boost from the 1990s' national governments, this time through the installation of "tourist brands" as a form of promotion of certain localities. The role of the Ministry of Tourism has been essential in the construction of the tourist imagery installing a diversity of touristic products abroad, influenced by private interests (Decasper and Servalli 2016). Transnational tourism has formulated Southern Andean Patagonia as an exotic landscape and ES of consumer value for bourgeois leisure. The definition of tourist brands integrates the local economy to the global market, orienting consumption trends and exchange value and focusing on market demand (Dimitriu 2002) which links it to the modern social imaginary identified by Buckles (2018).

The consequences of the pursuit of the modern social imaginary in Patagonia tourism varied among different social groups. The progressive improvement of transport, hosting, and other facilities and services allowed for a growing access of direct beneficiaries (tourists) to this region's attractions while enabling the flow of indirect benefits to tourist service providers (e.g., employers and employees in transport, hotel and gastronomy sectors, as well as other indirect beneficiaries).

On the other side, access to tourism benefits has had some negative influences on the environment and the social groups that depend on it. Private investments normally imply different mechanisms of exclusion to formerly public goods, and the tourism sector is no exception. Thus, private investments in tourism within Patagonia have threatened the access of local social groups to common goods and services provided by ecosystems or have led to the loss of such access. For example, one of the problems that are recurrently associated with the tourist appreciation of the region consists of the blocking of public access to the coasts of rivers and lakes or other natural areas of renowned beauty. Specifically, the free passage is limited, leaving these lands to the exclusive use of elites or those who are able to pay for its use (Abarzua and Di Nicolo 2018). Since free movement along these coasts is supported by the Argentinean and Chilean laws, these de facto blockades are a source of frequent conflicts.

In the following paragraphs, transformation and resistance forces emerging along different touristification processes illustrate the coexistence and relative influence of modern and postmodern imaginaries.

Residential projects or mismatches between values, expectations and behaviors Mega-residential projects aimed at promoting elite tourism have triggered resistance reactions to the consequent clearing of large tracts of native forests for large buildings and the localized loss of the wild Patagonian seal. Such is the experience in El Bolsón (Llosa 2019) where a ski village is to be constructed near the small ski resort in Cerro Perito Moreno. This urbanization includes a protected area which has been declared Biosphere Reserve by UNESCO and, as Llosa (2019) puts it, confronts two imaginaries, one in which these developments are seen as necessary to boost the economy through tourism (modern social imaginary), whereas others perceive it as an environmental threat which puts at risk a different way of living which is more in tune with nature and relies on sustainability (postmodern social imaginary).

A similar case is that of Villa La Angostura, a town surrounded by a national park in the Neuquén province, whose distinctive hallmark is its small size, with leisure cabins scattered across a matrix of native forests. This touristic destination offers a unique investment opportunity against the economic fluctuations of the country as its land value has been constantly increasing over the decades in spite of the cycles of economic recession (Gluch 2019). Since the mid-1980s, the population has increased at a constant pace due to a series of migratory waves from the country's urban centers, in search of a better lifestyle in contact with nature as well as new labor opportunities (TRECC 2007). A private initiative to build an urbanization project adjacent to the ski slopes in Cerro Bayo was confronted by NGOs and other environmental interest groups until the project was cancelled (Svampa and Viale 2014; Gluch 2019). This has resulted in a diversity of opposing actors, which has configured a heterogeneous society with contradictory interests. The town's main stakeholders are confronted today by the tensions of opposing development models, one that threatens natural resources and the future of tourism as an economic activity that relies on them by accepting or promoting town growth and development and another that resists such transformations, aiming to protect nature assets and the "mountain village" hallmark. Both parties are standing on different theoretical imaginaries. One, represented by the environmental associations, emphasizes the idea of sustainability as its main objective, trying to preserve the system in its present state and protect the touristic image of the town, which it intends to promote. The other group, consisting of the professional associations together with the local government, claims that change is inevitable and is more confident in the buffer effect of the proximity of the national park for securing the pursuit of tourism as an economic activity (Gluch 2019). Because of the way in which power is distributed among the stakeholders, those who can directly intervene in the regulation and management of the locality's natural resources are only a few, and the decisions made in terms of governance affect all beneficiaries. The implementation of these policies (e.g., granting of exceptions to those who do not follow them) is contested by the stakeholders nucleated within the environmental interest groups, who see a great risk in these actions as they do not share the development imaginary that lies at the bottom of these decisions (Gluch 2019).

Similar conflicting views around territorial development permeate the political and economic decisions taken in Aluminé, in the Neuquén province, where the Puel community has been dispossessed of their ancestral land, now highly valued by its beauty and touristic potential, and displaced into less fertile lands (Rodriguez 2015). Other examples of real estate speculation increasing land values occurred in coastal cities, such as Puerto Madryn, where local economy was inclined toward the tourist and service sector, resulting in disparities in terms of land access, creating social and ethnic urban segregation (Kaminker 2015).

In the case of Chile, we can see a similar situation, e.g., the region of Aysén, which until the 1980s was not part of the neoliberal project but was reconfigured in the 1990s as a "life reserve." Nature became globally valued as a business niche that ended up attracting corporate investments (Núñez et al. 2018). This has resulted in the construction of elite oriented fishing lodges, boutique hotels, and many private natural reserves. An example of these new interests, mentioned by Núñez et al. (2018), is the project "Patagonian land, Premium lands in the southern end of South America," which belongs to a binational company dedicated to health and tourism. The process of elite or high-income touristification has led to the privatization of large extensions of previously public areas. A particular aspect in Chile is the link between private conservation and touristification of nature, with large areas such as Pumalín, Tantauco, and Karukinka oriented toward nature-based tourism for higher-income level visitors.

Touristification and consumerism Modern tourism in Patagonia or anywhere in the world responds to psychosocial factors which are common to modern (and post-modern) consumerism behavior (Korstanje and Seraphin 2017). Focusing on the consumer as an agent, it is worth asking what individual and contextual factors shape our intentions and decisions to use the free time to join the torrent of mass tourism, compared to enrolling in more authentic experiences, distanced from corporatized tourism and closest to, for example, volunteering tourism for social and/or environmental purposes. From the available knowledge, it is difficult to disentangle the influence of imageries promoted from institutions, from those driven by shared preferences and even to dimension to what extent these mechanisms reinforce each other. Perhaps, autonomous modifications of consumer behavior are not sufficient as forces of resistance against the touristification process, or as transforming forces, where this process has been installed; perhaps the future that is coming demands new institutions compatible with more sustainable tourism (Higgins-Desbiolles 2010).

Synthesis of touristification versus independent trajectories Analyzed cases suggest that development of tourism in the Argentine side of Patagonia has been largely dictated by the views and discourses of powerful stakeholders who hold political and economic capital under the modern social imaginary. Private investments displace lower-income families to less-valued areas and advance over natural assets (e.g., resources and viewsheds). This posture has meant that other social imaginaries, held by local environmental interest groups or indigenous communities, have been relegated. As a common result, touristic destinations act as powerful attractors to the establishment of amenity migrants, promoting fast population growth, bolstering construction, and increasing land value and speculation. Such trajectories have been previously described in Argentina for Villa La Angostura (Merlos and Otero 2013; Paez and Otero 2014; Gluch 2019), San Martín de los Andes, Aluminé and Villa Pehuenia-Moquehue in Neuquén province (Impemba 2008; Svampa and Viale 2014; Valverde et al. 2015; Abarzua and Di Nicolo 2018), for San Carlos de Bariloche (Landriscini et al. 2019; Rodriguez 2019) and El Bolsón in Rio Negro province (Llosa 2016, 2019), and for Lago Puelo and El Hoyo of Epuyén in Chubut province (Crespo 2017).

4.4 Damming of Rivers for Hydroelectric Use

The first hydroelectricity plant in Chile and second in South America was built in 1897, but the expansion of hydropower as a source of transformation of the territory began after the postwar period and the Cold War (1940s to early 1980s). That momentum of the hydroelectric sector in Chile was aligned with the modern imaginary linked to the manipulation, control, and industrialization of nature, typical of modernity and functional to the sense of nation. Thus, the hydroelectric plants became "the temples of modernity" in the western imaginary of the 1960s (Purcell 2018).

This modern imaginary consolidated within a neoliberal institutional framework (1973–1990) that included reforms in both the water and electricity sectors (Prieto and Bauer 2012; Susskind et al. 2014) aiming to a larger participation of the market as the hegemonic approach toward natural resources governance (Prieto and Bauer 2012).

The generation of hydroelectricity in Chilean Patagonia forms part of the mining-energy-development trilogy, which reflects the interconnectivity between extreme northern regions where copper mining is mainly located, with the energy potential of the southern region. These constitute some of the keys ordering the territory and the new role of Patagonia within a country where development necessarily implies an increase in energy demand (Rodríguez et al. 2015). The expansion of hydropower as a modernizing imaginary is reinforced by the "green growth" discourse as a new version of the modernization ideas that promoted dams during the 1960s. In the green economy, hydropower plays a key role in both climate change mitigation (e.g., low carbon electricity source and as an enabler for other renewable energy sources) and climate adaptation and in a green energy transition (Hommes 2019; Schapper et al. 2020).

In Patagonia, a territory that until recently was considered the last frontier of capitalism (Mendoza et al. 2017), hydropower expansion has led to the confrontation between the imaginaries represented by different actors and different discourses; the most emblematic example is HydroAysén (companies Enel Chile and Colbún). In this context, the Patagonia Sin Represas (PWD) movement, Chile's largest environmental campaign, which enabled organizations and activists to find the common ground needed to collaborate (Schaeffer 2017), was raised through narratives on history, the environment, geography, and culture, and that served as the basis for rebuilding Patagonia-Aysén territory (Núñez et al. 2017). These elements, combined with a new political and social context in the country, help understand the halt of HidroAysén, one of the PDW movement's most important achievements (Schaeffer 2017).

The most powerful actors, including national and international environmental organizations, environmental entrepreneurs, anti-dam organizations and other local organizations, established in Aysén what Núñez et al. (2017) called an "ecopower". As a social force, this "ecopower", aimed at promoting consensus on the central role of Patagonia's natural capital for Chile's development, through legal reforms and institutions, the dissemination of an alternative society/nature relationship, the designation of areas for environmental conservation, and the alternative use of resources (Núñez et al. 2017). This ecopower re-imagined Patagonia as a "blessed land" and as an "extension of the hand of God", as a "reserve of natural and existential life, as a refuge from the aggressiveness and exhaustion of modernity and extractivist accumulation model" (Rodríguez et al. 2015: Urrutia et al. 2019).

In the case of Argentina; the hydroelectric projects are carried out by the state and private companies in the last three decades of the twentieth century, which produced the resettlement or forced relocation of many local populations that resided near them, e.g., El Chocón-Cerros Colorados, Alicurá, Piedra del Águila, and Pichi Picún Leufú over Limay and Neuquén rivers. These ventures, imbued with ideas of development and progress, aimed to control floods, increase irrigation areas and improve the national provision of electricity at preferential prices through renewable sources of energy. Mapuche communities, such as the Painemil in Cerros Colorados, those who lived in Pilquiniyeu del Limay in Rio Negro or the Ancatruz in the Neuquén province, were forced to leave behind their settlements. Small-scale producers experienced the same situation in the areas adjacent to the Colorado river, or even whole towns, such as Pichi Picún Leufú. Several of these relocations occurred under the government of the military dictatorship of 1966–1973, while others took place during the 1980's. However, in some cases as private ranches, like Alicurá, they were compensated for the loss of their lands (Radovich and Balazote 2005).

Compensations, however, presented disparities in terms of who was benefited and how, depending on who were recognized as landowners by the state or mere occupants of fiscal land. The Mapuche community who lived in Pilquiniyeu del Limay was given a greater extension of land and of superior quality to the ones which had to be sacrificed, by expropriating the "María Sofia" Ranch which was annexed to the reserve. While others, such as the rural population in Pichi Picún Leufú or the small-scale breeders around El Chocón, received meagre payments. Some rural settlers were forced to migrate toward precarious urban areas. Indigenous communities recognized by the state, like Pilcaniyeu del Limay, were successful in presenting collective land claims as a form of resistance. Others, however, as in Casa de Piedra, found resistance and collective protests repressed by threatening community leaders (Radovich and Balazote 2005). Strategic political and economic interests were imposed in terms of regional development, generating stress, loss of community ties, and a sense of belonging toward the land.

Similar situations occurred recently in Santa Cruz province, with the Kirchner-Cepernic dams, formerly known as Cóndor Cliff-La Barrancosa. The undertaking was awarded in August 2013 to the consortium formed by Electroingeniería S.A., China Gezhouba Group Company Ltd., and Hidrocuyo S.A., which has since been questioned by private lawyer associations, social movements, environmental organizations, and indigenous peoples due to the absence of an appropriate environmental feasibility study and lack of transparency in the decision process. In addition, the World Bank disqualified the Gezhouba Group in 2015 due to misconduct in three bank-financed projects in China related to water conservation, earthquake recovery, and flood management (Acevedo 2017; Mora 2018). In spite of strong popular opposition, the Argentine national and local government decided to pursue the endeavor and signed an agreement with the Chinese government. After the national elections of 2015, the Association of Environmental Lawyers of Patagonia filed a new appeal to the Supreme Court, which finally unanimously ordered the suspension of the project. As Mora (2018) and Mora-Motta (2018) suggest these social resistances show different perceptions and valuations around nature which collide with dominant discourses and imaginaries while evidencing inequalities in terms of the distribution of environmental degradation as well as the exclusion of society at large from the natural resource governance and land use.

During the 1990s, Latin-American indigenous groups gained international visibility, and new laws were passed opening discussions around these inclusion and exclusion practices. The defense of human rights, particularly those of marginal groups, gained importance in political debate (Kropff 2005; Carpinetti 2006). In Argentina, this process involved questioning the social imaginary through which the idea of nation had been built, in a land considered as devoid of indigenous communities, which had become extinct (Carpinetti 2006). This did not necessarily imply the disappearance of a derogatory hegemonic discourse, which denied these peoples of their rights and history, invisibilizing them in order to legitimize the prevalence of development over local interests (Kropff 2005). This allowed the state to enforce megaprojects over powerless popular sectors, small producers, and impoverished Mapuche populations (Radovich and Balazote 2000).

5 Discussion and Recommendations

The role of social imaginaries on nature's governance is an emerging topic with still few examples (Hommes et al. 2016; Gluch 2019; Archibald et al. 2020). In agreement with these studies, the cases revised in this chapter have allowed to highlight the role of NSI on the use and transformation of natural capital. NSI can be recognized as main factors underlying controversies regarding nature governance, whether influencing the shared perception of nature contributions to well-being, the shared expectations about behaviors (governance institutions), or the final decisions affecting natural capital. The relative importance of different mechanisms by which NSI influence SES trajectories (Fig. 19.2) may be inferred from the compiled information (Table 19.1).

The NSI promotes social cohesion in the face of the dispersion of interests through mechanisms that provide feedback and regulate the objectives and decisions of society in relation to nature, including "slow" variables such as values, beliefs, and norms that underlie practices. Our studied cases also illustrate that NSI can promote transformation mainly through mechanisms capable of eroding resistance against competing NSI, e.g., in Sect. 4.2, NSI with different roles were described in relation to industrial forestry. While recent attempts to expand the forestry industry in Tierra del Fuego (during the post-dictatorship periods in Argentina and Chile) were regulated by the free expression of values and beliefs shared by the local population, during the dictatorship of General Augusto Pinochet, *Pinus* spp. and *Eucalyptus* spp. plantations replaced large tracts of the native forest of Chile almost without restrictions, affecting landscapes and ES provision.

Our revision of study cases revealed that expressions of the postmodern imaginary are as diverse as complex, sometimes taking the form of resistance to forest transformation, which can lead to other forms of transformation such as touristification or green grabbing. Often, public policies that obey to opposing imaginaries are self-justifying in their positive consequences on the general well-being. Socio-environmental sustainability has already permeated the different discourses and does not exert a clear dividing line between the imaginaries, e.g., Trillium case suggests that governance of forests in the Tierra del Fuego is not merely based on scientific or technical knowledge nor on public perceptions or individual preferences. While more scientifically and economically than perceptually based decisions are claimed from one side, or more attention to economic opportunity costs are claimed from other sides, these arguments were overridden by the prevalence of environmentalists' NSI, clearly more prone to the application of precautionary principle.

While all the components of the components of the imaginaries' machinery are necessary for explaining NSI influence on socio-ecological systems, the relevance of shared behaviors is self-evident. Modern tourism in Patagonia or anywhere in the world responds to psychosocial factors which are common to modern (and post-modern) consumerism behavior (Korstanje and Seraphin 2017). Focusing on the consumer as an agent, it is worth asking what individual and contextual factors shape our intentions and decisions to use the free time to join the torrent of mass tourism, compared to enrolling in more authentic experiences, distanced from corporatized tourism and closest to, for example, volunteering tourism for social and/or environmental purposes. From the available knowledge, it is difficult to disentangle the influence of imageries promoted from institutions, from those driven by shared preferences, and even to dimension to what extent these mechanisms reinforce each other. Perhaps, autonomous modifications of consumer behavior are not sufficient as forces of resistance against the touristification process, or as transforming forces, where this process has been installed; perhaps the future that is coming

Table 19.1 Influence mechanisms of nature-society imaginaries (NSI) on socio-ecological systems (SES); some examples as illustrated by the analyzed controversies around governance of Patagonian SES

	Nature-society imaginaries: mechanisms of influence on SES		
	Setting preferences (Paths 1 and 2, Fig. 19.2)	Setting institutions (Path 3, Fig. 19.2)	Setting behaviors (Path 4, Fig. 19.2)
Modern NSI	Exaltation of progress, science, and technology, in opposition to the natural, the wild, the uncultivated, and the uncultured in the historical literature Exaltation of European culture and "industriousness"; invisibility of native peoples; "civilization or barbarism" Recreation and tourism as a human need and as consumerism objects Commercial advertising associated with tourism products for mass consumption	Enabling legislation for land and green grabs Enabling legislation and economic subsidies to the forest industry in detriment of native forests Holiday entitlement of formal workers Enabling legislation for water privatization and dam's construction	Psychosocial factors influencing offer and demand of tourism modalities and destinations
Postmodern NSI	In general, nature for its own sake over nature for its instrumental values. Examples: Native forest conservation is preferred over too much tourist infrastructure Native forests and tourism are preferred over industrial forestry Small, local or national forestry companies preferred over large foreign companies Free rivers and local access to water preferred over privatized water in dams for the sale of hydroelectric energy to large cities	Environmental NGO and participatory governance mechanisms active against industrial forestry and damming Environmental NGO and participatory governance mechanisms active against tourist expansion at the expense of urban and peri-urban forests Enactment of new laws enabling private conservation	Social learning, activism, and militancy in defense of native forests and water sources Environmentally friendly behaviors regarding water, energy consumption, and recreation Enrolment of private lands for conservation purposes

(continued)

Table 19.1 (continued)

	Nature-society imaginaries: mechanisms of influence on SES		
	Setting preferences (Paths 1 and 2, Fig. 19.2)	Setting institutions (Path 3, Fig. 19.2)	Setting behaviors (Path 4, Fig. 19.2)
Indigenous NSI	The land as the axis of individual and social development Good living with the land and not at the expense of the exploitation and transformation of the land The earth as an integral entity, not disagreeable in its natural resources Recognition of customary rights over individual rights only	Life in communities in rural contexts and indigenous organizations in urban contexts Recovery of traditions, language, and cultural identity Recovery of old institutions such as indigenous parliaments	Protest demonstrations and other expressions of resistance, sometimes illegal, sometimes violent Territorial claims, nonrecognition of government institutions and private property Psychosocial factors influencing decisions for not supporting indigenous claims, or for protesting, against indigenous claims

demands new institutions compatible with more sustainable tourism (Higgins-Desbiolles 2010).

As we said before, the particular case of the indigenous NSI cannot be separated from their more comprehensive worldview nor from their continued claim for their territory rights. Despite the adversity of the higher-ranking institutions for these claims, the success of the Mapuche NSI is directly reflected in the behaviors of resistance legitimized and promoted by their organizations and partly guaranteed by new institutions. Constitutional reforms in Argentina and Chile that recognize the ethnic and cultural preexistence of indigenous peoples have allowed Mapuche communities to legally confront provincial and national governments (Radovich 2013; Crespo and Tozzini 2013) with variable success. These groups have also engaged in the building of networks with other social actors. Such is the case of forums where indigenous and rural communities have converged with neighboring groups in Lago Puelo, Cholila, Epuyén, and El Hoyo, to prevent the building of a hydroelectric dam on Lake Lezama (Crespo and Tozzini 2013). This constitutes, as a member of a Mapuche community verbalized, "an alliance of the weak" (Gluch 2019), whereby common threats create the potential for joint action between groups that had previously ignored each other or seen themselves as antagonists, over the defense of the environment in the light of the recognition of the rights of marginalized groups.

The recognition of NSI as factors that affect the trajectory of SES opens to new research questions. For example, what kind of SES trajectories can be promoted by NSI changes? What is the importance of the type of governance and power distribution among social actors over the interplay between different NSI? How compatible is a governance based on both scientific knowledge and local and indigenous knowledge forms plus different social imaginaries? Is the free expression and interaction between these imaginaries capable of promoting more sustainable trajectories? Is it true that the more science and less social imaginaries, the better public decisions in terms of fairness and sustainability? In the face of increasing environmental impairment and social conflict, should imaginaries be changed? As stated by Stephenson Jr (2011), "imaginaries may be changed, but only if those espousing them are given reason to bring them to consciousness, reflect afresh on their foundations, and embrace an alternate conception."

While these questions are beyond the scope of this chapter, the cases reviewed offer a preliminary perspective on the tremendous challenge involved in any attempt to answer them through generalizations. The influence of NSI on governance is apparently idiosyncratic, but certainly, it cannot be neglected, and socio-ecological sustainability requires plenty of attention to how desirable imaginaries can be promoted. Although we could not answer to which imaginaries should be promoted, the important fact is that "social imaginaries can be changed by theories that penetrate and transform the social imaginary, and when this happens, people take up, improvise, or are inducted into new practices" (Taylor 2005). For example, in these times of the COVID-19 pandemic, when a very high percentage of economic activity and jobs which depend on tourism now rely on external economic help or are disappearing, social vulnerability of highly touristified destinations is evident. Therefore, is touristification or should it be the only imagined option of development? Probably it is time to explore alternative imaginaries which are, eventually, much closer to the attributes of the postmodern social imaginary, where a central role is assigned to the local economy and the needs of the local communities, where tourism is seen as a complementary income and focuses on an economy that develops at a slower pace and a different scale, more respectful for nature assets, allowing for a more sustainable development (e.g., Dimitriu 2002; Higgins-Desbiolles 2010).

In sum, the NSI that can be identified in Patagonia are not essentially different from those observed in the rest of the world. The dynamic character of these imaginaries is not exclusive to this region either. Conflicts previously reserved for small negotiating tables, and often silenced by the force of power, today permeate wider circles of society. The rise of new forms of communication and social networks is already playing a role in the evolution of NSI imaginaries that exceed local interactions and represent new forms of tele-coupling between societies and ecosystems separated in space and time.

Acknowledgments This chapter was financed by the National Agency of Scientific and Technological Promotion of Argentina (ANPCyT, PICT 2015-0672).

References

Abarzua FD, Di Nicolo CA (2018) Extractivismo en territorios del norte de la Patagonia. La frutihorticultura en los Valles de Río Negro y el turismo en Villa Pehuenia-Moquehue, Neuquén, CA, n.° 10, pp. 318–345

Acevedo M (2017) Represas sobre el Río Santa Cruz. Informe Integrar 105. Instituto de Integración Latinoamericana. Buenos Aires, Argentina

Agudelo PA (2011) (Des)hilvanar el sentido/los juegos de Penélope. Una revisión del concepto imaginario y sus implicaciones sociales. Uni-pluri/versidad 11(3). Universidad de Antioquia, Medellín, Colombia

Aizen H, Tam Muro C (1992) Historia de Bariloche. Museo de la Patagonia, Bariloche, Argentina. http://www.museodelapatagonia.nahuelhuapi.gov.ar/pdf/HISTORIA%20DE%20BARILOCHE.pdf

Aliaga C (2019) The land 'wars' in twenty-first century Patagonia. J Latin American Cult Stud 28(1):139–151

Appadurai A (1986) Introduction: commodities and the politics of value. In: Appadurai A (ed) The social life of things: commodities in cultural perspective. Cambridge University Press, Cambridge, UK, pp 3–63

Araos F, Anbleyth-Evans J, Riquelme W, Hidalgo C, Brañas F, Catalán E, Diestre F (2020) Marine indigenous areas: conservation assemblages for sustainability in Southern Chile. Coast Manag 48(4):289–307

Archibald JL, Anderson CB, Dicenta M, Roulier C, Slutz K, Nielsen EA (2020) The relevance of social imaginaries to understand and manage biological invasions in southern Patagonia. Biol Inv 22(11):3307–3323

Bandieri S (2009) Pensar una Patagonia con dos océanos: el proyecto de desarrollo de Ezequiel Ramos Mexía. Quinto Sol 13:47–71

Barrera MG, Huircapan SC, Foote S, Key JP (2019) Pueblo Mapuche. La representación de la nación a través de la producción discursiva en el Gulumapu. Anclajes 23(2):1–20

Bergesen AJ (1977) Neo-ethnicity as defensive political protest. Am Sociol Rev 42(5):823–825

Bertoncello R (2006) Turismo, territorio y sociedad. El 'mapa turístico de la Argentina'. América Latina: cidade, campo e turismo. Amalia Inés Geraiges de Lemos, Mónica Arroyo, María Laura Silveira. CLACSO, Consejo Latinoamericano de Ciencias Sociales, San Pablo

Bessera EM (2006) La Colonia Nahuel Huapi y los orígenes de la actividad turística en la región Andino-Patagónica. In: Historia de la Patagonia. Universidad Nacional del Comahue, Neuquén

Bizama G, Torrejón F, Aguayo M, Muñoz MD, Echeverría C, Urrutia R (2011) Pérdida y fragmentación del bosque nativo en la cuenca del río Aysén (Patagonia-Chile) durante el siglo XX. Revista de Geografía Norte Grande 49:125–138

Blanco DN, Mendes JM (2006) Aproximaciones al análisis de los conflictos ambientales en la Patagonia. Reflexiones de historia reciente 1980–2005. Ambient Soc 9(2):47–69

Braun AC, Troeger D, Garcia R, Aguayo M, Barra R, Vogt J (2017) Assessing the impact of plantation forestry on plant biodiversity: a comparison of sites in Central Chile and Chilean Patagonia. Glob Ecol Conserv 10:159–172

Buckles J (2018) Education, sustainability and the ecological social imaginary: connective education and global change. Palgrave Macmillan, Cham

Candia P (2013) Áreas Silvestres Protegidas Privadas Y De Pueblos Originarios: Integración De Diversos Actores. In: La Protección De Las Áreas Silvestres En Chile (Private Protected Areas and Indigenous Peoples: Integration of Various Actors in the Protection of Wildland in Chile). Justicia Ambiental, p 167

Carpinetti B (2006) Los Parques Nacionales y los Pueblos Indígenas: una historia de desencuentros. In: Derechos indígenas en el Parque Nacional Lanin… de la expulsión al comanejo. Buenos Aires, Administración de Parques Nacionales, pp 19–25

Clapp RA (1995) Creating competitive advantage: forest policy as industrial policy in Chile. Econ Geogr 71(3):273–296

Collado L (2001) Tierra del Fuego forest: analysis of their stratification through satellite images for the forest province inventory. Multequina 10:1–16

Crespo C (2017) Contornos de lo decible, exhibible y pensable. Los pueblos originarios en las políticas turístico-culturales en el noroeste de Chubut (Patagonia, Argentina). PASOS. Revista de Turismo y Patrimonio Cultural 15(3):533–544

Crespo C, Tozzini MA (2013) Trayectorias de relaciones en el marco de luchas étnico – Territoriales en la cordillera chubutense. Papeles de Trabajo N° 25. Centro de Estudios Interdisciplinarios en Etnolingüística y Antropología Socio-Cultural. Julio. ISSN 1852-4508

Decasper S, Servalli N (2016) Imaginarios turísticos: Argentina para el turista brasileño. Turismo y Sociedad 18(jun. 2016):43–60

Díaz E (1996) Qué es el Imaginario Social? In: Diaz E (ed) La Ciencia y el Imaginario Social. Bs. As, Biblos, pp 13–21

Díaz S, Demissew S, Carabias J, Joly C, Lonsdale M, Ash N, Bartuska A (2015) The IPBES Conceptual Framework—connecting nature and people. Curr Opin Environ Sustain 14:1–16

Di Giminiani P, Fonck M (2018) Emerging landscapes of private conservation: enclosure and mediation in southern Chilean protected areas. Geoforum 97:305–314

Dimitriu A (2002) Producir y consumir lugares: Reflexiones sobre la Patagonia como Mercancía. Eptic, Economía política de las tecnologías de la información y la comunicación. ISSN 1518-2487

Farris M, Martínez-Royo O (2019) El capitalismo del holding transnacional en el sector forestal chileno: la consolidación de una hegemonía territorial. Izquierdas 45:23–50

Gamondes Moyano I, Morgan RK, Martínez Pastur G (2016) Reshaping forest management in Southern Patagonia: a qualitative assessment. Sustain For 35(1):37–59

Gea-Izquierdo G, Pastur GM, Cellini JM, Lencinas MV (2004) Forty years of silvicultural management in southern Nothofagus pumilio primary forests. For Ecol Manage 201(2–3):335–347

Gluch MV (2019) Aportes de la noción de imaginario social al capital social: estudio de caso: Villa la Angostura, provincia de Neuquén (Master's thesis, Buenos Aires: FLACSO. Sede Académica Argentina). http://200.41.82.22/bitstream/10469/15783/2/TFLACSO-2019MVG.pdf

Gorenstein SM, Ortiz R (2016) La tierra en disputa: agricultura, acumulación y territorio en la Argentina reciente. Revista Latinoamericana de Estudios Rurales 1(2):1–26

Harambour A (2019) Soberanías fronterizas: Estados y capital en la colonización de Patagonia (Argentina y Chile, 1830–1922). Ediciones Universidad Austral de Chile

Hasbrouck A (1935) The conquest of the desert. Hisp Am Hist Rev 15(2):195–228

Heilmayr R, Echeverría C, Fuentes R, Lambin EF (2016) A plantation-dominated forest transition in Chile. Appl Geogr 75:71–82

Higgins-Desbiolles F (2010) The elusiveness of sustainability in tourism: the culture-ideology of consumerism and its implications. Tour Hosp Res 10(2):116–129

Hoffmann A (1998) La Tragedia del Bosque Chileno. Ocho Libros Editores Ltda, Santiago de Chile

Holmes G (2014) What is a land grab? Exploring green grabs, conservation, and private protected areas in southern Chile. J Peasant Stud 41(4):547–567

Holmes G (2015) Markets, nature, neoliberalism, and conservation through private protected areas in southern Chile. Environ Plan 47(4):850–866

Hommes L, Boelens R, Maat H (2016) Contested hydrosocial territories and disputed water governance: struggles and competing claims over the Ilisu Dam development in southeastern Turkey. Geoforum 71:9–20

Hommes L (2019) Desarrollo hidroeléctrico y reconfiguraciones territoriales históricas en la cuenca del Rímac, en Lima, Perú. Estudios atacameños 63:233–249

Impemba M (2008) Comunidades mapuche y expansionismo turístico. Estudio de caso: territorios en disputa en el Cerro Chapelco. III Jornadas de Historia de la Patagonia. San Carlos de Bariloche, 6–8 de noviembre de 2008

Jagoe ELA (2008) The end of the world as they knew it: writing experiences of the Argentine South. Associated University Press

Kaminker S (2015) Segregación residencial y proyectos de ciudad: Puerto Madryn como espacio en disputa. Vessuri, H. & Bocco, G. (Comps.) Conocimiento, paisaje, territorio. Procesos de cambio individual y colectivo. Río Gallegos : Universidad Nacional de la Patagonia Austral; Centro de Investigaciones en Geografía Ambiental-UNAM; Centro Nacional Patagónico; Universidad Nacional de Río Negro

Klepeis P, Laris P (2006) Contesting sustainable development in Tierra del Fuego. Geoforum 37(4):505–518

Korstanje M, Seraphin H (2017) Revisiting the sociology of consumption in tourism. In: The Routledge handbook of consumer behaviour in hospitality and tourism, pp 2–16

Kropff L (2005) Activismo mapuche en Argentina: trayectoria histórica y nuevas propuestas. In: CLACSO, Consejo Latinoamericano de Ciencias Sociales Pueblos indígenas, estado y democracia. Bs. As

Landriscini G, Civitaresi HM, del V Colino E (2019) Turismo, transformaciones territoriales y resiliencia. Evidencia de una ciudad intermedia patagónica. 1er Congreso Argentino de Desarrollo Territorial. 3ras Jornadas de Desarrollo Local Regional, las redes locales y el desafío de la innovación en una nueva etapa de la globalización. Villa María: Universidad Nacional de Villa María

Latorre JI, Pedemonte NR (2016) El conflicto forestal en territorio mapuche hoy. Ecología política 51:84–87

Llosa C (2016) Desarrollo y conflictividad socio-territorial en Patagonia. La integración de la Patagonia al mercado. El caso de El Bolsón, Río Negro ¿Del desenclave zonal al enclave turístico? Theomai, N° 34. Red Internacional de Estudios sobre Sociedad, Naturaleza y Desarrollo

Llosa C (2019) Transformaciones socio-territoriales en disputa. El caso del cerro Perito Moreno (Patagonia Argentina). Boletín geográfico 41(2):61–77

Lotz-Sisitka H (2010) Changing social imaginaries, multiplicities and 'one sole world': reading Scandinavian environmental and sustainability education research papers with Badiou and Taylor at hand. Environ Educ Res 16(1):133–142

Martínez Pastur G, Lencinas MV, Peri PL, Moretto A, Cellini JM, Mormeneo I, Vukasovic R (2007) Harvesting adaptation to biodiversity conservation in sawmill industry: Technology innovation and monitoring program. J Technol Manag Innov 2(3):58–70

Mastrangelo ME, Gavin MC, Laterra P, Linklater WL, Milfont TL (2014) Psycho-social factors influencing forest conservation intentions on the agricultural frontier. Conserv Lett 7(2):103–110

McGregor A (2004) Sustainable development and warm fuzzy feelings': discourse and nature within Australian environmental imaginaries. Geoforum 35(5):593–606

MEA (2005) Ecosystems and human Well-being: synthesis. Millennium ecosystem assessment. World resources institute. Island Press, Washington, DC

Meadows D (1999) Leverage points: places to intervene in a system. The Sustainability Institute, Hartland

Méndez LM (2016) "Turismo de elite y turismo social en el "Edén de América": El Parque Nacional Nahuel Huapi entre 1934 y 1955." Textos y contextos desde el sur, N°4, Vol II (2), julio

Mendoza M, Fletcher R, Holmes G, Ogden LA, Schaeffer C (2017) The Patagonian imaginary: natural resources and global capitalism at the far end of the world. J Lat Am Geogr 16(2):93–116

Merlos M, Otero A (2013) La producción del espacio en destinos post-turísticos. Caso: Villa La Angostura. Realidad. Tendencias y Desafíos en Turismo (CONDET), [S.l.] 11(1):79–94. ISSN 2545-6199

Meza LE (2009) Mapuche struggles for land and the role of private protected areas in Chile. J Lat Am Geogr 8(1):149–163

Montalba NR, Carrasco HN (2003) Modelo forestal chileno y conflicto indígena¿ Ecologismo cultural Mapuche? Ecología Política 26:63–77. P69

Montalba NR, Carrasco HN, Araya CJ (2005) Contexto económico y social de las plantaciones forestales en Chile. El caso de la comuna de Lumaco, Región de la Araucanía. Movimiento Mundial por los Bosques Tropicales, Montevideo

Mora S (2018) Resistencias sociales a la cooperación de China en infraestructura: las represas Kirchner-Cepernic en Argentina. Colombia Internacional 94:53–81

Mora-Motta A (2018) Plantaciones forestales en Chile: ¿hacia un modelo más sustentable? Gestión y Ambiente 21(2 Suppl):100–116

Moynihan DP, Glazer N (eds) (1975) Ethnicity: theory and experience, vol 531. Harvard University Press, Cambridge, MA
Murmis M, Murmis MR (2012) Land concentration and foreign land ownership in Argentina in the context of global land grabbing. Can J Dev Stud/Rev Can Eudes Dév 33(4):490–508
Núñez A, Aliste Almuna E, Bello Á, Osorio M (2017) Imaginarios geográficos, prácticas y discursos de frontera: Aisén-Patagonia desde el texto de la nación, Serie Geolibros, vol 25. Pontificia Universidad Católica de Chile, Instituto de Geografía
Nuñez A (2018) Capítulo 9: Geografías de lo-sublime y el proceso de turistificación en Aysen Patagonia. In: Araucania-Norpatagonia II La fluidez, lo disruptivo y el sentido de la frontera. Universidad Nacional de Río Negro (Argentina), pp 249–269. https://doi.org/10.4000/books.eunrn.1788
Núñez A, Aliste E, Bello Á, Astaburuaga JP (2018) Eco-extractivismo y los discursos de la naturaleza en Patagonia – Aysén: nuevos imaginarios geográficos y renovados procesos de control. Revista Austral de Ciencias Sociales 35:133–153
Paez MP, Otero A (2014) De los discursos a las representaciones e imaginarios en migrantes de amenidad. Caso de estudio Villa La Angostura, Neuquén. Realidad. Tendencias y Desafíos en Turismo (CONDET), Año XIV. Volumen 12. Noviembre
Pastor JMA, Mora VG (2013) El imaginario tecnológico de Domingo Faustino Sarmiento: representaciones y arquetipos de América (1845–1885). Anu Estud Am 70(2):673–697
Peri PL, Lencinas MV, Martínez Pastur G, Wardell-Johnson GW, Lasagno R (2013) Diversity patterns in the steppe of Argentinean Southern Patagonia: environmental drivers and impact of grazing. In: Morales Prieto MB, Traba Díaz J (eds) Steppe ecosystems: biological diversity, management and restoration, Series: Environmental research advances, Chapter 4. Nova Science Publishers, Hauppauge, NY, pp 73–95
Peri PL, Monelos L, Díaz B, Mattenet F, Huertas L, Bahamonde H, Rosas YM, Lencinas MV, Cellini JM, Martínez Pastur G (2019) Estado y usos de los bosques nativos de lenga, siempreverdes y mixtos en Santa Cruz: Base para su conservación y manejo. Editorial INTA-Consejo Agrario Provincial (CAP). 108 pp
Pintos JL (2005) "Comunicación, construcción de la realidad e imaginarios sociales". Utopía y Praxis Latinoamericana, vol. 10, núm. 29, abril-junio, 2005, pp. 37–65. Universidad del Zulia. Venezuela
Potschin-Young M, Haines-Young R, Görg C, Heink U, Jax K, Schleyer C (2018) Understanding the role of conceptual frameworks: reading the ecosystem service cascade. Ecosyst Serv 29:428–440
Prieto M, Bauer C (2012) Hydroelectric power generation in Chile: an institutional critique of the neutrality of market mechanisms. Water Int 37(2):131–146
Purcell F (2018) Imaginarios socioculturales de la hidroelectricidad en sudamérica 1945–1970. Atenea (Concepción) 518:97–116
Radovich JC, Balazote A (2000) Mapuches en Neuquén: conflictos en el orden económico y simbólico. In: El resignificado del desarrollo. UNIDA, Buenos Aires
Radovich JC, Balazote A (2005) Reasentamientos poblacionales en Norpatagonia, Argentina como consecuencia de la construcción de megarepresas hidroeléctricas. In: Taller "Acompañando al pueblo Huarpe". Universidad Nacional de, Cuyo. Mendoza
Radovich JC (2013) Los mapuches y el Estado neuquino: algunas características de la política indígena. RUNA XXXIV (1), pp 13–29. FFyL – UBA – ISSN 0325-1217 13
Raymond CM, Kenter JO (2016) Transcendental values and the valuation and management of ecosystem services. Ecosyst Serv 21:241–257
Rivera C, Vallejos-Romero A (2015) La privatización de la conservación en Chile: repensando la gobernanza ambiental. Bosque (Valdivia) 36(1):15–25
Robinson GJ (2013) Remembering the borderlands: traditional music and the post-frontier in Aisén, Chile. Ethnomusicology 57(3):455–484
Rodríguez JC, Gissi Barbieri N, Medina Hernández P (2015) Lo que queda de Chile: La Patagonia, el nuevo espacio sacrificable. Andamios 12(27):335–356

Rodríguez JC, Medina P, Reyes SE (2014) Territorio, paisaje y marketing global: Imaginarios en la construcción de la Patagonia como marca. Magallania (Punta Arenas) 42(2):109–123

Rodriguez MD (2015) Disputa por el territorio en un área de cordillera: Tensiones, negociaciones y conflictos entre instancias gubernamentales y población indígena mediadas por el turismo. Revista de Ciencias Sociales (Quilmes). Editorial Universidad Nacional de Quilmes: Bs.As. ISSN: 2347-1050

Rodriguez NJ (2019) Interacciones sociales en la conformación de ciudades turísticas de montaña. Bariloche. Patagonia Argentina (1991–2006). Revista CTS N°40, vol. 14. Febrero, pp.245.266

Rojas P, Mellado MA, Blanco-Wells G (2020) Sobrenaturaleza mapuche: extractivismo, seres no humanos y miedo en el Centro Sur de Chile. Revista Austral de Ciencias Sociales 38:7–30

Schapper A, Unrau C, Killoh S (2020) Social mobilization against large hydroelectric dams: a comparison of Ethiopia, Brazil, and Panama. Sustain Dev 28(2):413–423

Schaeffer C (2017) Democratizing the flows of democracy: Patagonia sin represas in the awakening of Chile's civil society. In: Social movements in Chile. Palgrave Macmillan, New York, pp 131–159

Salas C, Donoso PJ, Vargas R, Arriagada CA, Pedraza R, Soto DP (2016) The forest sector in Chile: an overview and current challenges. J For 114(5):562–571

Sepúlveda B, Guyot S (2016) Escaping the border, debordering the nature: protected areas, participatory management, and environmental security in Northern Patagonia (ie Chile and Argentina). Globalizations 13(6):767–786

Serenari C, Peterson MN, Leung YF, Stowhas P, Wallace T, Sills EO (2015) Private development-based forest conservation in Patagonia: comparing mental models and revealing cultural truths. Ecol Soc 20(3)

Serenari C, Peterson MN, Wallace T, Stowhas P (2017a) Private protected areas, ecotourism development and impacts on local people's well-being: a review from case studies in Southern Chile. J Sustain Tour 25(12):1792–1810

Serenari C, Peterson MN, Wallace T, Stowhas P (2017b) Indigenous perspectives on private protected areas in Chile. Nat Areas J 37(1):98–107

Skewes JC, Eugenia Solari M, Guerra D, Jalabert D (2012) Los paisajes del agua: naturaleza e identidad en la cuenca del río Valdivia. Chungará (Arica) 44(2):299–312

Smit M (2017) Environmental philanthropy. MSc Thesis: Leisure, Tourism & Environment. Wageningen University and Research Department of Environmental Sciences Cultural Geography Chair Group. https://edepot.wur.nl/430513. Last accessed 11 Nov 2020

Stephenson MO Jr (2011) Considering the relationships among social conflict, social imaginaries, resilience, and community-based organization leadership. Ecol Soc 16(1)

Susskind L, Kausel T, Aylwin J, Fierman E (2014) The future of hydropower in Chile. J Energy Nat Resour Law 32(4):425–481

Svampa M, Viale E (2014) Maldesarrollo. La Argentina del extractivismo y el despojo. Katz Editores, Buenos Aires. ISBN 978-987-1566-92-1

Taylor C (2003) Modern social imaginaries. Public Planet Books, Duke University Press

Taylor C (2005) Modern social imaginaries. Duke University Press, Durham, NC

Tecklin DR, Sepúlveda C (2014) The diverse properties of private land conservation in Chile: growth and barriers to private protected areas in a market-friendly context. Conserv Soc 12(2):203–217

Tomasello M, Carpenter M, Call J, Behne T, Moll H (2005) Understanding and sharing intentions: the origins of cultural cognition. Behav Brain Sci 28(5):675–691

TRECC (2007) Formulación del plan de ordenamiento urbano y actualización de los códigos de planeamiento urbano y edificación de la localidad de El Bolsón, Provincia de Río Negro, de la localidad de Villa La Angostura de la Provincia de Neuquén y estudios de funcionalidad de la Vega Plana de la localidad de San Martín de los Andes Provincia del Neuquén". Informe final Villa La Angostura Julio 2007. Programa de Mejora de la Competitividad del Sector Turismo en Áreas Piloto. Préstamo BID 1648/OC-AR. TRECC Consultores

Tricot T (2020) El movimiento del movimiento en Lumako: Pensándose como mapuche desde lo mapuche. Revista Trabajo Social 81:33–54

Urrutia S, Núñez A, Aliste E (2019) Naturaleza salvaje y agreste: Los imaginarios de la naturaleza en la construcción del camino longitudinal Austral, Chile 1976–1990. Magallania (Punta Arenas) 47(2):55–72

Valverde S, Maragliano G, Impemba M (2015) Expansionismo turístico, poblaciones indígenas Mapuche y territorios en conflicto en Neuquén, Argentina. PASOS, Revista de Turismo y Patrimonio Cultural 13(2. Special Issue):395–410

Vitar B (2010) Los caminos del Wallmapu (País mapuche). Disparidades. Revista de Antropología 65(1):255–288

Wagner JR, Berge E, Brara R, Bruns B, Caton SC, Donahue J, Swyngedouw E (2012) Water and the commons imaginary. Curr Anthropol 53(5):000–000

Zoomers A (2011) Introduction: rushing for land: equitable and sustainable development in Africa, Asia and Latin America. Development 54(1):12–20

Zorondo-Rodríguez F, Díaz M, Simonetti-Grez G, Simonetti JA (2019) Why would new protected areas be accepted or rejected by the public?: lessons from an ex-ante evaluation of the new Patagonia Park Network in Chile. Land Use Policy 89:104248

Chapter 20
The Challenges of Implementing Ecosystem Services in the Argentinean and Chilean Patagonia

L. Nahuelhual, C. Minaverry, P. Laterra, F. Henríquez, L. Delgado, and G. Martínez Pastur

Abstract Implementation means turning the ecosystem service (ES) recognition into incentives and institutions that will guide wise investments and actions to conserve natural capital, on a large scale. Like many other countries, Argentina and Chile face a discrepancy between the conceptual understanding of ES in science and the limited implementation. This chapter reveals implementation gaps within the Patagonian region of Argentina and Chile and the blind spots that ES research faces in order to support implementation. The introductory section provides an overview of ES research worldwide and, specifically, in Argentina and Chile. The second sec-

L. Nahuelhual (✉)
Instituto de Economía, Universidad Austral de Chile, Campus Isla Teja, Valdivia, Chile

Centro de Investigación Dinámica de Ecosistemas Marinos de Altas Latitudes, Universidad Austral de Chile, Campus Isla Teja, Valdivia, Chile

Fundación Bariloche, San Carlos de Bariloche, Río Negro, Argentina

C. Minaverry
Consejo Nacional de Investigaciones Científicas y Técnicas (CONICET), Instituto de Ecología y Desarrollo Sustentable (INEDES), Universidad Nacional de Luján, Luján, Buenos Aires, Argentina

P. Laterra
Fundación Bariloche, San Carlos de Bariloche, Río Negro, Argentina

Consejo Nacional de Investigaciones Científicas y Técnicas (CONICET), CCT Patagonia Norte, San Carlos de Bariloche, Río Negro, Argentina

F. Henríquez
Master in Community Development, University of California, Davis, CA, USA

L. Delgado
Escuela de Ciencias, Facultad de Ciencias, Universidad de Chile, Santiago, Chile

Fundación Centro Transdisciplinario de Estudios FES-Sistémicos, Santiago, Chile

G. Martínez Pastur
Consejo Nacional de Investigaciones Científicas y Técnicas, Centro Austral de Investigaciones Científicas y Técnicas, Ushuaia, Tierra del Fuego, Argentina

P. L. Peri et al. (eds.), *Ecosystem Services in Patagonia*, Natural and Social Sciences of Patagonia, https://doi.org/10.1007/978-3-030-69166-0_20

tion reviews the blind spots in current published research conducted in Patagonia. The third section reports examples of ES implementation at the state, private, and society levels. The last section revises the main challenges for ES implementation.

Keywords Environmental governance · Implementation gaps · Decision-making · Science-policy interface · Nature conservation

1 Introduction

The appeal of the ecosystem service (ES) concept lies on its potential of contributing to conservation by fostering an understanding of humans' dependence on nature for their survival and well-being (Kronenberg 2014). The focus on human-nature interactions differentiates the ES concept from narrower conservation perspectives such as biological conservation or natural resources management (Baker et al. 2013).

Research on ES grew exponentially since the publication of the Millennium Ecosystem Assessment in 2005 (Delgado and Marín 2015; Pauna et al. 2018; Perevochtchikova et al. 2019; Balvanera et al. 2020), although with noticeable differences across countries, ecosystems, and disciplines. Most research has been conducted by scientists in developed countries as the leading ES journals testify, marine ecosystems remain behind terrestrial ecosystems (Arkema et al. 2015), and social scientists other than economists have had a marginal involvement (Chaudhary et al. 2015). Despite these differences, the expectations of ES research are threefold: firstly, to raise awareness of the importance of the natural or partially modified environments in sustaining individual and social needs and in maintaining human health and well-being through the provision of ES; secondly, to provide a transparent and objective framework for understanding the trade-offs between human development and the conservation of natural systems (Beaumont et al. 2017); and finally, to promote the development of new instruments and institutions capable of simultaneously safeguarding nature and human well-being (Saarikoski et al. 2018).

In the last decade, the concept of ES has transcended the research sphere to generate an opening to transdisciplinary issues such as changes in conservation paradigms and participation of social actors and institutions (e.g., Virapongse et al. 2016), among others. In addition, countries have started implementing the ES concept in their legislations and environmental strategies and policies (Ruckelshaus et al. 2015). Implementation in this context has been defined as turning the ES recognition into incentives and institutions capable of guiding investments to conserve natural capital on a large scale (Daily and Matson 2008), and it varies widely across countries. Argentina and Chile, particularly, do not exhibit a significant implementation of ES, at least with regard to formal institutions and, alike other countries (Posner et al. 2016; Lautenbach et al. 2019; Matzek et al. 2019), they face a discrepancy between the considerable conceptual understanding of ES in science and the limited takeoff of the concept in practice. The absence of the concept in concrete operational decision-making contexts has been named "implementation gap" (Levrel et al. 2017).

The fact that most of the resource-use policy decisions still do not consider ES stems in part from an ineffective interface between ES science and policy. This limitation is reflected in the ongoing discussion on how the ES concept could be improved, mainstreamed, or operationalized (e.g., Fisher and Brown 2015; Olander et al. 2016; Costanza et al. 2017; Lautenbach et al. 2019). This discussion has at least three different components: (i) the specification of the ES concept itself, (ii) the available knowledge on ES for practical implementation of the concept, and (iii) the best practice for implementation of the concept (Lautenbach et al. 2019).

This chapter investigates implementation gaps in the Patagonian region of Argentina and Chile and the blind spots that ES research faces to support implementation. We also identify the main challenges to achieve implementation.

2 Implementation of Ecosystem Services in the Argentinean and Chilean Patagonia

It is important to notice that many cases of "implementation" referred to in literature are still in a research stage of development. In such cases, there is no real practical and regional application process involving stakeholders (Verburg et al. 2016), so we exclude them from this analysis.

2.1 State-Level Implementation

There is still a weak level of ES implementation in public decision-making of both Argentina and Chile. In most examples presented in Table 20.1, the use of ES is conceptual rather than instrumental. As defined by McKenzie et al. (2011), instrumental knowledge flows from scientists to rational decision-makers who make observable decisions on technical grounds. Conceptual knowledge broadens and deepens understanding, shapes thinking, and enables people to develop new beliefs and values. Strategic knowledge is used to support and promote a specific intervention or policy option or justify previously held beliefs and values.

In Argentina, the concept is mainly included in native forest policies. A significant example is the National Plan of Native Forests Restoration, which promotes restoration, recovery, and rehabilitation of native forests (Resolution 267/19 of the National Ministry of Environment). The plan includes the following aspects: (i) restoring the functional processes of native forests and biodiversity, (ii) reducing degradation factors and greenhouse gas emissions, and (iii) promoting environmental education concerning the importance of native forests.

Table 20.1 Ecosystem service implementation at the state level for three categories of policies: (a) national and subnational; (b) economic and fiscal incentives; (c) sectoral policies. Policy options are those in which ES have been implemented in developed countries (e.g., Europe and Australia) (see Ranganathan et al. 2008; Matzek et al. 2019)

Policy option	Examples from Argentina	Examples from Chile
National and subnational policies		
Mainstream ES into economic and development planning	(a) National Native Forests Law 26.331 (2007) includes a Payment for Ecosystem Services (PES) mechanism and is complemented by provincial regulations and the National Strategy of Biodiversity, the Plan of Action 2016–2020 and the National Plan of Forests and Climate Change of 2017 (REDD+) (b) National Law of protected marine areas 27.037 (2014)	(a) The 2020 Chile's Nationally Determined Contributions (NDC) includes instruments "to protect, maintain and increase natural carbon sinks, considering the multiple ecosystem services they provide" (b) The Environmental Accounting System National Plan envisions the incorporation of ES economic valuation (c) The National Landscape Restoration Plan for 2020–2030 considers to restore landscapes ES
Include investments in ES in government budgeting	National Native Forests Law 26.331 (2007) and climate change and restoration regulations propose and implement ES-based incentive mechanisms	No
Include ES in public protected areas management plans	National Law of Marine Protected Areas 27.037 (2014) states that the development of a management plan must have an ecosystem-based approach	Chile is pursuing the protection of several areas of Patagonia (terrestrial and marine), including their cities' wetlands by considering their ES or ecosystem flows, and planning to have a national wetland inventory by 2025 which acknowledges the concept
Economic and fiscal incentives		
Tax deductions and credits in ES transactions and investments	No	Activities oriented to manage and restore native forests are subsidized by law
Reduce perverse subsidies affecting ES	No	No
Set limits and establish trading systems for the use of ES	No	No

(continued)

Table 20.1 (continued)

Policy option	Examples from Argentina	Examples from Chile
Fund ES valuation and research for improving ES valuation methods	The National Strategy of Biodiversity and the Plan of Action 2016–2020 (National Ministry of Environment) included an item denominated "valuation of biodiversity" by incorporating ecological qualitative and quantitative valuation of ES and functions	The Ministry of Environment has pursued several studies on ES valuation, assessment, and indicators, for decision-making
Use procurement policies to focus demand on products and services that conserve ES	No	No
Support wetland banking schemes	No	No
Sectoral policies		
Include ES in sectoral policies and strategic environmental assessment	ES are indirectly or explicitly considered in different national, provincial, and municipal regulations (e.g., urban and provincial reserves, forest harvesting regulations, environmental restrictions for any human-regulated activity including tourism) such as the National Plan of Native Forests Restoration, the National Plan of Forests and Climate Change, and the Strategic Environmental Assessment in Andean Patagonia	(a) The Ministry of Environment must assess risks on ES under the regulation for the determination of minimum ecological water flow (b) The National Strategy of Biodiversity 2017–2030 aims to protect and restore biodiversity and its ES (c) The National Plan on Wetlands 2018–2022 aims to protect and restore wetlands' biodiversity and ES (d) The protection of species also has been based on ES research (e) The National Policy on Rural Development focuses partly on biodiversity and ES. The methodology to impose economic sanctions considers ES affected by noncompliance actions (f) Regarding strategic impact assessment, the government established the shoreline zoning of the Aysén region by recognizing areas of natural value and ES

(continued)

Table 20.1 (continued)

Policy option	Examples from Argentina	Examples from Chile
Use of zoning or easements to keep land available for priority ES	No	The government established the shoreline zoning of the Aysén region by recognizing areas of natural value and ES. A new law established conservation easements, which acknowledges the ES concept
Use of physical structures or technology to substitute ES	No	No
Use of regulating ES such as natural hazard protection or water filtration instead of built infrastructure	No	The authorities have invested in protecting and restoring the national reserve that houses the watershed which ensures water supply for the city of Valdivia in Los Ríos region
Establish ES-based certification schemes to encourage best management practices	There are (i) several voluntary norms and guides regarding environmental certification, (ii) ES payments (e.g., ISO 14.000 series and ISO 14.007, 14.008, etc.), and (iii) the Good Practices Manual for the management of Forests plantations in the North of Patagonia (2015) which includes ES	The Ministry of Environment should have a proposal for the establishment of a National Certification System for Biodiversity and Ecosystem Services, oriented to promote certification on activities, practices, or sites that contribute to ecosystem services provision
Introduce education or extension programs on good practices	National Strategy of Biodiversity, the Plan of Action 2016–2020, and the National Plan to fight against desertification and degradation of lands and mitigation of draughts (2019) – National Ministry of Environment	(a) The Ministry of Education included the concept of ES within the secondary education curriculum (b) The Ministry of Environment offered training to municipality officers on ES in mountains' biological corridors
Develop and encourage the use of products and methods that reduce dependence and impact on ES	No	(a) A recent legal scheme was designed to reduce, recycle, and reuse waste, which could eventually reduce impact on ES (b) The General Analysis of Economic and Social Impacts uses the ES concept as an indicator of impacts of environmental quality norms

In Chile, a promising space for implementation of ES is the Environmental Courts. Specifically, the third Environmental Court of Valdivia was created in 2012 to cover the jurisdictional territory comprised between the Ñuble and Magallanes regions. This court decides over illegality claims covering a range of environmental administrative decisions, mainly issued by agencies linked to the environmental impact assessment and enforcement systems and the Ministry of Environment and over environmental damage lawsuits. Since the beginning of its jurisdictional operation (Dec. 2013[1]), the third Environmental Court's rulings took into account ES. Currently, at least 20 of the Court's final rulings include the ES concept. An example is the rejection of the water quality standard for the Valdivia River Basin in 2016, considering (among other arguments) that the ES assessment involved in the cost-benefit analysis of the standard was superficial.

2.2 *Private-Level Implementation*

Private implementation of ES involves the use of ES knowledge by companies and other initiatives led by individuals or private organizations such as NGOs and foundations, different from society's organizations (e.g., the Natural Capital Project comprising The Nature Conservancy (TNC) and World Wildlife Fund (WWF)) (Hanson et al. 2012; TEEB 2012; BSR 2013; Hrabanski et al. 2013).

In the Patagonian region of Argentina and Chile, the uptake of the ES concept by the private sector has at least one decade of history, and the terms of use are commonplace among conservation NGOs, private landowners, and private firms. The most renowned NGOs such as World Wildlife Fund (WWF), Wildlife Conservation Society (WCS), The Nature Conservancy (TNC), and Pew Charitable Trust endorse the concept, and some of them have included ES in their strategic plans (e.g., WWF).

The participation of private landowners in conservation in Patagonia has been widely recognized as a way to complement insufficient public investments in conservation outside public protected areas (e.g., Rosas et al. 2019). Among private landowners, many ranchers have implemented conservation programs to protect biodiversity (e.g., Estancia El Condor y Los Toldos in Santa Cruz province, Argentina), which shows that public-private joint ventures are not only possible, but they also contribute to ES provision at the landscape level. Important conservation networks such as "Así Conserva Chile" or "Red Ecofluvial de la Patagonia" have also endorsed the concept within their conservation strategies and environmental education proposals.

Companies, in turn, have progressively included new concepts such as biodiversity and ES in their silvicultural plans, certification strategies, or mitigation proposals. Examples are Forestal Arauco in Chile and Argentina, Forestal Mininco in Chile, Kareken sawmill in Argentina, and Russfin sawmill in Chile (Martínez Pastur et al. 2007). The inclusion of new management criteria has helped these companies to decrease social concerns regarding their operations and to open market opportunities, allowing them to export products to many international markets with environmental restrictions (e.g., Europe). However, the use of these concepts by companies has not been exempt from criticism, regarding the factors influencing

[1] https://3ta.cl/tercer-tribunal-ambiental/conozca-al-tribunal/

their decisions to invest voluntarily in biodiversity and ES protection (Krause and Matzdorf 2019). Recent literature has coined the term "greenwashing" effect to define the use of nature-evoking elements, including ES, in advertisements to enhance a brand's ecological image (Parguel et al. 2015; Wilkinson 2014).

2.3 Community-Led Initiatives

Many community-based environmental projects or public-private projects recognize and use ES as a boundary object that helps integrate diverse forms of knowledge across social groups and organizational scales (Steger et al. 2018). These include action-research projects (e.g., BEST-p network project with participation of researchers from Chile and Argentina), environmental education projects (e.g., Environmental Protection Fund in Chile), or natural resources management projects (e.g., Bosque Modelo Panguipulli and Bosque Modelo Chiloé in Chile). The concepts used in research and development interventions have resulted problematic in some cases, particularly in indigenous communities, as the ES concept opposes their ontology of nature (Outeiro et al. 2015a; Nahuelhual et al. 2018). The use of the ES approach with indigenous communities requires the understanding of additional elements, such as intercultural dialogue, a comprehensive view of the territory, cultural principles regarding the care of nature, material and immaterial nature use systems (Albarracín-Álvarez et al. 2019), as well as the legal protection of their environmental and cultural rights (Minaverry and Martínez 2018).

In the case of Chile, a concrete experience carried on between the academia and rural communities in the Los Ríos region is the case of Rural Water Committees (e.g., Román et al. 2009), which are rural associations, based on the Chilean Law 19.418, organized to provide water to households within a community. They also deal with water shortage scenarios, common during summer, by collecting and maintaining rainwater or setting restrictions in water use (Delgado et al. 2013; Delgado and Marín 2016). Yet, a follow-up of these experiences reveals that the ES knowledge achieved during the execution of the projects (e.g., participatory research projects) is no longer used once projects end.

3 Blind Spots in Ecosystem Services Research in the Patagonian Region of Argentina and Chile

We analyzed some of the main features of the ES research developed to date in the Patagonian region, based on (i) a quick review of articles published since 2005 that explicitly mentioned ES in their title and abstract and (ii) the authors' own experience. In both countries, research has widened from an emphasis on ecological components of terrestrial ecosystems (mainly forests) to other various dimensions of ES and other ecosystems.

In the Argentinean Patagonia, terrestrial ecosystems are the most studied, and the most frequently assessed ES are provision services from grasslands and forests (e.g., Peri et al. 2016a; Martínez Pastur et al. 2013). These studies combine the assessment of economic and ecological values (e.g., Lindenmayer et al. 2012; Martínez Pastur et al. 2019), as well as the spatial assessment of ES supply and economic values using benefit transfer (e.g., Carreño and Viglizzo 2007). Other studies focus on the impact of invasive species (e.g., beavers) on natural capital and provision of ES (e.g., Henn et al. 2016; Huertas Herrera et al. 2020) and plants that decrease grazing potential (e.g., Huertas Herrera et al. 2018; Soler et al. 2019). One of the most studied topics are the trade-offs among different types of ES and biodiversity, under different management strategies and conservation planning within natural reserve networks (e.g., Peri et al. 2013, 2016b; Martínez Pastur et al. 2017; Rosas et al. 2019).

Over the past few years, the interest in other ES has greatly increased, considering cultural (e.g., Martínez Pastur et al. 2016a), regulation, and supporting ES, both in urban and natural landscapes (e.g., Zagarola et al. 2017; Peri et al. 2019). Most of these studies were initially conducted at local scales but quickly spanned to modelling at regional scales (e.g., Martínez Pastur et al. 2016b; Peri et al. 2017, 2018; Rosas et al. 2019).

Recent studies cover stakeholders' perspectives and institutional challenges facing ES provision and implementation in practice (Dick et al. 2018; Saarikoski et al. 2018), including policy instruments (e.g., Lorenzo et al. 2018; Martínez Pastur et al. 2020), governmental institutions (e.g., Gamondes Moyano et al. 2016; Turkelboom et al. 2018), effectiveness of natural reserves (e.g., Rosas et al. 2021), and research strategies and funding (e.g., Anderson et al. 2012; Soler et al. 2018). Emerging literature focuses on the benefits of ES, such as direct and indirect job opportunities (Laterra et al. 2019). Marine ES are far behind terrestrial ES, with only few studies in the recent years (e.g., Roldán et al. 2015; Martinetto et al. 2020).

In the Chilean Patagonia, terrestrial ecosystems are also the most studied, particularly temperate rainforests (e.g., Delgado et al. 2013; Nahuelhual et al. 2014; Rodríguez-Echeverry et al. 2018), including specific research conducted in protected natural areas (e.g., Durán et al. 2013). The best studied ES are the regulating services and among them, water regulation and nutrient retention (e.g., Lara et al. 2009; Little et al. 2015; Rodríguez-Echeverry et al. 2018; Alvarez-Garreton et al. 2019; Becerra-Rodas et al. 2019), followed by cultural services and among them, recreational ES (e.g., Nahuelhual et al. 2013, 2016). The ecological focus is still the main ES topic addressed; therefore, the most common types of assessments are ecological modeling and mapping of ES provisioning areas (e.g., Rodríguez-Echeverry et al. 2017), although the evaluation of social and legal dimensions of ES has recently begun to increase (e.g., Delgado and Marín 2016; Delgado et al. 2019; Pastén et al. 2018). Ecological and economic assessments usually evaluate a single or few ES (e.g., Núñez et al. 2006; Ponce et al. 2011; Barrena et al. 2014), and the inclusion of more elements such as benefits, values, and vulnerability is scarce (e.g.,

Laterra et al. 2016). For instance, synergy and trade-off analysis are rarely included (e.g., Lara et al. 2009; Benra et al. 2019). Mapping studies are mainly associated with specific localities or watersheds in the southern and austral regions, with a low occurrence in the Magallanes region. The application of the results to conservation and planning is generally limited, mainly due to the restricted scale of the studies (especially local) and their limited scope (few services evaluated, single scales, no trade-off analysis), which makes it difficult to apply them to concrete management and planning instruments (De la Barrera et al. 2015). As in Argentina, marine ES have been far less studied, with few examples in Chiloé and Magallanes (e.g., Outeiro and Villasante 2013; Outeiro et al. 2015b; Nahuelhual et al. 2017).

In Table 20.2 we synthesize the blind spots found in recent publications in Patagonia, which prevent ES research from being included in decision-making, which largely coincide with those reported for developed countries (e.g., Lautenbach et al. 2019). Blind spots are related to critical questions and key facets that characterize the "holistic ideal of ecosystem services research," namely, (i) socio-ecological validity of ecosystem data and models, (ii) assessment of trade-offs between ES, (iii) recognition of off-site effects, (iv) comprehensive and shrewd involvement of stakeholders, and (v) relevance and usability of study results for the operationalization of the ecosystem service concept in practice (Lautenbach et al. 2019).

Some blind spots are more critical than others, which depends on the decision that the ES research is intended to support. For example, decisions concerning local community development may require little consistency in selecting the appropriate ES and measures to use, because there is no need for comparison or tracking. In contrast, decisions supported with national funding, affected by national regulations, or concerning public lands or waters require a greater degree of consistency, especially if decision-makers need to make project comparisons, track performance, and assess returns on investments (Olander et al. 2016). Decision-making that aims at overcoming sectoral perspectives by integrating a socio-ecological systems approach would benefit from the comprehensive integration of ES supply, ES demand, and policy options (scenarios, trade-offs) affecting demand and/or supply of ES (Orenstein and Groner 2014; Wei et al. 2017). Kirchner et al. (2015), for example, demonstrate how the combination of stakeholder involvement, model integration, trade-off analysis, scenario analysis, and optimization can improve targeting of agro-environmental payment schemes to provide a more balanced and efficient supply of ES and promote rural development. The same authors sustain that if no policy instruments are tested, and/or if thematic, temporal, and spatial resolution and extent and scale of the assessment are not selected in accordance with decision-makers' requirements, specific policy recommendations cannot be generated or have poor chances of being applied.

Another neglected aspect is the capacity of scaling-up, to cover larger areas of land and seascape and encompass complex and extensive resource systems. This leads to the second necessary advance (besides coherently selecting and defining ES): the development of methodologies for ES assessments capable of crossing landscape units and be incorporated in decision-making among multiple jurisdictions and at government levels. Local-scale efforts can be more successful in

Table 20.2 Blind spots in ecosystem services research in the Patagonian region of Argentina and Chile

ES feature	Blind spots
ES definitions	The many definitions used to refer to ES (i.e., functions, processes, benefits, components of nature, aspects of ecosystems) may confound decision-makers and prevent comparisons. For example, some assessments of "ecosystem" services are reduced to functions performed by some of their components (e.g., pollination, biological control, soil infiltration)
ES typologies	The use of different ES typologies (or no specific typology) (e.g., MA, TEEB, CICES, other ad hoc) limits the application of results in decision-making at larger scales
Scope of ES assessments	The assessment of single or few ES limits trade-off analysis and the analysis of consequences of land and marine interventions on different ES beneficiaries The nonparticipatory selection of ES (usually defined by the researchers) can limit the scope and validity of assessments The restrictive nature of assessments, usually limited to ecological components, reduces the comprehension of the effects of land and marine management on the well-being of people
Scale (spatial and temporal)	The usually limited temporal and spatial scales of analysis (subregional case studies) restrict the application of results to decisions covering large extensions of land or marine ecosystems
Methods (ecological modeling, mapping, valuation, other assessments)	The lack of standardized and thorough methods prevents application to real-world decisions. Blind spots include: Trade-offs and scenario analyses are usually omitted Different indicators are used to measure the same ES (e.g., SWAT[a] versus ECOSER[b] to calculate water supply) Indicators are not field-validated Uncertainty of ES evaluations is not assessed Ecological thresholds are omitted Heterogeneity of ES values within society is disregarded Influences from power and access asymmetries are omitted from benefits assessment Exposition, sensitivity, and adaptability to loss of ES are not considered for ES-based land-use planning Appropriate assessment tools have not been developed
Missing links between assessments' goals and products	Despite being defined as "nature contributions to people," ES spatial assessments neglect current and future ES demands Spatial assessments focus on supply areas and hotspots rather than on areas prone to ES losses ES hotspots are proposed as priority conservation areas, regardless of the benefits that those hotspots can sustain and the number of people who benefit from them

[a]The Soil & Water Assessment Tool (SWAT) is a small watershed to river basin-scale model used to simulate the quality and quantity of surface and ground water and predict the environmental impact of land use, land management practices, and climate change (https://swat.tamu.edu/)
[b]ECOSER is a GIS-based collaborative tool developed to support decision-making about land use, as well as research and disciplinary integration and scientific collaboration regarding ecosystem functions (EF) and ecosystem services (ES), their capture by society, as well as the "socio-ecological vulnerability" (SEV) under different ES loss scenarios (http://eco-ser.com.ar)

engaging local communities, yet they lack the purview to drive larger-scale protection efforts or to manage ES in ways that incorporate multiple cumulative impacts that change as ecosystem types vary (Portman 2013).

Many of the blind spots are related to decisions made during the design of the studies. To overcome the blind spots identified here, careful experimental design is of key importance, including not only the design of field experiments but also the design of scenarios, stakeholder integration, and simulation experiments (Lautenbach et al. 2019).

4 Implementation Gaps and Challenges for an ES-Based Governance in Patagonia

Despite the progress in research, an important implementation gap prevails (Fig. 20.1). A first reason for this gap is the lack of understanding of the decision-making and policy context within the ES research community (Peixotto et al. 2019). Most scientific studies are not framed within specific decisions (public or private), although the majority of them aspire to support a decision, which in most cases is landscape or marine spatial planning. In other words, our research still does not take into account the context of use of the information that is being generated. Furthermore, researchers and decision-makers have different goals, attitudes toward information, languages, and perception of time, among others.

A second reason is the limited demand for ES research in both countries. In the case of the public sector, this demand is channeled fundamentally through specific

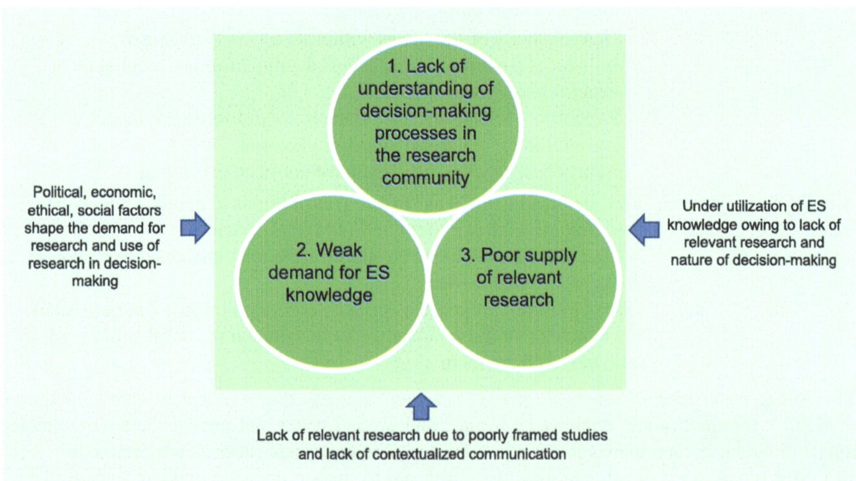

Fig. 20.1 Reasons that explain the ES knowledge decision-making gap. (Adapted from Bolo (2013; https://es.slideshare.net/Scinnovent/bridgingg-the-research-policy-gap-influencing-policy-changenairobi))

consultancies of the National Ministry of Environment and Sustainable Development and similar provincial offices in the case of Argentina and also through consultancies and competitive public funds from the Ministry of Environment (https://mma.gob.cl/servicios-ecosistemicos/) and Corporación Nacional Forestal (e.g., Native Forest Fund; https://www.conaf.cl/) in the case of Chile. It is obviously not an easy task for public administrators and researchers to move from a fragmented view of management toward a complex social-ecological system approach as ES require, which implies accepting (a) contextuality, (b) social participation, (c) uncertainty, and (d) the urgency of local problems (Delgado et al. 2019; Minaverry 2016).

A third reason is the poor supply of relevant knowledge, which is closely linked to the first reason. Poorly framed ES research has been associated with the lack of standards regarding terminology, methods, and reporting (Polasky et al. 2015; Olander et al. 2016, 2018; Jax et al. 2018; Nahuelhual et al. 2020). Mostly, this context requires clear research guidelines, for researchers to get involved in policy-based research, including their role among stakeholders, and the development of specific ES assessment tools (e.g., Jax et al. 2018).

In this context, ES implementation in Patagonia faces various challenges (Loft et al. 2015; Falk et al. 2018).

1. The existence of a set of disparate institutions (e.g., markets and hierarchies) operating at different administrative scales (see Table 20.1) and policy instruments (e.g., public protected areas, forest subsidies) which were not developed under a systemic logic but under a natural resource extraction rationale (e.g., water markets, watershed management, and fisheries management).
2. Property rights (e.g., private, communal, and customary) that have evolved to large levels of private concentration (Nahuelhual et al. 2018) and determine who benefits from ES and who does not.
3. Weak institutional frameworks, which among other things are reflected in the low budgets assigned to conservation and management agendas and the still low technical capacity of agencies for undertaking a system-based management approach, as the ES framework requires.
4. The existence of sectorial and fragmented interventions and agencies without an integrated vision of environmental problems. Legal and institutional fragmentation favors regulation overlaps, the ineffectiveness of environmental law, and citizens' participation and gives room for corruption, revealing the limitations of control agencies (Minaverry 2016).

At present, a diversity of different governance arrangements coexist, which have evolved around and on top of each other and not focused on ES conservation and human well-being (see Table 20.1). This may lead to uncoupling of different arrangements and/or the delivery of redundant and sometimes contradictory incentives (e.g., forest clearing bans versus forest subsidies) for the provision of a single or different ES, resulting in inefficiencies and trade-offs in their supply (e.g., Benra et al. 2019). Additionally, there is a difficulty in deciding which ES should be prioritized and for whom since stakeholders are heterogeneous in terms of demands, values, interest, and bargaining power (e.g., Zagarola et al. 2014; Mendoza 2016;

Mrotek et al. 2019). Furthermore, stakeholders have neither a minimum knowledge regarding ES and their regulations nor opportunities for binding participation in environmental issues. An equilibrium of the different social actors is key for representing the different interests in relation to ecosystems and their services. Argentina, for instance, has made modifications in the Constitution (1994) in order to improve regional citizen participation (Figueroa and Gutiérrez 2018).

Overcoming these challenges is not just about marginal improvements to existing structures but rather a modification in the actual objectives of environmental public policy, which implies incorporating new values, management visions, and new actors in a democratic and fair decision-making process. Doing so demands an approach that differs from the typically top-down, technocratic, and linear processes that characterize much of Argentina and Chile's policy-making. The top-down approach currently used in conservation and management policy in both countries is not well suited for the complexities and multiple definitions inherent to the ES concept, which would rather benefit from a more iterative decision-making and knowledge exchange among stakeholders (Ruckelshaus et al. 2015; Congreve and Cross 2019).

At a legal level, amendments are needed to overcome the current absence of an environmental code. It is also important to reorganize related environmental regulations, which in some cases contradict one another. This would also help solve the common problem of delays in the regulation of laws and their application in practice (Capaldo 2018) due to the dispersion of regulations. Likewise, efforts to harmonize legal and institutional frameworks should be undertaken in order to guarantee an effective protection of ES.

The incorporation of criminal sanctions should also be considered (such as the case of Spain, which enacted a criminal environmental responsibility law N° 26/2007), to complement the current environmental regulations which generally fix minor sanctions, unable to promote law-abiding conducts or environmental consciousness to prevent ecosystem damages. For example, in Argentina, fines or civil damage compensations for deforestation were set too low, and they are not functioning in practice (Minaverry 2016).

At a jurisprudential level, it is urgent to achieve a progress, but advancements depend on the existence of efficient regulation that protects ES, which so far is not enough as compared to the regulation of single resources such as water. Furthermore, the judges do not know the concept of ES.

The operationalization of the ES concept is as much a political process as an economic one, which is complicated due to the fact that academics, policy-makers, and society, in general, may not clearly understand the relationship between markets and institutions (Norgaard 2010). Although it behooves researchers to develop shared and pluralistic understandings, some degree of standardized methods for ES measurement and valuation are needed in order to move from theory to practice and address the previous challenges (Portman 2013; Polasky et al. 2015; Jax et al. 2018; Olander et al. 2018; Steger et al. 2018). Communication of ES research to society is also a key issue, such as society becomes supportive of biodiversity and ES regulations and public policies (Minaverry 2020). In this sense, the training of researchers and scholars to engage with society and decision-makers is also relevant, as well as the creation of networks between universities and public institutions aimed at the resolution of specific environmental problems (Ortega Uribe 2014; Minaverry 2020).

Acknowledgments L.N. thanks FONDAP Grant 15150003 and Fondecyt Grant 1190207 for financial support.

P.L. thanks Proyecto Agencia Nacional de Promoción Científica y Técnica (PICT 2015 – 0672 – Préstamo BID).

References

Albarracín-Álvarez OL, Novoa Mahecha D, Rodríguez Peña SM (2019) Elementos de enfoque y estudio de caso para abordar los servicios ecosistémicos en áreas protegidas de la Amazonia colombiana. Documentos de trabajo del Instituto Humboldt. Edición Cambio Climático 4(1):30–50

Alvarez-Garreton C, Lara A, Boisier JP, Galleguillos M (2019) The impacts of native forests and forest plantations on water supply in Chile. Forests 10(6):473

Anderson CB, Celis-Diez JL, Bond BJ, Martinez Pastur G, Little C, Armesto J, Ghersa CM, Austin AT, Schlichter T, Lara A, Carmona M, Chaneton EJ, Gutiérrez J, Rozzi R, Vanderbilt K, Oyarce G, Fernandez Alducin RJ (2012) Progress in creating a joint research agenda that allows networked long-term socio-ecological research in southern South America: addressing crucial technological and human capacity gaps limiting its application in Chile and Argentina. Austral Ecol 37(5):529–536

Arkema KK, Verutes GM, Wood SA, Clarke-Samuels C, Rosado S, Canto M, Rosenthal A, Ruckelshaus M, Guannnel G, Tofj FJ, Silver JM, Griffin R, Guerry AD (2015) Embedding ecosystem services in coastal planning leads to better outcomes for people and nature. Proc Natl Acad Sci 112(24):7390–7395

Baker J, Sheate WR, Phillips P, Eales R (2013) Ecosystem services in environmental assessment—help or hindrance? Environ Impact Assess Rev 40:3–13

Balvanera P, Pérez-Harguindeguy N, Perevochtchikova M, Laterra P, Cáceres DM, Langle-Flores A (2020) Ecosystem services research in Latin America 2.0: expanding collaboration across countries, disciplines, and sectors. Ecosyst Serv 42(101086):579–598

Barrena J, Nahuelhual L, Báez A, Schiappacasse I, Cerda C (2014) Valuing cultural ecosystem services: agricultural heritage in Chiloé island, southern Chile. Ecosyst Serv 7:66–75

Beaumont NJ, Mongruel R, Hooper T (2017) Practical application of the Ecosystem Service Approach (ESA): lessons learned and recommendations for the future. Int J Biodiver Sci Ecosyst Serv Manag 13(3):68–78

Becerra-Rodas C, Little C, Lara A, Sandoval J, Osorio S, Nimptsch J (2019) The role of streamside native forests on dissolved organic matter in forested and agricultural watersheds in northwestern Patagonia. Forests 10(7):595

Benra F, Nahuelhual L, Gaglio M, Gissi E, Aguayo M, Jullian C, Bonn A (2019) Ecosystem services tradeoffs arising from non-native tree plantation expansion in southern Chile. Landsc Urban Plan 190:103589

BSR (Business for Social Responsibility) (2013) Private sector uptake of ecosystem services concepts and frameworks. The current state of play. Available at https://www.bsr.org/en/our-insights/report-view/private-sector-uptake-of-ecosystem-services-concepts-and-frameworks

Capaldo G (2018) Eficacia del Derecho Ambiental. En: Gestión sustentable del agua en ecosistemas, comp. por Capaldo, G. Eudeba, Buenos Aires, pp 53–54

Carreño L, Viglizzo EF (2007) Provisión de servicios ecológicos y gestión de los ambientes rurales en Argentina (No. 333.7614). Instituto Nacional de Tecnología Agropecuaria, Argentina

Chaudhary S, McGregor A, Houston D, Chettri N (2015) The evolution of ecosystem services: a time series and discourse-centered analysis. Environ Sci Pol 54:25–34

Congreve A, Cross ID (2019) Integrating ecosystem services into environmental decision-making. J Appl Ecol 56(3):494–499

Costanza R, De Groot R, Braat L, Kubiszewski I, Fioramonti L, Sutton P, Farber S, Grasso M (2017) Twenty years of ecosystem services: how far have we come and how far do we still need to go? Ecosyst Serv 28:1–16

Daily GC, Matson PA (2008) Ecosystem services: from theory to implementation. Proc Natl Acad Sci 105(28):9455–9456

Delgado LE, Sepúlveda MB, Marín VH (2013) Provision of ecosystem services by the Aysén watershed, Chilean Patagonia, to rural households. Ecosyst Serv 5:102–109

Delgado LE, Marín VH (2015) Ecosystem services: where on earth? Ecosyst Serv 14:24–26

Delgado LE, Marín VH (2016) Well-being and the use of ecosystem services by rural households of the Río Cruces watershed, southern Chile. Ecosyst Serv 21:81–91

Delgado, L.E., Tironi-Silva A & Marín V.H. (2019) Sistemas socio-ecológicos y servicios ecosistémicos: modelos conceptuales para el Humedal del Río Cruces (Valdivia, Chile). Chile. Cerda, C.L., Silva-Rodriguez, E., & Briceño, C. Naturaleza en sociedad: Una mirada a la dimensión humana de la conservación de la biodiversidad. Editorial Ocho Libros, Santiago, Chile. 177–205

De la Barrera F, Bachmann-Vargas P, Tironi A (2015) La investigación de servicios ecosistémicos en Chile: una revisión sistemática. Investigaciones Geográficas 50:3–18

Dick J, Turkelboom F, Woods H, Iniesta-Arandia I, Primmer E, Saarela SR et al (2018) Stakeholders' perspectives on the operationalisation of the ecosystem service concept: results from 27 case studies. Ecosyst Serv 29:552–565

Durán AP, Casalegno S, Marquet PA, Gaston KJ (2013) Representation of ecosystem services by terrestrial protected areas: Chile as a case study. PLoS One 8(12):e82643

Falk T, Spangenberg JH, Siegmund-Schultze M, Kobbe S, Feike T, Kuebler D, Settele J, Vorlaufer T (2018) Identifying governance challenges in ecosystem services management–conceptual considerations and comparison of global forest cases. Ecosyst Serv 32:193–203

Figueroa LM, Gutiérrez RA (2018) Enfrentados por el ambiente: incidencia de las coaliciones sociedad-estado en la protección de bosques nativos. In: Gutiérrez RA (ed) Construir el ambiente: sociedad, estado y políticas ambientales en Argentina. Teseo, Buenos Aires, pp 103–166

Fisher JA, Brown K (2015) Reprint of "Ecosystem services concepts and approaches in conservation: just a rhetorical tool?". Ecol Econ 117:261–269

Gamondes Moyano IM, Morgan RK, Martinez Pastur G (2016) Reshaping forest management in southern Patagonia: a qualitative assessment. J Sustain For 35(1):37–59

Hanson C, Ranganathan J, Iceland C, Finisdore J (2012) Integrating ecosystem services into business performance systems. Available at https://www.wri.org/publication/nature-performance

Henn JJ, Anderson CB, Pastur GM (2016) Landscape-level impact and habitat factors associated with invasive beaver distribution in Tierra del Fuego. Biol Invasions 18(6):1679–1688

Hrabanski M, Bidaud C, Le Coq JF, Méral P (2013) Environmental NGOs, policy entrepreneurs of market-based instruments for ecosystem services? A comparison of Costa Rica, Madagascar and France. Forest Policy Econ 37:124–132

Huertas Herrera AH, Cellini JM, Barrera M, Lencinas MV, Pastur GM (2018) Environment and anthropogenic impacts as main drivers of plant assemblages in forest mountain landscapes of Southern Patagonia. For Ecol Manag 430:380–393

Huertas Herrera AH, Lencinas MV, Toro Manríquez M, Miller JA, Martínez Pastur G (2020) Mapping the status of the North American beaver invasion in the Tierra del Fuego archipelago. PLoS One 15(4):e0232057

Jax K, Furman E, Saarikoski H, Barton DN, Delbaere B, Dick J, Duke G, Görg C, Gómez-Baggethun E, Harrison PA, Maes J, Pérez-Soba M, Saarela SR, Turkelboom F, van Dijk J, Watt AD (2018) Handling a messy world: lessons learned when trying to make the ecosystem services concept operational. Ecosyst Serv 29:415–427

Kirchner M, Schmidt J, Kindermann G, Kulmer V, Mitter H, Prettenthaler F, Rüdisser J, Schauppenlehner T, Schönhart M, Strauss F, Tappeiner U, Tasser E, Schmid E (2015) Ecosystem services and economic development in Austrian agricultural landscapes—the impact of policy and climate change scenarios on trade-offs and synergies. Ecol Econ 109:161–174

Krause MS, Matzdorf B (2019) The intention of companies to invest in biodiversity and ecosystem services credits through an online-marketplace. Ecosyst Serv 40:101026

Kronenberg J (2014) Environmental impacts of the use of ecosystem services: case study of birdwatching. Environ Manag 54(3):617–630

Lara A, Little C, Urrutia R, McPhee J, Álvarez-Garretón C, Oyarzún C, Soto D, Donoso P, Nahuelhual L, Pino M, Arismendi I (2009) Assessment of ecosystem services as an opportunity for the conservation and management of native forests in Chile. For Ecol Manag 258(4):415–424

Laterra P, Barral P, Carmona A, Nahuelhual L (2016) Focusing conservation efforts on ecosystem service supply may increase vulnerability of socio-ecological systems. PLoS One 11(5):e0155019

Laterra P, Nahuelhual L, Gluch M, Sirimarco X, Bravo G, Monjeau A (2019) How are jobs and ecosystem services linked at the local scale? Ecosyst Serv 35:207–218

Lautenbach S, Mupepele AC, Dormann CF, Lee H, Schmidt S, Scholte S, Seppelt R, van Teeffelen A, Verhagen W, Volk M (2019) Blind spots in ecosystem services research and challenges for implementation. Reg Environ Chang:1–22

Levrel H, Cabral P, Feger C, Chambolle M, Basque D (2017) How to overcome the implementation gap in ecosystem services? A user-friendly and inclusive tool for improved urban management. Land Use Policy 68:574–584

Lindenmayer DB, Franklin JF, Lõhmus A, Baker SC, Bauhus J, Beese W, Brodie A, Kiehl B, Kouki J, Martínez Pastur G, Messier C, Neyland M, Palik B, Sverdrup-Thygeson A, Volney J, Wayne A, Gustafsson L (2012) A major shift to the retention approach for forestry can help resolve some global forest sustainability issues. Conserv Lett 5(6):421–431

Little C, Cuevas JG, Lara A, Pino M, Schoenholtz S (2015) Buffer effects of streamside native forests on water provision in watersheds dominated by exotic forest plantations. Ecohydrology 8(7):1205–1217

Loft L, Mann C, Hansjürgens B (2015) Challenges in ecosystem services governance: multi-levels, multi-actors, multi-rationalities. Ecosyst Serv 16:150–157

Lorenzo C, Kelly J, Pastur GM, Saavedra FE, Lencinas MV (2018) How are Argentina and Chile facing shared biodiversity loss? Int Environ Agreem 18(6):801–810

Martinetto P, Alemany D, Botto F, Mastrángelo M, Falabella V, Acha EM, Antón G, Bianchi A, Campagna C, Cañete G, Filippo P, Iribarne O, Laterra P, Martínez P, Negri R, Piola AR, Romero SI, Santos D, Saraceno M (2020) Linking the scientific knowledge on marine frontal systems with ecosystem services. Ambio 49(2):541–556

Martínez Pastur GM, Lencinas MV, Peri P, Moretto A, Cellini JM, Mormeneo I, Vukasovic R (2007) Harvesting adaptation to biodiversity conservation in sawmill industry: technology innovation and monitoring program. J Technol Manag Innov 2(3):58–70

Martínez Pastur G, Peri PL, Lencinas MV, Cellini JM, Barrera M, Soler R, Ivancich H, Mestre L, Moretto A, Anderson C, Pulido F (2013) La producción forestal y la conservación de la biodiversidad en los bosques de Nothofagus en Tierra del Fuego y Patagonia Sur. In: Donoso P, Promis A (eds) Silvicultura en bosques nativos: Avances en la investigación en Chile, Argentina y Nueva Zelanda. Universidad Austral de Chile, Valdivia, Chile, pp 155–179. Capítulo 8, (ISBN 978–956-7173-32-7)

Martínez Pastur G, Peri PL, Lencinas MV, García-Llorente M, Martín-López B (2016a) Spatial patterns of cultural ecosystem services provision in Southern Patagonia. Landsc Ecol 31(2):383–399

Martínez Pastur G, Peri PL, Soler RM, Schindler S, Lencinas MV (2016b) Biodiversity potential of Nothofagus forests in Tierra del Fuego (Argentina): tool proposal for regional conservation planning. Biodivers Conserv 25(10):1843–1862

Martínez Pastur G, Peri PL, Huertas Herrera A, Schindler S, Díaz-Delgado R, Lencinas MV, Soler R (2017) Linking potential biodiversity and three ecosystem services in silvopastoral managed forest landscapes of Tierra del Fuego, Argentina. Int J Biodiver Sci Ecosyst Ser Manag 13(2):1–11

Martínez Pastur G, Rosas YM, Toro Manríquez T, Huertas Herrera A, Miller JA, Cellini JM, Barrera MD, Peri P, Lencinas MV (2019) Knowledge arising from long-term research of variable retention harvesting in Tierra del Fuego: where do we go from here? Ecol Process 8(1):24

Martínez Pastur G, Schlichter T, Matteucci SD, Gowda JH, Huertas Herrera A, Toro Manríquez M, Lencinas M, Cellini J, Peri P (2020) Synergies and trade-offs of national conservation policy and agro-forestry management over forest loss in Argentina during the last decade. In: Latin America in times of global environmental change. Springer, Cham, Switzerland, pp 135–155. (ISBN 978-3-030-24253-4)

Matzek V, Wilson KA, Kragt M (2019) Mainstreaming of ecosystem services as a rationale for ecological restoration in Australia. Ecosyst Serv 35:79–86

McKenzie E, Irwin F, Ranganathan J, Hanson C, Kousky C, Bennett K et al (2011) Incorporating ecosystem services in decisions. In: Natural capital: theory and practice of mapping ecosystem services. Oxford University Press, Oxford, UK, pp 339–355

Mendoza M (2016) Educational policing: park rangers and the politics of the green (E) state in Patagonia. J Latin Amer Carib Anthrop 21(1):173–192

Minaverry C, Martínez A (2018) Los servicios ecosistémicos culturales y su interrelación con el Derecho Ambiental. El caso de Argentina. 56° Congreso Internacional de Americanistas, Memoria del 56° Congreso Internacional de Americanistas, VoluMEN 12: Estudios Sociales, Manuel Alcántara, Mercedes García Montero y Francisco Sánchez López (Coord.), Editorial Universidad de Salamanca, 534–545

Minaverry C (2016) El Derecho Penal Ambiental a la luz de un reciente precedente jurisprudencial sobre desmontes ilegales en la Provincia de Salta, Argentina, Erreius eonline, Editorial Errepar, publicado el 12/08/2016

Minaverry C (2020) El reconocimiento de los servicios ecosistémicos culturales en el ámbito científico del Derecho Ambiental y de las Ciencias Sociales. Aplicación de sus contribuciones al caso de la protección jurídica de los bosques nativos en Norpatagonia argentina y chilena. Revista Lex de la Facultad de Derecho y Ciencias Políticas 18(25):109–137

Mrotek A, Anderson CB, Valenzuela AE, Manak L, Weber A, Van Aert P, Malizia M, Nielsen EA (2019) An evaluation of local, national and international perceptions of benefits and threats to nature in Tierra del Fuego National Park (Patagonia, Argentina). Environ Conserv 46(4):326–333

Nahuelhual L, Carmona A, Lozada P, Jaramillo A, Aguayo M (2013) Mapping recreation and ecotourism as a cultural ecosystem service: an application at the local level in Southern Chile. Appl Geogr 40:71–82

Nahuelhual L, Carmona A, Aguayo M, Echeverria C (2014) Land use change and ecosystem services provision: a case study of recreation and ecotourism opportunities in southern Chile. Landsc Ecol 29(2):329–344

Nahuelhual L, Ochoa FB, Rojas F, Díaz GI, Carmona A (2016) Mapping social values of ecosystem services: what is behind the map? Ecol Soc 21(3)

Nahuelhual L, Vergara X, Kusch A, Campos G, Droguett D (2017) Mapping ecosystem services for marine spatial planning: recreation opportunities in Sub-Antarctic Chile. Mar Policy 81:211–218

Nahuelhual L, Saavedra G, Henríquez F, Benra F, Vergara X, Perugache C, Hasen F (2018) Opportunities and limits to ecosystem services governance in developing countries and indigenous territories: the case of water supply in Southern Chile. Environ Sci Pol 86:11–18

Nahuelhual L, Vergara X, Bozzeda F, Campos G, Subida MD, Outeiro L, Lillasante S, Fernández M (2020) Exploring gaps in mapping marine ecosystem services: a benchmark analysis. Ocean Coast Manag 192:105193

Norgaard RB (2010) Ecosystem services: from eye-opening metaphor to complexity blinder. Ecol Econ 69(6):1219–1227
Núñez D, Nahuelhual L, Oyarzún C (2006) Forests and water: the value of native temperate forests in supplying water for human consumption. Ecol Econ 58(3):606–616
Olander L, Urban D, Johnstone RJ, van Houtven G, Kagan J (2016) Proposal for increasing consistency when incorporating ecosystem services into decision making. Duke University. National Ecosystem Services Partnership Policy Brief, (16-01)
Olander LP, Johnston RJ, Tallis H, Kagan J, Maguire LA, Polasky S, Urban D, Boyd J, Wainger L, Palmer M (2018) Benefit relevant indicators: ecosystem services measures that link ecological and social outcomes. Ecol Indic 85:1262–1272
Orenstein DE, Groner E (2014) In the eye of the stakeholder: changes in perceptions of ecosystem services across an international border. Ecosyst Serv 8:185–196
Ortega Uribe T (2014) Estudios transdisciplinarios en socio-ecosistemas: reflexiones teóricas y su aplicación en contextos latinoamericanos. Investigación Ambiental, VI 2:109–122
Outeiro L, Villasante S (2013) Linking salmon aquaculture synergies and trade-offs on ecosystem services to human wellbeing constituents. Ambio 42(8):1022–1036
Outeiro L, Gajardo C, Oyarzo H, Ther F, Cornejo P, Villasante S, Ventine LB (2015a) Framing local ecological knowledge to value marine ecosystem services for the customary sea tenure of aboriginal communities in southern Chile. Ecosyst Serv 16:354–364
Outeiro L, Häussermann V, Viddi F, Hucke-Gaete R, Försterra G, Oyarzo H, Kosiel K, Villasante S (2015b) Using ecosystem services mapping for marine spatial planning in southern Chile under scenario assessment. Ecosyst Serv 16:341–353
Parguel B, Benoît-Moreau F, Russell CA (2015) Can evoking nature in advertising mislead consumers? The power of 'executional greenwashing'. Int J Advert 34(1):107–134
Pastén R, Olszynski M, Hantke-Domas M (2018) Does slow and steady win the race? Ecosystem services in Canadian and Chilean environmental law. Ecosyst Serv 29:240–250
Pauna VH, Picone F, Le Guyader G, Buonocore E, Franzese PP (2018) The scientific research on ecosystem services: a bibliometric analysis. Ecol Quest 29(3):53–62
Peixotto J, Godfrid J, Stevenson H (2019) La difusión del concepto de servicios ecosistémicos en la Argentina: alcances y resistencias. Revista SAAP: Sociedad Argentina de Análisis Político 13(2):313–340
Perevochtchikova M, Flores JÁH, Marín W, Flores AL, Bueno AR, Negrete IAR (2019) Systematic review of integrated studies on functional and thematic ecosystem services in Latin America, 1992–2017. Ecosyst Serv 36:100900
Peri PL, Lencinas MV, Martínez Pastur G, Wardell-Johnson GW, Lasagno R (2013) Diversity patterns in the steppe of Argentinean southern Patagonia: environmental drivers and impact of grazing. In: Prieto MBM, Díaz JT (eds) Steppe ecosystems: biological diversity, management and restoration, Series: Environmental research advances, Chapter 4. Nova Science Publishers, Hauppauge, New York, pp 73–95. (ISBN 978-1-62808-298-2)
Peri PL, Ladd B, Lasagno RG, Pastur GM (2016a) The effects of land management (grazing intensity) vs. the effects of topography, soil properties, vegetation type, and climate on soil carbon concentration in Southern Patagonia. J Arid Environ 134:73–78
Peri P, Caballe G, Hansen N, Bahamonde HA, Lencinas MV, Von Müller A, Ormaechea SG, Gargaglione VB, Soler E, Rosina M, Sarasola MM, Rusch VE, Borrelli LB, Fernández ME, Gyenge J, Tejera LE, Lloyd CE, Martínez Pastur G (2016b) Silvopastoral systems under native forest in Patagonia Argentina. In: Peri PL, Dube F, Varella A (eds) Silvopastoral systems in Southern South America, Series: Advances in agroforestry 11, Chapter 6. Springer, Bern, Switzerland, pp 117–168
Peri PL, López DR, Rusch V, Rusch G, Rosas YM, Martínez Pastur G (2017) State and transition model approach in native forests of Southern Patagonia (Argentina): linking ecosystem services, thresholds and resilience. Int J Biodiver Sci Ecosyst Serv Manag 13(2):105–118

Peri PL, Rosas YM, Ladd B, Toledo S, Lasagno RG, Martínez Pastur G (2018) Modelling soil carbon content in South Patagonia and evaluating changes according to climate, vegetation, desertification and grazing. Sustainability 10(2):438

Peri PL, Rosas YM, Ladd B, Toledo S, Lasagno RG, Martínez Pastur G (2019) Modeling soil nitrogen content in South Patagonia across a climate gradient, vegetation type, and grazing. Sustainability 11(9):2707

Polasky S, Tallis H, Reyers B (2015) Setting the bar: standards for ecosystem services. Proc Natl Acad Sci 112(24):7356–7361

Ponce RD, Vásquez F, Stehr A, Debels P, Orihuela C (2011) Estimating the economic value of landscape losses due to flooding by hydropower plants in the Chilean Patagonia. Water Resour Manag 25(10):2449

Portman ME (2013) Ecosystem services in practice: challenges to real world implementation of ecosystem services across multiple landscapes–a critical review. Appl Geogr 45:185–192

Posner SM, McKenzie E, Ricketts TH (2016) Policy impacts of ecosystem services knowledge. Proc Natl Acad Sci 113(7):1760–1765

Ranganathan J, Raudsepp-Hearne C, Lucas N, Irwin F, Zurek M, Bennett K, Ash N, West P (2008) Ecosystem services: a guide for decision makers. World Resources Institute, Washington, DC, 11(1)

Rodríguez-Echeverry J, Echeverría C, Oyarzún C, Morales L (2018) Impact of land-use change on biodiversity and ecosystem services in the Chilean temperate forests. Landsc Ecol 33(3):439–453

Rodríguez-Echeverry J, Echeverría C, Oyarzún C, Morales L (2017) Spatial congruence between biodiversity and ecosystem services in a forest landscape in southern Chile: basis for conservation planning. Bosque 38(3):495–506

Roldán VA, Villasante S, Outeiro L (2015) Linking marine and terrestrial ecosystem services through governance social networks analysis in Central Patagonia (Argentina). Ecosyst Serv 16:390–402

Román B, Nahuelhual L, Morey F (2009) Programa de educación ambiental para comités de agua potable rural en el sur de Chile. Educere 13(45):523–529

Rosas YM, Peri PL, Bahamonde HA, Cellini JM, Barrera MD, Huertas Herrera A, Lencinas MV, Marínez Pastur G (2019) Trade-offs between management and conservation for the provision of ecosystem services in the southern Patagonian forests. In: Stanturf J (ed) Achieving sustainable management of boreal and temperate forests. Burleigh Dodds Science Publishing, Cambridge, UK. Chapter 6, 33 pp. (ISBN 178-676-2927)

Rosas YM, Peri PL, Carrasco J, Lencinas MV, Pidgeon AM, Politi N, Martinuzzi, Martínez Pastur G (2021) Improving potential biodiversity and human footprint in *Nothofagus* forests of Southern Patagonia through the spatial prioritization of their conservation values. In: Shit PK, Pourghasemi H, Das P, Bhunia GS (eds) Spatial modelling in forest resources management, Springer-Nature: Environmental Science and Engineering. Apartado: Rural livelihood and sustainable development. Springer, Cham, Switzerland, pp 441–471. (ISBN 978-3-030-56541-1)

Ruckelshaus M, McKenzie E, Tallis H, Guerry A, Daily G, Kareiva P, Polasky S, Ricketts T, Bhagabati N, Wood S, Bernhardt J (2015) Notes from the field: lessons learned from using ecosystem service approaches to inform real-world decisions. Ecol Econ 115:11–21

Saarikoski H, Primmer E, Saarela SR, Antunes P, Aszalós R, Baró F, Berry P, Garcia Blanko G, Goméz-Baggethun E, Carvalho L, Dick J, Dunford R, Hanzu M, Harrison PA, Izakovicova Z, Kertész M, Kopperoinen L, Köhler B, Langemeyer J, Lapola D, Liquete C, Luque S, Mederly P, Niemelä J, Palomo I, Martinez Pastur G, Peri P, Preda E, Priess JA, Santos R, Schleyer C, Turkelboom F, Vadineanu A, Verheyden W, Vikström S, Young J (2018) Institutional challenges in putting ecosystem service knowledge in practice. Ecosyst Serv 29:579–598

Soler R, Peri PL, Bahamonde H, Gargaglione V, Ormaechea S, Herrera AH et al (2018) Assessing knowledge production for agrosilvopastoral systems in South America. Rangel Ecol Manag 71(5):637–645

Soler R, Rumpf SB, Schindler S, Martinez Pastur G, Barrera M, Cellini JM, Pérez Flores M, Essl F, Rabitsch W, Lencinas MV (2019) Twelve-year dynamics of alien and native understorey plants following variable retention harvesting in Nothofagus pumilio forests in Southern Patagonia. For Ecol Manag 449:117447

Steger C, Hirsch S, Evers C, Branoff B, Petrova M, Nielsen-Pincus M, Wardropper C, van Riper CJ (2018) Ecosystem services as boundary objects for transdisciplinary collaboration. Ecol Econ 143:153–160

TEEB (The Economics of Ecosystems and Biodiversity) (2012) In: Wittmer H, Gundimeda H (eds) The economics of ecosystems and biodiversity in local and regional policy and management. Routledge

Turkelboom F, Leone M, Jacobs S, Kelemen E, García-Llorente M, Baró F, Termansen M, Barton DN, Berry P, Stange E, Thoonen M, Kalóczkai A, Vadineanu A, Castro AJ, Czúcz B, Röckmann C, Wurbs D, Odee D, Preda E, Gómez-Baggethun E, Rusch GM, Martínez Pastur G, Palomo I, Dick J, Casaer J, van Dijk J, Priess JA, Langemeyer J, Mustajoki J, Kopperoinen L, Baptist MJ, Peri P, Mukhopadhyay R, Aszalós R, Roy SB, Luque S, Rusch V (2018) When we cannot have it all: ecosystem services trade-offs in the context of spatial planning. Ecosyst Serv 29:566–578

Verburg R, Selnes T, Verweij P (2016) Governing ecosystem services: national and local lessons from policy appraisal and implementation. Ecosyst Serv 18:186–197

Virapongse A, Brooks S, Metcalf EC, Zedalis M, Gosz J, Kliskey A, Alessa L (2016) A social-ecological systems approach for environmental management. J Environ Manag 178:83–91

Wei H, Fan W, Wang X, Lu N, Dong X, Zhao Y, Ya X, Zhao Y (2017) Integrating supply and social demand in ecosystem services assessment: a review. Ecosyst Serv 25:15–27

Wilkinson K (2014) Payment for 'ecosystem services' and the 'green economy': green-washing or something new? J Human Rights Environ 5(2):168–191

Zagarola JPA, Anderson CB, Veteto JR (2014) Perceiving Patagonia: an assessment of social values and perspectives regarding watershed ecosystem services and management in southern South America. Environ Manag 53(4):769–782

Zagarola JPA, Martínez Pastur G, Lopez ME, Anderson CB (2017) Assessing the effects of urbanization on streams in Tierra del Fuego. Ecol Austral 27:45–54

Chapter 21
Natural Capital and Local Employment in Argentine Patagonia

Pedro Laterra, Laura Nahuelhual, Ximena Sirimarco, Adrián Monjeau, Mariana Gluch, and Gonzalo Bravo

Abstract Employment is a largely neglected indirect benefit in ecosystem services (ES) assessments, despite its relevance in countries with abundant rural landscapes. We present an analysis of the spatial connections between landscape composition (as a proxy of natural capital and ES) and employment level as measured by the number of people employed in tourism and gastronomy, crop production, animal production, forestry, and fishery production in the Argentinean Patagonia. We assessed landscape composition through land cover analysis within a buffer around the main city of each municipality, and we classified the municipalities in three groups according to their proximity to main national highways at Patagonia (R3, R22, R40): (i) R3 with dominance of rangelands and ocean coasts, (ii) R22 with dominance of rangelands and irrigated crops (mainly apples, pears, grapes, toma-

P. Laterra (✉) · A. Monjeau
Fundación Bariloche, San Carlos de Bariloche, Río Negro, Argentina

Consejo Nacional de Investigaciones Científicas y Técnicas (CONICET), CCT Patagonia Norte, San Carlos de Bariloche, Río Negro, Argentina

L. Nahuelhual
Instituto de Economía, Universidad Austral de Chile Centro de Investigación: Dinámica de Ecosistemas Marinos de Altas Latitudes Instituto Milenio en Socio-Ecología Costera, Valdívia, Chile

X. Sirimarco
Facultad de Ciencias Agrarias Universidad Nacional de Mar del Plata, Balcarce, Provincia de Buenos Aires, Argentina

M. Gluch
Fundación Bariloche, San Carlos de Bariloche, Río Negro, Argentina

G. Bravo
Fundación Bariloche, San Carlos de Bariloche, Río Negro, Argentina

Sede Andina Universidad Nacional de Río Negro, San Carlos de Bariloche, Río Negro, Argentina

P. L. Peri et al. (eds.), *Ecosystem Services in Patagonia*, Natural and Social Sciences of Patagonia, https://doi.org/10.1007/978-3-030-69166-0_21

toes, and lucerne), and (iii) R40 with a combination of rangelands, forests, and lakes. Finally, we estimated the portion of employment in each economic subsector attributable to different land covers within each municipality group using generalized additive models and specific employment multiplier coefficients. Our results reveal that land covers explain approximately half of the variation in inter-municipality employment in the different economic subsectors. We found strong differences in employment level among municipality groups, as well as among economic subsectors within groups, which we discussed considering the particular importance of different land cover types, as a measure of natural capital and ES, for employment generation.

Keywords Ecosystem services · Indirect benefits · Benefits distribution · Employment multipliers · Tourism · Agriculture

1 Introduction

Natural resources and ecosystem services (ES) are a part of the real wealth of nations (OCDE 2011; Andersen et al. 2018). They contribute toward fiscal revenue, income, and poverty reduction. Sectors related to natural resources use provide jobs and are often the basis of livelihoods in poor communities (OCDE 2011).

The relationship between economic development and abundance of natural resources has received extensive academic and policy interest. Academic research, however, is dominated by the disciplines of economics, and the related fields of management and business mostly focused on the notion of the "resource curse" which emphasizes governance problems related to corruption and different aspects of monetary policy in natural resource-abundant countries (Manzano and Gutiérrez 2019).

Whereas the ES approach is essentially about connecting the natural capital of ecosystems and human benefits, most ES assessments fail to integrate the benefits that accrue to second-order or indirect beneficiaries through the generation of income and employment along ES value chains (Daw et al. 2011; Rawlins et al. 2018). Instead, they focus only on the benefits that accrue to "first-order" or direct beneficiaries, through use, consumption, and/or enjoyment of ES (Martín-López et al. 2014; Jacobs et al. 2016; Pascual et al. 2017).

Basic employment is a clear and vital requirement for human well-being. While the act of employment may have little relationship with ES, numerous job types are directly related to these. For example, employment involved with agriculture and food production (e.g., apple picking), food distribution (Daily et al. 1998), forestry (Daily et al. 1997), green architecture and design (Jackson 2003), and environmental protection (Daily 2000) has a dependence upon basic ecosystem services

(Summers et al. 2012). Ecosystem services sustain agriculture, which feeds the world and provides thousands of jobs in developing countries, e.g., by regulating the water balance, stabilizing the microclimate, and pollinating crops. An estimated three out of four jobs that make up the global workforce are either heavily or moderately dependent on water. This implies that water shortages and access limitations to water and sanitation would limit economic growth and job creation in the next decades (WWAP 2016). Ecosystem services also sustain nature-based tourism, which is one of the fastest-growing economic activities in Patagonia, both in Chile and Argentina, generating thousands of jobs. If these jobs are not considered in the evaluation of policies, decisions made can have detrimental effects on human well-being. Furthermore, metrics such as employment can provide stronger arguments for nature conservation than current measures of economic or social values of ES accruing only to first-order beneficiaries (Laterra et al. 2019).

In this chapter, we focus on the relationships between capital natural and jobs related to economic sectors that rely on natural resources, namely, tourism, crop production, animal production, poultry, beekeeping, forestry, hunting, fishery production, and associated activities, which by definition are mediated by ecosystem services supply and capture. Quantifying the importance of natural capital and ES in employment creation is a complex matter and often can only be illustrated with reference to examples. Sound estimates exist for certain industries, such as agriculture and forestry, but not for others such as small-scale fishery production or tourism, which have a large informal component, particularly in developing countries. Some efforts include the extended version of the input-output (I-O) models in order to incorporate the influence of several types of ES (e.g., Grêt-Regamey and Kytzia 2007), the characterization of the local circulation of money in forestry operations using network analysis and Markov chains (e.g., Kelly et al. 2014), the estimation of income and employment multiplier effects of the economic activity in different sectors (Briassoulis 1991; Miller and Blair 2009), and the application of the ES value chain framework (Rawlins et al. 2018).

This chapter is based on a previous study by Laterra et al. (2019) that comprised Argentina as a whole and their main ecoregions. In this opportunity, we offer a disaggregated analysis of the Argentinean Patagonia to illustrate the extent to which natural capital (as reflected by landscape composition) influences local employment at municipality scale and how these influences vary within the region.

We build on the premise that ES mediate the relationships between natural capital and human benefits (Potschin-Young et al. 2018). However, since ES may not be directly observable constructs, here we interpret the relationships between natural capital (as represented by landscape composition) and employment as the contribution of ES to indirect benefits (Fig. 21.1). This "black box" approach (in the sense that ES are not directly observed and measured) limits the understanding of the complex mechanisms that explain how nature matters to the dynamics of employment in nature-based industries and sectors (e.g., circuits of natural capital accumu-

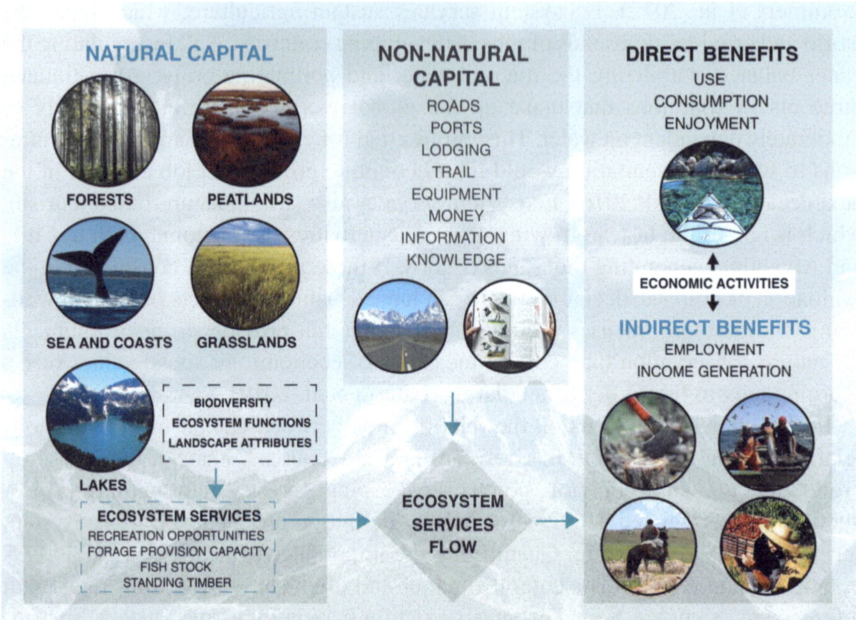

Fig. 21.1 The path from natural capital to jobs due to ecosystem services (observed variables in this chapter, in blue), mediated by the supply and capture of ecosystem services (unobserved variables, in black)

lation) (see Boyd et al. 2001). However, it has two important advantages in our case: (i) it does not require a disaggregated quantification and mapping of ES, and (ii) it allows centering the discussion on the management of ecosystems rather than on the management of individual ES or bundles of them. This approach is also more in line with previous research that has directly related natural resources stocks and flows with development and employment generation.

2 Materials and Methods

We collected secondary information on employment from different economic subsectors and landscape composition in terms of relative cover of main vegetation and land-use types at municipality scales. Then we explored the degree to which employment can be explained by landscape composition, across three municipality groups belonging to three contrasting zones of the Argentine Patagonia.

2.1 Study Site: General Description, Zoning, and Selection of Sampling Units

The Argentinean Patagonia occupies a wide climate range (2500 to less than 200 mm $yr.^{-1}$ of annual precipitation), but it is distinguished by a cold and semi-arid climate in most of the territory, which prevents extensive crop production, as well as by the presence of extractive activities, such as mining and oil production (Paruelo et al. 1998). It also includes a relatively small portion of the Andean-Patagonian Forest region along the Andes mountain system, shared with Chile, which has a dominant forest cover, mostly within the 2000–1000 mm $yr.^{-1}$ annual precipitation range (Kitzberger 2012). Due to its large extension, its low human population density, and its high portion of original habitat retained (intactness), this region is recognized as one of the main wilderness areas in the world (Mittermeier et al. 2003). Most of the area is covered by rangelands (steppes, grasslands, and shrublands), except for the Andean zone, characterized by a mix of temperate forests, rangelands and abundance of water bodies (mostly rivers and lakes), and some irrigated valleys like the Valley of Río Negro, the most important agricultural area within the region. In terms of tourism, Patagonia exhibits the highest hotel occupation levels in relation to their population size, as compared to the country. Lakes, dams, water reservoirs, and rivers constitute important tourist attractions in this region. The touristic activity concentrates on the municipalities near the region's national parks, along the Patagonian forests, as well as on the seacoast.

The geographic heterogeneity of our study site raises the difficulty of controlling many other factors, which may explain the level of employment, besides natural capital. In order to reduce such heterogeneity, we performed separate analysis for three major zones: Coastal Patagonia, dominated by rangelands and marine coasts; the main irrigated valley (Valley of Río Negro) dominated by rangelands, perennial crops, and orchards; and the Andean Patagonia covered by combinations of rangelands, mountain forests, and lakes. Three highways respectively traverse these

Table 21.1 Groups of selected municipalities, elaborated from the 2010 census data (INDEC 2010)

	Municipal groups		
	Route 3	Route 22	Route 40
Number of selected municipalities	32	31	53
Mean population size	17,620	11,431	3755
Number of employees in tourism and gastronomy	5811	1301	2776
Number of employees in agriculture	4928	8807	3458
Number of employees in animal production, beekeeping, poultry, forestry, and hunting	4366	1507	2245
Number of employees in fishery production	1584	33	37

The number of employees is the average of employees for each economic subsector within the municipal group

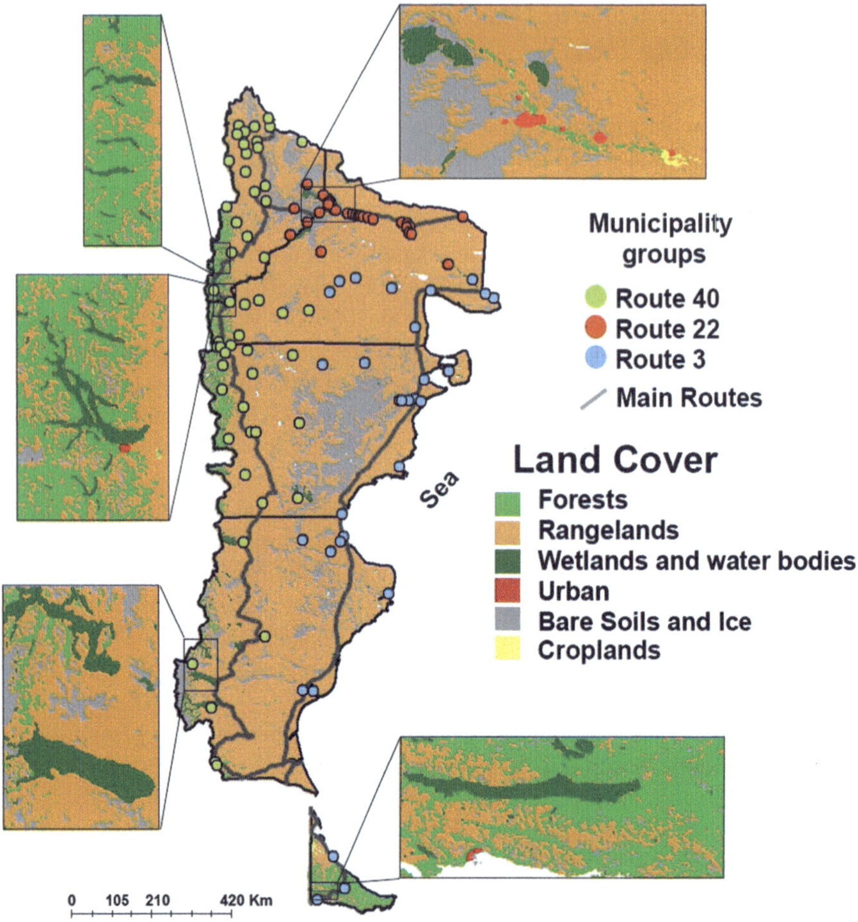

Fig. 21.2 Land cover categories of Argentine Patagonia. Thick lines indicate the three main highways, and colored circles indicate the location of the main cities of each of the municipalities selected along those highways. Some zoomed areas of wetlands and water bodies are shown in boxes

major zones: the National Route 3 (R3), National Route 22 (R22), and National Route 40 (R40) (Table 21.1 and Fig. 21.2).

Our sampling units consisted of all the municipalities within the study site, excluding those with more than 100,000 inhabitants in order to avoid the "large-city" phenomena. The selected municipalities were classified into one of three groups according to their relative proximity to the three abovementioned highways.

2.2 Procedures

Methodological procedures involved three steps (see Laterra et al. 2019 for details): (i) description of landscape composition within a buffer around each of the municipalities' main city, (ii) estimation of the portion of employment (number of people) in each economic subsector attributable to the landscape composition within each municipality group, and (iii) the estimation of overall employment at municipality scale attributed to ES, by considering propagation effects on the rest of the economic subsectors.

Data on landscape composition came from SERENA land cover map for Latin America and the Caribbean (Blanco et al. 2013), based on daily MODIS-Terra surface reflectance data (MOD09GA of collection 5) for the year 2008. Different categories of original cover data were merged into six land cover classes, namely, forests (merging all forest types), rangelands (merging all savannas, shrublands, and grassland types), wetlands and other water bodies (excluding marine ecosystems), croplands, ice (merging ice and snow and bare soil covers), and sea. We assessed land covers for a circular buffer of 20 km radius around the municipalities' main cities, which represents half of the mean distance from the head city center to the municipality's edge and controls the shape and size of the municipalities. This scale showed the best explanatory capacity in a previous study (Laterra et al. 2019) in agreement with previous reports on suitable ES flow propagation area (e.g., Fisher et al. 2009; Serna-Chavez et al. 2014; Schröter et al. 2018). Landscape composition is one of the main data sources used to construct ES indicators (Nahuelhual et al. 2015), thus we are confident in its suitability as a proxy of ES supply.

We calculated employment at municipality level based on the last National Population Census of 2010 (INDEC 2010) for four economic subsectors: (i) tourism and gastronomy (hereafter, tourism); (ii) agricultural production; (iii) livestock, beekeeping, poultry, forestry, and hunting (hereafter, animal production and forestry); and (iv) fishery production. The level of aggregation of census data does not allow for the discrimination between the levels of employment in gastronomy and those in the tourism sector, or between nature-based tourism and other types of tourism at the scale of the analysis. This is not an impediment to answer our research questions, as it is not our intention to discriminate between different types of tourism.

In order to explore the relationships between the employment (response variable) and landscape composition, we used a series of generalized additive models (GAM, Hastie and Tibshirani 1987), which do not assume any specific structure for the relationships. We specified GAM models using a Poisson error distribution and a log link. In order to avoid overfitting and spuriously high levels of explained variation, we limited the complexity of the fitted curves to two effective degrees of freedom. We modeled employment in absolute terms (number of jobs at municipality level) and omitted one of the land cover categories (percent of urban areas plus bare soil) from models in order to avoid collinearity. To avoid estimating some large and difficult-to-interpret constants, we forced the curves through the origin.

We considered the propagation effects of employment in the four subsectors of economy on the rest of employments, that is, the number of jobs that these four i sectors activate in the remaining economic sectors, by using multiplicative factors (EMF_i) estimated by Laterra et al. (2019). EMF_i represents the overall number of expected employed persons per each employed person in the economic subsector i, so it is an indirect measure of the full effect of natural capital into the economy of the Patagonian region. We then estimated landscape composition influences on employment through the overall multiplier factor corrected by the portion of the employment in the i economic subsector (E_i) explained by landscape composition (EP_i), as reflected by the R^2 values of the corresponding fitted model, so.

$$IB_i = (E_i * EP_i) + (E_i * EP_i * EMF_i) \tag{1}$$

where IB_i is the total number of jobs explained by landscape composition through the i economic subsector. Given the variation in population sizes between municipalities, for comparison purposes, IB was relativized as follows:

$$RIB_i = IB_i / E_i = EP_i * (1 + EMF_i) \tag{2}$$

where RIB_i is the number of employments attributable to landscape composition per each employment in the i economic subsector.

Table 21.2 Generalized additive models (GAM) of number of employees in different economic subsectors in response to landscape composition (excluding urban area) for the different municipality groups

Municipality group	Dependent variables (Employees by economic subsector)	N	R^2	GAM coefficients[a] Land cover categories: Forests	Rangelands	Surface water bodies	Croplands	Sea
Route 3	Tour	32	0.34	**7.11**	**3.94**	**−5.19**	**27.02**	**3.44**
	Agr		0.25	**5.88**	**3.88**	**−11.06**	**42.99**	**4.05**
	Anp		0.33	**5.98**	**3.81**	**−12.35**	**4.24**	**3.88**
	Fish		0.48	**2.04**	**2.32**	**2.19**	**0.56**	**4.58**
Route 22	Tour	31	0.50	**2.37**	**1.33**	**5.74**	**12.21**	–[b]
	Agr		0.77	**8.11**	**3.26**	**3.23**	**4.37**	–[b]
	Anp		0.67	**5.59**	**2.40**	**0.48**	**3.23**	–[b]
	Fish		0.83	−6.53	**0.52**	**10.13**	−1.80	–[b]
Route 40	Tour	53	0.78	**2.86**	**2.10**	**5.80**	**−1.51**	–[b]
	Agr		0.64	**5.05**	**2.33**	**0.49**	**−0.62**	–[b]
	Anp		0.23	**1.81**	**2.60**	**2.69**	**11.71**	–[b]
	Fish		0.67	**2.62**	**−1.63**	**1.79**	−1.16	–[b]

[a]Significant ($p < 0.05$) correlation coefficients are shown in bold letter
[b]Sea cover is absent within this data set
Tour tourism, *Agr* agricultural production, *Anp* animal production and forestry, *Fish* fishery production

3 Results

3.1 *GAM Models of Employment: Landscape Composition Relationships*

According to GAM, a significant portion of inter-municipal variation of employment in the selected economic subsectors was related to landscape composition, within all the different municipality groups and within the complete data set (Table 21.2). Although a minimum freedom degree was adopted to avoid GAM overfitting, adjusted R^2 values indicate that the 12 explored models were able to explain more than 20% variation in specific subsector employments, and half of them were able to explain more than 60% of that variation.

The explanatory capacity of the models was highly variable among different employment sources and regions (see R^2 values in Table 21.2). For example, while only 34% and 50% of variation in employment in tourism along R3 and R22, respectively, was explained by landscape composition (mostly agriculture cover), it reached 78% for the group of municipalities along R40, where it was mainly associated to water bodies (mostly Andean lakes) combined with temperate forests and negatively associated to croplands. It is consistent with main attractions of those

Fig. 21.3 Images from different Patagonian landscapes: (**a**) seaside localities only found along R3, (**b**) the main irrigated valley along R22, (**c**) the typical combination of clear lakes and dense native forests along R40, and (d) the dominant rangeland matrix that is common to R3, R22, and R40 landscapes

touristic corridors, consisting in orchard farms, vineyards, and wineries in R3 or the natural beauties of R40 (Fig. 21.3).

Employment within the agriculture subsector was better explained by landscape composition within the R22 and R40 than within the R3 municipality groups (Table 21.2). Along R22, jobs in agriculture were more influenced by forest cover than cropland cover, probably because the presence of parks, riparian forests, and cultivated forest curtains is indicative of fruit production, which is the most labor-intensive activity in the zone, in contrast with fodder crops, which are less associated to forest cover and less labor-intensive activity. We have not found a reasonable explanation for the importance of the employment level linked to forest cover in the agriculture subsector along R40 municipalities.

Animal production on rangelands typically consists of extensive sheep raising and, to a much lesser extent, cow raising, which have low labor demand. Thus, it is not surprising that rangeland cover represented a medium weight within models of employment in animal production and forestry and mainly related to forest cover along R22 and R3, where planted forests are indicative of cities with intensive production of small animals (chickens, pigs), slaughterhouses, and beekeeping. However, we do not have a reasonable explanation for the high weight shown by cropland cover on this model for the R40 data set.

Employment in fishery production was clearly related to marine environments along R3, probably due to commercial fishing activity in various coastal locations, and it is also related to surface water bodies along the R22, where the Río Negro sustains both sport and commercial (frequently illegal) fishing. Employment in the fishery subsector along R40 is related to both forests and surface water bodies (mountain rivers and lakes) as indicative of the importance of sport fishing, an activity based on introduced salmonid species.

3.2 Propagation of Nature-Based Employments into the Rest of the Economy

Total employment due to ES (Fig. 21.3d) represents an important portion of the employment in the selected economic subsectors within Argentine Patagonia. Overall, the employment in selected economic subsectors explained by landscape composition was between 50% and 100% higher than the unexplained employment within the R22 and R40 municipality groups and 50% lower in the case of R3 municipalities (Fig. 21.3d). Furthermore, employment resulting from the multiplicative effect of the ES-related employment on the rest of the economy represented approximately 40% of total IB for R3 and R40 and 70% for R22. Employment in fishery production was not included in this analysis because a reliable employment multiplier factor was not available for this subsector.

Employment due to ES showed different patterns when it is disaggregated into different economic subsectors. For example, while in the R40 municipalities the contribution of employment in tourism due to ES was 3.5 times higher than the

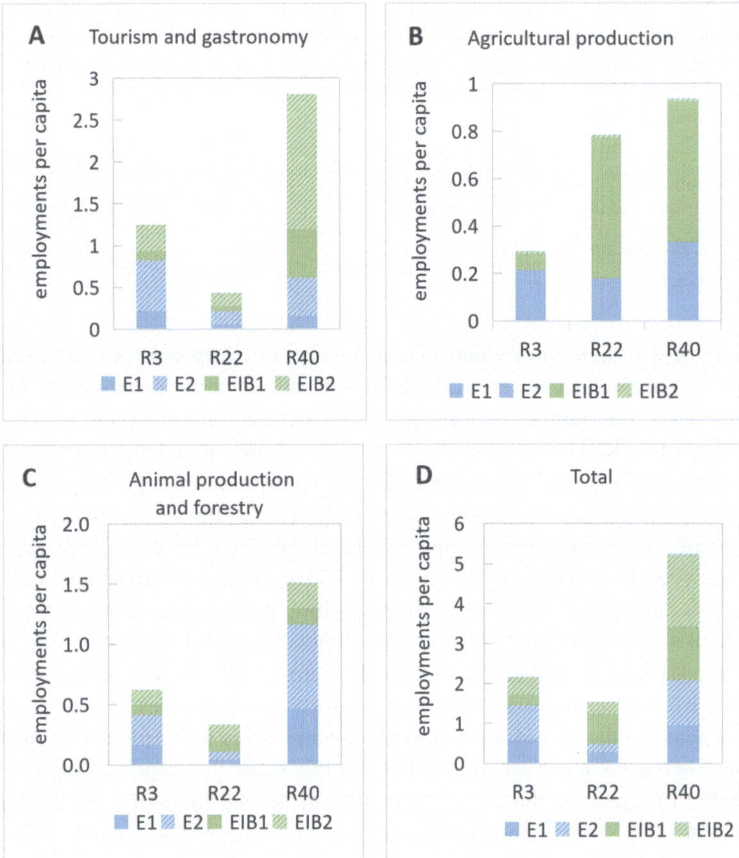

Fig. 21.4 Number of employments per capita in tourism and gastronomy (tourism, **a**), agricultural production (**b**); animal production, forestry, beekeeping, poultry, and hunting (animal production, **c**); and the three economic subsectors (**d**), within different municipality groups along main highways and landscapes (R3, R22, and R40). Employments are disaggregated into those explained by ecosystem services (indirect benefits, IB) and those due to other sources (E). In turn, IB are disaggregated into the employments directly related to landscape composition (IB1) and those that result from the multiplicative effect of IB1 on the rest of the economy (IB2)

contribution of employment from the same subsector due to other sources, employment in animal production and forestry due to ES was more than three times lower due to other sources (compare Fig. 21.4a, b). Within the agricultural subsector, there were important contributions of employment due ES in the R22 and R40 groups (Fig. 21.4b) in comparison to the overall pattern (Fig. 21.4d). This was associated with the high dependence on irrigated areas of the labor-intensive agricultural systems in the R22 group of municipalities and with the better soil and climatic conditions for agricultural production in forest than in rangeland-covered zones in the case of the R40 group. It is also worth noting the relatively low contribution of ES

Table 21.3 Number of employments (IB) explained by landscape composition per each employment in the *i* economic subsector (relative indirect benefits, RIB), for the main economic subsectors linked to local ecosystem services: tourism (including both tourism and gastronomy), agricultural production, and animal production (including livestock production, forestry, beekeeping, and hunting)

Economic subsector	RIB		
	R3	R22	R40
Tourism	1.29	1.90	2.96
Agricultural production	0.26	0.79	0.65
Animal production	0.83	1.70	0.58

to employment in animal production and forestry in comparison to other employment sources, especially for the R3 and the R40 municipality groups (Fig. 21.4c).

In synthesis, the number of jobs due to ES per each employment in the *i* economic subsector (RIB) was relatively high for the tourism and gastronomy subsector, and this was especially noticeable in the R40 municipality group, where for each employment in tourism, landscape composition explained approximately three employments in the overall economy (Table 21.2). Thus, for example, each of the 104 employees in tourism registered in the 2010 census for the municipality of Epuyén was supporting 2.96 jobs in the remaining economic subsectors (including the tourism subsector) due to landscape composition within the main city buffer, that is, approximately 308 employments. Therefore, 22% of the jobs in tourism is explained by landscape composition in the R40 municipalities (Table 21.3), and they propagate to the rest of the economy of San Carlos de Bariloche according to the same EMF used in the calculus of RIB for this subsector, representing 189 additional jobs. In short, of the 497 jobs explained by the tourism subsector in this municipality, 62% is explained by landscape composition that is due to ES supply at the local level.

4 Discussion and Recommendations for Policy-Makers

This work shows the relevant links between employment level in four selected economic subsectors and environment composition in the surrounding landscapes and seascapes in the Argentine Patagonia. Our models suggest that landscape composition, and therefore the capacity to provide ES, is important for local jobs where economic subsectors like tourism crop production, animal production, fishery, and forestry are at least partly controlled by surrounding land covers. For example, the R22 municipality group is characterized by general aridity. Here, the observed relation between croplands and surface water bodies (as indicators of ES) for one side, and number of jobs in tourism, reflected how the specific ES of water supply support the propagation of employment, through the rise of agricultural production and recreation opportunities. In this municipality group, the forestry industry is negligible, and employments in forestry are registered in the census with animal production, not with agriculture. Therefore, the strong contribution of forest cover to employments in agriculture may be better reflecting the covariation of water supply with forest cover and agriculture employment, rather than a causal influence of forest cover on agriculture employment.

Accounting for the way in which ecosystems and their natural capital sustain indirect benefits such as employment is a necessary step in the analysis of benefits and costs of policies and management actions that can alter terrestrial and marine ecosystems. Demonstrating the importance of natural capital and ES in sustaining employment can engage policy-makers and other social actors in creating new structures and institutions aimed at protecting ES (Cowling et al. 2008; Monjeau 2010; Cáceres et al. 2016).

Furthermore, consequences of economic multipliers on employment levels are rarely assessed at a local scale. The relationships between landscape composition and local indirect benefits have been mostly restricted to studies on "environmental incomes" based on primary data gathered at household scale (Vedeld et al. 2007; Angelsen et al. 2014).

The relations between landscape composition and employment varied within Patagonia (Fig. 21.4). The R22 and R40 municipality groups owe a high portion of their employment in the analyzed economic subsectors to surrounding landscape composition. However, while in R22 municipalities employment related to landscape composition was mainly mediated by agricultural production, in the R40 municipalities, it was mediated by tourism. Therefore, ES contribution to employment was moderate and more balanced among the economic subsectors in the R3 group.

Our results show that differences in ES contributions to overall employment among the municipality groups have two main causes. Firstly, the contribution landscape composition to employment through particular economic subsectors depends on the geographic context (e.g., tourism employment due to landscape composition is 78% for R40 group, but 34% for R3 group (Table 21.2)). Secondly, the relative contribution of particular land cover types to employment in an economic subsector varies among municipality groups. Thus, for example, cover of surface water bodies showed the main positive contribution to tourism employment in R22 and R40 municipalities but the most negative influence for the R3 group. Cropland cover showed the main influence in this subsector for the R3 and R22 groups but the most negative influence for the R40 group (Table 21.2). Since the relative influence IB to the rest of the economy varies among economic subsectors, those municipalities with highest ES contribution to the subsector with highest RIB values (e.g., R40 group, Tables 21.2 and 21.3) will probably show the highest overall IB values (Fig. 21.4d).

Despite the fact that our results did not necessarily reflect causal relationships between natural capital and employment, they can provide useful insights for the design of public policies. First, the results indicated how land-use decisions for improving the overall employment level should consider the conservation or conversion of different land cover types, depending on the geographical context (e.g., conversion of rangelands to croplands should benefit tourism employment for R3 and R22 groups, but for R40 the reverse is true (Table 21.2)). Results also suggested that the influence from different land covers on ES supply, and then on the employment in particular economic subsector, propagates to the rest of the local economy, thus reinforcing other criteria defining land-use policies and taking (or not) particular land-use decisions (Paruelo et al. 2014).

Different sources of mismatching or loose relationships between landscape composition and local employment level have been discussed by Laterra et al. (2019) for Argentina as a whole and may also be applicable to the present analyses. For example, extra-local influences on employments may explain large employment levels in the tourism subsector of some coastal cities of R3 group due to their relative proximity to large cities (although the inclusion of distance factor in the GAM models did not improve their fit levels); therefore, the transport of agricultural products from R22 to R40 municipalities prevents the employment in local agricultural and animal production. The creation of casinos along the R22 also illustrates how local natural attractiveness by constructed amenities contributes to the loss of natural capital-tourism employment relationships.

The predictive power of landscape composition for explaining employment may also be limited by methodological constraints discussed in a previous work (Laterra et al. 2019). For example, landscape composition may not be a good proxy of specific ES sustaining economic activities like attributes of the cultural landscape (e.g., sites of paleontological or archaeological interest). Therefore, landscape composition account for ES supply, but it cannot account for access to ES, which depends on the supporting facilities from near cities and/or protected areas, among others. On the other hand, an arbitrary buffer of 20 km ratio may not properly represent the scale of access to all ES types, either by neglecting relevant areas of ES provision outside the buffer or by taking into account ES that cannot be reached or captured by local population or visitors. More importantly, and in agreement with our objectives, the results only reflect the relevance of local ES on local employments, so their utilization as indicative of influences on total employment is misleading and should take into account the non-estimated influences on extra-local employment.

In synthesis, we can arrive to the following conclusions:

(i) The use of employment models to evaluate the indirect benefits derived from ES is a promising approach for improving policy evaluation, which nonetheless requires further development, so great caution must be exercised in interpretation.
(ii) Linear models based on employment multipliers may be insufficient to predict the consequences of the ES supply of a landscape on employment, since there may be nonlinear changes in the supply and demand of inputs. Therefore, we recommend restricting the application of our ES-employment relationship model to marginal changes in the supply of ES where linear responses are expected, but far from the thresholds (where nonlinear responses can occur).
(iii) Our approach is limited to understanding the local coupling between ES and employment, and it would not be correct to extrapolate to the national scale. For example, the RIB due to local farmland ecosystems increases with the

number of indirect beneficiaries (e.g., tractor drivers, mechanics, supply vendors, etc.) living in the corresponding municipality but neglects indirect beneficiaries living in large distant cities (e.g., tractor manufacturers), used in the food, feed, fertilizer, and pesticide industries, among others.

We hope that the quantitative assessment of the indirect benefits derived from ES will complement conventional economic valuations. Our results provide additional support to nature-based climate solutions, since the conservation of a landscape not only is a source of carbon sequestration but also conserves biodiversity and is a generator of employment. This fact can be a powerful tool to increase public awareness about the contribution of nature to society and with it encourages the prioritization of environmental policies of the political sector.

An important future research avenue is trying to include employment quality into the assessments, using indicators such as the rate of wage and salaried workers and the rate of non-vulnerable employment in total employment. Therefore, instead of simply assessing if natural resources create jobs, we can assess if they create good-quality jobs. Another addition would be the inclusion of other less traditional nature-based industries and sectors for which data is limited, such as knowledge-based (e.g., research and technology) and artistic industries.

Acknowledgments This study was funded by grants Fondo para la Investigación Científica y Tecnológica FONCYT (ANPCYT) PICT 15-0672.

References

Angelsen A, Jagger P, Babigumira R, Belcher B, Hogarth NJ, Bauch S, Börner J, Smith-Hall C, Wunder S (2014) Environmental income and rural livelihoods: a global-comparative analysis. World Dev 64:S12–S28

Andersen AD, Marín A, Simensen EO (2018) Innovation in natural resource-based industries: a pathway to development? Introduction to special issue. Innov Dev 8(1):1–27

Blanco PD, Colditz RR, Saldaña GL, Hardtke LA, Llamas RM, Mari NA, Fisher A, Caride C, Aceñolaza PG, del Valle HF, Lillo-Saavedra M, Coronato FR, Opazo SA, Morelli F, Anya JA, Sione WF, Zamboni P, Barrena Arroyo V (2013) A land cover map of Latin America and the Caribbean in the framework of the SERENA project. Remote Sens Environ 132:13–31

Boyd W, Prudham WS, Schurman RA (2001) Industrial dynamics and the problem of nature. Soc Nat Resour 14(7):555–570

Briassoulis H (1991) Methodological issues: tourism input-output analysis. Ann Tour Res 18(3):485–495

Cáceres DM, Silvetti F, Díaz S (2016) The rocky path from policy-relevant science to policy implementation—a case study from the South American Chaco. Curr Opin Environ Sustain 19:57–66

Cowling RM, Egoh B, Knight AT, O'Farrell PJ, Reyers B, Rouget M, Roux DJ, Welz A, Wilhelm-Rechman A (2008) An operational model for mainstreaming ecosystem services for implementation. Proc Natl Acad Sci 105(28):9483–9488

Daily GC, Alexander S, Ehrlich PR, Goulder L, Lubchenco J, Matson PA, Mooney HA, Postel S, Schneider SH, Tilman D, Woodwell GM (1997) Ecosystem services: benefits supplied to

human societies by natural ecosystems. In: Grifo F, Rosenthal J (eds) Biodiversity and human. Island Press, Washington, DC
Daily G, Dasgupta P, Bolin B, Crosson P, Du Guerny J, Ehrlich P, Folke C, Jansson AM, Bengt-Owe J, Kautsky N, Kinzig A, Levin S, Mäller KG, Pinstrup-Andersen P, Sinilasco D, Walker B (1998) Food production, population growth, and the environment. Science 281:1291–1292
Daily GC (2000) Management objectives for the protection of ecosystem services. Environ Sci Pol 3:333–339
Daw T, Brown K, Rosendo S, Pomeroy R (2011) Applying the ecosystem services concept to poverty alleviation: the need to disaggregate human well-being. Environ Conserv 38:370–379
Fisher B, Turner RK, Morling P (2009) Defining and classifying ecosystem services for decision making. Ecol Econ 68(3):643–653
Grêt-Regamey A, Kytzia S (2007) Integrating the valuation of ecosystem services into the input–output economics of an Alpine region. Ecol Econ 63(4):786–798
Hastie T, Tibshirani R (1987) Generalized additive models: some applications. J Am Stat Assoc 82(398):371–386
INDEC (2010) WebINDEC – Población/Censos/Censo 2010 [WWW Document]. WebINDEC – Poblac Censos Censo 2010. https://www.indec.gov.ar/. Last accessed 10 Oct 2017
Jacobs S, Dendoncker N, Martín-López B, Barton DN, Gomez-Baggethun E, Boeraeve F, McGrath FL, Vierikko K, Geneletti D, Sevecke KJ (2016) A new valuation school: integrating diverse values of nature in resource and land use decisions. Ecosyst Serv 22:213–220
Jackson LE (2003) The relationship of urban design to human health and condition. Landsc Urban Plan 64(4):191–200
Kelly G, Cooper A, Pinkerton E (2014) Quantifying Equity with Messrs. Markov, Lorenz and Gini: A Case Study of Dunster, British Columbia. J Rural Comm Dev 9(3):142–156
Kitzberger T (2012) Ecotones as complex arenas of disturbance, climate, and human impacts: the trans-Andean forest-steppe ecotone of northern Patagonia. In: Ecotones between forest and grassland. Springer, New York, NY, pp 59–88
Laterra P, Nahuelhual L, Gluch M, Sirimarco X, Bravo G, Monjeau A (2019) How are jobs and ecosystem services linked at the local scale? Ecosyst Serv 35:207–218
Manzano O, Gutiérrez JD (2019) The subnational resource curse: theory and evidence. Extr Ind Soc 6(2):261–266
Martín-López B, Gómez-Baggethun E, García-Llorente M, Montes C (2014) Trade- offs across value-domains in ecosystem services assessment. Ecol Indic 37:220–228
Miller RE, Blair PD (2009) Input-output analysis: foundations and extensions. Cambridge University Press. 737 p
Mittermeier RA, Mittermeier CG, Brooks TM, Pilgrim JD, Konstant WR, da Fonseca GA, Kormos C (2003) Wilderness and biodiversity conservation. Proc Natl Acad Sci 100(18):10309–10313
Monjeau A (2010) Conservation crossroads and the role of hierarchy in the decision- making process. Natureza Conservação 8(2):112–119
Nahuelhual L, Laterra P, Villarino S, Mastrángelo M, Carmona A, Jaramillo A, Barral P, Burgos N (2015) Mapping of ecosystem services: missing links between purposes and procedures. Ecosyst Serv 13:162–172
OCDE (2011) The economic significance of natural resources: key points for reformers in eastern Europe, Caucasus and Central Asia. 40 p. Available at http://www.oecd.org/env/outreach/2011_AB_Economic%20significance%20of%20NR%20in%20EECCA_ENG.pdf
Paruelo JM, Beltrán A, Jobbágy E, Sala OE, Golluscio RA (1998) The climate of Patagonia: general patterns and controls on biotic processes. Ecol Austral 8(02):085–101
Paruelo JM, Jobbágy EG, Laterra P, Dieguez H, García Collazo MA, Panizza A (eds) (2014) Ordenamiento Territorial: Conceptos, Métodos y Experiencias. FAO, MAGyP, FAUBA. 576 p. ISBN 978-92-5-308619-1. http://www.fao.org/3/a-i4195s.pdf
Pascual U, Balvanera P, Díaz S, Pataki G, Roth E, Stenseke M, Watson RT, Başak Dessane E, Islar M, Kelemen E, Maris V, Quaas M, Subramanian SM, Wittmer H, Adlan A, Ahn S, Al-Hafedh YS, Amankwah E, Asah ST, Berry P, Bilgin A, Breslow SJ, Bullock C, Cáceres D, Daly-Hassen

H, Figueroa E, Golden CD, Gómez-Baggethun E, González-Jiménez D, Houdet J, Keune H, Kumar R, Ma K, May PH, Mead A, O'Farrell P, Pandit R, Pengue W, Pichis-Madruga R, Popa F, Preston S, Pacheco-Balanza D, Saarikoski H, Strassburg BB, van den Belt M, Verma M, Wickson F, Yagi N (2017) Valuing nature's contributions to people: the IPBES approach. Curr Opin Environ Sustain 26:7–16

Potschin-Young M, Haines-Young R, Görg C, Heink U, Jax K, Schleyer C (2018) Understanding the role of conceptual frameworks: Reading the ecosystem service cascade. Ecosyst Serv 29:428–440

Rawlins JM, De Lange WJ, Fraser GC (2018) An ecosystem service value chain analysis framework: a conceptual paper. Ecol Econ 147:84–95

Schröter M, Koellner T, Alkemade R, Arnhold S, Bagstad KJ, Erb KH, Frank K, Kastner T, Kissinger M, Liu J, López-Hoffman L, Maes J, Marques A, Martín-López B, Meyer C, Schulp CJE, Thober J, Wolff S, Bonn A (2018) Interregional flows of ecosystem services: concepts, typology and four cases. Ecosyst Serv 31:231–241

Serna-Chavez HM, Schulp CJE, Van Bodegom PM, Bouten W, Verburg PH, Davidson MD (2014) A quantitative framework for assessing spatial flows of ecosystem services. Ecol Indic 39:24–33

Vedeld P, Angelsen A, Bojö J, Sjaastad E, Berg GK (2007) Forest environmental incomes and the rural poor. Forest Policy Econ 9(7):869–879

Summers JK, Smith LM, Case JL, Linthurst RA (2012) A review of the elements of human well-being with an emphasis on the contribution of ecosystem services. Ambio 41(4):327–340

WWAP (United Nations World Water Assessment Programme) (2016) The United Nations world water development report 2016: water and jobs. UNESCO, Paris, France

Chapter 22
Ecosystem Services in Patagonia: A Synthesis and Future Directions

Laura Nahuelhual, Guillermo Martínez Pastur, and Pablo L. Peri

Abstract In this closing chapter, we provide a synthesis of the Patagonian ecosystems and ES that were more frequently addressed in the preceding chapters, along with the main transformations and associated drivers. We also synthetize the research gaps and the recommendations provided by the authors and delineate future directions for ES research in Patagonia. Natural and human-induced drivers have modeled and remodeled Patagonian landscapes continuously. The chapters in this book describe recent landscape transformations and the major human-derived impacts on biodiversity and provision of ecosystem services (ES). The chapters also discuss implications of these changes for human well-being and provide recommendations for decision-making.

Keywords Patagonian landscapes · Terrestrial and marine ecosystems · Land-use change · Nature conservation · Human well-being

L. Nahuelhual (✉)
Instituto de Economía, Universidad Austral de Chile Centro de Investigación: Dinámica de Ecosistemas Marinos de Altas Latitudes Instituto Milenio en Socio-Ecología Costera, Valdívia, Chile
e-mail: lauranahuel@uach.cl

G. Martínez Pastur
Consejo Nacional de Investigaciones Científicas y Técnicas, Centro Austral de Investigaciones Científicas y Técnicas, Ushuaia, Tierra del Fuego, Argentina

P. L. Peri
National Agricultural Technology Institute, Río Gallegos, Argentina

P. L. Peri et al. (eds.), *Ecosystem Services in Patagonia*, Natural and Social Sciences of Patagonia, https://doi.org/10.1007/978-3-030-69166-0_22

1 Introduction

Fifteen years after the Millennium Ecosystem Assessment (MEA 2005), the concept of ecosystem services (ES) is widely recognized, both scientifically and politically (Bouwma et al. 2018). In parallel, the science related to the evaluation of ES and their contribution to human well-being has had an exponential expansion (Delgado and Marín 2015; Pauna et al. 2018; Perevochtchikova et al. 2019; Balvanera et al. 2020). Nonetheless, most ES research continues to focus on the biophysical quantification of ES flows or supply, over social dimensions (Lautenbach et al. 2019). The research carried out in the preceding chapters (Fig. 22.1) also reveals this trend.

To give continuity to the MEA, in 2012, 118 countries signed as members of the Intergovernmental Platform on Biodiversity and Ecosystem Services (IPBES), including Argentina and Chile, with the mission of assessing the state of biodiversity and ES. IPBES presented its Regional Assessment Report on Biodiversity and Ecosystem Services for the Americas in 2018 (IPBES 2018) and its Global Assessment in 2019 (IPBES 2019), which confirmed the decreasing trend of biodiversity and ES reported by MEA in 2005. Among key messages, the America's IPBES Report stated that "many aspects of quality of life are improving at regional and sub-regional scales, but that the majority of countries are using nature more intensively than the global average and exceeding nature's ability to renew the contributions it makes to quality of life" (A4, p. 10). As a result, biodiversity and ecosystem conditions in many parts of the Americas are declining, resulting in a reduction in nature's contributions to people's quality of life.[1] Namely, "65% of nature's contributions to people in all units of analysis are declining, with 21% declining strongly" (B1, p. 12). In the case of Argentinean and Chilean Patagonia, the report documents an increase in the amount of temperate forests and woodlands habitat, but an increase in habitat degradation along with the decrease of native species diversity and the increase of alien and invasive species. The trends are similar for Patagonian grasslands and peatlands, but with a decrease in habitat amount.

Thus, the ES approach, despite its increasing popularity, has failed to reverse the loss of ecosystems' natural capital (Levrel et al. 2017) as neither have the approaches focused on the management of natural resources and the conservation of biological diversity (MEA 2005; IPBES 2019). The reasons for this outcome are diverse and

[1] Nature's contributions to people (NCP) are all the contributions, both positive and negative, of living nature (e.g., diversity of organisms, ecosystems, and their associated ecological and evolutionary processes) to the quality of life for people. Quality of life is understood in IPBES as the achievement of a fulfilled human life, a notion which may vary strongly across different societies and groups within societies. It is a context-dependent state of individuals and human groups, comprising aspects such as access to food, water, energy and livelihood security, health, good social relationships and equity, security, cultural identity, and freedom of choice and action. "Living in harmony with nature," "living-well in balance and harmony with Mother Earth," and "human well-being" are examples of different perspectives on a "good quality of life."

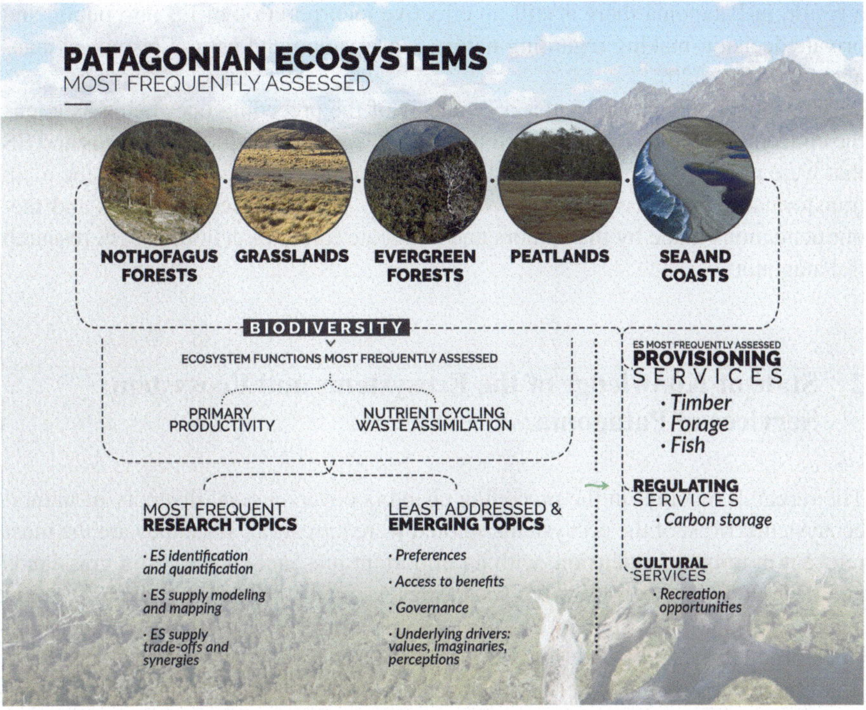

Fig. 22.1 Ecosystems, ecosystem services, and type of assessments covered in the book's chapters

range from structural ones, such as the capitalist production model, to operational ones, such as the gaps between ES science and decision-making.

Capitalism, as a political-economic system, depends on the expropriation and exploitation of natural resources and the environment, creating social inequalities and environmental degradation on an ever-increasing scale (O'Connor 1991; Moore 2017). This has been called the "second contradiction of capitalism" designated as "the absolute general law of environmental degradation under capitalism" (O'Connor 1991).

On the other hand, various studies have described and exemplified the causes of science-policy gaps in the field of ES (Saarikoski et al. 2018). Among them are (i) the rapid proliferation of definitions, conceptual frameworks, approaches, and evaluation models of ES within the scientific community, which has made it difficult to understand and implement the concept (Polasky et al. 2015), (ii) the complexity that the concept represents for decision-makers (Posner et al. 2016), and (iii) the different ontologies regarding nature and conservation (Dick et al. 2018).

The research in the precedent chapters shows significant advances in ES knowledge but also reveals the challenges we still have for bridging ES knowledge and decision-making. Additionally, the legal and institutional systems of Argentina and Chile do not always offer opportunities for the incorporation of the ES approach. As a result, in Patagonia there is still no effective incorporation of ES into public and private decision-making regarding nature conservation and natural resources management (Chap. 20).

This closing chapter provides a synthesis of the preceding ones using as a lens the conceptual framework presented in Chap. 1. We focus on the ecosystems and ES that were more frequently addressed in the preceding chapters, along with the main transformations and associated drivers. We also synthetize research gaps and recommendations made by the authors and delineate future directions for ES research in Patagonia.

2 State of Knowledge of the Ecosystems and Ecosystem Services of Patagonia

The research reported in the preceding chapters covers a great diversity of natural ecosystems. Most of the ecosystems studied were terrestrial, since they are the most relevant in terms of extension, with an important presence of studies in grasslands and deciduous forests of *Nothofagus* species, as well as mixed evergreen. In these ecosystems, the most studied ES were provisioning ES, specifically forage, livestock production, and wood (Fig. 22.1).

Aquatic ecosystems were less represented, particularly freshwater ecosystems. The most frequently addressed topics focused on the quantification and modeling of ecosystem functions and intermediate and final ES as well as the mapping of ES bundles (Chaps. 2, 3, 4, 5, 6, 7, 8, 10, 12, and 17). Additionally, some of these studies considered different trade-offs and synergies between biodiversity and ES such as fauna biodiversity and livestock production (Chap. 6) or gains in biodiversity from marine conservation vs. fair distribution of benefits across different social actors (Chap. 13) and between ES such as wood and water regulation, under pastoral and forest management scenarios (Chaps. 4 and 6). These studies included different scales of analysis from forest stands to provinces and larger territories, using a variety of spatial indicators and quantification and mapping techniques. On the contrary, issues associated with the demand side of ES, specifically the evaluation of benefits and beneficiaries, were less covered or superficially included. The exceptions were Chap. 15 that analyzed the supply and coproduction of marine ES and their distribution across direct and indirect beneficiaries and Chap. 16 that compared ES and benefits obtained from ES by rural households in two watersheds.

As emerging research topics that depart from traditional ES supply assessments, the following can be highlighted: (i) studies that analyzed individual preferences and contributions of ES to well-being (Chaps. 13 and 14) and social imaginaries of nature-human relations influencing stakeholder's decisions regarding the use of ecosystems (Chap. 19); (ii) distributive studies, which looked at the result of the interaction between social actors, institutions, and natural capital (Chaps. 15 and

18); (iii) studies that associated ES and long-term sustainability (Chaps. 9 and 11); and (iv) studies that looked at indirect benefits arising from ES (Chap. 21).

From the conceptual and methodological point of view, there are "blind spots" related to critical questions that characterize the "holistic ideal of ecosystem services research" (Lautenbach et al. 2019). Table 22.1 summarizes five criteria that help typify these blind spots, which largely coincide with the diagnosis presented in Chap. 20 for previous studies conducted in the Patagonian region. The first criterion is social-ecological validity, meaning that measurements, modeling and monitoring of ecosystem functions and the social dimension related to ES supply and demand are close to the phenomenon measured. For example, the capacity of a forest to regulate water flow is not closely represented by the change in stream or river flows. Likewise, the well-being obtained from forest firewood is not closely accounted for by the per capita consumption of firewood.

The chapters that measured, modeled or monitored ES still have the challenge to meet this first criterion. Most studies still use proxies and GIS-based models as opposed to process-based models and statistical models that are capable of capturing ES supply, demand, and interactions.

The second criterion is the analysis of trade-offs, which is a crucial step for identifying promising management options (e.g., White et al. 2012). In general, the chapters showed an increasing recognition of the importance of trade-offs, but when trade-offs were indeed assessed, the most frequent approach was the use of a simple map overlay to assess bundles of ES or trade-offs between them (e.g., Chap. 4). The assessments did not use optimization approaches, analysis of the trade-offs at different scenarios or management alternatives, or statistical analysis of survey data including trait-based analysis (e.g., Hevia et al. 2017).

The third criterion is the assessment of off-site effects, also called peri-couplings (ES flows between contiguous social-ecological systems) and tele-couplings (ES flows between distant social-ecological systems). From the perspective of regional

Table 22.1 Blind spots in ES research as reported in the book's chapters

Criteria	Blind spots
Socio-ecological validity of ecosystem data and models	Single or few ES are assessed Modeling approaches used do not account for feedbacks, nonlinear effects, or spatial and temporal variability Lack of integration between biophysical and social aspects of ES Use of proxies and models without interactions Disciplinary studies; little interdisciplinary research
Trade-offs analyses	Few ES are considered Policy scenarios are generally not considered Trade-offs are simple ES overlays
Recognition of off-site effects or tele-couplings	ES demand aspects are not included Data is limited
Involvement of stakeholders	Stakeholders are only partially engaged in research design
Relevance and usability of study results	Critical decision-making problems are not detected

and global sustainability, it is important that place-based ES assessments do not overlook effects on other social-ecological systems (Liu et al. 2013). Without consideration of such off-site effects, there is considerable risk for the spatial spillover rebound effect (Maestre Andres et al. 2012), meaning that policies intending to protect biodiversity or ES in one place can have negative impacts on biodiversity, ES, or well-being in another place. This topic was partially considered in only one chapter of the book (Chap. 15).

The fourth criterion is stakeholders' involvement, which can build stronger links between science, policy, and society and ensure that research addresses real-world needs (Menzel and Teng 2010). Despite this recognition, ES research in Patagonia seems to be mostly driven by researchers' own interests, which might explain the low level of stakeholder engagement. In chapters that did engage stakeholders, their participation was limited to identifying and prioritizing ES (Chaps. 13 and 14).

The last criterion is the relevance and usability of research results. Addressing single aspects of ES assessments improves our knowledge on specific aspects of the conceptual framework in Chap. 1. However, the ES research most relevant for decision-making is the integrated assessment of multiple ES supply and demand linked with societal needs (Verburg and Selnes 2014; Beaumont et al. 2017; Lautenbach et al. 2019). Decision-making that aims to overcome sectoral views (e.g., forest management; water management) by integrating the components and interactions of coupled social-ecological systems needs to pursue the integration of ES supply and demand. This is a common blind spot in most chapters that included ES quantification and mapping, with the exception of those that simultaneously tackled ES supply, demand, and/or policies (Chaps. 15 and 18). Cash et al. (2003) concluded that the effectiveness of scientific information in societal decision-making is related to three main characteristics: saliency (relevance to decision-making), legitimacy (fair and unbiased information production that also respects stakeholders' values), and credibility (scientific adequacy). Failure to achieve these criteria can at least partially explain why ES knowledge has not been sufficiently implemented in decision-making in Patagonia (Chap. 20).

There are several reasons for blind spots: (i) perceived importance of the ES categories only determined by researchers and/or stakeholders, (ii) different research background of study leaders, and (iii) financial, logistic, and scientific challenges to assess ES supply, demand, and interactions in the territory. The blind spots summarized in Table 22.1 apply to those studies that focused on measuring, mapping, and monitoring ES. However, the chapters in this book also made important conceptual contributions such as the role of social imaginaries in nature conservation and ES (Chap. 19) and distributive justice issues (Chaps. 13 and 18).

3 Drivers of Change in the Socio-ecological Systems of Patagonia

In both sides of Patagonia, Argentina and Chile, the ecosystem transformations reported in the chapters occurred mostly in private lands. In the case of marine ecosystems, the status of the marine space as a public good or a "common pool resource" resulted in much more complex social-ecological interactions and transformations such as those described in Chaps. 12, 13, and 15.

The direct driver most frequently described in terrestrial ecosystems was land-use change and, specifically, the loss and degradation of forests due to urban expansion or the extraction of firewood and timber and the degradation of grasslands by overgrazing (Chaps. 3, 7, and 9). Another direct driver of great effect on the ecosystems of Patagonia was the expansion of invasive species such as the beaver, nonnative tree species, and salmon, which produce several adverse impacts on the natural ecosystems, but, at the same time, can generate services that benefit specific groups of people. For example, beavers modify most of riparian natural forests but at the same time generate ponds that are used by livestock in some areas or generate earnings for local tourism (Chap. 10). The expansion of nonnative tree species such as *Pinus radiata* and *Eucalyptus* sp., mostly in the northern Chilean Patagonia and of murrayana pine (*P. contorta*), ponderosa pine (*P. ponderosa*), and Oregon pine (*Pseudotsuga menziesii*) in Northern Patagonia on the Argentine side (Sarasola et al. 2006), while sustaining the timber industry, has been reported to have a myriad of negative effects on biodiversity and ES (Franzese et al. 2017; Corley et al. 2018). These controversial trade-offs and synergies are difficult to reconcile and present serious challenges to decision-makers.

Among the indirect or underlying drivers, the chapters described sociopolitical and, to a lesser extent, cultural drivers (values, beliefs, norms, and perceptions shared by people). Indirect drivers encompass the forces influencing private and public decision-making such as stakeholders' imaginaries and levels of education and knowledge and governance modes (Fig. 22.2), among others. These factors in turn influence the institutional arrangements for ecosystem management, as well as property rights over ES.

Both MEA (2005) and IPBES (2018) have recognized "weak governance" among the underlying drivers of ecosystem change. In Argentina and Chile, ecosystems and biodiversity are managed mostly under centralized governance arrangements, which are not aligned with the paradigm shift that the ES approach entitles. Although there are environmental policies and instruments (see Chap. 20) that aim to reduce pressure on biodiversity and ES, they have often not been effectively coordinated to achieve their objectives. Furthermore, subordination of environment to economic policies results in trade-offs and inequities in distribution of benefits (IPBES 2018) that cannot be fixed or prevented under current nature governance approaches.

The chapters generally focused on one or few drivers, but we have yet to make progress in studying the synergistic effects of multiple drivers on landscape

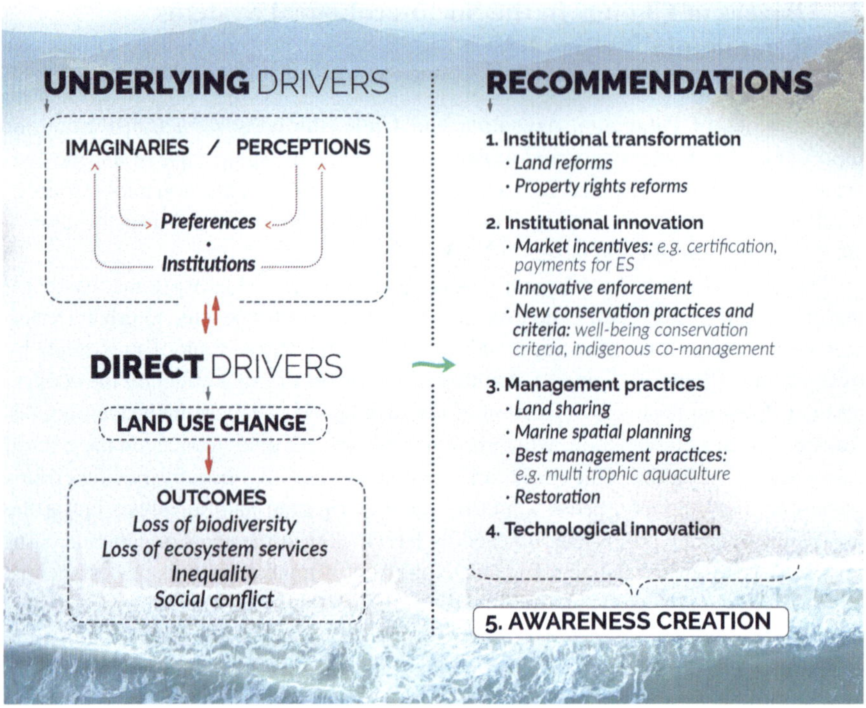

Fig. 22.2 Underlying and direct drivers of ecosystem change in Patagonia and recommendations in this book

transformations. When these transformations occurred on private lands, the trade-offs reported were usually those between provision ES and the other ES (Chap. 2). In these cases, private decisions largely responded to economic incentives (e.g., subsidies to agriculture or nonnative tree plantations) and were influenced by indirect external drivers, such as trade globalization. Specifically, while provisioning services have an exchange value in markets, regulating and cultural ES generally do not (with exception of carbon emission reductions and recreation opportunities); therefore private decisions do not take them into account. The exception to this rule occurs under other institutions such as private protected areas, whose objectives are not primarily economic (Chaps. 13 and 18).

Changes in ecosystems occur through the interaction between these multiple drivers, at different spatial and temporal scales. For instance, global trade (e.g., wood pulp, meat, salmon) places challenges to national, regional, and local governance, regulations, and management practices on ecosystems and their services, enhancing good practices but worsening the damage caused by poor practices. Increased trade can accelerate degradation of ES in commodity-exporting countries such as Argentina and Chile, if their policy, regulatory, and management systems are inadequate or weak.

Additionally, changes in ecosystems influence drivers in complex ways. For example, invasive species such as the beaver, which have led to the enormous destruction of forests in Tierra del Fuego, has been reimagined as an icon of regional identity, which creates conflicts when it comes to their management through authorized hunting. However, the benefits that this species could generate are well below the environmental and social costs it produces. Finally, altered ecosystems create new opportunities and constraints, induce institutional changes in response to degradation (e.g., restoration policies) and resource scarcity, and lead to social effects such as changes in employment.

4 Recommendations

The chapters provided recommendations to address direct and indirect drivers (Fig. 22.2), which are in line with those guidelines established in the IPBES assessments at both global and regional levels. IPBES guidelines include the implementation of specific public policies, behavioral change, improved technology, effective governance arrangements, education and awareness programs, scientific research, monitoring and evaluation, adequate finance arrangements, and supporting documentation and capacity-building. Specifically, the recommendations made in the chapters can be grouped into the following categories:

(i) Institutional transformation. An institutional transformation involves the change in the formal "rules of the game." This includes profound reforms to land ownership in the case of Chile, where very few owners own most of the lands and forests (Chap. 18). This inequality means that only large farms can sustain the provision of ES, consolidating distributive inequalities (Benra and Nahuelhual 2019). Along the same lines and in the case of marine ecosystems, a more equitable distribution of ES involves profound changes to access regimes and rights to marine resources (Chaps. 13 and 15), which in the case of Chile requires transformations to the Constitution, which are expected to occur in the coming years with a new constituent process. Deep social conflicts and efforts to secure purposive change are likely to demand strong civil society organization response if new imaginaries (Chap. 19) are to be discerned and effectively shared in ways that encourage sustained dialogue and the development of new social understandings (Stephenson 2011).

    Materializing institutional changes does require not only marginal improvements to existing structures but also a modification of the actual objectives of environmental and development public policy, which implies incorporating new values, management visions, and new actors in a democratic and fair decision-making process. Doing so demands an approach that differs from the typically top-down, technocratic, and linear processes that characterize much of Argentina and Chile policy-making.

At a legal level, amendments are needed to overcome the current absence of an environmental code. It is also important to reorganize related environmental regulations, which in some cases contradict one another. This would also help solving the common problem of delays in the regulation of laws and their application in practice (Capaldo 2018) due to the dispersion of regulations. Likewise, efforts to harmonize legal and institutional frameworks should be undertaken in order to guarantee an effective protection of ES (Chap. 22).

(ii) Institutional innovation. Innovation often comprises technical processes and organizational changes in production and marketing, sometimes products, but rarely institutions. New formal or informal institutions emerge as responses to the shocks and stresses induced by market, social, and policy changes (Buttoud et al. 2011). For example, ES applications in the territory may incorporate innovation based on the creation, exchange, and application of new ideas into marketable goods and nonmarketable services, leading both to the success of an enterprise and the advancement of society (Boisvert et al. 2013).

Some chapters provide recommendations that involve institutional innovations. One example is the change in conservation criteria for marine protected areas (Chap. 13) as the result of increasing conflicts between local communities and public protected areas, whose creation obeys the command and control approach to governance. Traditionally, the creation and management of terrestrial and marine public protected areas has been guided by the protection of the so-called "objects of conservation", which are exclusively biological (e.g., endangered or iconic species or habitats). However, the new conservation standards for the achievement of the Sustainable Development Goals (SDG) require the inclusion of social criteria such as the recognition of livelihoods and the distribution of the benefits of conservation. The incorporation of "objects of well-being" in new conservation planning strategies (e.g., open standards for conservation) represents an opportunity to implement the ES approach in protected areas management through the identification of ES and key benefits for particular beneficiaries (Brain et al. 2020).

Another example is the reorientation of existing policies in order to improve the distribution of ES benefits. For example, the focus of regional tourism policies in Chilean Patagonia has been continuously placed on the generation of economic profits rather than on allowing local inhabitants to get to know their territory and natural beauties or on generating small-scale enterprises with local identity. Redirecting tourism and related economic policies toward local tourism markets can not only improve the distribution of benefits from ES but also increase local resilience in the face of events such as the coronavirus pandemic experienced during 2020–2021 (Chap. 15).

(iii) Management practices. The sustainable provision of ES to society is the main target of different ecosystem management approaches. In the precedent chapters, these management practices focus on two main areas: (i) the effect of specific practices on the provision of one or several ES and (ii) the interaction, trade-offs, and synergies between ES under different management scenarios.

These chapters mostly focused on provisioning ES, especially on timber obtained from deciduous *Nothofagus* forests and described the advantages of different silviculture alternatives (e.g., silvopastoral systems, firewood extraction schemes) (Chaps. 2, 4, 5, and 6). These management practices highlight the need of multipurpose objectives that promote several ES with broader social benefits. In addition, these proposals promote conservation within managed stands (land-sharing strategy) as opposed to those that only secure ES provision or biodiversity in natural reserve networks (land-sparing strategy). Sustainable management of forest stands is based on maintaining some legacies or natural values in the managed landscapes, which include not only monetary values (e.g., provision of timber or cattle) but also other ES (e.g., regulation, supporting, cultural) and biodiversity. Grasslands and shrublands were also analyzed in this book, where the equilibrium between the livestock and the natural ecosystem maintenance was the main challenge to face desertification, soil erosion, and the impacts of climate change (Chaps. 3 and 7). Management and conservation of peatlands were also described (Chap. 8). Finally, Chap. 9 proposed restoration practices to recover the losses of natural capital in the managed areas where sustainable management was not achieved.

The different chapters proposing management recommendations showed that under certain circumstances, it is possible to reach an equilibrium among social, ecological, and economic criteria in the management and conservation of the natural ecosystems in Patagonia. They also showed the importance of the private sector in elaborating these proposals (Chaps. 6, 8, 12, 14, and 15). The need for better policies to reach sustainable management of particular ecosystems was clear in the recommendations of the different chapters, which should aim at reducing some pressures to ecosystems (e.g., intensive salmon production; Chap. 12) or at promoting specific back-to-nature practices (e.g., retention forestry or harvesting based on gap creation; Chap. 4).

A pending task is the application of "adaptive management," through theoretical and modeling approaches that can accommodate the complexities of real-world problems and embrace uncertainty through innovative experimentation and monitoring approaches (Keith et al. 2011).

(iv) Technology innovation. Modern nature conservation science operates at the frontier of technology. Innovative technology can enhance biodiversity conservation using a variety of technological options including big data, drones, artificial intelligence, and technological processes. Only one chapter formally proposed "eco-innovations" as one of the recommended solutions to the impacts of salmon farming on marine ecosystems (Chap. 12). At a global level, the salmon industry has made efforts to reduce the impacts by incorporating technology to mitigate environmental pressures, thus improving feed digestibility, food composition, and feeding technology. As a result, a reduction in nutrient excess, food waste, and deposition in sediments has been observed. A measure, not well explored yet in aquaculture in Patagonia, is the integrated multi-trophic aquaculture (IMTA). This proposal combines three trophic levels such as shellfish, finfish and but the scale of mariculture needed to mitigate pollution is sometimes unrealistic (Chap. 12).

Chapter 11 addressed forest restoration, which ranks very high among the options to recover impaired ecosystems. Whereas the chapter proposed important criteria for targeting forest restoration areas, a next step would be to assess implementation strategies, including benefits and costs of alternative options. Innovations in restoration usually rely on technological tools requiring high investments. However, there are also opportunities for making better use of the existing funds and for low-cost solutions (Brancalion and Van Melis 2017). Finally, technological solutions should take into account that ES coproduction in a given territory (e.g., spatial and temporal scales) is a complex multilayer process involving a variety of biophysical factors in interaction with a diversity of actors with different backgrounds, interests, and different spheres of influence.

Awareness creation and education. The ES has many strengths regarding awareness and education. In example, it increases awareness of the extent of human dependence on the environment, it promotes the integration between the natural and social sciences and helps acknowledging stakeholder knowledge, it helps understanding the impacts of environmental change and environmental policy on human well-being, and it contributes toward the achievement of sustainable society-ecosystems relationships (Bull et al. 2016).

The chapters in this book explicitly or implicitly support awareness creation and education for promoting nature conservation and management. Creating awareness implies making people more conscious of the benefits from nature and the relation between ES and well-being, including risk reduction. Making people aware of a greater number of ES may encourage them to design habitat management that better balances the provision of conflicting services (Richards et al. 2017). Awareness also concerns sustainability issues, which leads to changes in human consumption patterns and facilitates a transition toward less material- and energy-intensive activities. As stated in the IPBES America's Report, this implies, among others, a significant reduction in the consumption of meat and eggs as well as reduced wastage, which leads to less agricultural production and thus the reduction of the associated biodiversity loss.

Two chapters showed that people tend to recognize a large variety of ES and benefits derived from them (Chaps. 13 and 14). While provisioning services were more easily acknowledged, social actors also appreciated spiritual values. Awareness of regulating and supporting services, including those that were important for maintaining the stability and productivity of agroecosystems, was generally low.

Yet, it is important to recognize that there is still much distance between awareness and action. The high awareness-low priority dichotomy or also called behavior-impact gap (Csutora 2012) is principally due to the ineffectiveness of communication strategies. Some conventional awareness-raising approaches, such as fear creation, moralizing, and information provision, are insufficient in drawing positive behavior changes from the public on environmental issues (Chen 2016). In some cases, these approaches may create undesired effects, such as denial or anxiety. Therefore, alternative approaches are needed to close the gap, such as a deeper restructuring of the socioeconomic determinants of life (e.g., imaginaries covered in Chap. 19), including the culture of consumption (Csutora 2012).

5 From the Global Environmental Agenda to Local Research Directions

The international society expresses itself in norms, values, and institutions and in "anything that interferes with human activities beyond domestic jurisdiction" (Wight 1966). However, there is no single expression (either a norm or institution) that bears the title of "official" or "exclusive representative" of the international agenda (Martínez Reyes 2014). The international (global) agenda is a heterogeneous group of issues that are constantly discussed on the list of goals to be achieved.

An international agenda constitutes a "road map" where nations ultimately decide how to achieve a particular set of goals. In the case of the environment and sustainable development, these agendas are usually not legally binding, and therefore failure to comply does not imply sanctions, and the goals are generally renewed under new names. For example, the United Nations Millennium Development Goals (MDG) agenda (1990–2015) was not met and was replaced by the United Nations SDG agenda (2015–2030). The MDG environmental targets on which the world failed most roundly were "reversal of the loss of environmental resources" and a "reduction of biodiversity loss." Likewise, a decade ago, the world agreed to 20 biodiversity targets (Aichi Targets 2020) none of which was met by 2020.

Allegedly, the most integrative international agenda is the UN-SDG since it includes all issues relevant to the world and nations. The SDG are framed as a universal project, with substantial institutional monitoring mechanisms aimed at ensuring the successful implementation of aligned policies. Nonetheless, SDG have been criticized for being inconsistent, difficult to quantify, implement, and monitor. Some scholars suggest that there exists a potential inconsistency in the SDGs, particularly between the socioeconomic development and the environmental sustainability goals (Swain 2018). Other scholars show that the SDG agenda may be aimed in part at undermining political struggles that aspire for more socially just and ecologically sustainable approaches to development. Weber (2017) shows that the SDG framework is deeply aligned with the rules and regulations of key international development institutions, such as the World Trade Organization (WTO) and its highly contentious policies.

The SDG 13, 14, and 15 associated with climate action, life below water and life on land, have a strong link to biodiversity conservation and ES, in line with the Convention on Biological Diversity agenda and IPBES guidelines. The online information available for Argentina and Chile on SDG monitoring shows different levels of progress (https://dashboards.sdgindex.org/profiles). While Argentina reports moderate improvements for SDG 13, 14, and 15, Chile exhibits a decreasing trend in SDG 15 (Argentina shows stagnation). This reveals that Chile is failing to protect, restore, and promote sustainable use of terrestrial ecosystems, sustainably manage forests, reduce desertification, and reverse land degradation and biodiversity loss.

The achievement of the SDG implies a commitment by governments to change course and to leave "inertial" policies behind. Five years after the SDS adoption, both governments have failed to translate the proclaimed transformative vision of the 2030 Agenda into real policies. However, it is important to recognize that the implementation of the 2030 Agenda is not just a matter of better policies or science. The effectiveness of the political reforms requires holistic changes in power structures and depends on the existence of strong, democratic, and transparent public institutions at the regional, national, and international levels (Glass and Newig 2019).

Additionally, in 2020, the pandemic increased poverty and hunger, and revealed the weaknesses of health and education systems and global cooperation. The global recession caused by the COVID-19 response is alarming and has made researchers question whether the SDG are fit for the post-pandemic time.

The challenges and the low level of progress exhibited by Argentina and Chile in SDG 13, 14, and 15 (as well as other SDG) had been evidenced in the chapters of the book: (i) the change from land use to urban land, industrial plantations, and croplands continues to impair natural ecosystems, biodiversity, and ES, and (ii) the current management practices have a limited impact in reversing these trends as long as population and consumption continue to increase. The same is true in ocean ecosystems although uncertainties are even greater.

General directions to address sustainability have been proposed by several authors (see Franco et al. 2019) and include environmental education, increased participation of economic and noneconomic interest groups in proposing relevant policy actions, policy-making and implementation coherence, adaptive governance, and democratic institutions (Glass and Newing 2019). Most importantly, because of the complex human-environment interactions, most environmental challenges require fundamental changes in attitudes and behaviors from governments, industry, and individuals. While most people think sustainability is an important problem, they are often unresponsive, seem slow to act, do not always understand, and often deny environmental imperatives, creating substantial social and psychological barriers (Soron 2010).

Given the size of the task, the role of science (and technology) is obviously limited, for several reasons. Firstly, current mechanistic, reductionist science is inherently incapable of providing the complete and accurate information, which is required to successfully address environmental problems. Even under new research paradigms as sustainability sciences, there are fundamental bounds to our ability to design sustainable transformation pathways based on evidence. Human-environment systems remain highly complex and difficult, or impossible, to map fully. Causes and effects are often hard to distinguish and context dependent (Voulvoulis and Burgman 2019). Stakeholders frequently disagree about problems and solutions. In such cases, decision-makers must navigate ways forward based on careful consideration of risks, uncertainty, and issues of social justice. Precautionary measures or interventions may be advisable even if cause-and-effect relationships are not fully established. Secondly, both the conservation of mass principle and the second law of thermodynamics dictate that most remediation technologies, while successful in solving specific pollution problems, can cause inevitable negative environmental impacts elsewhere or in the future. Thirdly, it is intrinsically impossible for extractive (e.g., large-scale timber extraction or livestock operations) or industrial processes (e.g., industrial salmon farming) to have zero environmental impacts. Fourthly, most environmental problems and their solutions have more to do with political decisions (e.g., income distribution and poverty reduction policies with north-south cooperation) than with environmental science or technology innovation.

In this state of affairs, it is necessary to reflect on two linked questions: (i) how much more ES research is necessary to achieve sustainability in Patagonia and (ii) on which components of the conceptual framework proposed in Chap. 1. Chapter 20 and the present chapter make a contribution toward answering the second question by means of synthetizing research gaps. As to the first question, whichever the answer, we believe that a great effort is needed to share and systematize the existing ES knowledge. Thus, a true innovation might be research synthesis, which is still incipient in ES in Argentina and Chile. Research synthesis is the integration of existing knowledge and research findings pertinent to an issue, in order to increase the generality and applicability of those findings and to develop new knowledge through the process of integration. Synthesis is promoted as "an approach that deals with the challenge of information overload, delivering products that further our understanding of problems and distil relevant evidence for decision-making" (Wyborn et al. 2018). Yet, different ontological positions, epistemological positions, paradigms of inquiry, foundational theories, and philosophies and methodologies can make the synthesis a daunting initiative (Sandelowski et al. 2006).

Despite the challenges and faced with the urgency to meet sustainability targets, ES research needs to move to evidence syntheses or integration of research findings derived from systematic reviews of empirical research in targeted research areas to answer specific research questions addressing specific practice problems. Such evidence syntheses might truly have the potential to increase the utility of ES research and the effectiveness of practice.

References

Balvanera P, Pérez-Harguindeguy N, Perevochtchikova M, Laterra P, Cáceres DM, Langle-Flores A (2020) Ecosystem services research in Latin America 2.0: expanding collaboration across countries, disciplines, and sectors. Ecosyst Serv 42(101086):579–598

Beaumont NJ, Mongruel R, Hooper T (2017) Practical application of the Ecosystem Service Approach (ESA): lessons learned and recommendations for the future. Int J Biodiv Sci Ecosyst Serv Manag 13(3):68–78

Benra F, Nahuelhual L (2019) A trilogy of inequalities: land ownership, forest cover and ecosystem services distribution. Land Use Policy 82:247–257

Boisvert V, Méral P, Froger G (2013) Market-based instruments for ecosystem services: institutional innovation or renovation? Soc Nat Resour 26(10):1122–1136

Bouwma I, Schleyer C, Primmer E, Winkler KJ, Berry P, Young J et al (2018) Adoption of the ecosystem services concept in EU policies. Ecosyst Serv 29:213–222

Brain MJ, Nahuelhual L, Gelcich S, Bozzeda F (2020) Marine conservation may not deliver ecosystem services and benefits to all: insights from Chilean Patagonia. Ecosyst Serv 45:101170

Brancalion PH, van Melis J (2017) On the need for innovation in ecological restoration1. Ann Mo Bot Gard 102(2):227–236
Bull JW, Jobstvogt N, Böhnke-Henrichs A, Mascarenhas A, Sitas N, Baulcomb C et al (2016) Strengths, weaknesses, opportunities and threats: a SWOT analysis of the ecosystem services framework. Ecosyst Serv 17:99–111
Buttoud G, Kouplevatskaya-Buttoud I, Slee B, Weiss G (2011) Barriers to institutional learning and innovations in the forest sector in Europe: markets, policies and stakeholders. Forest Policy Econ 13(2):124–131
Capaldo G (2018) Eficacia del Derecho Ambiental. En: Gestión sustentable del agua en ecosistemas, comp. por Capaldo, G., Buenos Aires: Eudeba, 53–54
Cash DW, Clark WC, Alcock F, Dickson NM, Eckley N, Guston DH et al (2003) Knowledge systems for sustainable development. Proc Natl Acad Sci 100(14):8086–8091
Chen MF (2016) Impact of fear appeals on pro-environmental behavior and crucial determinants. Int J Advert 35(1):74–92
Corley JC, Dimarco RD, Fischbein D, Lantschner MV, Martínez AS, Masciocchi M et al (2018) A synthesis on the impact of non-native conifer plantations on ant and beetle diversity in northwestern Patagonia. South For J For Sci 80(4):285–291
Csutora M (2012) One more awareness gap? The behaviour–impact gap problem. J Consum Policy 35(1):145–163
Delgado LE, Marín VH (2015) Ecosystem services: where on earth? Ecosyst Serv 14:24–26
Dick J, Turkelboom F, Woods H, Iniesta-Arandia I, Primmer E, Saarela SR et al (2018) Stakeholders' perspectives on the operationalisation of the ecosystem service concept: results from 27 case studies. Ecosyst Serv 29:552–565
Franco IB, Chatterji T, Derbyshire E, Tracey J (2019) Actioning the global goals for local impact. Springer Nature Singapore Pte Ltd., Singapore
Franzese J, Urrutia J, García RA, Taylor K, Pauchard A (2017) Pine invasion impacts on plant diversity in Patagonia: invader size and invaded habitat matter. Biol Invasions 19(3):1015–1027
Glass LM, Newig J (2019) Governance for achieving the Sustainable Development Goals: how important are participation, policy coherence, reflexivity, adaptation and democratic institutions? Earth Syst Gov 2:100031
Hevia V, Martín-López B, Palomo S, García-Llorente M, de Bello F, González JA (2017) Trait-based approaches to analyze links between the drivers of change and ecosystem services: synthesizing existing evidence and future challenges. Ecol Evol 7(3):831–844
IPBES (2018) Summary for policymakers of the regional assessment report on biodiversity and ecosystem services for the Americas of the Intergovernmental Science-Policy Platform on Biodiversity and Ecosystem Services. Rice J, Seixas CS, Zaccagnini ME, Bedoya-Gaitán M, Valderrama N, Anderson CB, Arroyo MTK, Bustamante M, Cavender-Bares J, Diaz-de-Leon A, Fennessy S, García Márquez JR, Garcia K, Helmer EH, Herrera B, Klatt B, Ometo JP, Rodríguez Osuna V, Scarano FR, Schill S, Farinaci JS (Eds.). IPBES Secretariat, Bonn, Germany. 41 pp
IPBES (2019) Summary for policymakers of the global assessment report on biodiversity and ecosystem services of the Intergovernmental Science-Policy Platform on Biodiversity and Ecosystem Services. Díaz S, Settele J, Brondízio ES, Ngo HT, Guèze M, Agard J, Arneth A, Balvanera P, Brauman KA, Butchart SHM, Chan KMA, Garibaldi LA, Ichii K, Liu J, Subramanian SM, Midgley GF, Miloslavich P, Molnár Z, Obura D, Pfaff A, Olasky S, Purvis A, Razzaque J, Reyers B, Roy Chowdhury R, Shin YJ, Visseren-Hamakers IJ, Willis KJ, Zayas CN (Eds.). IPBES Secretariat, Bonn, Germany. 56 pp
Keith DA, Martin TG, McDonald-Madden E, Walters C (2011) Uncertainty and adaptive management for biodiversity conservation. Biol Conserv 144(4):1175–1178
Lautenbach S, Mupepele AC, Dormann CF et al (2019) Blind spots in ecosystem services research and challenges for implementation. Reg Environ Chang 19:2151–2172. https://doi.org/10.1007/s10113-018-1457-9
Levrel H, Cabral P, Feger C, Chambolle M, Basque D (2017) How to overcome the implementation gap in ecosystem services? A user-friendly and inclusive tool for improved urban management. Land Use Policy 68:574–584

Liu J, Hull V, Batistella M, DeFries R, Dietz T, Fu F, Hertel TW, Izaurralde RC, Lambin EF, Li S, Martinelli LA, McConnell WJ, Moran EF, Naylor R, Ouyang Z, Polenske KR, Reenberg A, de Miranda Rocha G, Simmons CS, Verburg PH, Vitousek PM, Zhang F, Zhu C (2013) Framing sustainability in a telecoupled world. Ecol Soc 18(2):26. https://doi.org/10.5751/ES-05873-180226
Maestre Andres S, Mir LC, van den Bergh JC, Ring I, Verburg PH (2012) Ineffective biodiversity policy due to five rebound effects. Ecosyst Serv 1(1):101–110
Martínez Reyes E (2014) Notas sobre la agenda internacional: Desarrollo¿ para qué? Revista de El Colegio de San Luis 4(7):274–282
MEA (2005) Ecosystems and human well-being: synthesis. Island, Washington, DC
Menzel S, Teng J (2010) Ecosystem services as a stakeholder-driven concept for conservation science. Conserv Biol 24(3):907–909
Moore JW (2017) The Capitalocene, Part I: on the nature and origins of our ecological crisis. J Peasant Stud 44(3):594–630
O'Connor J (1991) On the two contradictions of capitalism. Capital Nat Social 2(3):107–109
Pauna VH, Picone F, Le Guyader G, Buonocore E, Franzese PP (2018) The scientific research on ecosystem services: a bibliometric analysis. Ecol Quest 29(3):53–62
Perevochtchikova M, Flores JÁH, Marín W, Flores AL, Bueno AR, Negrete IAR (2019) Systematic review of integrated studies on functional and thematic ecosystem services in Latin America, 1992–2017. Ecosyst Serv 36:100900
Polasky S, Tallis H, Reyers B (2015) Setting the bar: standards for ecosystem services. Proc Natl Acad Sci 112(24):7356–7361
Posner SM, McKenzie E, Ricketts TH (2016) Policy impacts of ecosystem services knowledge. Proc Natl Acad Sci 113(7):1760–1765
Richards DR, Warren PH, Maltby L, Moggridge HL (2017) Awareness of greater numbers of ecosystem services affects preferences for floodplain management. Ecosyst Serv 24:138–146
Saarikoski H, Primmer E, Saarela SR, Antunes P, Aszalós R, Baró F et al (2018) Institutional challenges in putting ecosystem service knowledge in practice. Ecosyst Serv 29:579–598
Sandelowski M, Voils CI, Barroso J (2006) Defining and designing mixed research synthesis studies. Res Sch 13(1):29
Sarasola MM, Rusch VE, Schlichter TM, Ghersa CM (2006) Invasión de coníferas forestales en áreas de estepa y bosques de ciprés de la cordillera en la Región Andino Patagónica. Ecol Austral 16(2):143–156
Soron D (2010) Sustainability, self-identity and the sociology of consumption. Sustain Dev 18:172–181
Stephenson Jr, MO (2011) Considering the relationships among social conflict, social imaginaries, resilience, and community-based organization leadership. Ecol Soc 16(1):34. http://www.ecologyandsociety.org/vol16/iss1/art34/
Swain RB (2018) A critical analysis of the sustainable development goals. In: Handbook of sustainability science and research. Springer, Cham, pp 341–355
Verburg RW, Selnes T (2014). The ecosystem services approach as an instrument for action: on the experiences in the United Kingdom, Belgium and the Netherlands (No. 30). Wettelijke Onderzoekstaken Natuur & Milieu
Voulvoulis N, Burgman MA (2019) The contrasting roles of science and technology in environmental challenges. Crit Rev Environ Sci Technol 49(12):1079–1106
Weber H (2017) Politics of 'leaving no one behind': contesting the 2030 sustainable development goals agenda. Globalizations 14(3):399–414. https://doi.org/10.1080/14747731.2016.1275404
White C, Halpern BS, Kappel CV (2012) Ecosystem service tradeoff analysis reveals the value of marine spatial planning for multiple ocean uses. Proc Natl Acad Sci 109(12):4696–4701
Wight M (1966) Western values in international relations. In: Jackson R, Sorensen G (eds) Introduction to international relations: theories and approaches. Oxford University Press, Nueva York
Wyborn C, Louder E, Harrison J, Montambault J, Montana J, Ryan M et al (2018) Understanding the impacts of research synthesis. Environ Sci Pol 86:72–84

Index

P. L. Peri et al. (eds.), *Ecosystem Services in Patagonia*, Natural and Social Sciences of Patagonia, https://doi.org/10.1007/978-3-030-69166-0

The manufacturer's authorised representative in the EU is Springer Nature Customer Service Centre GmbH, Europaplatz 3, 69115 Heidelberg, Germany. If you have any concerns regarding our products, please contact ProductSafety@springernature.com

Printed and bound by CPI Group (UK) Ltd, Croydon, CR0 4YY
23/09/2026
02221532-0013